Jean-Claude Falmagne · Jean-Paul Doignon

Learning Spaces

Interdisciplinary Applied Mathematics

Springer

Prof. Dr. Jean-Claude Falmagne
University of California, Irvine
Department of Cognitive Sciences
Institute of Mathematical Behavioral Sciences
92697-5100 Irvine California
USA
jcf@uci.edu

Prof. Dr. Jean-Paul Doignon
Université Libre de Bruxelles
Département de Mathématique, c.p. 216
Bd du Triomphe
1050 Bruxelles
Belgium
doignon@ulb.ac.be

ISBN 978-3-642-01038-5 e-ISBN 978-3-642-01039-2
DOI 10.1007/978-3-642-01039-2
Springer Heidelberg Dordrecht London New York

Library of Congress Control Number: 2010933254

Cover design: KuenkelLopka GmbH

Printed on acid-free paper

Springer is part of Springer Science+Business Media (www.springer.com)

Learning Spaces

To Cheryl,
with my best regards
Jean-Claude

Preface

This book is a much enlarged second edition of "Knowledge Spaces", by Jean-Paul Doignon and Jean-Claude Falmagne, which appeared in 1999. Chapters 2, 10, 16, 17 and 18 are new, and several of the other chapters have been extensively revised[1]. The reasons for the change of title and the extended content are explained below. As much of our earlier preface remains valid, we reproduce the useful parts here.

The work reported in these pages began during the academic year 1982–83. One of us (JCF) was on sabbatical leave at the University of Regensburg. For various reasons, the time was ripe for a new joint research subject. Our long term collaboration was thriving, and we could envisage an ambitious commitment. We decided to build an efficient machine for the assessment of knowledge—for example, that of students learning a scholarly subject. We began at once to work out the theoretical components of such a machine. Until then, we had been engaged in topics dealing mostly with geometry, combinatorics, psychophysics, and especially measurement theory. The last subject has some bearing on the content of this book. A close look at the foundations of measurement, in the sense that this term has in the physical sciences, may go a long way toward convincing you of its limited applicability. It seemed to us that in many scientific areas, from chemistry to biology and especially the behavioral sciences, theories must often be built on a very different footing than that of classical physics. Evidently, the standard physical scales such as time, mass, or length may always be used in measuring aspects of phenomena. But the substrate proper to these other sciences may very well be of a fundamentally different nature. In short, nineteenth century physics is a bad example. This is not always understood. There was in fact a belief, shared by many nineteenth century scientists, that for an academic endeavour to be called a 'science', it had to resemble classical physics in critical ways. In particular, its basic observations had to be quantified in terms of measurement scales in the exact sense of classical physics.

Prominent advocates of that view were Francis Galton, Karl Pearson and William Thomson Kelvin. Because that position is still influential today, with a detrimental effect on fields such as 'psychological measurement', which is relevant to our subject, it is worth quoting some opinions in detail. In Pearson's biography of Galton (Pearson, 1924, Vol. II, p. 345), we can find the following definition:

> "**Anthropometry**, *or the art of measuring the physical and mental faculties of human beings, enables a shorthand description of any individual by measuring a small sample of his dimensions and qualities. These will sufficiently define his bodily proportions, his massiveness,*

[1] The content of the book is summarized in Section 1.5 on page 12.

> *strength, agility, keenness of sense, energy, health, intellectual capacity and mental character, and will substitute concise and exact* **numerical**[2] *values for verbose and disputable estimates*[3]*.*"

For scientists of that era, it was hard to imagine a non-numerical approach to precise study of an empirical phenomenon. Karl Pearson himself, for instance—commenting on a piece critical of Galton's methods by the editor of the *Spectator*[4]—, wrote

> "*There might be difficulty in ranking Gladstone and Disraeli for "Candour", but few would question John Morley's position relative to both of them in this quality. It would require an intellect their equal to rank truly in scholarship Henry Bradshaw, Robertson Smith and Lord Acton, but most judges would place all three above Sir John Seeley, as they would place Seeley above Oscar Browning. After all, there are such things as brackets, which only makes the statistical theory of ranking slightly less simple in the handling*" (Pearson, 1924, Vol. II, p. 345).

In other words, measuring a psychical attribute such as 'candour' only requires fudging a bit around the edges of the order relation of the real numbers, making it either, in current terminology, a 'weak order' (cf. 1.6.7 in Chapter 1) or perhaps a 'semiorder' (cf. Problems 9 and 10 in Chapter 4).

As for Kelvin, his position on the subject is well-known, and often summarized in the form: "*If you cannot measure it, then it is not science.*" (In French: "*Il n'y a de science que du mesurable.*") The full quotation is:

> "*When you can measure what you are speaking about, and express it in numbers, you know something about it; but when you cannot measure it, when you cannot express it in numbers, your knowledge is of a meager and unsatisfactory kind: it may be the beginning of knowledge, but you are scarcely, in your thoughts, advanced to the stage of* **science**, *whatever the matter may be*" (Kelvin, 1889).

Such a position, which equates precision with the use of numbers, was not on the whole beneficial to the development of mature sciences outside of physics. It certainly had a costly impact on the assessment of mental traits. For instance, for the sake of scientific precision, the assessment of mathematical

[2] Our emphasis.

[3] This excerpt is from an address on "Anthropometry at Schools" given in 1905 by Galton at the London Congress of the Royal Institute for Preventive Medicine. The text was published in the *Journal of Preventive Medicine*, Vol. XIV, pp. 93–98, London, 1906.

[4] The *Spectator*, May 23, 1874. The editor of the *Spectator* was taking Galton to task for his method of ranking applied to psychical character. He used 'candour' and 'power of repartee' as examples.

knowledge was superseded in the US by the measurement of mathematical aptitude using instruments directly inspired from Galton via Alfred Binet in France. They are still used today in such forms as the S.A.T.[5], the G.R.E. (*Graduate Record Examination*), and other similar tests.

In the mind of Galton and his followers, the numerical measurement of mental traits was to be a prelude to the establishment of sound, predictive scientific theories in the spirit of those used so successfully in classical physics. The planned constructions, however, never went much beyond the measurement stage[6].

Of course, we are enjoying the benefits of hindsight. In all fairness, there were important mitigating circumstances affecting those who uphold the cause of numerical measurement as a prerequisite to science. For one thing, the appropriate mathematical tools were not yet available for different conceptions. More importantly, the 'Analytical Engine' of Charles Babbage was still a dream, and close to another century had to pass before the appearance of computing machines capable of handling the symbolic manipulations that would be required.

The material of this book represents a sharp departure from other approaches to the assessment of knowledge. Its mathematics is in the spirit of current research in combinatorics. No attempt is made to obtain a numerical representation[7]. We start from the concept of a possibly large but essentially discrete set of 'units of knowledge.' In the case of elementary algebra, for instance, one such unit might be a particular type of algebra problem. The full set of questions may contain several hundred such problems. Two key concepts are: the 'knowledge state', a subset of problems that some individual is capable of solving correctly, and the 'knowledge structure', which is a distinguished collection of knowledge states. For beginning algebra, a useful knowledge structure may contain several million feasible knowledge states.

An important difference between the psychometric approach and that expounded in this book concerns the choice of problems representing a particular curriculum, such as beginning algebra. In our case, there is no essential restriction. The problems involved in any assessment can be chosen in a pool covering the entire curriculum. By contrast, the problems selected in the construction of a psychometric test must satisfy a criterion of homogeneity: they

[5] Interestingly, the meaning of the acronym S.A.T. was changed a few years ago by Educational Testing Service from '*Scholastic Aptitude Test*' to '*Scholastic Assessment Test*', suggesting that a different philosophy on the part of the test makers was considered. Today, it appears that 'S.A.T.' has become an abbreviation, a mnemonic without any intended meaning.

[6] Sophisticated mathematical theories can certainly be found in some areas of the behavioral sciences, but they do not generally rely on 'psychological measurement.'

[7] For example, in the form of one or more measurement scales quantifying some 'aptitudes.'

must contribute to the estimation of a numerical score. Any problem that does not satisfy this criterion—for which there is a technical statistical definition—may be rejected, even though it may be an essential part of the curriculum.

If a test is used as part of the assessment of students competence, with potential consequences not only for the students but also for the teacher or the school, the practice of teaching-to-the-test is unavoidable. In that respect, the difference between the two approaches is critical. With assessments based on the knowledge space methodology, teaching-to-the-test is beneficial[8] because the questions of the tests are essential parts of the curriculum. When psychometric tests are used, the practice may result in a distortion of the educational process, as is often argued and criticized.

The title "Knowledge Spaces" given to the first edition was consonant with our initial goal of building a machine for assessing knowledge. It later turned out that the resulting instrument could form the core component of a teaching engine, for the sensible reason that ascertaining the exact knowledge state of a student in a scholarly subject is the essential step toward educating the student in that subject. However, the change of focus from assessing to teaching prompted the development of a particular kind of knowledge space, called a 'learning space', hence the title of this new edition of our book. The difference between the names is not merely skin-deep. It is motivated by a re-axiomatization of the concept. Learning spaces are specified by two simple, pedagogically inescapable axioms. They are presented and discussed in Chapter 2. (Chapter 1 contains a general, non technical introduction to our subject.)

The two concepts of knowledge state and knowledge structure give rise to various lattice-theoretical developments motivated by the empirical application intended for them. This material is presented in Chapters 2-8. The concept of a learning space has been generalized in the form of an algebraic system called a medium. The connections between media and learning spaces are spelled out in Chapter 10.

The behavioral nature of the typical empirical observations—the responses of human subjects to questions or problems—practically guarantees noisy data. Also, it is reasonable to suppose that all the knowledge states (in our sense) are not equally likely in a population of reference. This means that a probabilistic theory had to be forged to deal with these two kinds of uncertainties. This theory is expounded in Chapters 11 and 12. Chapters 9, 13 and 14 are devoted to various practical schemes for uncovering an individual's knowledge state by sophisticated questioning. Chapters 15 and 16 tackle the complex problems of constructing knowledge spaces and learning spaces.

For a real-life demonstration of a system based on the concepts of this book, we direct the reader to `http://www.aleks.com` where various full-

---

[8] It is also relevant that, in standard applications of knowledge space theory, such as the `ALEKS` system (see page 10), there are typically no multiple choice problems.

scale programs involving both an assessment module and a learning module are available, covering mathematics and science subjects (see page 10 for a more detailed description). Chapter 17 is devoted to an investigation of the `ALEKS` assessment software from the standpoint of its validity, that is, the accuracy of its predictions. The final chapter 18 contains a list of open problems.

Many worthwhile developments could not be included here. There is much on-going research, especially in two European centers: the University of Padua, with Francesca Cristante, Luca Stefanutti and their colleagues, and the University of Graz by the research team of Dietrich Albert. However, we had to limit our coverage. Further theoretical concepts and results can be found in chapters of two edited volumes, by Albert (1994) and Albert and Lukas (1998). The second one also contains some applications to various domains of knowledge. Current references on knowledge spaces can be obtained at

`http://wundt.kfunigraz.ac.at/hockemeyer/bibliography.html`

thanks to Cord Hockemeyer, who maintains a searchable database.

Our enterprise, from the first idea to the completion of *Knowledge Spaces*, the first edition of this monograph, took 17 years, during which Falmagne benefited from major help in the form of several grants from the National Science Foundation at New York University. JCF also ackowledges a grant from the Army Research Institute (to New York University). He spent the academic year 1987-88 at the Center for Advanced Study in the Behavioral Sciences in Palo Alto. JPD, as a Fulbright grantee, was a visitor at the Center for several months, and substantial progress on our topic was made during that period. Another major grant from the National Science Foundation to JCF at the University of California, Irvine, was instrumental for the development of the educational software `ALEKS` (which belongs to UCI and is licenced to ALEKS Corporation[9]). We thank all these institutions for their financial support.

Numerous colleagues, students and former students were helpful at various stages of our work. Their criticisms and suggestions certainly improved this book. We thank especially Dietrich Albert, Biff Baker, Eric Cosyn, Charlie Chubb, Chris Doble, Nicolas Gauvrit, Cord Hockemeyer, Yung-Fong Hsu, Geoffrey Iverson, Mathieu Koppen, Kamakshi Lakshminarayan, Wil Lampros, Damien Lauly, Arnaud Lenoble, Josef Lukas, Jeff Matayoshi, Bernard Monjardet, Cornelia Müller-Dowling, Louis Narens, Misha Pavel, Michel Regenwetter, Selim Rexhap, Ragnar Steingrimsson, Ching-Fan Seu, Nicolas Thiéry, Vanessa Vanderstappen, Hassan Uzun and Fangyun Yang. We also benefited from the remarks of students from two Erasmus courses given by JPD (Leuven, 1989, and Graz, 1998).

Some special debts must be ackowledged separately. One is to Duncan Luce, for his detailed remarks on a preliminary draft of the first edition, many of which led us to alter some aspects of our text. Chris Doble's carefully read part of the present edition and his comments were also very useful to us.

[9] JCF is the Chairman and a co-founder of ALEKS Corporation.

As mentioned at the beginning of this preface, JCF spent the all important gestation period of 1982-83 at the University of Regensburg, in the stimulating atmosphere of Professor Jan Drösler's team there. This stay was made possible by a Senior US Scientist Award to Falmagne from the von Humboldt Foundation. The role of Drösler and his colleagues and of the von Humboldt Foundation is gratefully recognized here. During the initial software development phase at University of California at Irvine, Steve Franklin's friendly criticisms and cooperation have been invaluable to JCF. During the preparation of the revised edition, JPD's work was partially supported by an ARC (*Actions de Recherche Concertées*), fund of the *Communauté française de Belgique*.

We are indebted to Wei Deng for her uncomprimising chase for the typographical errors and solecisms committed in an early draft of the first edition. As the second edition largely relies on the first one, the present work also gained from Wei's pointed remarks. Because this is a human enterprise, the reader will surely uncover some remaining incongruities, for which we accept all responsibilities. We have also benefited from the kind efficiency of Mrs. Glaunsinger, Mrs. Fischer and Dr. Engesser, all from Springer-Verlag who much facilitated the production phase of this work.

Many of the figures in this new edition have been manufactured in `tikz`, the LaTeX-based graphic software created by Till Tantau. We thank him for making this powerful instrument available.

Finally, but not lastly, we thank again our respective spouses, Dina and Monique, for their kind and unwavering support.

Jean-Claude Falmagne
Irvine, CA

Jean-Paul Doignon
Watermael-Boitsfort, Belgium

May 31, 2010

Contents

1 Overview and Basic Mathematical Concepts

A student is facing a teacher, who is probing his[1] knowledge of high school mathematics. The student, a new recruit, is freshly arrived from a foreign country, and important questions must be answered. To which grade should the student be assigned? What are his strengths and weaknesses? Should the student take a remedial course in some subject? Which topics is he ready to learn? The teacher will ask a question and listen to the student's response. Other questions will then be asked. After a few questions, a picture of the student's state of knowledge will emerge, which will become increasingly sharper in the course of the examination.

However refined the questioning skills of the teacher may be, some important aspects of his task are not *a priori* beyond the capability of a clever machine. Imagine a student sitting in front of a computer terminal. The machine selects a problem and displays it on the monitor. The student's response is recorded, and the database—which keeps track of the set of all feasible knowledge states[2] consistent with the responses given so far—is updated. The next question is selected so as to maximize, in some appropriate sense, the expected information in the student's response. The goal is to focus as fast as possible on some knowledge state capable of explaining all the responses.

The purpose of this book is to expound a mathematical theory for the construction of such an assessment routine. We shall also describe some related probabilistic computer algorithms and some applications.

One reason why a machine might conceivably challenge a human examiner resides in the poor memory of the latter. No matter how the concept of a 'knowledge state in high school mathematics' is defined, a comprehensive list of such states will contain millions of entries[3]. The human mind is not especially suitable for fast and accurate scanning of such large databases. We forget, confuse and distort routinely. A fitting comparison is chess. We also

[1] In most instances, we shall conform to current standards in that the fictitious characters appearing in our story will remain nameless and genderless. Otherwise, whenever needed, we will use 'she' and 'her' in even chapters and reserve 'he' and 'his' for odd chapters.

[2] By a 'feasible knowledge state', we mean a knowledge state that can conceivably occur in the population of reference.

[3] An example is discussed in Chapter 17.

J.-C. Falmagne, J.-P. Doignon, *Learning Spaces*,
DOI 10.1007/978-3-642-01039-2_1, © Springer-Verlag Berlin Heidelberg 2011

have there, before any typical move, a very large number of possibilities to consider. A few years ago, the top human players were still capable of challenging the immense superiority of the machine as a scanning device. Today, however, some of the best chess programs, such as *Rybka*[4], have beaten human world class champions in blitz tournament. Few people believe that the human edge, if any still exists, could be long lasting[5].

Our developments will be based on a few commonsensical concepts and the mathematical ones built upon them. They are introduced informally in our next section.

1.1 Main Constructs

1.1.1 The questions and the domain. We envisage a field of knowledge that can be parsed into a set of questions, each of which has a correct response. An instance of a question in high school algebra is

[P1] *What are the roots of the equation* $3x^2 + \frac{11}{2}x - 1 = 0$?

We shall consider a basic set of such questions, called the 'domain', that is large enough to give a fine-grained, representative coverage of the field. In high school algebra, this means a set containing at least several hundred questions. Let us avoid any misunderstanding. Obviously, we are not especially interested in a student's capability of solving [P1] with the particular values 3, $\frac{11}{2}$ and -1 indicated for the coefficients. Rather, we want to assess the student's capability of solving all quadratic equations of that kind. In this book, the label 'question' (we also say 'problem', or 'item') is reserved for a class of test units differing from each other by the choice of some numbers in specified classes, or possibly also by the particular phrasing of a word problem. In that sense, [P1] is an instance of the question

[P2] *Express the roots of the equation* $\alpha x^2 + \beta x + \gamma = 0$ *in terms of* α, β *and* γ.

When the machine tests a student on question [P2], the numbers α, β and γ are selected in some specified manner. Practical considerations may enter in such a selection. For example, one may wish to choose α, β and γ in such a way that the roots of the equation can be expressed conveniently as simple fractions or decimal numbers. Nevertheless, a student's correct response to [P1] is revealing of his mastery of [P2]. Objections to our choice of fundamental concepts are sometimes made, which we shall address later in this chapter.

4 See the website: `http://rybkachess.com`.

5 Besides, the world class chess players capable of challenging the best chess programs have gone through a long and punishing learning process, during which poor moves were immediately sanctioned by a loss of a piece—and of some self-esteem. No systematic effort is made toward preparing human examiners with anything resembling the care taken in training a good chess player.

1.1.2 The knowledge states. The 'knowledge state' of an individual is represented in our approach by the subset of questions in the domain that he is capable of answering correctly in ideal conditions. This means that he is not working under time pressure or impaired by emotional turmoil of any kind. In reality, careless errors arise. Also, the correct response to a question may occasionally be guessed by a student lacking any real understanding of the question asked. (This is certainly the case when a 'multiple choice' format is used, but it may also happen in other situations.) In general, an individual's knowledge state is thus not directly observable, and has to be inferred from the responses to the questions. The connections between the knowledge state and the actual responses are explored in Chapters 11, 12, 13, and 14 where the probabilistic aspects of the theory are elaborated.

1.1.3 The knowledge structure. In our experience, for any non-trivial domain, the number of feasible knowledge states tends to be quite large. In an experiment reported in Chapter 17, for example, the number of knowledge states obtained for a domain containing around 300 questions in beginning algebra (often called 'Algebra 1' in the US) is on the order of several million. A working list of states can be obtained by interviewing educators using an automated questioning technique called '`QUERY`' (cf. 1.1.9 and Chapter 15; see Koppen and Doignon, 1990; Koppen, 1993). Such a list can then be amended by a statistical analysis of students' responses in the course of assessments.

Several million feasible states may seem to be an excessively large number of possibilities to sort out. However, it is but a minuscule fraction of the set of all 2^{300} subsets of the domain. The collection of all the knowledge states captures the organization of the knowledge and will be referred to as the 'knowledge structure.' Figure 1.1 displays a miniature example of a knowledge structure for the domain

$$Q = \{a, b, c, d, e\}.$$

In this case, the number of states is small enough that a graphic representation is possible. Considerably more complex examples are discussed later in this book[6]. The graph in Figure 1.1 represents the knowledge structure

$$\mathcal{K} = \{\varnothing, \{a\}, \{b\}, \{a,b\}, \{a,d\}, \{b,c\}, \{a,b,c\}, \\ \{a,b,d\}, \{b,c,d\}, \{a,b,c,d\}, Q\}. \tag{1.1}$$

This knowledge structure contains 11 states. The domain Q and the empty set $\varnothing$, the latter symbolizing complete ignorance, are among them. The arcs of the graph represent the *covering relation* of set inclusion: an arc linking a state K to a state K' located to its right in the graph means that $K \subset K'$ (where $\subset$ denotes the strict inclusion), and that there is no state K'' such that $K \subset K'' \subset K'$. Such a graphic representation is often used. When scanned from left to right, it suggests a learning process: at first, a student knows

---

[6] In particular, in Chapter 17.

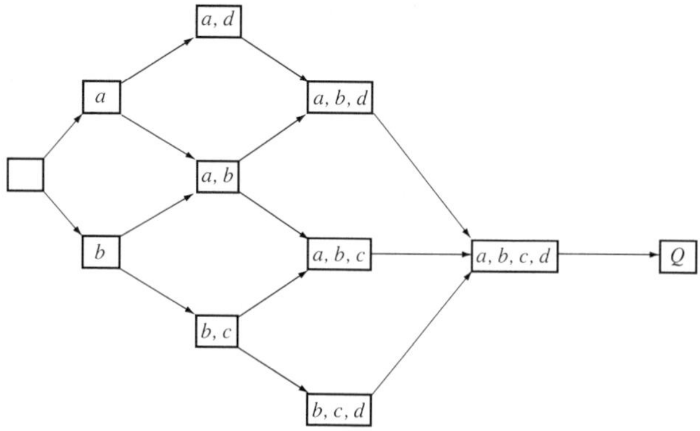

Figure 1.1. The knowledge structure of Equation (1.1).

nothing at all about the field, and is thus in state $\varnothing$, which is represented by the empty box on the left of the figure. He may then gradually progress from state to state, following one of the paths in Figure 1.1, until a complete mastery of the topic is achieved in state Q. This idea suggests the central mechanisms of the theory, which are outlined in our next subsection.

The knowledge structure of Figure 1.1 and Equation (1.1) is not artificial. It was obtained empirically, and successfully tested on a large number of subjects, in the framework of a probabilistic model. The questions $a, \ldots, e$ were problems in elementary Euclidean geometry. The text of these questions can be found in Figure 12.1 on page 216. This work is due to Lakshminarayan (1995) and we discuss it in Chapter 12.

1.1.4 Learning spaces. The step-by-step learning process outlined above suggests the two core axioms of the theory. These axioms concern a knowledge structure $\mathcal{K}$ on a domain Q; we suppose thus that both $\varnothing$ and Q are states of $\mathcal{K}$. We only give informal statements of these axioms here (for a mathematical formulation, see Section 2.2 on page 26).

1. LEARNING SMOOTHNESS. *If the state K of the learner is included in some other state L, then the learner can reach state L by mastering the missing items one by one.*

   This implies that step-by-step learning is feasible.

2. LEARNING CONSISTENCY. *Suppose that a learner in state K is capable of mastering some new item q. Then, any learner in some state L including K either has already mastered q, or is also capable of mastering it.*

   In short: knowing more does not prevent from learning something new.

A knowledge structure satisfying the these two axioms is a 'learning space.' These axioms seem *a priori* reasonable from a pedagogical standpoint. They have strong, useful consequences.

1.1.5 The fringes of a knowledge state. A knowledge state may sometimes be quite large. In beginning algebra, for example, the knowledge state of a student may contain 200 items. Representing a student's mastery of beginning algebra by such a long list is not helpful. Fortunately, the axioms of a learning space enable the exact specification of any knowledge state by its two 'fringes', which are almost always much smaller sets. Intuitively, the 'outer fringe' $K^{\mathcal{O}}$ of a knowledge state K is the set of items that a student is ready to learn. In set-theoretic terms, $K^{\mathcal{O}}$ contains all those items q such that $K \cup \{q\}$ is also a state. (The student can thus progress from state K to state $K \cup \{q\}$ simply by mastering q.) The 'inner fringe' $K^{\mathcal{I}}$ is the complementary concept, namely: $K^{\mathcal{I}}$ contains exactly those items which, when removed from K, yield another state of the learning space[7]. One interpretation of the inner fringe is that it contains the most advanced items in a student's state. In a learning space, any knowledge state is defined by its two fringes. The formal definition of the fringes and the relevant theorem are 4.1.6 and 4.1.7.

As an illustration, the two fringes of the knowledge state $\{b, c, d\}$ in the learning space displayed in Figure 1.1 are

$$\{b, c, d\}^{\mathcal{O}} = \{a\} \quad \text{and} \quad \{b, c, d\}^{\mathcal{I}} = \{d\}.$$

The economy realized in replacing the state $\{b, c, d\}$ by its fringes $\{a\}$ and $\{d\}$ is trivial in this case, which is not typical.

1.1.6 Knowledge spaces. We turn to another important consequence of the two axioms of a learning space. Examining Figure 1.1 reveals that the family $\mathcal{K}$ of states satisfies a powerful property: if K and K' are any two states in $\mathcal{K}$, then $K \cup K'$ is also a state. In mathematical lingo: the family $\mathcal{K}$ is 'closed under finite union.' A knowledge structure satisfying 'closure under union' is called a 'knowledge space.' This property results from the two axioms: any learning space is a knowledge space[8] (cf. Theorem 2.2.4). The closure under union plays a central role. It allows, for example, to summarize exhaustively any learning space by a distinguished subfamily of its states called the 'base' of the learning space. In many cases, this subfamily can be considerably smaller than the learning space itself, which may be of critical importance for some purposes, such as the economical storage of a learning space in the memory of a computer.

1.1.7 Wellgradedness. Still another consequence of the two axioms of a learning space is important in this work, for example in connection with some assessment procedures. These procedures are designed to uncover the knowledge state of a student (see Chapters 13 and 14). Consider the two states $\{a, d\}$ and $\{b, c, d\}$ of the knowledge struture $\mathcal{K}$ of Equation (1.1). They differ by exactly three items, namely a, b and c (for this reason, we will say that

[7] Obviously, the outer fringe of the domain and the inner fringe of the empty state $\varnothing$ are both empty.

[8] But not conversely. Can you think of a counterexample?

$\{a, d\}$ and $\{b, c, d\}$ are at distance 3 from one another). It happens that the state $\{a, d\}$ can be transformed into the state $\{b, c, d\}$ by a sequence of three elementary steps, namely along the 'path' of the states

$$\{a, d\}, \quad \{a, b, d\}, \quad \{a, b, c, d\}, \quad \{b, c, d\}.$$

Such a path is 'tight' because each of its states differs from the next one (if any) in exactly one item. It can be checked that a similar property holds for any two states in $\mathcal{K}$, whatever their distance is. For this reason, we say that the knowledge structure $\mathcal{K}$ is 'well-graded.' The notion of wellgradedness is defined in 2.2.2 and investigated in Chapter 4. Wellgradedness is an essential property because it guarantees that any state can be summarized by its fringes, without loss of information (see Theorem 4.1.7).

1.1.8 Surmise functions and clauses. Knowledge spaces can be regarded as a generalization of quasi orders (i.e. reflexive and transitive relations; see 1.6.3, 1.6.4 and 1.6.6). Indeed, according to a classical result of Birkhoff (1937), any family of sets closed under both union and intersection can be recoded as a quasi order, and vice versa. Moreover, the correspondence is one-to-one (cf. Theorem 3.8.3). A similar representation can be obtained for families of sets closed only under union, but the representing concept is not a quasi order, nor even a binary relation. Rather, it is a function σ associating, to each item q in Q, a family $\sigma(q)$ of subsets of Q. The family $\sigma(q)$ has the following interpretation: if a student has mastered question q, that student must also have mastered all the items in at least one of the sets in $\sigma(q)$. One may think of a particular element of $\sigma(q)$ as a possible set of prerequisites for q. This is consonant with the view that there may be more than one way to achieve the mastery of any particular question q. This interpretation leads to impose certain natural conditions on the function σ, which will be referred to as a 'surmise function.' The elements of $\sigma(q)$ for a particular item q will be called the 'clauses' for q. We may also call them 'backgrounds' of q, or 'foundations' for q. In the example of Figure 1.1, a subject having mastered item d must also have mastered either at least item a, or at least items b and c. So, Item d has two clauses: we have

$$\sigma(d) = \{\{a, d\}, \{b, c, d\}\}.$$

Notice that these two clauses for d contain d itself. By convention, any clause for an item contains that item. In other words, any item is a prerequisite for itself. (This property generalizes the reflexivity of a partial order. In 5.1.2, we also discuss a condition on the surmise function generalizing transitivity.)

In the special case where the family of states is closed under both union and intersection, any question q has a unique clause. In this case, the surmise function σ is essentially a quasi order (i.e. becomes a quasi order after a trivial change of notation): the unique clause for an item q contains q itself, plus all the items which are either equivalent to q or precede q in the quasi order.

1.1.9 Entailment relations and human expertise. We will also use another representation of a knowledge space, which has an important function in practical applications. Human experts (such as practiced teachers) may possess critical information concerning the knowledge states which are feasible in some empirical situation. However, straightforward queries would not work. We cannot realistically ask a teacher to give a complete list of the feasible states, with the hope of getting a useful response[9]. Fortunately, a recoding of the concept of knowledge space is possible, which leads to a more fruitful approach. Consider asking a teacher queries of the type

> [Q1] *Suppose that a student has failed items $q_1, \ldots, q_n$. Do you believe this student would also fail item q_{n+1}? You may assume that chance factors, such as lucky guesses and careless errors, play no role in the student's performance, which reflects his actual mastery of the field.*

(We reserve the label '[Q0]' for the special case where $n = 1$; see Chapter 7.) The set of positive responses to *all* queries of that kind, for a given domain Q, define a binary relation $\mathcal{P}$ pairing subsets of Q with elements of Q. Thus, a positive response to query [Q1] is coded as

$$\{q_1, \ldots, q_n\} \, \mathcal{P} \, q_{n+1}.$$

It can be shown that if the relation $\mathcal{P}$ satisfies certain natural conditions, then it uniquely specifies a particular knowledge space. The relation $\mathcal{P}$ is then called an 'entailment.' Algorithms based on that relation have been written, which are instrumental when querying an expert. The most widely used algorithm is the '`QUERY`' routine. In applying one of these algorithms, it is supposed that an expert relies on an implicit knowledge space to respond to queries of the type [Q1]. The output of the procedure is the personal knowledge space of the expert. We discuss these questioning procedures in Chapter 15 and 16.

1.1.10 Remarks. a) Obviously, a teacher's responses to questions of type [Q1] may be unreliable if n is large, or even with $n = 1$. To alleviate this problem Cosyn and Thiéry (2000) (cf. also Heller, 2004) developed a sophisticated version of the `QUERY` routine, called '`PS-QUERY`.' The '`PS`' in `PS-QUERY` stands for 'pending status' and signals the fact that the implementation of an expert's response is delayed until it is confirmed by a later response. If a second response conflicts with the earlier one, both of them are discarded.

b) In practice, only a very small fraction of the queries of the form [Q1] must be asked. The procedure typically terminates with $n \leq 5$. Actually, even with just $n = 1$ the procedure will deliver all the states of the knowledge space. However, if the procedure is stopped at that point, many fictitious states will remain which would be eliminated by further questioning. The resulting knowledge space must then be pruned down by a statistical analysis of student's data based, for example, on the elimination of states with low probability or by other methods (cf. Chapter 15).

---

[9] Besides, as mentioned earlier, there may be millions of states in some cases.

c) The QUERY and other similar routines, with or without the help of a statistical analysis of students' data, only guarantee that a knowledge space will emerge, which will not necessarily be a learning space[10]. Since obtaining a learning space is often the goal, a relevant modification of the QUERY procedure is presented in Chapter 16. The new procedure implements only the QUERY responses that will produce a learning space. Efficient tests are provided to this aim. Another method to build learning spaces is due to Eppstein, Falmagne, and Uzun (2009) and is briefly discussed in the same chapter. The idea is to first build a knowledge space and then supplement it by carefully chosen additional states turning the space into a learning space.

As mentioned in 1.1.2, in realistic situations, subjects sometimes fail problems that they fully understand, or provide amazingly correct responses to problems that they do not understand at all. That is, the knowledge states are not directly observable. The usual way out of such difficulties is a probabilistic approach.

1.1.11 Probabilistic knowledge structures. Probabilities may enter in two ways in the theory. We first suppose that to each knowledge state K is attached a number $P(K)$, which can be interpreted as the probability of finding, by a random sample in the population of reference, a subject in state K. The relevant structure $(Q, \mathcal{K}, P)$, where Q is the domain and $\mathcal{K}$ the set of states, is called a 'probabilistic knowledge structure.' Next, we shall introduce, for any subset R of the domain Q and any state K in $\mathcal{K}$, a conditional probability $r(R, K)$ that a subject in state K would provide correct responses to all the items in R, and only to those items. Any subset R of the domain Q is a 'response pattern.' The overall probability $\rho(R)$ of observing a response pattern R can be computed from the weighted sum

$$\rho(R) = \sum_{K \in \mathcal{K}} r(R, K)P(K).$$

Formal definitions of these probabilistic concepts will be found in Chapters 11 and 12 where specific forms of the functions P and r are also investigated. Empirical tests are also described.

1.1.12 Assessment procedures. The machinery of knowledge states, (probabilistic) structures, and learning spaces provides the foundation for a number of algorithmic assessment procedures. The goal of such a procedure is to uncover, as efficiently as possible, the knowledge state of a student by asking appropriate questions from the domain. The most widely used algorithm, which is discussed in Chapter 13, proceeds by setting up an *a priori* likelihood function on the set of all states. This likelihood function is then updated after each response of the student by a Bayesian operator. The choice of the next question is based on the current likelihood function. The output of the

[10] An exception is the questioning data restricted to the case $n = 1$ in [Q1]. As pointed out, however, this may result in a large number of fictitious states.

assessment algorithm is a knowledge state which best represents the student's competence in the domain. As such a result may be unwieldy—we made that point in 1.1.5—this output is then routinely transformed into the fringes of the selected state.

A practical implementation of an automated system built on the concepts of this book is outlined in Section 1.3.

1.2 Possible Limitations

The concept of a knowledge state, which is at the core of our work, is sometimes criticized on the grounds that it trivializes an important idea. In the minds of some critics, the concept of a 'knowledge state' should cover much more than a set of questions fully mastered by a student. It should contain many other features related to the student's current understanding of the material, such as the type of errors that he is likely to make. A reference to the work of Van Lehn (1988) is often made in this connection.

For the most part, such criticisms come from a misconception of the exact status of the definition of 'knowledge state' in our work. An important aspect of this work has consisted in developing a formal language within which critical aspects of knowledge assessment could be discussed and manipulated. 'Knowledge state' is a defined concept in that formal language. We make no claim that the concept of a 'knowledge state' in our sense captures all the cognitive features that one might associate with such a word, any more than in topology, say, the concept of a compact set captures the full physical intuition that might be evoked by the adjective "compact." As far as we know, the concept of a knowledge state has never been formally defined in the literature pertaining to computerized learning. There was thus no harm, we believed, in appropriating the term, and giving it a formal status. On the other hand, it is certainly true that if some solid information is available regarding refined aspects of the students' performance, such information could be taken into account in the assessment. There are various ways this can be achieved.

As a first example, we take the type of error mechanisms discussed by Van Lehn (1988). Suppose that for some or all of our questions, such error mechanisms have been elucidated. In other words, erroneous responses are informative, and can be attributed to specific faulty rules that the student applies. One possibility would be to analyze or redesign our domain (our basic set of questions) in such a manner that a knowledge state itself would involve the diagnosis of error mechanisms. After all, if a student routinely applies some faulty rule, this rule should be reflected in the pattern of responses to suitably chosen questions. A description of the knowledge state could then include such error mechanisms.

More generally, a sophisticated description of the knowledge states could be obtained by a precise and detailed tagging of the items. For example, it is possible to associate, to each item in the domain, a detailed list of information featuring entries such as: the part of the field to which this item belongs (e.g.: calculus, derivatives; geometry, right triangles), the type of problem (e.g.: word problem, computation, logical reasoning), the expected grade at which this item should be learned, the concepts used in the formulation or in the solution of this item, the most frequent type of misconceptions, etc.

When an assessment algorithm has uncovered the state of some student, these tags can be used to prepare, via various manipulations, a comprehensive description of the state in everyday language. We give a mathematical development of this idea in Chapter 6.

1.3 A Practical Application: The ALEKS System

The computer educational system ALEKS provides on the Internet[11] a bilingual (English-Spanish) educational environment with two modules: an *assessment module* and a self-paced *learning module* with many tools. The name 'ALEKS' is an acronym for 'Assessment and Learning in Knowledge Spaces.' The assessment module is based on the work described in this monograph. Its working version implements one of the continuous Markov procedures presented in Chapter 13. It relies on knowledge structures built with the techniques of Chapters 7, 15 and 16, refined by extensive analyses of student data. The system currently[12] covers all of grade 3-12 mathematics including precalculus and geometry, plus a few other topics such as elementary chemistry, accounting and undergraduate statistics. The assessment is comprehensive in the sense that the set of all possible questions covers the whole curriculum in the topic. An 'answer editor' permits the student to enter the responses in a style imitating a paper and pencil method. In contrast with most standardized tests, almost all the questions have open responses[13]. At the end of the assessment, the system delivers a detailed report describing the student's accomplishments, making recommendations for further learning, and giving immediate access to the learning module of the system.

While using the learning module, the student may request an 'explanation' of any problem proposed by the system. A variety of dedicated calculators and a mathematics dictionary are available on-line. The dictionary provides definitions of all the technical terms and is accessed by clicking on any highlighted word. Both the report and the learning module of the system are programmable, so that they can be made consistent with the educational standards of every state in the US. The current default standards of the system are the California Standards. A user-friendly tool allows the teacher to modify these standards, if necessary.

---

[11] See www.aleks.com.

[12] In the spring of 2010.

[13] In the few cases where a multiple choice question is used, the number of possible responses offered to the student is large.

Other systems exist, which are also based on the concepts of knowledge structures and knowledge states in the sense of this book and similar in spirit to the ALEKS system. One is the system created by Cornelia Dowling and her colleagues (see, e.g. Dowling, Hockemeyer, and Ludwig, 1996). Several systems have been developed by Dietrich Albert and his team, in cooperation with project partners. We must mention in particular RATH (Hockemeyer, Held, and Albert, 1998), APeLS (Conlan, Hockemeyer, Wade, and Albert, 2002), iClass (Albert, Nussbaumer, and Steiner, 2008), ELEKTRA (Kickmeier-Rust, Marte, Linek, Lalonde, and Albert, 2008), and MedCAP (Hockemeyer, Nussbaumer, Lövquist, Aboulafia, Breen, Shorten, and Albert, 2009). Other relevant references are Albert (1994); Albert and Held (1994); Albert and Hockemeyer (1997); Albert and Lukas (1999); Desmarais and Pu (2005); Desmarais, Fu, and Pu (2005); and Pilato, Pirrone, and Rizzo (2008).

1.4 Potential Applications to Other Fields

Even though our theoretical developments have been primarily guided by a specific application in education, the basic concepts of knowledge structure, knowledge space and learning spaces are very general, and have potential uses in superficially quite different fields. A few examples are listed below.

1.4.1 Failure analysis. Consider a complex device, such as a telephone interchange (or a computer, or a nuclear plant). At some point, the device's behavior indicates a failure. The system's administrator (or a team of experts), will perform a sequence of tests to determine the particular malfunction responsible for the difficulty. Here, the domain is the set of observable signs. The states are the subsets of signs induced by all the possible malfunctions.

1.4.2 Medical diagnosis. A physician examines a patient. To determine the disease (if any), the physician will check which symptoms are present. As in the preceding example, a carefully designed sequence of verifications will take place. Thus, the system is the patient, and the state is a subset of symptoms specifying his medical condition. For an early example of a computerized medical diagnosis system, see Shortliffe and Buchanan (1975) or Shortliffe (1976). As a web search would indicate, the literature on this topic is vast and quickly expanding.

1.4.3 Pattern recognition. A pattern recognition device analyzes a visual display to detect one of many possible patterns, each of which is defined by a set of specified features. Consider a case in which the presence of features is checked sequentially, until a pattern can be identified with an acceptable risk of error. In this example, the system is a visual display, and the possible patterns are its states. For a first contact with the vast literature on pattern recognition, consult, for instance, Duda and Hart (1973) and Fu (1974).

1.4.4 Axiomatic systems. Let E be a collection of well-formed expressions in some formal language, and suppose that we also have a fixed set of derivation rules. Consider the relation $\mathfrak{I}$ on the set of all subsets of E, with the following interpretation: we write $A \mathfrak{I} B$ if all the expressions in B are derivable from the expressions in A by application of the derivation rules. We call any $K \subseteq E$ a *state* of $\mathfrak{I}$ if $B \subseteq K$ whenever $A \subseteq K$ and $A\mathfrak{I}B$. It is easily shown that the collection $\mathcal{L}$ of all states is closed under intersection; that is, $\cap\mathcal{F} \in \mathcal{L}$ for any $\mathcal{F} \subseteq \mathcal{L}$ (See Problem 2 in Chapter 3). Notice that this constraint is the dual of that defining a knowledge space, in the sense that the set

$$\overline{\mathcal{L}} = \{Z \in 2^E \,|\, \overline{Z} \in \mathcal{L}\}$$

is closed under union.

1.5 On the Content and Organization of this Book

A synopsis of some basic, standard mathematical concepts and notation will be found in the next section of this chapter. This synopsis contains entries such as 'binary relation', 'partial order', 'chains', 'Hausdorff maximal principle', etc. In writing this book, we had in mind a reader with a minimum mathematical background corresponding roughly to that of a mathematics major: e.g. a three semester sequence of calculus, a couple of courses in algebra, and a couple of courses in probability and statistics. However, a reader equipped with just that may find the book rough going, and will have to muster up a fair amount of patience and determination. To help all readers, problem solving exercises are provided at the end of each chapter.

Chapters 2–10 are devoted to algebraic aspects of the theory. They cover the main concepts, learning spaces, knowledge spaces, and a variety of ancillary topics, such as: surmise functions, entailments, and the concept of an 'assessment language' (Chapter 9). The axioms for a learning space turn out to be equivalent to those defining an 'antimatroid' in combinatorics. We devote part of Chapter 2 to the relationship between learning spaces and (union-closed) antimatroids[14]. The connection between learning spaces and the semigroups of transformations called 'media' is discussed in Chapter 10.

Chapters 11 and 12 deal with probabilistic knowledge structures. These two chapters develop a number of stochastic learning models describing the successive transitions, over time, between the knowledge states. Chapter 13 and 14 are devoted to various stochastic algorithms for assessing knowledge. Chapters 15 and 16 describes some procedures for the construction of knowledge spaces and learning spaces in a practical situation. Chapter 17 describes an evaluation of the validity and reliability of an assessment by the `ALEKS` system, based on a very extensive data set. We end up in Chapter 18 with a list of open problems which we believe to be of interest. For convenience of

[14] Two dual versions of the concept of antimatroid are in used in the combinatoric literature: union-closed and intersection-closed.

reference, we give on pages 379-396 a glossary of the standard mathematical symbols used and of the new formal concepts introduced in this book. Fairly extensive definitions of these concepts can be found there.

Chapters are divided into sections and subsections. A single lexicographic system is used. The title

"**1.4 Potential Applications to Other Fields**"

on page 11 is that of the fourth section of Chapter 1. This section is organized into four 'subsections.' The first subsection in Section 1.4 is entitled:

"**1.4.1 Failure Analysis.**"

Thus, when a number "**n.m.p**" is used for a subsection, the "**n**" denotes the chapter number and the "**m**" and "**p**" are the section and subsection numbers within that chapter. The most frequent titles for subsections will be "Examples", "Definition", "Theorem", and "Remark(s)[15]." Minor results may sometimes appear as remarks. Especially difficult chapters, sections, subsections and exercises are marked by a star. We also use a star to indicate a part of the book that can be omitted at first reading without loss of continuity.

Some printing conventions should be remembered. We put single quotation marks around a term used in a technical sense, but yet undefined. (Only the first occurence of the term is so marked.) A word or phrase mentioned, but not used, is flanked by double quotation marks. (This does not apply to mathematical symbols, however.) A slanted font is used for quotations, for technical terms in their definition, and for the text of theorems or lemmas. The preceding pages contain many applications of all these conventions.

1.6 Basic Mathematical Concepts and Notation

1.6.1 Set theory, relations, mappings. Standard set-theoretic notation will be used throughout. We sometimes write $+$ for the disjoint union. Set inclusion is denoted by $\subseteq$ and proper (or strict) inclusion by $\subset$. The size, or cardinality, or cardinal number of a set Q is denoted by $|Q|$. The collection of all subsets of Q, or *power set* of Q, is symbolized by 2^Q. An element (x, y) of the Cartesian product $X \times Y$ is often abbreviated as xy. A *relation* $\mathcal{R}$ from a set X to a set Y is a subset of $X \times Y$. The *complement* of $\mathcal{R}$ (*with respect to* $X \times Y$) is the relation $\overline{\mathcal{R}} = (X \times Y) \setminus \mathcal{R}$, also from X to Y. The specification "with respect to $X \times Y$" is usually omitted when no ambiguity can arise. The phrase "with respect to" is occasionally abbreviated by the acronym "w.r.t." In the same vein, "l.h.s." and "r.h.s." are shorthand for the left hand side and the right hand side, respectively, of an equation or a logical formula. The *converse* of $\mathcal{R}$ is the relation

$$\mathcal{R}^{-1} = \{yx \in Y \times X \,|\, x\mathcal{R}y\}$$

from Y to X.

[15] The style of this introductory chapter is not representative of the rest of the book.

A *mapping*, or function, f from the set X to the set Y is a relation from X to Y such that for any $x \in X$, there exists exactly one $y \in Y$ with $x\, f\, y$; we then write $y = f(x)$. This mapping is *injective* if $f(x) = f(x')$ implies $x = x'$ for all $x, x' \in X$. It is *surjective* if for any $y \in Y$ there is some $x \in X$ with $y = f(x)$. Finally, f is *bijective*, or a *one-to-one correspondence*, (with 'one-to-one' often abbreviated as 1–1), if f is both injective and surjective. The converse of a bijective function is called its *inverse*.

A detailed treatment of relations can be found, for example, in Suppes (1960) or Roberts (1979, 1984). Some basic facts and concepts are recalled in the following few subsections.

1.6.2 Relative product. The *(relative) product* of two relations $\mathcal{R}$ and $\mathcal{S}$ is denoted by

$$\mathcal{R}\mathcal{S} = \{xy \,|\, \exists z: \; x\mathcal{R}z \wedge z\mathcal{S}y\},$$

in which $\exists$ stands for "there exists." When $\mathcal{R}$ and $\mathcal{S}$ are explicitly given as relations from X to Y and from Z to W respectively, the element z in the above formula is to be taken in $Y \cap Z$, and the relation $\mathcal{R}\mathcal{S}$ is from X to W. It is easy to check that the product operation is associative: for any three relations $\mathcal{R}$, $\mathcal{S}$ and $\mathcal{T}$, we have $(\mathcal{R}\mathcal{S})\mathcal{T} = \mathcal{R}(\mathcal{S}\mathcal{T})$. So, the parentheses play no useful role. Accordingly, we shall write $\mathcal{R}_1\mathcal{R}_2 \cdots \mathcal{R}_k$ for the product of k relations $\mathcal{R}_1, \mathcal{R}_2, \ldots, \mathcal{R}_k$. For any relation $\mathcal{R}$ from a set X to itself and any positive integer n, we write $\mathcal{R}^n$ for the nth *(relative) power* of the relation $\mathcal{R}$, that is

$$\mathcal{R}^n = \underbrace{\mathcal{R}\mathcal{R} \cdots \mathcal{R}}_{\text{n times}}.$$

By convention, $\mathcal{R}^0$ denotes the *identity relation* on the ground set X, which will always be specified explicitly or by the context. For a relation $\mathcal{R}$ from X to Y, we set $\mathcal{R}^0 = \{xx \,|\, x \in X \cup Y\}$.

1.6.3 Properties of relations. The symbol $\neg$ is read "not" and denotes the logical negation. A relation $\mathcal{R}$ on a set X is

reflexive (on X)	when $x\mathcal{R}x$ for all $x \in X$;
symmetric (on X)	when $x\mathcal{R}y$ implies $y\mathcal{R}x$ for all $x, y \in X$;
asymmetric (on X)	when $x\mathcal{R}y$ implies $\neg(y\mathcal{R}x)$ for all $x, y \in X$;
antisymmetric (on X)	if ($x\mathcal{R}y$ and $y\mathcal{R}x$) implies $x = y$ for all $x, y \in X$.

1.6.4 Transitive closure. A relation $\mathcal{R}$ is *transitive* if whenever $x\mathcal{R}y$ and $y\mathcal{R}z$, we also have $x\mathcal{R}z$. More compactly, in the relative product notation of 1.6.2, $\mathcal{R}$ is transitive if $\mathcal{R}^2 \subseteq \mathcal{R}$. The *transitive closure* (or more precisely *reflexo-transitive closure*) of a relation $\mathcal{R}$ is the relation $t(\mathcal{R})$ defined by

$$t(\mathcal{R}) = \mathcal{R}^0 \cup \mathcal{R}^1 \cup \cdots \cup \mathcal{R}^k \cup \cdots = \cup_{k=0}^{\infty} \mathcal{R}^k.$$

1.6.5 Equivalence relations and set partitions. An *equivalence relation* $\mathcal{R}$ on a set X is a reflexive, transitive, and symmetric relation on X. It corresponds exactly to a *partition* of X, that is, a family of nonempty subsets of X which are pairwise disjoint and whose union is X: these subsets, or *classes* of the equivalence relation $\mathcal{R}$, are all the subsets of X of the form $\{x \in X \,|\, x\mathcal{R}z\}$, for some $z \in X$.

1.6.6 Quasi orders, partial orders. A *quasi order* on a set X is any relation which is transitive and reflexive on X, as for instance the relation $\leq$ on the set of all real numbers. A set equipped with a quasi order is *quasi ordered*. An antisymmetric quasi order is a *partial order*. A *strict partial order* on a set X is an irreflexive and transitive relation on X.

Any quasi order $\mathcal{P}$ on X gives rise to the equivalence relation $\mathcal{P} \cap \mathcal{P}^{-1}$. A partial order $\mathcal{P}^*$ is obtained on the set X^* of all equivalence classes by setting $C\,\mathcal{P}^*\,C'$ if $c\mathcal{P}c'$ for some (and thus for all) c in C and c' in C'. The partially ordered set $(X^*, \mathcal{P}^*)$ is called the *reduction* of the quasi ordered set $(X, \mathcal{P})$. An element x in a quasi order $(X, \mathcal{P})$ is *maximal* when $x\mathcal{P}y$ implies $y\mathcal{P}x$, for all $y \in X$. It is a *maximum* if it is maximal and moreover $y\mathcal{P}x$ for all $y \in X$. *Minimal elements* and *minimum* are similarly defined. A partially ordered set can have at most one maximum and one minimum (Problem 12).

1.6.7 Weak orders, linear orders. A *weak order* $\mathcal{P}$ on the set X is a quasi order on X which is *complete*, in the sense that for all $x, y \in X$ we have $x\mathcal{P}y$ or $y\mathcal{P}x$. The reduction of a weak order is a *linear*, or *simple*, or *total order*.

1.6.8 Covering relation, Hasse diagram. In a partially ordered set $(X, \mathcal{P})$, the element x is *covered* by the element y when $x\mathcal{P}y$ with $x \neq y$ and moreover $x\mathcal{P}t\mathcal{P}y$ implies $x = t$ or $t = y$. The *covering relation* or *Hasse diagram* of $(X, \mathcal{P})$ is the relation $\breve{\mathcal{P}}$ containing all the pairs xy with y covering x. When X is infinite, the Hasse diagram of $(X, \mathcal{P})$ may be empty even though $\mathcal{P}$ itself is not empty. When X is finite, the Hasse diagram of $(X, \mathcal{P})$ provides a comprehensive summary of $\mathcal{P}$ in the sense that the transitive closure of the Hasse diagram of $(X, \mathcal{P})$ is equal to $\mathcal{P}$. In fact, in this case, the Hasse diagram of $(X, \mathcal{P})$ is the smallest relation having its transitive closure equal to $\mathcal{P}$ (cf. Problem 13).

1.6.9 Graphs. When the set X of a partial order $(X, \mathcal{P})$ is small, the Hasse diagram of $\mathcal{P}$ can be conveniently displayed by a 'graph' drawn according to the following conventions: the elements of X are represented by points on a page, with an ascending *edge* from x to y when x is covered by y. Such a graph is called a *directed graph* or *digraph*, because the edges have an orientation (upward in this case)[16]. Formally, there is a 1-1 correspondence between the Hasse diagram $\breve{\mathcal{P}}$ and the collection of all *(directed) edges* of its representing

[16] In some cases, the drawing may be rotated for convenience, so that an edge may be drawn left-right rather than down-up.

graph. The points of such a graph are refered to as *vertices*. The edges of a directed graph are sometimes called *arcs*.

More generally, the language of relations is coextensive with that of graphs, with the latter applying when a geometrical representation is serviceable.

1.6.10 Chain, Hausdorff Maximality Principle. A *chain* in a partially ordered set $(X, \mathcal{P})$ is any subset C of X such that $c\mathcal{P}c'$ or $c'\mathcal{P}c$ for all $c, c' \in C$ (in other words, the order induced by $\mathcal{P}$ on C is linear). In several proofs belonging to starred material, the Hausdorff maximality principle is invoked to establish the existence of a maximal element in a partially ordered set. This principle is equivalent to Zorn's Lemma and states that a quasi ordered set $(X, \mathcal{P})$ admits a maximal element whenever all its chains are bounded above, that is: for any chain C in X, there exists b in X with $c\mathcal{P}b$ for all $c \in C$. For details on the Hausdorff Maximality Principle and related conditions, see Dugundji (1966) or Suppes (1960).

1.6.11 Basic number sets. The following notation is used for the basic sets:

$\mathbb{N}$, the set of natural numbers (excluding 0);
$\mathbb{Z}$, the set of integer numbers;
$\mathbb{Q}$, the set of rational numbers;
$\mathbb{R}$, the set of real numbers;
$\mathbb{R}^+$, the set of (strictly) positive real numbers.

A set S is *countable* if there exists an injective function $S \to \mathbb{N}$. (Such a function may be surjective.)

We denote by $]x, y[= \{z \in \mathbb{R} \,|\, x < z < y\}$ the real open intervals, and by $]x, y]$, $[x, y[$ and $[x, y]$ the half open and closed intervals.

1.6.12 Metric spaces. A mapping $d : X \times X \to \mathbb{R}$ is called a *distance* on X if it satisfies the following three conditions for all $x, y, z \in X$:

(1) $d(x, y) \geq 0$, with $d(x, y) = 0$ iff $x = y$ (i.e., d is *positive definite*);
(2) $d(x, y) = d(y, x)$ (i.e., d is *symmetric*);
(3) $d(x, y) \leq d(x, z) + d(z, y)$ (the *triangular inequality*).

A *metric space* is a set equipped with a distance. As an example, let E be any finite set. The *symmetric difference distance* or *canonical distance* on the power set of E is defined by setting, for $A, B \in 2^E$:

$$d(A, B) = |A \triangle B| \tag{1.2}$$

(cf. Problem 7). Here, $A \triangle B = (A \setminus B) \cup (B \setminus A)$ denotes the *symmetric difference* of the sets A and B.

1.6.13 Probabilistic and statistical concepts. Whenever needed, we use the standard techniques of probability theory, stochastic processes and statistics. The symbol $\mathbb{P}$ denotes the probability measure of the probability space under consideration. Some statistical techniques dealing with goodness-of-fit tests are briefly reviewed in Chapter 11.

1.7 Original Sources and Main References

A first pass at developing a mathematical theory for knowledge structures was made by Doignon and Falmagne (1985). A rather technical follow up paper is Doignon and Falmagne (1988). The stochastic aspects of the theory were initially presented in Falmagne and Doignon (1988a,b). A comprehensive description of our program, intended for non mathematicians, is given in Falmagne, Koppen, Villano, Doignon, and Johannesen (1990). Short introductions to knowledge space theory are contained in Doignon and Falmagne (1987) and Falmagne (1989b). A more leisurely paced text is Doignon (1994a).

The application in the schools and universities that began around 1999 in the form of the internet based software `ALEKS` provided the impetus for further developments, especially in the form of statistical analyses of students' data. General sources for such results are Falmagne, Cosyn, Doignon, and Thiéry (2006a) and Cosyn, Doble, Falmagne, Lenoble, Thiéry, and Uzun (2010). These works can also be regarded as up-to-date, non-technical introductions to the topic. In the course of these applications, an important theoretical change of focus took place, already alluded to in our preface and earlier in this chapter. Originally, the essential axiom of the system was the closure under finite union: if K and L are knowledge states, then so is $K \cup L$. In the framework of finite knowledge structures, this assumption defines the knowledge spaces. While adopting such a rule can be convincingly demonstrated to be sound, its pedagogical justification is not *a priori* overwhelming[17]. Another, quite different reason for a changed outlook lies in the need for a convenient (economical, evocative) description of the knowledge state of a student at the end of an assessment. In this respect, Doignon and Falmagne (1997) showed that wellgradedness is the key concept, because it allows the description of any knowledge state by its fringes. A knowledge state may contain many items. Providing the user—a teacher, for example—with a full list may not be helpful, and summarizing such a large set of items in a meaningful way is a challenge. These considerations, added to the fact that the closure under union axiom was not persuasive, led Falmagne to propose a re-axiomatization of the theory in the guise of a learning space (cf. 1.1.4), which is more compelling from a pedagogical standpoint and now forms the core

[17] The rationale given in Doignon and Falmagne (1999) was that if two students, whatever their respective knowledge states are, collaborate intensely, then one of them may end up mastering all the items originally mastered by either of them.

of the theory. In a learning space, any knowledge state can be summarized uniquely by its fringes (cf. 1.1.5). The definition of "fringe" and the relevant theorem are 4.1.6 and 4.1.7. Recently, Cosyn and Uzun (2009) have shown that a knowledge structure is a learning space if and only if it is a knowledge space satisfying a condition of 'wellgradedness.' This work is discussed and expanded on in Chapter 2.

Since its inception in 1985, the work on knowledge spaces and learning spaces came to the attention of a number of other researchers, who provided their own contributions to the development of the field. Dietrich Albert and his team in Austria, Cornelia Dowling, Ivo Duntsch, Günther Gediga, Jurgen Heller and Ali Ünlü in Germany, Mathieu Koppen in Holland, and Francesca Cristante, Luca Stefanutti and their colleagues at the University of Padua in Italy, must be mentioned in that category. More specific references pertaining to particular aspects of our work and that of others are given in the last section of each chapter of this book. An extensive database on knowledge spaces, with hundreds of titles, is maintained by Cord Hockemeyer at the University of Graz: `http://wundt.uni-graz.at/kst.php` (see also Hockemeyer, 2001).

As indicated in Subsections 1.4.1 to 1.4.4, our results are potentially applicable to other fields, such as computerized medical diagnosis, pattern recognition or the theory of feasible symbologies (for the latter, see Problem 14 of Chapter 3).

The literature on computerized medical diagnosis is vast and quickly expanding. We only mention here the well-known early example of Shortliffe and Buchanan (1975) and Shortliffe (1976). For pattern recognition, we refer the reader to Duda and Hart (1973) and Fu (1974). For symbology theory, the reader may consult Jameson (1992).

There are obvious similarities between knowledge assessment in the framework of learning spaces as developed in this book, and the technique known as 'tailor testing' in psychometrics. In both situations, subjects are presented with a sequence of well-chosen questions, and the goal is to determine, as accurately and as efficiently as possible, their mastery of a given field. There is, however, an essential difference in the theoretical foundations of the two approaches. In psychometric theory, it is assumed that the responses to the items reflect primarily the subject's ability with respect to some intellectual traits, and most often, just one such trait. In constructing the test, the psychometrician chooses and formats the items so as to minimize the impact of other determinants of the subject's performance, such as schooling, culture, or knowledge in the sense of this book. In fact, the primary or sole aim of such psychometric tests is to measure the subject's abilities on some numerical scale. The ubiquitous I.Q. test is the exemplary case of this enterprise. Accordingly, the models underlying a 'tailor testing' procedure tend to be simple numerical structures, in which a subject's ability is represented as a real number[18]. Psychometric models are developed either in the guise of *Clas-*

[18] Or possibly as a real vector with a small number of dimensions.

sical Test Theory (CTT) for which a standard source is still Lord and Novick (1974) (but see also Wainer and Messick, 1983), or of the more recent *Item Response Theory* (IRT) (cf., for example, Nunnally and Bernstein, 1994). For tailor testing, the reader may consult Lord (1974), Weiss (1983), or Wainer, Dorans, Eignor, Flaugher, Green, Mislevy, Steinberg, and Thissen (2000). A few authors have proposed to escape the unidimensionality of classical 'tailor testing.' As an example, Durnin and Scandura (1973) developed tests under the assumption that the subject performs a well-defined, algorithmic treatment of tasks to be solved.

Subsection 1.4.4 provides examples of families of subsets that are closed under intersection. In a sense which will be made precise in Definition 2.2.2, these families are dual to knowledge spaces. There is a vast mathematical literature concerned with the general study of such families, referred to as 'closure spaces', 'abstract convexities', and a few other technical terms.

More detailed comments on relationships of our work with that of others can be found in the sources sections at the end of each chapter.

Problems

The first few problems below cannot be solved rigorously without using the formal apparatus—definitions, axioms—introduced in the following chapters. The reader should rely on the intuitive conception of a learning space developed in this chapter to analyze the problem and attempt a formalization. We propose such exercises as a useful preparation for the rest of this volume.

The other problems are meant to help the reader refresh his command of the concepts and notation recalled in Section 1.6. A reader experiencing difficulties with such problems should study basic texts such as Suppes (1960) or Roberts (1979, 1984).

1. For each of the following pairs, check whether it forms (i) a knowledge structure; (ii) a knowledge space; (iii) a learning space.
   a) $\big(\{a,b,c,d\},\ \big\{\{a\},\ \{a,b\},\ \{a,b,c\},\ \{a,b,c,d\}\big\}\big)$;
   b) $\big(\{a,b,c,d\},\ \big\{\varnothing,\ \{a\},\ \{a,b\},\ \{a,b,c\},\ \{a,b,c,d\}\big\}\big)$;
   c) $\big(\{a,b,c,d\},\ \big\{\varnothing,\ \{a\},\ \{b\},\ \{a,b\},\ \{a,b,c\},\ \{a,b,c,d\}\big\}\big)$;
   d) $\big(\{a,b,c,d\},\ \big\{\varnothing, \{a\}, \{c\}, \{a,b\}, \{c,d\}, \{a,b,c\}, \{b,c,d\}, \{a,b,c,d\}\big\}\big)$;
   e) $\big(\{a,b,c,d\},\ \big\{\varnothing,\ \{a\},\ \{c\},\ \{a,b\},\ \{a,c\},\ \{c,d\},\ \{a,b,c\},\ \{b,c,d\},\ \{a,b,c,d\}\big\}\big)$;
   f) $\big(\{a,b,c,d\},\ \big\{\varnothing,\ \{a\},\ \{c\},\ \{a,b\},\ \{c,d\},\ \{a,b,c\},\ \{a,c,d\},\ \{b,c,d\},\ \{a,b,c,d\}\big\}\big)$.

2. For each of the following pairs, check whether it forms: (i) a knowledge structure; (ii) a knowledge space; (iii) a learning space. Let n and k be two natural numbers, with $0 \leq k \leq n$, and let Q be a set of size n. Consider the various possibilities for the respective values of k and n.
   a) $\big(Q, \{\varnothing, Q\}\big)$;
   b) $\big(Q, \{K \in 2^Q \mid |K| \leq k\} \cup \{Q\}\big)$;
   c) $\big(Q, \{K \in 2^Q \mid |K| \geq k\} \cup \{\varnothing\}\big)$;
   d) $\big(Q, \{K \in 2^Q \mid |K| \text{ is even}\} \cup \{Q\}\big)$.

3. By the learning smoothness axiom of learning spaces (cf. 1.1.4) if two states K and L satisfy $K \subset L$, then the number of states between them is finite. Considering this observation, can a learning space be infinite?

4. Are there knowledge spaces that are not learning spaces? If you believe so, provide a counterexample.

5. Suppose that each of the states in a knowledge structure $\mathcal{K}$ is specified by its fringes (meaning: no other state in $\mathcal{K}$ has the same fringes). Is $\mathcal{K}$ necessarily a learning space, a knowledge space? Justify your response by formal arguments or counterexamples. (If needed, go to Doignon and Falmagne, 1997, for clarification, or see 4.1.8 in this book.)

6. (Continuation.) Can you guess the defining property guaranteeing that the states of a knowledge structure are uniquely specified by their fringes? (Otherwise, see 4.1.8).

7. Verify that the symmetric difference $|X \triangle Y|$ distance between finite sets X and Y satisfies Conditions (1), (2) and (3) of Definition 1.6.12.

8. Prove that, if R, S, T and M are relations, then the two following hold:
$$S \subseteq M \quad \Longrightarrow \quad RST \subseteq RMT \tag{1.3}$$
$$R(S \cup T) \subseteq RS \cup RT. \tag{1.4}$$

9. Using the implication (1.3), give a short proof of the fact that the relative product of two transitive relations is transitive.

10. Prove that the collection $\mathfrak{P}$ of all the partial orders on a given finite set is closed under intersection: that is, if $\mathcal{P}$ and $\mathcal{Q}$ are in $\mathfrak{P}$, then so is $\mathcal{P} \cap \mathcal{Q}$.

11. Spell out the relationship between (reflexive) partial orders and irreflexive partial orders.

12. Prove that a partial order can have at most one maximum element. Provide an example in which neither maximum nor minimum elements exist.

13. Let $\mathcal{P}$ be a strict partial order on a finite set. Suppose that $\mathcal{H}$ is a relation whose transitive closure $t(\mathcal{H})$ is equal to $\mathcal{P}$. Verify that we have then $\mathcal{H} \supseteq \breve{\mathcal{P}}$, with $\breve{\mathcal{P}}$ the covering relation of $\mathcal{P}$. (So, the phrase 'the Hasse diagram of $(X, \mathcal{P})$ is the smallest relation having its transitive closure equal to $\mathcal{P}$' makes sense in 1.6.8.)

14. Let $\breve{\mathcal{P}}$ be the Hasse diagram of a partial order $\mathcal{P}$. Is it true that $\breve{\mathcal{P}}$ is countable if and only if $\mathcal{P}$ is countable? Prove your answer.

15. When is the Hasse diagram of a nonempty partial order empty? Can we have an infinite partial order with a nonempty Hasse diagram? What if the partial order is uncountable?

16. Suppose that $d(\breve{\mathcal{P}}, \breve{\mathcal{Q}}) = n$ for some partial orders $\mathcal{P}$ and $\mathcal{Q}$ on a same finite set Q, where d is the symmetric difference distance defined by (1.2). Is it true that there exists a sequence of partial orders on Q, say $\mathcal{P} = \mathcal{P}_0$, $\mathcal{P}_1$, $\ldots$, $\mathcal{P}_n = \mathcal{Q}$ such that $d(\breve{\mathcal{P}}_{j-1}, \breve{\mathcal{P}}_j) = 1$ for $j = 1, \ldots, n$. Prove or give a counterexample. (In other terms, is the collection of all Hasse diagrams on Q a well-graded family in the sense of 1.1.7?)

2

Knowledge Structures and Learning Spaces

Suppose that some complex system is assessed by an expert, who checks for the presence or absence of some revealing features. Ultimately, the state of the system is described by the subset of features, from a possibly large set, which are detected by the expert. This concept is very general, and becomes powerful only on the background of specific assumptions, in the context of some applications. We begin with the combinatoric underpinnings of the theory formalizing this idea.

2.1 Fundamental Concepts

2.1.1 Example. (*Knowledge structures in education.*) A teacher is examining a student to determine, for instance, which mathematics courses would be appropriate at this stage of the student's career, or whether the student should be allowed to graduate. The teacher will ask one question, then another, chosen as a function of the student's response to the first one. After a few questions, a picture of the student's knowledge state will emerge, which will become increasingly more precise in the course of the examination. By 'knowledge state' we mean here the set of all problems that the student is capable of solving in ideal conditions[1]. The next few definitions provides a rigorous framework.

2.1.2 Definition. A *knowledge structure* is a pair $(Q, \mathcal{K})$ in which Q is a nonempty set, and $\mathcal{K}$ is a family of subsets of Q, containing at least Q and the empty set $\varnothing$. The set Q is called the *domain* of the knowledge structure. Its elements are referred to as *questions* or *items* and the subsets in the family $\mathcal{K}$ are labeled *(knowledge) states*. Occasionally, we shall say that $\mathcal{K}$ is a *knowledge structure on* a set Q to mean that $(Q, \mathcal{K})$ is a knowledge structure. The specification of the domain can be omitted without ambiguity since we have $\cup\mathcal{K} = Q$.

[1] We assume, for the time being, that there are no careless errors or lucky guesses.

J.-C. Falmagne, J.-P. Doignon, *Learning Spaces*,
DOI 10.1007/978-3-642-01039-2_2, © Springer-Verlag Berlin Heidelberg 2011

2.1.3 Example. Consider the domain $U = \{a, b, c, d, e, f\}$ equipped with the knowledge structure

$$\mathcal{H} = \{\varnothing, \{d\}, \{a, c\}, \{e, f\}, \{a, b, c\}, \{a, c, d\}, \{d, e, f\}, \\ \{a, b, c, d\}, \{a, c, e, f\}, \{a, c, d, e, f\}, U\}. \tag{2.1}$$

As illustrated by this example, we do not assume that all subsets of the domain are states. The knowledge structure $\mathcal{H}$ contains eleven states out of sixty-four possible subsets of U.

2.1.4 Definition. Let $\mathcal{F}$ be a family of sets. We denote by $\mathcal{F}_q$ the collection of all sets in $\mathcal{F}$ containing q. In the knowledge structure $\mathcal{H}$ of Example 2.1.3, we have, for instance

$$\mathcal{H}_a = \{\{a, c\}, \{a, b, c\}, \{a, c, d\}, \{a, b, c, d\}, \{a, c, e, f\}, \{a, c, d, e, f\}, U\},$$
$$\mathcal{H}_e = \{\{e, f\}, \{d, e, f\}, \{a, c, e, f\}, \{a, c, d, e, f\}, U\}.$$

Items a and c carry the same information relative to $\mathcal{H}$ in the sense that they are contained in the same states: any state containing a also contains c, and vice versa. In other terms, we have $\mathcal{H}_a = \mathcal{H}_c$. From a practical viewpoint, any individual whose state contains item a has necessarily mastered item c, and vice versa. Thus, in testing the acquired knowledge of a subject, only one of these two questions must be asked. Similarly, we also have $\mathcal{H}_e = \mathcal{H}_f$.

2.1.5 Definition. In a knowledge structure $(Q, \mathcal{K})$, the set of all the items contained in the same states as item q is denoted by q^* and is called a *notion*; we thus have

$$q^* = \{r \in Q \,|\, \mathcal{K}_q = \mathcal{K}_r\}.$$

The collection Q^* of all notions is a partition of the set Q of items. When two items belong to the same notion, we shall sometimes say that they are *equally informative*. In such a case, the two items form a pair in the equivalence relation on Q associated to the partition Q^*.

In Example 2.1.3, we have the four notions

$$a^* = \{a, c\}, \quad b^* = \{b\}, \quad d^* = \{d\}, \quad e^* = \{e, f\},$$

forming the partition $U^* = \{\{a, c\}, \{b\}, \{d\}, \{e, f\}\}$.

A knowledge structure in which each notion contains a single item is called *discriminative*. A discriminative knowledge structure can always be manufactured from any knowledge structure $(Q, \mathcal{K})$ by forming the notions, and constructing the knowledge structure $\mathcal{K}^*$ induced by $\mathcal{K}$ on Q^* via the definitions

$$K^* = \{q^* \,|\, q \in K\} \qquad (K \in \mathcal{K})$$
$$\mathcal{K}^* = \{K^* \,|\, K \in \mathcal{K}\}.$$

Note that as $\varnothing$, $Q \in \mathcal{K}$ and $\varnothing^* = \varnothing$, we have $\varnothing, Q^* \in \mathcal{K}^*$.

The knowledge structure $(Q^*, \mathcal{K}^*)$ is called the *discriminative reduction* of $(Q, \mathcal{K})$. Since this construction is straightforward, we shall often simplify matters and suppose that a particular knowledge structure under consideration is discriminative.

2.1.6 Example. We construct the discriminative reduction of the knowledge structure $(U, \mathcal{H})$ of Example 2.1.3 by setting

$$a^* = \{a, c\}, \quad b^* = \{b\}, \quad d^* = \{d\}, \quad e^* = \{e, f\};$$
$$U^* = \{a^*, b^*, d^*, e^*\};$$
$$\mathcal{H}^* = \big\{\varnothing, \{d^*\}, \{a^*\}, \{e^*\}, \{a^*, b^*\}, \{a^*, d^*\}, \{d^*, e^*\}, \{a^*, b^*, d^*\}, \{a^*, e^*\}, \{a^*, d^*, e^*\}, U^*\big\}.$$

Thus, $(U^*, \mathcal{H}^*)$ is formed by aggregating equally informative items from U. The graph of the discriminative reduction $\mathcal{H}^*$ is displayed in Figure 2.1. (The graph of a knowledge structure was introduced in 1.1.3.)

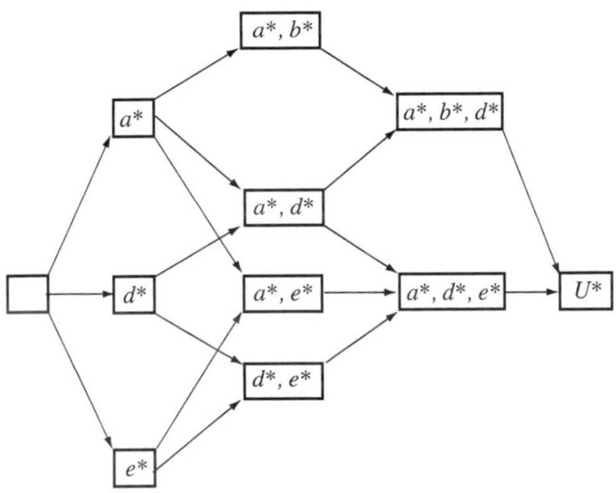

Figure 2.1. The discriminative reduction of the knowledge structure $\mathcal{H}$ of Eq. (2.1).

2.1.7 Definition. A knowledge structure $(Q, \mathcal{K})$ is called *finite* (respectively *essentially finite*), if Q (respectively $\mathcal{K}$) is finite. A similar definition holds for *countable* (respectively *essentially countable*) knowledge structures.

Typically, knowledge structures encountered in education are essentially finite. They may not be finite however: at least conceptually, some notions may contain a potentially infinite number of equally informative questions. Problems 3 and 4 require the reader to show that any knowledge structure $\mathcal{K}$ has the same cardinality as its discriminative reduction $\mathcal{K}^*$, and that the knowledge structure $(Q, \mathcal{K})$ is essentially finite if and only if Q^* is finite.

As suggested by these first few definitions, our choice of terminology is primarily guided by Example 2.1.1, which has also motivated many of our theoretical developments. However, as illustrated by Examples 1.4.1 to 1.4.4 our results are potentially applicable to very different fields.

An important special case of a knowledge structure arises when the family of states is a learning space.

2.2 Axioms for Learning Spaces

2.2.1 Definition. A knowledge structure $(Q, \mathcal{K})$ is called a *learning space* if it satisfies the two following conditions.

[L1] LEARNING SMOOTHNESS. For any two states K, L such that $K \subset L$, there exists a finite chain of states

$$K = K_0 \subset K_1 \subset \cdots \subset K_p = L \tag{2.2}$$

such that $|K_i \setminus K_{i-1}| = 1$ for $1 \leq i \leq p$ and so $|L \setminus K| = p$.

Intuitively, in pedagogical language: *If the state K of the learner is included in some other state L then the learner can reach state L by mastering the missing items one at a time.*

In the sequel, we refer to a chain (2.2) as an *L1–chain* from K to L.

[L2] LEARNING CONSISTENCY. If K, L are two states satisfying $K \subset L$ and q is an item such that $K + \{q\} \in \mathcal{K}$, then $L \cup \{q\} \in \mathcal{K}$.

In short: *Knowing more does not prevent learning something new.*

Notice that any learning space is finite. Indeed, Condition [L1] applied to the two states $\varnothing$ and Q implies that Q is a finite set.

From the pedagogical standpoint of Example 2.1.1, both of these axioms seem sensible. This mathematical object occurs in another field, however. In the combinatoric literature, a learning space is sometimes referred to as an 'antimatroid', which is then typically defined by different (but equivalent) axioms (e.g. Korte, Lovász, and Schrader, 1991). As we shall see in Definition 2.2.2, this structure is a family of sets closed under union and satisfying a particular 'accessibility' condition. Originally, however, the label 'antimatroid' was attached to their dual structures, namely, to families closed under intersection (see in particular Edelman and Jamison, 1985; Welsh, 1995; Björner, Las Vergnas, Sturmfels, White, and Ziegler, 1999). As the dates of the last two references suggest, this dual usage is still current. For an overview of the origins and the many avatars of the 'antimatroid' concept, we refer the reader to Monjardet (1985).

Our next definition formalizes various properties ensuing from Axioms [L1] and [L2]. We have encountered two of them earlier, namely, closure under union and wellgradedness (cf. 1.1.6 and 1.1.7). Another one is that of a '(union-closed) antimatroid.' Theorem 2.2.4 spells out the relationships among all these concepts.

2.2.2 Definition. A family $\mathcal{K}$ is *closed under union* when $\cup\mathcal{F} \in \mathcal{K}$ whenever $\mathcal{F} \subseteq \mathcal{K}$. This implies $\varnothing \in \mathcal{K}$, because by convention the union of the empty subfamily is the empty set. When a family is closed under union, we will sometimes say for short that it is *union-closed*, or even $\cup$-*closed*. When the family $\mathcal{K}$ of a knowledge structure $(Q, \mathcal{K})$ is union-closed, we call $(Q, \mathcal{K})$ a *knowledge space*; we may also say equivalently, that $\mathcal{K}$ is a *knowledge space*.

On occasion, we also use the phrase *closed under finite union*. When applied to a family $\mathcal{K}$, it means that for any K and L in $\mathcal{K}$, the set $K \cup L$ is also in $\mathcal{K}$. Note that, in such a case, the empty set does not necessarily belong to the family $\mathcal{K}$.

The *dual* of a knowledge structure $\mathcal{K}$ on Q is the knowledge structure $\overline{\mathcal{K}}$ containing all the complements of the states of $\mathcal{K}$, that is, the family

$$\overline{\mathcal{K}} = \{K \in 2^Q \,|\, Q \setminus K \in \mathcal{K}\}.$$

Thus, $\mathcal{K}$ and $\overline{\mathcal{K}}$ have the same domain. It is clear that if $\mathcal{K}$ is a knowledge space, then $\overline{\mathcal{K}}$ is an *intersection-closed* knowledge structure, that is, $\cap\mathcal{F} \in \overline{\mathcal{K}}$ whenever $\mathcal{F} \subseteq \overline{\mathcal{K}}$, with $\varnothing, Q \in \overline{\mathcal{K}}$.

We recall (from 1.6.12) that the canonical distance d between two finite sets A and B is defined by counting the number of elements in their symmetric difference $A \triangle B$:

$$d(A, B) = |A \triangle B| = |(A \setminus B) \cup (B \setminus A)|. \tag{2.3}$$

A family of sets $\mathcal{F}$ is *well-graded* or a *wg-family* if, for any two distinct sets $K, L \in \mathcal{F}$, there is a finite sequence of states $K = K_0, K_1, \ldots, K_p = L$ such that $d(K_{i-1}, K_i) = 1$ for $1 \leq i \leq p$ and moreover $p = d(K, L)$. We call the sequence of sets (K_i) a *tight path* from K to L. It is clear that a well-graded knowledge structure is finite and discriminative (Problem 2).

A family $\mathcal{K}$ of subsets of a finite set $Q = \cup\mathcal{K}$ is an *antimatroid* if it is closed under union and satisfies the following axiom:

[MA] If K is a nonempty subset of Q belonging to the family $\mathcal{K}$, then there is some q in K such that $K \setminus \{q\} \in \mathcal{K}$.

We call the sets in $\mathcal{K}$ states, and we also say then that the pair $(Q, \mathcal{K})$ is an antimatroid. It is clear that $(Q, \mathcal{K})$ is then a discriminative knowledge structure. A family $\mathcal{K}$ satisfying Axiom [MA] is said to be *accessible* or *downgradable* (see Doble, Doignon, Falmagne, and Fishburn, 2001, for the latter term).

The following result allows us to trim some proofs.

2.2.3 Lemma. *The two following conditions are equivalent for a* $\cup$*-closed family of sets* $\mathcal{K}$*:*

(i) $\mathcal{K}$ *is well-graded;*
(ii) *for any two sets* K *and* L *such that* $K \subset L$*, there is a tight path from* K *to* L*.*

PROOF. It is clear that (i) implies (ii). Suppose that (ii) is true. For any two distinct sets K and L, there exists a tight path $K = K_0 \subset K_1 \subset \cdots \subset K_q = K \cup L$ and another tight path $L = L_0 \subset L_1 \subset \cdots \subset L_p = K \cup L$. Reversing the order of the sets in the latter tight path and redefining $K_{q+1} = L_{p-1}$, $K_{q+2} = L_{p-2}$, ..., $K_{q+p} = L = L_0$, we get the tight path $K = K_0$, K_1, ..., $K_{q+p} = L$, with $|K \triangle L| = q + p$. □

When applied to knowledge structures, the wellgradedness property is a strengthening of [L1]: any L1–chain is a special kind of tight path.

All three of the conditions introduced in Definition 2.2.2—∪-closure, wellgradedness, and accessibility—hold in any learning space. In fact, we have the following result.

2.2.4 Theorem. *For any knowledge structure $(Q, \mathcal{K})$, the following three conditions are equivalent.*

(i) $(Q, \mathcal{K})$ *is a learning space.*
(ii) $(Q, \mathcal{K})$ *is an antimatroid.*
(iii) $(Q, \mathcal{K})$ *is a well-graded knowledge space.*

The equivalence of (i) and (iii) was established by Cosyn and Uzun (2009). Clearly, under each of the three conditions, the knowledge structure $(Q, \mathcal{K})$ is discriminative. Note in passing that this result also holds under a substantially weaker form of Axiom [MA] (cf. Condition (iii) in Theorem 5.4.1)[2].

PROOF. (i) ⇒ (ii). Suppose that $(Q, \mathcal{K})$ is a learning space. Thus, Q is necessarily finite. Axiom [MA] results immediately from the fact that, for any state K, there is an L1-chain from $\varnothing$ to K. Turning to the ∪-closure, we take any two states K and L in $\mathcal{K}$ and suppose that neither of them is empty or a subset of the other (otherwise ∪-closure holds trivially). Since $\varnothing \subset L$, Axiom [L1] implies the existence of an L1–chain $\varnothing \subset \{q_1\} \subset \cdots \subset \{q_1, \ldots, q_n\} = L$.

Let $j \in \{1, \ldots, n\}$ be the first index with $q_j \notin K$. If $j > 1$; we have

$$\{q_1, \ldots, q_{j-1}\} \subset K, \text{ and } \{q_1, \ldots, q_{j-1}\} + \{q_j\} \in \mathcal{K}. \tag{2.4}$$

By Axiom [L2] and with $q_j \in L$, we get $K + \{q_j\} \in \mathcal{K}$ with $K + \{q_j\} \subseteq K \cup L$. A similar argument applies with $\varnothing \subset K$ in (2.4) if $j = 1$. Applying induction yields $K \cup L \in \mathcal{K}$.

(ii) ⇒ (iii). Only the wellgradedness must be established. We use Lemma 2.2.3. Take any two states K, L with $K \subset L$ (with possibly $K = \varnothing$). Repeated applications of Axiom [MA] to state L gives us a sequence of states $L_0 = L, L_1, \ldots, L_k = \varnothing$ such that $q_{i-1} \in L_{i-1}$ and $L_i = L_{i-1} \setminus \{q_{i-1}\}$ for $i = 1, \ldots, k$. Let j be the largest index such that $q_j \notin K$ (there must be such an index since $K \subset L$). We obtain $K \subset K \cup \{q_j\} = K \cup L_j \subseteq L$. Replacing K with $K \cup \{q_j\}$ and using induction we see that the condition in Lemma 2.2.3 is satisfied, and so the wellgradedness of $(Q, \mathcal{K})$.

---

[2] We give yet another characterization of learning spaces in Theorem 11.5.3.

(iii) $\Rightarrow$ (i). Axiom [L1] results from the wellgradedness condition. Suppose that $K \subset L$ for two states K and L and that $K + \{q\}$ is also a state. By $\cup$-closure, the set $(K + \{q\}) \cup L = L \cup \{q\}$ is also a state; so, [L2] holds. □

2.2.5 Remarks. The concept of a well-graded knowledge space was investigated by Falmagne and Doignon (1988b). In the early stage of our work, knowledge spaces were at the focus of our developments. From a pedagogical standpoint, they were motivated by the following argument.

Consider the case of two students engaged in extensive interactions for a long time, and suppose that their initial knowledge states with respect to a particular body of information are K and L. At some point, one of these students could conceivably have acquired the joint knowledge of both. The knowledge state of this student would then be $K \cup L$. Obviously, there is no certainty that this will happen. However, requiring the existence of a state in the structure to cover this case may be reasonable.

Some may find such an argument only moderately convincing. As for the wellgradedness condition, its *a priori* justification is far from obvious. Yet, the two conditions are equivalent to [L1]-[L2]. In fact, both the $\cup$-closed and wellgradedness conditions do play critical roles, but their pedagogical imports are subtle. We will see in Chapter 3 that the $\cup$-closed condition makes it possible to summarize any knowledge space by its 'base'[3], which is a typically much smaller subfamily of the knowledge space. In view of the very large size of the knowledge structures encountered in practice, this feature is precious because it facilitates computation. As for the wellgradedness condition, it guarantees that any state can be faithfully represented by its two 'fringes' which are also comparatively much smaller sets[4] (cf. Theorem 4.1.7).

We devote Chapters 3 and 4 to a detailed discussion of knowledge spaces and well-graded knowledge structures, respectively.

As a preparation for our next section, a weakening of some of the concepts of Theorem 2.2.4 is in order.

2.2.6 Definition. A family $\mathcal{F}$ of subsets of a nonempty set Q is a *partial knowledge structure* if it contains the set $Q = \cup\mathcal{F}$. The discriminative concept introduced in Definition 2.1.5 also applies in the partial case. We do not assume that $|\mathcal{F}| \geq 2$. We also call the sets in $\mathcal{F}$ *states*. A partial knowledge structure $\mathcal{F}$ is a *partial learning space* if it satisfies Axioms [L1] and [L2]. A family $\mathcal{F}$ is *partially* $\cup$*-closed* if for any nonempty subfamily $\mathcal{G}$ of $\mathcal{F}$, we have $\cup\mathcal{G} \in \mathcal{F}$. (Contrary to the $\cup$-closure condition, partial $\cup$-closure does not imply that the empty set belongs to the family.) A *partial knowledge space* $\mathcal{F}$ is a partial knowledge structure that is partially $\cup$-closed.

---

[3] This concept is defined and investigated in Section 3.4.

[4] In practical applications of the concept of a learning space, the fringes summarize—without loss of information—a knowledge state in a way that may be more meaningful, for the teacher and the student, than a full listing of all the items mastered.

The equivalence (i) $\Leftrightarrow$ (iii) in Theorem 2.2.4 ceases to hold in the case of partial structures. Indeed, we have the following result.

2.2.7 Lemma. *Any well-graded partially $\cup$-closed family is a partial learning space. The converse implication is false.*

PROOF. Let $\mathcal{K}$ be a well-graded partially $\cup$-closed family. Axiom [L1] is a special case of the wellgradedness condition. If $K \subset L$ for two sets K and L in $\mathcal{K}$ and $K + \{q\}$ is in $\mathcal{K}$, then the set $(K + \{q\}) \cup L = L \cup \{q\}$ is in $\mathcal{K}$ by partial $\cup$-closure, and so [L2] holds. The example below disproves the converse. $\square$

2.2.8 Example. The family of sets

$$\mathcal{L} = \{\{a\}, \{c\}, \{a, b\}, \{b, c\}, \{a, b, c\}\}$$

is a partial learning space. It is the union of the two chains

$$\{a\} \subset \{a, b\} \subset \{a, b, c\}, \qquad \{c\} \subset \{b, c\} \subset \{a, b, c\}$$

with $\cup\mathcal{L}$ as the only common state. However, $\mathcal{L}$ is neither $\cup$-closed nor well-graded. The knowledge structure $\mathcal{L}' = \{\varnothing\} \cup \mathcal{L}$ does not satisfy [L1] since we have $\varnothing \subset \{a\}$ with $\varnothing + \{c\}$ as a state of $\mathcal{L}'$, but $\{a\} \cup \{c\}$ is not a state.

2.3 The nondiscriminative case*

The axiomatics given in the previous section for learning spaces and well-graded knowledge spaces imply that their models are always discriminative knowledge structures (in the sense of Definition 2.1.5). It is straightforward to adapt the axioms so as to cover nondiscriminative structures. We state here the modified axioms and review some of their consequences, without going into much detail.

2.3.1 Definition. In the case of structures that may not be discriminative, we must modify the concept of distance between two states in the structure. Rather than counting the number of items by which two states differ, we count here the number of notions. Remember from 2.1.5 that q^* denotes the notion containing the item q, and that for any state K, we set $K^* = \{q^* \mid q \in K\}$.

Suppose that $(Q, \mathcal{K})$ is an essentially finite knowledge structure. Let K and L be two states in $(Q, \mathcal{K})$. The *essential distance* between K and L is defined by

$$e(K, L) = |K^* \mathbin{\triangle} L^*|.$$

We can verify that the function $e : \mathcal{K} \times \mathcal{K} \to \mathbb{R}$ is a distance in the usual sense (cf. 1.6.12).

2.3.2 Definition. A knowledge structure $(Q, \mathcal{K})$ is called a *quasi learning space* if it satisfies the two following conditions.

[L1*] QUASI LEARNING SMOOTHNESS. For any two states K, L such that $K \subset L$ there exists a chain of $1 + p$ states

$$K = K_0 \subset K_1 \subset \cdots \subset K_p = L \tag{2.5}$$

with $p = e(K, L)$ and $K_i = K_{i-1} + \{q_i^*\}$ for some $q_i \in Q$, $1 \leq i \leq p$.

In the sequel, we refer to a chain (2.2) as a *quasi L1–chain* from K to L.

[L2*] QUASI LEARNING CONSISTENCY. If K, L are two states satisfying $K \subset L$ and q is an item such that $K + \{q^*\} \in \mathcal{K}$, then $L \cup \{q^*\} \in \mathcal{K}$.

Our next definition introduces a nondiscriminative variant of the wellgradedness condition.

2.3.3 Definition. A family of sets $\mathcal{F}$ is *quasi well-graded* or a *qwg-family* if, for any two distinct states $K, L \in \mathcal{F}$, there exists a finite sequence of states $K = K_0$, K_1, $\ldots$, $K_p = L$ such that $e(K_{i-1}, K_i) = 1$ for $1 \leq i \leq p$ and moreover $p = e(K, L)$. We call the sequence of sets (K_i) a *quasi tight path* from K to L. A quasi well-graded knowledge structure is essentially finite (Problem 16).

We leave to the reader the verification of the following result, which extends the equivalence (i) $\Leftrightarrow$ (iii) in Theorem 2.2.4 (see Problem 9) .

2.3.4 Theorem. *For any knowledge structure $(Q, \mathcal{K})$, the following two conditions are equivalent.*

(i) *$(Q, \mathcal{K})$ is a quasi learning space.*
(ii) *$(Q, \mathcal{K})$ is a quasi well-graded knowledge space.*

We will not pursue here the extension of the theory to structures which may not be discriminative. In any event, discriminative reduction is always at hand to generate a discriminative structure from a nondiscriminative one.

2.4 Projections

As we argued before, an empirical learning space can be very large, numbering millions of states. The concept of a 'projection' discussed in this section provides a way of parsing such a large structure into meaningful components. Moreover, when the learning space concerns a scholarly curriculum such as high school algebra, a projection may provide a convenient instrument for the programming of a placement test.

The key idea is that if $\mathcal{K}$ is a learning space on a domain Q, then any proper subset Q' of Q defines a learning space $\mathcal{K}_{|Q'}$ on Q' which is in some sense consonant with $\mathcal{K}$. We call $\mathcal{K}_{|Q'}$ a 'projection' of $\mathcal{K}$ on Q', a terminology consistent with that used for media by Cavagnaro (2008) and Eppstein, Falmagne, and Ovchinnikov (2008). (We discuss the relationship between media and learning spaces in Chapter 10.) Moreover, this construction defines a partition of $\mathcal{K}$ such that each equivalence class is a subfamily of $\mathcal{K}$ satisfying two key properties of a learning space, namely wellgradedness and $\cup$-closure. Actually, it is possible to choose Q' so that each of these classes is essentially (via a trivial transformation) either a learning space consistent with $\mathcal{K}$ or the singleton $\{\varnothing\}$. These results, which are mostly due to Falmagne (2008), are presented in this section.

2.4.1 Definition. Suppose that $(Q, \mathcal{K})$ is a partial knowledge structure with $|Q| \geq 2$, and let Q' be any proper nonempty subset of Q. Define a relation $\sim_{Q'}$ on $\mathcal{K}$ by

$$K \sim_{Q'} L \iff K \cap Q' = L \cap Q' \tag{2.6}$$

$$\iff K \triangle L \subseteq Q \setminus Q'. \tag{2.7}$$

Thus, $\sim_{Q'}$ is an equivalence relation on $\mathcal{K}$. When the context specifies the subset Q', we sometimes use the shorthand $\sim$ for $\sim_{Q'}$ in the sequel. The equivalence between the right hand sides of (2.6) and (2.7) is easily verified (cf. Problem 11). We denote by $[K]$ the equivalence class of $\sim$ containing K, and by $\mathcal{K}_{\sim} = \{[K] \mid K \in \mathcal{K}\}$ the partition of $\mathcal{K}$ induced by $\sim$. We may also say for short that such a partition is *induced* by the set Q'. In the sequel, we always assume that $|Q| \geq 2$, so that $|Q'| \geq 1$.

2.4.2 Definition. Let $(Q, \mathcal{K})$ be a partial knowledge structure and take any nonempty proper subset Q' of Q. The family

$$\mathcal{K}_{|Q'} = \{W \subset Q' \mid W = K \cap Q' \text{ for some } K \in \mathcal{K}\} \tag{2.8}$$

is called the *projection* of $\mathcal{K}$ on Q'. We thus have $\mathcal{K}_{|Q'} \subseteq 2^{Q'}$. Depending on the context, we may also refer to $\mathcal{K}_{|Q'}$ as a *substructure* of $\mathcal{K}$. Each set $W = K \cap Q'$ with $K \in \mathcal{K}$ is called the *trace* of the state K on Q'. Example 2.4.3 shows that the sets in $\mathcal{K}_{|Q'}$ may not be states of $\mathcal{K}$. For any state K in $\mathcal{K}$ and with $[K]$ as in Definition 2.4.1, we define the family

$$\mathcal{K}_{[K]} = \{M \subseteq Q \mid M = L \setminus \cap[K] \text{ for some } L \sim K\}. \tag{2.9}$$

(If $\varnothing \in \mathcal{K}$, we thus have $\mathcal{K}_{[\varnothing]} = [\varnothing]$.) The family $\mathcal{K}_{[K]}$ is called a *Q'-child*, or simply a *child* of $\mathcal{K}$ when the set Q' is made clear by the context. As shown by our next example, a child of $\mathcal{K}$ may take the form of the singleton $\{\varnothing\}$ and we may have $\mathcal{K}_{[K]} = \mathcal{K}_{[L]}$ even when $K \not\sim L$. The set $\{\varnothing\}$ is called the *trivial* child. We refer to $\mathcal{K}$ as the *parent* structure.

2.4.3 Example. Consider the learning space

$$\mathcal{F} = \{\varnothing,\{b\},\{c\},\{a,b\},\{a,c\},\{b,c\},\{b,d\},\{a,b,c\},\{a,b,d\},\{b,c,d\},\{b,c,e\},$$
$$\{b,d,f\},\{a,b,c,d\},\{a,b,c,e\},\{b,c,d,e\},\{b,c,d,f\},\{b,c,e,f\},$$
$$\{a,b,d,f\},\{a,b,c,d,e\},\{a,b,c,d,f\},\{a,b,c,e,f\},\{b,c,d,e,f\},$$
$$\{a,b,c,d,e,f\},\{a,b,c,d,e,f,g\}\}. \tag{2.10}$$

The domain of this learning space is thus the set $Q = \{a,b,c,d,e,f\}$. The inclusion graph of $\mathcal{F}$ is pictured by the grey parts of the diagram of Figure 2.2.

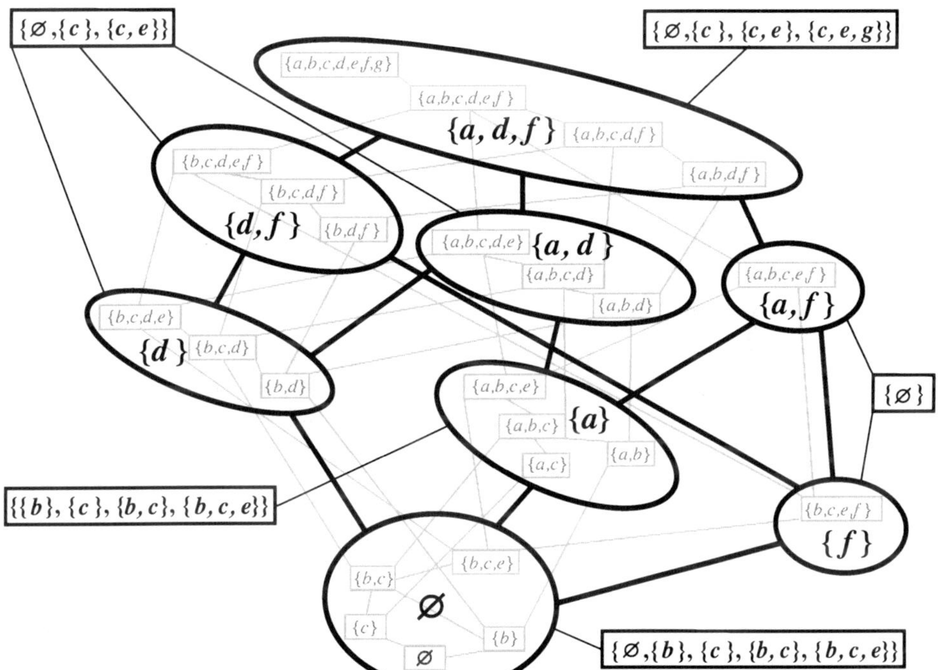

Figure 2.2. In grey, the inclusion graph of the learning space $\mathcal{F}$ of Equation (2.10). Each oval surrounds an equivalence class $[K]$ (in grey) and a particular state (in black) of the projection $\mathcal{F}_{|\{a,d,f\}}$ of $\mathcal{F}$ on $Q' = \{a,d,f\}$, signaling a 1-1 correspondence $\mathcal{F}_{\sim} \to \mathcal{F}_{|\{a,d,f\}}$ (cf. Lemma 2.4.5(ii)). Via the defining equation (2.9), the eight equivalence classes produce five children of $\mathcal{F}$, which are represented in the five black rectangles at the edge of the figure. One of these children is the singleton set $\{\varnothing\}$ (thus, the trivial child), and the others are learning spaces or partial learning spaces (cf. the projection Theorems 2.4.8 and 2.4.12).

The sets marked in black in the eight ovals of the figure represent the states of the projection $\mathcal{F}_{|\{a,d,f\}}$ of $\mathcal{F}$ on the set $\{a, d, f\}$. It is clear that $\mathcal{F}_{|\{a,d,f\}}$ is a learning space[5]. Each of these ovals also surrounds the inclusion subgraph corresponding to an equivalence class of the partition $\mathcal{F}_{\sim}$. This is consistent with Lemma 2.4.5(ii) below, according to which there is a 1-1 correspondence between $\mathcal{F}_{\sim}$ and $\mathcal{F}_{|\{a,d,f\}}$. In this example, the 'learning space' property is transmitted to the children: not only is $\mathcal{F}_{|\{a,d,f\}}$ a learning space, but also any child of $\mathcal{F}$ is a learning space or a partial learning space. Indeed, we have

$$\begin{aligned}
\mathcal{F}_{[\{b,c,e\}]} &= \{\varnothing, \{b\}, \{c\}, \{b,c\}, \{b,c,e\}\},\\
\mathcal{F}_{[\{a,b,c,e\}]} &= \{\{b\}, \{c\}, \{b,c\}, \{b,c,e\}\},\\
\mathcal{F}_{[\{b,c,d,e\}]} = \mathcal{F}_{[\{b,c,d,e,f\}]} = \mathcal{F}_{[\{a,b,c,d,e\}]} &= \{\varnothing, \{c\}, \{c,e\}\},\\
\mathcal{F}_{[\{a,b,c,d,e,f,g\}]} &= \{\varnothing, \{c\}, \{c,e\}, \{c,e,g\}\}\\
\mathcal{F}_{[\{b,c,e,f\}]} = \mathcal{F}_{[\{a,b,c,e,f\}]} &= \{\varnothing\}.
\end{aligned}$$

These five children are represented in the five black rectangles in Figure 2.2.

Theorem 2.4.8 shows that wellgradedness is inherited by the children of a learning space. These children are also partially $\cup$-closed. In the particular case of this example, just adding the set $\varnothing$ to the child not containing it already, that is, to the child $\mathcal{F}_{[\{a,b,c,e\}]}$, would result in all the children being learning spaces or trivial. This is **not** generally true. The situation is clarified by Theorem 2.4.12.

2.4.4 Remark. The concept of projection for learning spaces is closely related to the concept bearing the same name for media introduced by Cavagnaro (2008). The Projection Theorems 2.4.8 and 2.4.12, the main results of this section, could be derived via similar results concerning the projections of media (cf. Theorem 2.11.6 in Eppstein et al., 2008). This would be a detour, however. The route followed here is direct.

In the next two lemmas, we derive a few consequences of Definition 2.4.2.

2.4.5 Lemma. *The following two statements are true for any partial knowledge structure* $(Q, \mathcal{K})$.

(i) *The projection* $\mathcal{K}_{|Q'}$, *with* $Q' \subset Q$, *is a partial knowledge structure. If* $(Q, \mathcal{K})$ *is a knowledge structure, then so is* $\mathcal{K}_{|Q'}$.

(ii) *The function* $h : [K] \mapsto K \cap Q'$ *is a well defined bijection of* $\mathcal{K}_{\sim}$ *onto* $\mathcal{K}_{|Q'}$.

PROOF. (i) Both statements follow from $\varnothing \cap Q' = \varnothing$ and $Q \cap Q' = Q'$.

(ii) That h is a well defined function is due to (2.6). Clearly, $h(\mathcal{K}_{\sim}) = \mathcal{K}_{|Q'}$ by the definitions of h and $\mathcal{K}_{|Q'}$. Suppose that, for some $[K], [L] \in \mathcal{K}_{\sim}$, we have $h([K]) = K \cap Q' = h([L]) = L \cap Q' = X$. Whether or not $X = \varnothing$, this entails $K \sim L$ and so $[K] = [L]$. □

[5] This property holds in general. Notice that we have here the special case in which $\mathcal{F}_{|\{a,d,f\}}$ is the power set of $\{a, d, f\}$. However, it is not generally true that for any learning space $(Q, \mathcal{K})$ and $Q' \subset Q$, we have $\mathcal{K}_{|Q'} = 2^{Q'}$.

2.4.6 Lemma. *Let $\mathcal{K}$ be any $\cup$-closed family, with $Q = \cup\mathcal{K}$ not necessarily in $\mathcal{K}$, and take any $Q' \subset Q$. The following three statements are then true.*

(i) $K \sim_{Q'} \cup[K]$ *for any* $K \in \mathcal{K}$;

(ii) $\mathcal{K}_{|Q'}$ *is a $\cup$-closed family. If $\mathcal{K}$ is a knowledge space, so is* $\mathcal{K}_{|Q'}$.

(iii) *The children of $\mathcal{K}$ are also partially $\cup$-closed.*

For knowledge spaces, Lemma 2.4.6(ii) was obtained by Doignon and Falmagne (1999, Theorem 1.16 on p. 25, in which the term 'substructure is used instead of 'projection').

PROOF. (i) As $\cup[K]$ is the union of states of $\mathcal{K}$, we get $\cup[K] \in \mathcal{K}$. We must have $K \cap Q' = (\cup[K]) \cap Q'$ because $K \cap Q' = L \cap Q'$ for all $L \in [K]$; so $K \sim \cup[K]$.

(ii) Any subfamily $\mathcal{H}' \subseteq \mathcal{K}_{|Q'}$ is associated to the family

$$\mathcal{H} = \{H \in \mathcal{K} \,|\, H' = H \cap Q' \text{ for some } H' \in \mathcal{H}'\}.$$

As $\mathcal{K}$ is $\cup$-closed, we have $\cup\mathcal{H} \in \mathcal{K}$, yielding $Q' \cap (\cup\mathcal{H}) = \cup\mathcal{H}' \in \mathcal{K}_{|Q'}$. If $\mathcal{K}$ is a knowledge space, then $Q \in \mathcal{K}$, which implies $Q' \in \mathcal{K}_{|Q'}$. Thus $\mathcal{K}_{|Q'}$ is a knowledge space.

(iii) Take $K \in \mathcal{K}$ arbitrarily. We must show that $\mathcal{K}_{[K]}$ is $\cup$-closed. For any $\mathcal{H} \subseteq \mathcal{K}_{[K]}$ we define the associated family

$$\mathcal{H}^\dagger = \{H^\dagger \in \mathcal{K} \,|\, H^\dagger \sim K,\, H^\dagger \setminus \cap[K] \in \mathcal{H}\}.$$

So, $\mathcal{H}^\dagger \subseteq [K]$, which gives $L \cap Q' = K \cap Q'$ for any $L \in \mathcal{H}^\dagger$. Since $\mathcal{K}$ is $\cup$-closed, we have $\cup\mathcal{H}^\dagger \in \mathcal{K}$. We thus get $(\cup\mathcal{H}^\dagger) \cap Q' = K \cap Q'$ and $\cup\mathcal{H}^\dagger \sim K$.

The $\cup$-closure of $\mathcal{K}_{[K]}$ follows from the string of equalities

$$\cup\mathcal{H} = \cup_{H^\dagger \in \mathcal{H}^\dagger}(H^\dagger \setminus \cap[K]) = \cup_{H^\dagger \in \mathcal{H}^\dagger}(H^\dagger \cap (\overline{\cap[K]}) = (\cup_{H^\dagger \in \mathcal{H}^\dagger} H^\dagger) \setminus \cap[K]$$

which gives $\cup\mathcal{H} \in \mathcal{K}_{[K]}$ because $K \sim \cup\mathcal{H}^\dagger \in \mathcal{K}$.

Example 2.4.7 shows that the reverse implications in (ii) and (iii) do not hold. □

2.4.7 Example. Consider the projection of the knowledge structure

$$\mathcal{G} = \big\{\varnothing, \{a\}, \{b\}, \{c\}, \{a,b\}, \{a,c\}, \{a,b,c\}\big\},$$

on the subset $\{c\}$. We thus have the two equivalence classes $[\{a,b\}]$ and $[\{a,b,c\}]$, with the projection $\mathcal{G}_{|\{c\}} = \{\varnothing, \{c\}\}$. The two $\{c\}$-children are $\mathcal{G}_{[\varnothing]} = \big\{\varnothing, \{a\}, \{b\}, \{a,b\}\big\}$ and $\mathcal{G}_{[\{c\}]} = \big\{\varnothing, \{b\}\big\}$. Both $\mathcal{G}_{[\varnothing]}$ and $\mathcal{G}_{[\{c\}]}$ are well-graded and $\cup$-closed, and so is $\mathcal{G}_{|\{c\}}$. However, $\mathcal{G}$ is not $\cup$-closed since $\{b,c\}$ is not a state.

We state the first of our two projection theorems.

2.4.8 Theorem. *Let $(Q, \mathcal{K})$ be a learning space, with $|Q| = |\cup\mathcal{K}| \geq 2$. The following two properties hold for any proper nonempty subset Q' of Q.*

(i) *The projection $\mathcal{K}_{|Q'}$ of $\mathcal{K}$ on Q' is a learning space.*
(ii) *The children of $\mathcal{K}$ are well-graded and partially $\cup$-closed families.*

Note that we may have $\mathcal{K}_{[K]} = \{\varnothing\}$ in (ii) (cf. Example 2.4.3).

PROOF. (i) Since is a learning space, $\mathcal{K}_{|Q'}$ is a knowledge structure by Lemma 2.4.5(i). We prove that Axiom [L1] holds for $\mathcal{K}_{|Q'}$. Assume K, $L \in \mathcal{K}_{|Q'}$ with $K \subset L$. Then, there exist $\tilde{K}$ and $\tilde{L}$ in $\mathcal{K}$ such that $K = \tilde{K} \cap Q'$ and $L = \tilde{L} \cap Q'$. As $\mathcal{K}$ is a learning space, there is a $L1$–chain from $\tilde{K}$ to $\tilde{K} \cup \tilde{L}$, say $\tilde{K} = K_0$, K_1, ..., $K_q = \tilde{K} \cup \tilde{L}$. Then $K = K_0 \cap Q'$, $K_1 \cap Q'$, ..., $K_q \cap Q' = L$ yields a $L1$–chain from K to L in $\mathcal{K}_{[K]}$ after deleting from the sequence any set identical to a previous set. Axiom [L2] also holds for $\mathcal{K}_{|Q'}$. Indeed, take K, $L \in \mathcal{K}_{|Q'}$ and $q \in Q'$ with $K \subset L$ and $K \cup \{q\} \in \mathcal{K}_{|Q'}$. There exist $\tilde{K}$, $\tilde{L}$, M in $\mathcal{K}$ such that $K = \tilde{K} \cap Q'$, $L = \tilde{L} \cap Q'$ and $K \cup \{q\} = M \cap Q'$. So, we have $L \cup \{q\} = (\tilde{L} \cup M) \cap Q'$, thus $L \cup \{q\} \in \mathcal{K}_{|Q'}$.

(ii) Take any child $\mathcal{K}_{[K]}$ of $\mathcal{K}$. By Lemma 2.4.6(iii), $\mathcal{K}_{[K]}$ is a partially $\cup$-closed family. Axiom [L1] and the argument in the proof of Lemma 2.2.3 imply that $[K]$ is well-graded. The wellgradeness of $\mathcal{K}_{[K]}$ follows easily. □

2.4.9 Remark. In Example 2.4.3, we had a situation in which the non trivial children of a learning space were either themselves learning spaces, or would become so by the addition of the set $\{\varnothing\}$. This happens if and only if the subset Q' of the domain defining the projection satisfies the condition spelled out in the next definition.

2.4.10 Example. Take the learning space

$$\mathcal{K} = \{\varnothing, \{a\}, \{d\}, \{a,b\}, \{a,d\}, \{c,d\}, \\ \{a,b,c\}, \{a,b,d\}, \{a,c,d\}, \{b,c,d\}, \{a,b,c,d\}\},$$

with domain $Q = \{a,b,c,d\}$. We set $Q' = \{c\}$ and $K = \{c,d\}$. Then

$$[K] = \{\{c,d\}, \{a,b,c\}, \{a,c,d\}, \{b,c,d\}, \{a,b,c,d\}\}$$

and, as $\cap[K] = \{c\}$,

$$\mathcal{K}_{[K]} = \{\{d\}, \{a,b\}, \{a,d\}, \{b,d\}, \{a,b,d\}\}.$$

Clearly, the child $\mathcal{K}_{[K]}$ is not a learning space and even $\mathcal{K}_{[K]} \cup \{\varnothing\}$ is not: for instance, there is no tight path from $\varnothing$ to $\{a,b\}$. The reason lies in a feature of $[K]$: the element $\{a,b,c\}$ is a minimal element of $[K]$ covering $\cap[K]$ while at the same time $\{a,b,c\} \setminus \cap[K]$ contains more than one element.

2.4.11 Definition. Suppose that $(Q, \mathcal{K})$ is a partial knowledge structure, with $|Q| \geq 2$. A subset $Q' \subset Q$ is *yielding* if for any state L of $\mathcal{K}$ that is minimal for inclusion in some equivalence class $[K]$, we have $|L \setminus \cap[K]| \leq 1$. We recall that $[K]$ is the equivalence class containing K in the partition of $\mathcal{K}$ induced by Q' (cf. Definition 2.4.1). For any non trivial child $\mathcal{K}_{[K]}$ of $\mathcal{K}$, we call $\mathcal{K}^+_{[K]} = \mathcal{K}_{[K]} \cup \{\varnothing\}$ a *plus child* of $\mathcal{K}$.

2.4.12 Theorem. *Suppose that $(Q, \mathcal{K})$ is a learning space with $|Q| \geq 2$, and let Q' be a proper nonempty subset of Q. The two following conditions are then equivalent.*

(i) *The set Q' is yielding.*
(ii) *All the plus children of $\mathcal{K}$ are learning spaces*[6].

Problem 13 asks the reader to investigate whether any learning space always has at least one non trivial child.

PROOF. (i) $\Rightarrow$ (ii). By Lemma 2.4.6(iii), we know that any non trivial child $\mathcal{K}_{[K]}$ is $\cup$-closed. This implies that the associated plus child $\mathcal{K}^+_{[K]}$ is a knowledge space. We use Lemma 2.2.3 to prove that such a plus child is also well-graded. Suppose that L and M are states of $\mathcal{K}^+_{[K]}$, with $\varnothing \subseteq L \subset M$ and, for some positive integer n, $d(L, M) = n$. We have two cases.

CASE 1. Suppose that $L \neq \varnothing$. Then both L and M are in $\mathcal{K}_{[K]}$. As $\mathcal{K}_{[K]}$ is well-graded by Theorem 2.4.8(ii), there exists a tight path

$$L = L_0 \subset L_1 \subset \cdots \subset L_n = M.$$

Since $\varnothing \subset L_0$, this tight path lies entirely in the plus child $\mathcal{K}^+_{[K]}$.

CASE 2. Suppose now that $L = \varnothing$. In view of what we just proved, we only have to show that, for any nonempty $M \in \mathcal{K}^+_{[K]}$, there is a singleton set $\{q\} \in \mathcal{K}^+_{[K]}$ with $q \in M$. By definition of $\mathcal{K}^+_{[K]}$, we have $M = M^\dagger \setminus \cap[K]$ for some $M^\dagger \in [K]$. Take a minimal state N in $[K]$ such that $N \subseteq M^\dagger$ and so $N \setminus \cap[K] \subseteq M$. Since Q' is yielding, we get $|N \setminus \cap[K]| \leq 1$. If $|N \setminus \cap[K]| = 1$, then $N \setminus \cap[K] = \{q\} \subseteq M$ for some $q \in Q$ with $\{q\} \in \mathcal{K}^+_{[K]}$. Suppose that $|N \setminus \cap[K]| = 0$. Thus $N \setminus \cap[K] = \varnothing$ and N must be the only minimal set in $[K]$, which implies that $\cap[K] = N$. By the wellgradedness of $\mathcal{K}$, there exists some $q \in M^\dagger$ such that $M^\dagger \supseteq N + \{q\} \in \mathcal{K}$. We have in fact $N + \{q\} \in [K]$ since $q \in M \setminus N$ implies $N \cap Q' = (N + \{q\}) \cap Q'$. We thus get

$$(N + \{q\}) \setminus \cap[K] = (N + \{q\}) \setminus N = \{q\} \subseteq M \quad \text{with} \quad \{q\} \in \mathcal{K}^+_{[K]}.$$

We have proved that in both cases, the tight path from L to M exists. The plus child $\mathcal{K}^+_{[K]}$ is thus well-graded. Applying Theorem 2.2.4, we conclude that $\mathcal{K}^+_{[K]}$ is a learning space.

[6] Note that we may have $\varnothing \in \mathcal{K}_{[K]}$, in which case $\mathcal{K}^+_{[K]} = \mathcal{K}_{[K]}$ (cf. Example 2.4.3).

(ii) $\Rightarrow$ (i). Let L be a minimal element in the equivalence class $[K]$, where $K \in \mathcal{K}$. Then $\cap[K] \subseteq L$. If equality holds, we have $|L \setminus \cap[K]| = 0$. If $\cap[K] \subset L$ holds, then $\varnothing$ and $L \setminus \cap[K]$ are distinct elements in the plus child $\mathcal{K}^+_{[K]}$. By the wellgradedness of $\mathcal{K}^+_{[K]}$, there is a tight path from $\varnothing$ to $L \setminus \cap[K]$ in $\mathcal{K}^+_{[K]}$. Because L is minimal in $[K]$ and distinct from $\cap[K]$, we see that $L \setminus \cap[K]$ must be a singleton. Hence $|L \setminus \cap[K]| = 1$. □

2.4.13 Remark. The theory of learning spaces provides the combinatoric foundation for various knowledge assessment algorithms. As we discussed in 1.1.12, the goal of an assessment algorithm is to uncover the knowledge state of a student by a sequence of well chosen questions. Two quite different classes of stochastic assessment algorithms are described in Chapters 13 and 14. However, empirically constructed learning spaces are typically so large, with knowledge states numbering millions, that a straightforward application of an assessment algorithm is not always feasible. In such cases, the result of this section may be useful. For example, they pave the way to a two-step assessment in a learning space $(Q, \mathcal{K})$ which is serviceable in those cases in which $\mathcal{K}$ is very large. The first step uses a projection $\mathcal{K}_{|Q'}$ on a suitable—in particular, yielding—subset $Q' \subset Q$. This step ends up with a state $W \subseteq Q'$ of the projection $\mathcal{K}_{|Q'}$, with $W = K \cap Q'$ for some $K \in \mathcal{K}$. The second step is an assessment on the Q'-child $\mathcal{K}_{[K]}$ of $\mathcal{K}$, leading to some state $M = L \setminus \cap[K]$ of $\mathcal{K}_{[K]}$, for some state L of $\mathcal{K}$. The state L can then be taken as the final state obtained for the assessment. If the learning space $\mathcal{K}$ is extremely large, an n-phase assessment along these lines is also feasible in principle.

An objection to this procedure is that it does not feature any mechanism permitting a correction, during Step 2, of any assessment error made in Step 1. The state $W = K \cap Q'$ selected by Step 1 is taken for granted and defines the child $\mathcal{K}_{[K]}$. The assessment in the space $\mathcal{K}_{[K]}$ only amounts to selecting one among the states that are $\sim_{Q'}$ equivalent to K. A more flexible procedure is discussed in Section 13.7[7].

2.5 Original Sources and Related Works

As indicated in Chapter 1, the theory of knowledge spaces was initiated by Doignon and Falmagne (1985). Most of the early work was focused on the axiom of closure under union, in the finite case. From a pedagogical standpoint, there were some weaknesses in this approach. For one thing, the $\cup$-closure condition may not be convincing for an educator, at least a *priori*. For another, the state resulting from an assessment in a knowledge space has no natural,

[7] Another approach would rely on having two or more assessments pursued simultaneously rather that successively, with possible interplay between them. While this possibility is intriguing, we do not expand on this idea here.

economical representation in a style that would be useful to the teacher or the student.

The wellgradedness condition was introduced by Falmagne and Doignon (1988b) to palliate the latter defect. Under this condition, a meaningful representation of any knowledge state is feasible in the form of the two 'fringes' of that state (see Definition 4.1.6). The outer fringe spells out the items that the student is ready to learn, and the inner fringe contains all those items signaling the 'high points' in a student's state. However, the resulting concept of a well-graded knowledge space, although mathematically appropriate, was still suffering from a lack of an immediate pedagogical justification.

The axioms [L1] and [L2] axioms were later proposed by Falmagne to Eric Cosyn and Hasan Uzun as offering a more compelling basis for the theory. In a recent paper, Cosyn and Uzun (2009) proved that for a knowledge structure $\mathcal{K}$, the conditions [L1] and [L2] were in fact equivalent to the hypothesis that $\mathcal{K}$ is a well-graded knowledge space. This result is recalled as the equivalence (i) $\Leftrightarrow$ (iii) of Theorem 2.2.4.

The concept of a projection was already present in our original monograph 'Knowledge Spaces' in the form of a 'substructure' of a knowledge structure (see Doignon and Falmagne, 1999, Theorem 1.16 and Definition 1.17). What is new[8] in Section 2.4, which closely follows Falmagne (2008), is the extension of the results to learning spaces, and, most importantly, the analysis of a knowledge structure $\mathcal{K}$ into a projection $\mathcal{K}_{|Q'}$ and its satellite components in the form of the children $\mathcal{K}_{[K]}$ defined in 2.4.2. These results expand earlier work by Cosyn (2002), who also defined a partition of the knowledge structure, which he called 'coarsening.' However, his partition was chosen arbitrarily and did not arise from an equivalence relation defined by (2.6) via a subset Q' of the domain. By contrast, the definition of a projection given by Cavagnaro (2008) is conceptually quite similar to ours, but applies to media, which are semigroups of transformations rather than families of sets. As we pointed out in Remark 2.4.4, a precise connection between media and learning spaces exists, which will be delineated in Chapter 10.

Problems

1. Construct the discriminative reduction of the knowledge structure

$$\mathcal{K} = \big\{\varnothing, \{a,c,d\}, \{b,e,f\}, \{a,c,d,e,f\}, \{a,b,c,d,e,f\}\big\}.$$

2. Verify that any well-graded knowledge structure is discriminative. Why is a well-graded family of sets not necessarily discriminative?

[8] Besides the name 'projection' replacing 'substructure.'

3. Do we have $|\mathcal{K}| = |\mathcal{K}^*|$ for any knowledge structure $\mathcal{K}$? Prove your answer. (The '$*$' notation is as in Definition 2.1.4.)

4. Show that a knowledge structure $(Q, \mathcal{K})$ is essentially finite if and only if Q^* is finite.

5. Construct a drawing representing the knowledge structure $\mathcal{H}$ of Example 2.1.3 in the style of Figure 2.1 for Example 2.1.6.

6. Consider the following axiom generalizing the closure under union.
   [JS] For any subfamily of states $\mathcal{F}$ in a knowledge structure $(Q, \mathcal{K})$, there exists a unique minimal state $K \in \mathcal{K}$ such that $\cup\mathcal{F} \subseteq K$.
   (Under this axiom, $\mathcal{K}$ is thus a 'join semi lattice' with respect to inclusion.) Construct a finite example in which this axiom is not satisfied.

7. Is it true that if one set is finite in a well-graded family, then all sets of that family are finite? Give a proof or a counterexample.

8. Prove that if a knowledge structure $(Q, \mathcal{K})$ is discriminative, then so is its projection $\mathcal{K}_{|Q'}$ on a subset $Q' \subset Q$, but that the converse does not hold.

9. Prove Theorem 2.3.4.

10. Consider the following modification of Axiom [L1] for a knowledge structure $(Q, \mathcal{K})$.
    [L1'] If K and L are two states with $K \subset L \neq Q$, then there is a chain of states $K = K_0 \subset K_1 \subset \cdots \subset K_n = L$ with $K_i = K_{i-1} + \{q_i\}$ for $1 \leq i \leq n$ and $|L \setminus K| = n$.

    Suppose that $(Q, \mathcal{K})$ satisfies [L1'] and [L2]. Could the domain Q be uncountable? Prove your answer.

11. Let $(\mathcal{Q}, \mathcal{K})$ be a knowledge structure and let Q' be any proper subset of Q. With the equivalence classes $[K]$ defined as in 2.4.1, prove the following two statements:

$$K \triangle L \subseteq Q \setminus Q' \iff K \cap Q' = L \cap Q' \tag{2.11}$$
$$(\cap[K]) \cap Q' = K \cap Q'. \tag{2.12}$$

12. Describe the components $\mathcal{F}_{[\{a,b\}]}$ and $\mathcal{F}_{[\varnothing]}$ in the example of Figure 2.2.

13. Is it true that any learning space has at least one non trivial child, either (i) for some subset of the domain; or (ii) for a given subset of the domain (cf. Theorem 2.4.12).

14. Two knowledge structures $(Q, \mathcal{L})$ and $(Q^\dagger, \mathcal{K}^\dagger)$ are *isomorphic* if there exists a 1-1 correspondence $f : Q \to Q^\dagger$ such that for all $K \subseteq Q$, we have $K \in \mathcal{K}$ if and only if $f(K) \in \mathcal{K}^\dagger$. Prove that $(Q, \mathcal{L})$ is a learning space (resp. knowledge space) if and only if $(Q', \mathcal{K}')$ is a learning space (resp. knowledge space).

15. Is a partial learning space necessarily finite? How about a partial knowledge space?
16. Show that a quasi well-graded knowledge structure is essentially finite (cf. 2.3.3).
17. Prove by two counterexamples that the Axioms [L1] and [L2] independent.
18. What knowledge structures satisfy both [L1] and $\cup$-closure?

3

Knowledge Spaces

3.1 Outline

We have learned from Theorem 2.2.4 that any learning space is a knowledge space, that is, a knowledge structure closed under union. The $\cup$-closure property is critical for the following reason. Certain knowledge spaces, and in particular the finite ones, can be faithfully summarized by a subfamily of their states. To wit, any state of the knowledge space can be generated by forming the union of some states in the subfamily. When such a subfamily exists and is minimal for inclusion, it is unique and is called the 'base' of the knowledge space. In some cases, the base can be considerably smaller than the knowledge space, which results in a substantial economy of storage in a computer memory. The extreme case is the power set of a set of n elements, where the 2^n knowledge states can be subsumed by the family of the n singleton sets. This property inspires most of this chapter, beginning with the basic concepts of 'base' and 'atoms' in Sections 3.4 to 3.6. Other features of knowledge spaces are also important, however, and are dealt with in this chapter.

In the next section, we mention in passing the problem of building a knowledge space in practice. The key idea is to code the information concerning the structure in the form of a relation $\mathcal{R}$ on the power set of the domain, with the following interpretation: $A\mathcal{R}B$ holds if failing all the items in A implies failing all those in B. Such a relation defines a unique knowledge space. It can be obtained either by querying experts, or from assessment statistics. This topic is systematically expanded in Chapters 7, 15 and 16.

To any knowledge space $\mathcal{K}$ corresponds its 'dual', that is, the family containing all the complements of states of $\mathcal{K}$. This interplay is of interest because the duals of knowledge spaces belong to an important family of mathematical structures called 'closure spaces.' Much is known regarding 'closure spaces' that has a useful translation in our context. These structures are introduced Section 3.3.

This chapter also covers two situations in which a particular quasi order on the domain plays a special role. In Section 3.7, we introduce the 'surmise relation', a quasi order which is also called the 'precedence relation.' It holds between two items r and q when the mastery of r can be inferred from that

J.-C. Falmagne, J.-P. Doignon, *Learning Spaces*,
DOI 10.1007/978-3-642-01039-2_3, © Springer-Verlag Berlin Heidelberg 2011

of q. In this case, the quasi order is built from the knowledge structure (which need not be a knowledge space). In the second situation, which we discuss in Section 3.8, the quasi order on the domain can be primary. It then defines a particular kind of knowledge space in which the family of states is also closed under intersection. The relevant Theorem 3.8.3 is due to Birkhoff (1937). An extension to general knowledge spaces is given in Theorem 5.2.5.

3.2 Generating Knowledge Spaces by Querying Experts

A concept akin to knowledge spaces, but mathematically quite different, was introduced in 1.1.9 under the name of 'entailment.' We indicated there that entailments can be used to construct a knowledge space by querying an expert without actually asking him to provide an explicit list of all the knowledge states. Recalling the relevant passage of Chapter 1, imagine that an experienced teacher is asked, in a systematic way, questions of the following type:

[Q1] *Suppose that a student under examination has just provided wrong responses to all the items $q_1, \ldots, q_n$. Is it practically certain that this student will also fail item q_{n+1}? We assume that careless errors and lucky guesses are excluded*[1].

The responses to all such questions define a relation $\mathcal{R}$ on $2^Q \setminus \{\varnothing\}$, with the following interpretation: for any two nonempty sets A, B of items, we have

$$A\mathcal{R}B \quad \text{if and only if} \quad \begin{cases} \text{from the failure of all the items in } A \\ \text{we infer the failure of all the items in } B. \end{cases} \tag{3.1}$$

It turns out that any such relation $\mathcal{R}$ on $2^Q \setminus \{\varnothing\}$ specifies a unique knowledge space. The definition of its states is given in the next theorem, which extends the discussion to sets of arbitrary cardinality.

3.2.1 Theorem. *Suppose that Q is a nonempty set, with $\mathcal{R}$ a relation on $2^Q \setminus \{\varnothing\}$. Let $\mathcal{S}$ be the family of all the subsets K of Q satisfying the condition:*

$$K \in \mathcal{S} \iff \big(\forall (A, B) \in \mathcal{R} : \ A \cap K = \varnothing \ \Rightarrow \ B \cap K = \varnothing\big). \tag{3.2}$$

Then $\mathcal{S}$ contains $\varnothing$ and Q, and is closed under union.

PROOF. Take $\mathcal{F} \subseteq \mathcal{S}$ and suppose that $A\mathcal{R}B$, with $A \cap (\cup\mathcal{F}) = \varnothing$. We obtain: $A \cap K = \varnothing$, for all K in $\mathcal{F}$. Using (3.2), we derive that we must also have $B \cap K = \varnothing$ for all K in $\mathcal{F}$. This gives $B \cap (\cup\mathcal{F}) = \varnothing$ and thus $\cup\mathcal{F} \in \mathcal{S}$. Because the r.h.s. of (3.2) is trivially satisfied for $\varnothing$ and for Q, both must be in $\mathcal{S}$. □

[1] In practice, we suppose also that n is small, say $n \leq 5$. We shall see is Chapter 15 that this assumption is empirically justified. We recall that the label [Q0] is reserved for the special case where $n = 1$.

The knowledge space $(Q, \mathcal{S})$ in Theorem 3.2.1 need not be closed under intersection (see our next example). The relation $\mathcal{R}$ provides a powerful tool for constructing knowledge spaces in practice, whether one is willing to rely on human expertise or on statistics of student data yielding essentially the same type of information (see Remark 3.2.3). We devote Chapters 7 and 15 to this topic. The 1-1 correspondence between the collection of entailments on a set and the collection of knowledge spaces on the same set is established in Theorem 7.1.5. The relation $\mathcal{R}$ is also instrumental for building learning spaces, as we demonstrate in Chapter 16.

3.2.2 Example. With $Q = \{a, b, c\}$, suppose that $\mathcal{R}$ contains the single pair $(\{a, b\}, \{c\})$. Thus, both $\{a, c\}$ and $\{b, c\}$ are knowledge states in the knowledge space $(Q, \mathcal{S})$; however, their intersection $\{c\}$ is not a knowledge state.

3.2.3 Remark. Human expertise is not the only way of capitalizing on results such as Theorem 3.2.1 for constructing knowledge spaces. In cases where sizable sets of student data are available, one can also rely on conditional probabilities of failing to solve some problems. Specifically, the relation $\mathcal{R}$ of (3.1) could be constructed via the formula

$$A\mathcal{R}B \iff \mathbb{P}(\text{failing all the items in } B \,|\, \text{all the items in } A \text{ are failed}) > \alpha,$$

where $\mathbb{P}$ denotes the probability measure, and α is a suitably chosen parameter. This probability could be estimated from relative frequencies computed from students' data. This avenue is a realistic one since as mentioned in Footnote 1, in practice, the size of the set B need not be large.

However, because our primary interest is learning spaces, having succeeded in building a knowledge space gets us only part of the way. How can we optimally engineer—either by the addition of well chosen missing states or by some other technique—the wellgradedness of a knowledge space that is not a learning space? Solving this problem requires some new tools and results and we have to postpone the relevant developments. We devote Section 4.5 and Section 16.3 to these questions.

3.3 Closure Spaces

We recall from 2.2.2 that the dual of a knowledge structure $(Q, \mathcal{K})$ is the knowledge structure $\overline{\mathcal{K}}$ containing all the complements of states in $\mathcal{K}$, that is

$$\overline{\mathcal{K}} = \{K \in 2^Q \,|\, Q \setminus K \in \mathcal{K}\}.$$

Thus, $\mathcal{K}$ and $\overline{\mathcal{K}}$ have the same domain.

3.3.1 Definition. By a *collection* on Q we mean a family $\mathcal{K}$ of subsets of a *domain* Q. We often then write $(Q, \mathcal{K})$ to denote the collection. Note that a collection may be empty. A collection $(Q, \mathcal{L})$ is a *closure space* when the family $\mathcal{L}$ contains Q and is closed under intersection. This closure space is *simple* when $\varnothing$ is in $\mathcal{L}$. Thus, a collection $\mathcal{K}$ of subsets of a domain Q is a knowledge space on Q if and only if the dual structure $\overline{\mathcal{K}}$ is a simple closure space.

Examples of closure spaces abound in mathematics.

3.3.2 Examples. Let $\mathbb{R}^3$ be the set of all points of a 3-dimensional Euclidean space, and let $\mathcal{L}$ be the family of all affine subspaces (that is: the empty set, all the singletons sets, the lines, the planes and $\mathbb{R}^3$ itself). Then $\mathcal{L}$ is closed under intersection. Another example is the family of all convex subsets of $\mathbb{R}^3$.

These two examples of closure spaces are only instances of general classes: replace 3-dimensional Euclidean space by any affine space over an (ordered) skew field. Moreover, other classes of examples can be found in almost all branches of mathematics (e.g. by taking subspaces of a vector space, subgroups of a group, ideals in a ring, closed subsets of a topological space). Our next example comes from another discipline.

3.3.3 Example. In Example 1.4.4, we considered the collection L of all the well-formed expressions in some formal language, together with a fixed set of derivation rules, and a relation $\mathcal{I}$ on the set of all subsets of L, defined by: $A \mathcal{I} B$ if all the expressions in B are derivable from the expressions in A by application of the derivation rules. A knowledge structure can be obtained by calling any $K \subseteq L$ a *state* of $\mathcal{I}$ if $B \subseteq K$ whenever $A \subseteq K$ and $A\mathcal{I}B$. It is easily shown that the collection $\mathcal{L}$ of all states is closed under intersection; that is, $\cap\mathcal{F} \in \mathcal{L}$ for any $\mathcal{F} \subseteq \mathcal{L}$ (see Problem 2).

Closure spaces are sometimes called 'convex structures[2].' We record below an obvious construction.

3.3.4 Theorem. *Let $(Q, \mathcal{L})$ be a closure space. Then any subset A of Q is included in a unique element of $\mathcal{L}$, denoted as A', which is minimal for inclusion in $\mathcal{L}$; we have for $A, B \in 2^Q$,*

(i) $A \subseteq A'$;
(ii) $A' \subseteq B'$ *when* $A \subseteq B$;
(iii) $A'' = A'$.

Conversely, any mapping $2^Q \to 2^Q : A \mapsto A'$ which satisfies Conditions (i) *to* (iii) *is obtained from a unique closure space on Q; this establishes a one-to-one correspondence between those mappings and the closure spaces on Q. Moreover, $\varnothing' = \varnothing$ if and only if $\varnothing \in \mathcal{L}$.*

[2] For references, see the Sources section 3.9 of this chapter.

Proof. The intersection of all elements of $\mathcal{L}$ that include some $A \in 2^Q$ is an element of $\mathcal{L}$, and this intersection is the smallest possible element of $\mathcal{L}$ that includes A (by definition, Q itself is an element of $\mathcal{L}$, and includes A). Proving Conditions (i) to (iii) is straightforward, and left to the reader (Problem 11). Conversely, given a mapping $2^Q \to 2^Q : A \mapsto A'$ satisfying Conditions (i) to (iii), we set $\mathcal{L} = \{A \in 2^Q \mid A' = A\}$. It is easily verified that $\mathcal{L}$ is closed under intersection. Moreover, the mapping $A \mapsto A'$ is obtained from $\mathcal{L}$ via the above construction. It is now easy to prove the existence of the one-to-one correspondence mentioned in the statement; we also leave the verification of this fact to the reader. □

3.3.5 Definition. In the notation of Theorem 3.3.4, A' is called the *closure* of the set A (in the closure space $(Q, \mathcal{L})$).

3.4 Bases and Atoms

3.4.1 Definition. The *span* of a family of sets $\mathcal{G}$ is the family $\mathcal{G}'$ containing any set which is the union of some subfamily of $\mathcal{G}$. In such a case, we write $\mathbb{S}(\mathcal{G}) = \mathcal{G}'$ and we say that $\mathcal{G}$ *spans* $\mathcal{G}'$. By definition $\mathbb{S}(\mathcal{G})$ is thus $\cup$-closed. A *base* of a $\cup$-closed family $\mathcal{F}$ is a minimal subfamily $\mathcal{B}$ of $\mathcal{F}$ spanning $\mathcal{F}$ (where 'minimal' is meant with respect to set inclusion: if $\mathbb{S}(\mathcal{H}) = \mathcal{F}$ for some $\mathcal{H} \subseteq \mathcal{B}$, then $\mathcal{H} = \mathcal{B}$). By a standard convention, the empty set is the union of the empty subfamily of $\mathcal{B}$. Thus, since the base is minimal, the empty set never belongs to a base.

It is clear that a state K belonging to some base $\mathcal{B}$ of $\mathcal{K}$ cannot be the union of other elements of $\mathcal{B}$. Also, a knowledge structure has a base only if it is a knowledge space.

3.4.2 Theorem. *Let $\mathcal{B}$ be a base for a knowledge space $(Q, \mathcal{K})$. Then $\mathcal{B} \subseteq \mathcal{F}$ for any subfamily $\mathcal{F}$ of states spanning $\mathcal{K}$. Consequently, a knowledge space admits at most one base.*

Proof. Let $\mathcal{B}$ and $\mathcal{F}$ be as in the hypotheses of the theorem, and suppose that $K \in \mathcal{B} \setminus \mathcal{F}$. Then, $K = \cup\mathcal{H}$ for some $\mathcal{H} \subseteq \mathcal{F}$. Since $\mathcal{B}$ is a base, any state in $\mathcal{H}$ is a union of some states in $\mathcal{B}$. This implies that K is a union of sets in $\mathcal{B} \setminus \{K\}$, negating the minimality property of a base. The uniqueness of a base is now obvious. □

Some knowledge spaces have no base.

3.4.3 Example. The collection $\mathcal{O}$ of all the open sets of $\mathbb{R}$ is a knowledge space. It is spanned by the family $\mathcal{I}_1$ of all the open intervals with rational endpoints, as well as by the family $\mathcal{I}_2$ of all open intervals with irrational endpoints. If $\mathcal{O}$ had a base $\mathcal{B}$, Theorem 3.4.2 would imply that $\mathcal{B} \subseteq \mathcal{I}_1 \cap \mathcal{I}_2 = \varnothing$, which is absurd. Thus, $\mathcal{O}$ has no base (in the sense of Definition 3.4.1). In the finite case, however, a base always exists.

3.4.4 Theorem. *Any essentially finite knowledge space has a base.*

Indeed, since the number of states is finite, there must be a minimal spanning subfamily of states, that is, a base.

The next definition will be useful for the specification of the base when it exists. Notice that we do not restrict ourselves to the essentially finite case.

3.4.5 Definition. Let $\mathcal{F}$ be a nonempty family of sets. For any $q \in \cup\mathcal{F}$, an *atom at* q is a minimal set of $\mathcal{F}$ containing q. A set $X \in \mathcal{F}$ is called an *atom* if it is an atom at q for some $q \in \cup\mathcal{F}$.

Note that this meaning of the term 'atom' is different from that used in lattice theory (see e.g. Birkhoff, 1967; Davey and Priestley, 1990).

3.4.6 Example. In the space $\mathcal{K} = \{\varnothing, \{a\}, \{a,b\}, \{b,c\}, \{a,b,c\}\}$, the state $\{b,c\}$ is an atom at b and also an atom at c. There are two atoms at b, namely $\{a,b\}$ and $\{b,c\}$. There is only one atom at a, which is $\{a\}$. (However, a also belongs to the atom $\{a,b\}$, but the state $\{a,b\}$ is not an atom at a.)

The knowledge space of Example 3.4.3 has no atoms. On the other hand, in an essentially finite knowledge structure, there is at least one atom at every item.

Another characterization of the atoms in a knowledge space is given below.

3.4.7 Theorem. *A state K in a knowledge space $(Q, \mathcal{K})$ is an atom if and only if $K \in \mathcal{F}$ for any subfamily of states $\mathcal{F}$ satisfying $K = \cup\mathcal{F}$.*

PROOF. (Necessity.) Suppose that K is an atom at q, and that $K = \cup\mathcal{F}$ for some subfamily $\mathcal{F}$ of states. Thus, q must belong to some $K' \in \mathcal{F}$, with necessarily $K' \subseteq K$. We must have $K = K'$, since K is a minimal state containing q. Thus, $K \in \mathcal{F}$.

(Sufficiency.) If K is not an atom, for each $q \in K$, there must be some state $K'(q)$ with $q \in K'(q) \subset K$. With $\mathcal{F} = \{K'(q) \,|\, q \in K\}$, we thus have $K = \cup\mathcal{F}$, and $K \notin \mathcal{F}$. □

3.4.8 Theorem. *Suppose a knowledge space has a base. Then this base is formed by the collection of all the atoms.*

PROOF. Let $\mathcal{B}$ be the base of a knowledge space $(Q, \mathcal{K})$, and let $\mathcal{A}$ be the collection of all the atoms. (We do not assume that there is an atom at every item.) We have to show that $\mathcal{A} = \mathcal{B}$. If some $K \in \mathcal{B}$ is not an atom, then, for every $q \in K$, there exists a state $K'(q)$ with $q \in K'(q) \subset K$. But then $K = \cup_{q \in K} K'(q)$ and we cannot have $K \in \mathcal{B}$. (Since each $K'(q)$ is a union of states in $\mathcal{B}$, we see that K is a union of other states in $\mathcal{B}$.) Thus, K must be an atom for at least one item. Every element of the base is thus an atom, and we have $\mathcal{B} \subseteq \mathcal{A}$. Conversely, take any $K \in \mathcal{A}$. Then, $K = \cup\mathcal{F}$ for some $\mathcal{F} \subseteq \mathcal{B}$. By Theorem 3.4.7, we have $K \in \mathcal{F} \subseteq \mathcal{B}$. Thus, $\mathcal{A} = \mathcal{B}$. □

Even when the base exists, there may not be an atom at every item.

3.4.9 Example. Define $\mathcal{G} = \{[0, \frac{1}{n}] \mid n \in \mathbb{N}\} \cup \{\varnothing\}$. Then $([0,1], \mathcal{G})$ is a knowledge space, with a base consisting of all the states except $\varnothing$; every item has an atom, except 0. Note that $([0,1], \mathcal{G})$ is not discriminative. However, its discriminative reduction $([0,1]^*, \mathcal{G}^*)$ (cf. Definition 2.1.5) provides a similar counterexample. (It has a base but no atom at 0^*.)

The importance of the base as a compact summary of a knowledge space prompts the search for efficient algorithms for constructing that base and for generating the states from the base. Two such algorithms are sketched in the next two sections.

3.5 An Algorithm for Constructing the Base

We assume that the domain Q of the knowledge space is finite, with $|Q| = m$ and $|\mathcal{K}| = n$. By Theorem 3.4.8, the base of a knowledge space is formed by all the atoms. Recall from Definition 3.4.5 that an atom at q is a minimal state containing q. A simple algorithm for building the base, due to Dowling (1993b) and described below, is grounded on this definition of an atom.

3.5.1 Sketch of Algorithm. List the items arbitrarily as $q_1, \ldots, q_m$. List the states as $K_1, \ldots, K_n$ in such a way that $K_i \subset K_k$ implies $i < k$ for $i, k \in \{1, \ldots, n\}$. (Thus, list the states according to nondecreasing size, and arbitrarily for states of the same size.) Form an $n \times m$ array $T = (T_{ij})$ with the rows and columns representing the states and items, respectively; thus, the rows are indexed from 1 to n and the columns from 1 to m. At any step of the algorithm, a cell of T contains one of the symbols '$*$', '$+$' or '$-$'. Initially, set T_{ij} to $*$ if state K_i contains item q_j; otherwise, set T_{ij} to $-$. The algorithm inspects rows $i = 1, \ldots, n$ and tranforms any value $*$ in a cell (i,j) into $+$ whenever the following condition is satisfied: there exists an index p such that $1 \leq p < i$, state K_p contains item q_j, and $K_p \subset K_i$. When this is done, the atoms are the states K_i for which row i still contains at least one $*$.

3.5.2 Example. Take the space $\mathcal{K} = \{\varnothing, \{a\}, \{a,b\}, \{b,c\}, \{a,b,c\}\}$ from Example 3.4.6. The initial array T is shown on the left of Table 3.1. From the final value of T on the right, we conclude that the base is $\{\{a\}, \{a,b\}, \{b,c\}\}$.

Table 3.1. The initial and final values of the array T in Example 3.5.2.

	a	b	c
$\varnothing$	$-$	$-$	$-$
$\{a\}$	$*$	$-$	$-$
$\{a,b\}$	$*$	$*$	$-$
$\{b,c\}$	$-$	$*$	$*$
$\{a,b,c\}$	$*$	$*$	$*$

	a	b	c
$\varnothing$	$-$	$-$	$-$
$\{a\}$	$*$	$-$	$-$
$\{a,b\}$	$+$	$*$	$-$
$\{b,c\}$	$-$	$*$	$*$
$\{a,b,c\}$	$+$	$+$	$+$

It is easily checked that Algorithm 3.5.1 also works when provided with a spanning family instead of the space itself (see Problem 4).

The description given for this algorithm is intended for an initial encoding of a space or of a spanning family $\mathcal{F}$ as an $n \times m$ array with each cell (i, j) indicating whether state K_i contains item q_j. It is not difficult to redesign the search for atoms in the case of a different encoding of the space (for instance, as a list of states with each state being a list of items).

3.5.3 An Algorithm for Generating a Space from its Base. The solution described below owes much to that of Dowling (1993b). However, by clarifying some of the underlying ideas, we improve both the principle and the efficiency of the algorithm.

We consider a base $\mathcal{B}$ that contains p states, with $\mathcal{B} = \{B_1, \ldots, B_p\}$. The states of the corresponding knowledge space are manufactured by a sequential procedure based on considering increasingly larger subfamilies of the base. We set $\mathcal{G}_0 = \{\varnothing\}$ and for $i = 1, \ldots, p$, we define $\mathcal{G}_i$ as the space spanned by $\mathcal{G}_{i-1} \cup \{B_i\}$. This is the general scheme, but some care is required to ensure efficiency. Clearly, at any step i of the algorithm, the new states created by taking the span of $\mathcal{G}_{i-1} \cup \{B_i\}$ are all of the form $G \cup B_i$ with $G \in \mathcal{G}_{i-1}$. However, some states formed by taking the union of B_i with some states in $\mathcal{G}_{i-1}$ may already exist in $\mathcal{G}_i$. A straightforward application of this scheme would require verifying for each newly generated state whether it was obtained before. As in general the number n of states can grow exponentially as a function of p, such verifications may be prohibitive. Accordingly, we want to form $G \cup B_i$ only when this union delivers a state not created before (whether at the current step, or earlier). Here is the crucial point: among all states G from $\mathcal{G}_{i-1}$ producing a state $K = G \cup B_i$, there is a unique maximum one that we denote by M. We thus have $K = M \cup B_i$, and moreover $K = G \cup B_i$ for $G \in \mathcal{G}_{i-1}$ implies $G \subseteq M$. The existence and uniqueness of M follow from the fact that $\mathcal{G}_{i-1}$ is closed under union. Condition (ii) in the result below provides a manageable characterization of the state M which is the key component of the algorithm. In this theorem, we consider the situation in which an arbitrary subset B of a domain Q is added to the base $\mathcal{D}$ of a knowledge space $\mathcal{G}$ on Q.

3.5.4 Theorem. *Let $(Q, \mathcal{G})$ be a knowledge space with base $\mathcal{D}$, and take $M \in \mathcal{G}$ and $B \in 2^Q$. The following two conditions are equivalent:*

(i) $\forall G \in \mathcal{G}: \quad M \cup B = G \cup B \Rightarrow G \subseteq M;$

(ii) $\forall D \in \mathcal{D}: \quad D \subseteq M \cup B \Rightarrow D \subseteq M.$

PROOF. (i) $\Rightarrow$ (ii). If $D \subseteq M \cup B$ for some $D \in \mathcal{D}$, we get $M \cup B = (M \cup D) \cup B$. As $M \cup D \in \mathcal{G}$, our hypothesis implies $M \cup D \subseteq M$, that is $D \subseteq M$.

(ii) $\Rightarrow$ (i). If $M \cup B = G \cup B$ with $G \in \mathcal{G}$, there exists a subfamily $\mathcal{E}$ of $\mathcal{D}$ such that $G = \cup\mathcal{E}$. For $D \in \mathcal{E}$, we have $D \subseteq M \cup B$, hence by our hypothesis $D \subseteq M$. We conclude that $G \subseteq M$. □

Returning to our discussion of the algorithm, we now have a way of generating, at the main stage i, only distinct elements $G \cup B_i$: it suffices to take such a union exactly when G from $\mathcal{G}_{i-1}$ satisfies the condition

$$\forall D \in \{B_1, \ldots, B_{i-1}\}: \quad D \subseteq G \cup B_i \implies D \subseteq G. \tag{3.3}$$

We must also avoid generating a state $G \cup B_i$ belonging to $\mathcal{G}_{i-1}$ (that is, a state that was generated at some earlier main stage). To this effect, notice that for $G \in \mathcal{G}_{i-1}$ satisfying (3.3), we have $G \cup B_i \in \mathcal{G}_{i-1}$ if and only if $B_i \subseteq G$.

3.5.5 Sketch of Algorithm. Let $\mathcal{B} = \{B_1, \ldots, B_p\}$ be the base of some knowledge space $\mathcal{K}$ on Q to be generated by the algorithm. Initialize $\mathcal{G}$ to $\{\varnothing\}$. At each step $i = 1, 2, \ldots, p$, perform the following:

(1) Initialize $\mathcal{H}$ to $\varnothing$.

(2) For each $G \in \mathcal{G}$, check whether
$B_i \nsubseteq G$ and $\forall D \in \{B_1, \ldots, B_{i-1}\}: D \subseteq G \cup B_i \Rightarrow D \subseteq G$.
If the condition holds, add $G \cup B_i$ to $\mathcal{H}$.

(3) When all G's from $\mathcal{G}$ have been considered, replace $\mathcal{G}$ with $\mathcal{G} \cup \mathcal{H}$. (This terminates step i.)

The family $\mathcal{G}$ obtained after step p is the desired space $\mathcal{K}$.

3.5.6 Example. For the base $\mathcal{B} = \{\{a\}, \{a, b\}, \{b, c\}\}$, Table 3.2 displays the successive values of $\mathcal{G}$.

Table 3.2. The successive values of $\mathcal{G}$ in Example 3.5.6.

Main stage	base element	states in $\mathcal{G}$
initialization		$\varnothing$
1	$\{a\}$	$\varnothing, \{a\}$
2	$\{a, b\}$	$\varnothing, \{a\}, \{a, b\}$
3	$\{b, c\}$	$\varnothing, \{a\}, \{a, b\}, \{b, c\}, \{a, b, c\}$

3.5.7 Example. Here is another example, with $\mathcal{B} = \{\{a\}, \{b, d\}, \{a, b, c\}, \{b, c, e\}\}$. In Table 3.3, we only show on each line the base element considered at this main stage together with the additional state(s) produced.

Table 3.3. The successive values of $\mathcal{H}$ in Example 3.5.7.

Base element	states in $\mathcal{H}$
initialization	$\varnothing$
$\{a\}$	$\{a\}$
$\{b, d\}$	$\{b, d\}, \{a, b, d\}$
$\{a, b, c\}$	$\{a, b, c\}, \{a, b, c, d\}$
$\{b, c, e\}$	$\{b, c, e\}, \{b, c, d, e\}, \{a, b, c, e\}, \{a, b, c, d, e\}$

3.5.8 Remarks. a) Algorithm 3.5.5 can be applied to any spanning subfamily $\mathcal{F}$ of a knowledge space $\mathcal{K}$ to produce that space (see Problem 5). In the case of a spanning subfamily $\mathcal{F}$ which is not the base, we recommend however to first construct the base $\mathcal{B}$ of $\mathcal{K}$ by applying Algorithm 3.5.1 to $\mathcal{F}$. Algorithm 3.5.5 can then be applied to $\mathcal{B}$ to produce $\mathcal{K}$.

b) A few words about the efficiency of Algorithm 3.5.5 are in order. Experiments show that the execution time of a computer program implementing this algorithm may be affected by the order in which the base states (or spanning states) are listed. No 'best' rule seems to emerge about a plausible, optimal way of encoding the base states. On the other hand, an improvement is obtained with many data sets by the following variant of Algorithm 3.5.5. At each step i, build the union U of all B_j's with $0 < j < i$ and $B_j \subset B_i$. When a $G \in \mathcal{G}$ is taken into consideration, first check whether $U \subseteq G$. If $U \not\subseteq G$, the verification of step (2) in Algorithm 3.5.5 can be skipped, because the condition cannot hold. Dowling's original algorithm relies heavily on such unions U. This algorithm is also sensitive to the selected ordering. Our modified version performs usually faster by 10% to 30%.

c) On the theoretical side, the complexity of Algorithm 3.5.5 (in the sense of Garey and Johnson, 1979) is good. Because the cardinality n of the family $\mathcal{K}$ of sets spanned by a base $\mathcal{B}$ containing p states in a domain Q of m items can grow exponentially with p, we analyze the complexity in terms of m, p and n together. Algorithm 3.5.5 has execution time in $O(n \cdot p^2 \cdot m)$, in other words there exist natural numbers m_0, p_0 and n_0 and a positive real number c such that execution on a domain of size $m \geq m_0$ with a base of size $p \geq p_0$ producing a space of size $n \geq n_0$ will always take less than $c \cdot n \cdot p^2 \cdot m$ steps (see Problem 12).

3.6 Bases and Atoms: The Infinite Case*

The results on bases and atoms are straightforward in the case of essentially finite knowledge structures. As shown by Example 3.4.3, however, there is no guarantee that atoms exist in the infinite case. There is one type of infinite structures in which the base always exists, namely, the so-called 'finitary' case defined below. The term 'finitary' comes from the theory of closure spaces (see the Sources section 3.9 at the end of this chapter).

3.6.1 Definition. A knowledge structure $\mathcal{K}$ is *finitary* when the intersection of any chain of states in $\mathcal{K}$ is a state. We also say that $\mathcal{K}$ is *granular* if for any state K containing some item q, there is an atom at q that is included in K. Obviously, any essentially finite knowledge structure is both finitary and granular. Another example is given below.

3.6.2 Example. Let $(V, \mathcal{S})$ be a knowledge structure, where V is a real vector space and $\mathcal{S}$ is the family of all the subspaces of V. Then its dual knowledge structure $(V, \overline{\mathcal{S}})$ is a finitary and granular knowledge space.

3.6.3 Theorem. *Any finitary knowledge structure is granular. The converse does not hold, even for knowledge spaces.*

PROOF. Consider the collection $\mathcal{F}$ of states that are included in a state K and contain an item q, ordered by inclusion. By Hausdorff's Maximal Principle, $\mathcal{F}$ must include at least one maximal chain $\mathcal{C}$ (cf. 1.6.10). If the knowledge structure is finitary, $\cap\mathcal{C}$ is a state and an atom at q. Example 3.6.5 establishes the second statement. □

As an immediate consequence, we have:

3.6.4 Corollary. *The span of the family $\mathcal{A}$ of all the atoms of a finitary knowledge structure $\mathcal{K}$ necessarily includes $\mathcal{K}$.*

The next example is that of a granular knowledge space which is not finitary.

3.6.5 Example. Consider the following subsets of $[0, 2]$:

$$\{0\} \cup \left[\frac{1}{k}, \frac{2}{k}\right], \qquad \text{for } k \in \mathbb{N}.$$

Since none of these subsets includes any other one, their collection forms the base $\mathcal{B}$ of a granular knowledge space $\mathcal{K}$ (in this case, any state in the base is an atom at each of the items it contains). On the other hand, $\mathcal{K}$ is not finitary. Since $[0, \frac{2}{k}] = \{0\} \cup \left(\bigcup_{j=k}^{\infty}[\frac{1}{j}, \frac{2}{j}]\right)$, the family of intervals $[0, \frac{2}{k}]$, for $k \in \mathbb{N}$, constitutes a chain in $\mathcal{K}$ whose intersection $\{0\}$ is not in $\mathcal{K}$.

Together with Theorem 3.6.3, the following result shows that any finitary knowledge space has a base.

3.6.6 Theorem. *Any granular knowledge space has a base.*

PROOF. Consider the collection $\mathcal{B}$ of all the atoms in a granular knowledge space $(Q, \mathcal{K})$. By Definition 3.6.1, any knowledge state in $\mathcal{K}$ is the union of all the atoms that it includes. Thus the family $\mathcal{B}$ spans $\mathcal{K}$, and it is clearly minimal with this property. □

A knowledge space may have a base without being granular. This happens in Example 3.4.9: there is no atom at 0.

We now study the condition of closure under intersection from the point of view of the atoms of the knowledge space.

3.6.7 Theorem. *A knowledge space $\mathcal{K}$ closed under intersection has exactly one atom at each item q, which is specified by $\cap\mathcal{K}_q$. Moreover, a granular knowledge space $\mathcal{K}$ having exactly one atom at each item is necessarily closed under intersection.*

PROOF. The assertions in the first sentence are obvious. Suppose that the knowledge space $\mathcal{K}$ has exactly one atom at each item and let $\mathcal{F}$ be a subfamily of $\mathcal{K}$. If $\cap\mathcal{F} = \varnothing$, then $\cap\mathcal{F}$ is a state. Otherwise, take any $q \in \cap\mathcal{F}$, and let $K(q)$ be the unique atom at q. For all $K \in \mathcal{F}$, we must have $K(q) \subseteq K$ since, by granularity, there is an atom at q included in K. Hence $K(q) \subseteq \cap\mathcal{F}$, and because $\mathcal{K}$ is a knowledge space, we get $\cap\mathcal{F} = \cup_{q\in\cap\mathcal{F}} K(q) \in \mathcal{K}$. □

The following example shows that the granularity assumption cannot be dropped in the second statement of Theorem 3.6.7.

3.6.8 Example. Take the knowledge space $\mathcal{K}$ on $\mathbb{R}$ with base

$$\left\{ \left[0, \frac{1}{n}\right] \,\middle|\, n \in \mathbb{N} \right\} \cup \{\,]-\infty, 0],\ \mathbb{R}\,\}.$$

Then for each $r \in \mathbb{R}$ there is a unique atom at r. However, the intersection of the states $]-\infty, 0] \cap [0, 1] = \{0\}$ does not belong to $\mathcal{K}$.

3.6.9 Corollary. *A granular knowledge space $\mathcal{K}$ is closed under intersection if and only if there is exactly one atom at each item.*

A more systematic study of knowledge spaces closed under intersection is contained in the next section.

3.7 The Surmise Relation

An important part of this book concerns the analysis, within the framework of knowledge structures, of the possible ways of learning the material in a domain Q. This leads naturally to study the concept of a 'predecessor' of an item. Intuitively, an item r is a predecessor of an item q if r is never mastered after q, either for logical or historical reasons. The next definition formalizes the intuitive idea that the predecessors of some item q are the items contained in *all* the states containing q.

3.7.1 Definition. Let $(Q, \mathcal{K})$ be a knowledge structure, and let $\precsim$ be a relation on Q defined by

$$r \precsim q \iff r \in \cap\mathcal{K}_q. \tag{3.4}$$

The relation $\precsim$ will be called the *surmise relation* or sometimes the *precedence relation* of the knowledge structure. (The usage of the two terminologies is discussed in Remark 3.7.3.) When $r \precsim q$ holds, we say that r is *surmisable* from q, or that r *precedes* q. If moreover $q \precsim r$ does not hold, then we write $r \prec q$ and say that r *strictly precedes* q.

Notice the equivalence:

$$r \precsim q \iff \mathcal{K}_r \supseteq \mathcal{K}_q, \tag{3.5}$$

which holds for any knowledge structure $\mathcal{K}$ and any items q, r in its domain. We leave the verification of this fact to the reader (cf. Problem 6). The equivalence (3.5) immediately implies the following result.

3.7.2 Theorem. *The surmise relation of a knowledge structure is a quasi order. When the knowledge structure is discriminative, this quasi order is a partial order.*

By abuse of language, we will occasionally make reference to the Hasse diagram of a knowledge structure $\mathcal{K}$, to mean the Hasse diagram of the surmise relation of the discriminative reduction $\mathcal{K}^* = \{K^* \mid K \in \mathcal{K}\}$ (cf. 2.1.4).

3.7.3 Remark. Two viewpoints can be taken with regard to the relation $\precsim$. One is that of inference: if $r \precsim q$, then the mastery of r can be surmised from that of q. The other one is that of learning: $r \precsim q$ means that r is always mastered before or at the same time as q, either for logical reasons or because this is the custom in the population of reference. As an illustration, consider the following two questions in European history:

Question q:	*Who was the prime minister of Great Britain just before World War II?*
Question r:	*Who was the prime minister of Great Britain during World War II?*

Today, anybody knowing that the answer to question q is 'Neville Chamberlain' would also know that the next prime minister was Winston Churchill. In our terms, this means that any state containing q would also contain r, that is, $r \precsim q$. Obviously, logic plays no role in this dependency, which only relies on the structure of the collection of states, which itself is a reflection of the population of subjects under consideration[3]. In many cases, however, especially in mathematics or science, the formula $r \precsim q$ will mean that, for logical reasons, r must be mastered before or at the same time as q.

In Chapter 5, we shall discuss a generalization of the concept of a surmise relation formalizing the following natural idea: to any item q in the structure is attached a collection of possible learning backgrounds (that is, sets of items) preparing a student for the mastery of q.

We examine an example of a surmise relation.

3.7.4 Example. Consider the knowledge structure

$$\begin{aligned}\mathcal{G} = \{\varnothing, \{a\}, \{b\}, \{a,b\}, \{b,c\}, \{a,b,c\}, \{b,c,e\},\\ \{a,b,c,e\}, \{a,b,c,d\}, \{a,b,c,d,e\}\}.\end{aligned} \tag{3.6}$$

It is easy to verify that $\mathcal{G}$ is a discriminative knowledge space (cf. 2.1.4 and 2.2.2). The surmise relation $\precsim$ of $\mathcal{G}$ is thus a partial order. The Hasse diagram of $\precsim$ is given in Figure 3.1. We leave to the reader to work out the details of the construction of $\precsim$ from the knowledge structure $\mathcal{G}$.

[3] The idea is that in such a population, it is very unlikely to find some individual knowing the answer to Question q who would not also know the answer to Question r; so, the possibility can be ignored.

As an illustration, notice that

$$c \in \{a, b, c, d\} \cap \{a, b, c, d, e\} = \cap \mathcal{G}_d,$$

and so

$$c \precsim d.$$

On the other hand, $a \notin \{b, c, e\}$. Thus $a \notin \mathcal{G}_e$, yielding $\neg(a \precsim e)$: in the Hasse diagram, there is no broken line descending from e to a.

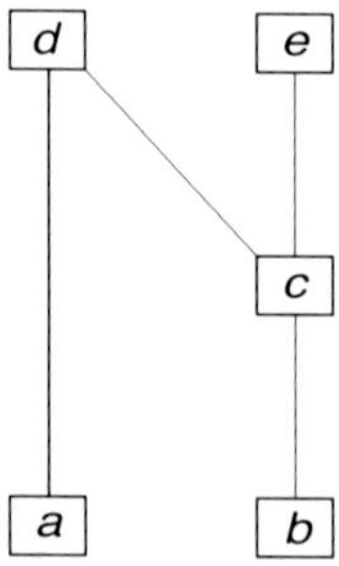

Figure 3.1. Hasse diagram of the surmise relation of the knowledge structure $\mathcal{G}$ specified by Equation (3.6).

The surmise relation offers a compact summary of the information contained in a knowledge structure, especially when the domain is finite with a small number of elements. It must be realized, however, that some information might be missing: different knowledge structures may have the same surmise relation.

For instance, the knowledge structure $\mathcal{G}' = \mathcal{G} \setminus \{\{b, c\}\}$ has the same surmise relation as $\mathcal{G}$. This raises the question: When is a knowledge structure fully described by its surmise relation? A well-known result of Birkhoff (1937), labeled as Theorem 3.8.3 in the next section, answers the question.

3.8 Quasi Ordinal Spaces

3.8.1 Definition. A knowledge space closed under intersection is called a *quasi ordinal space.* The chief reason for this terminology is that, in such a case, the knowledge structure is characterized by a quasi order, namely, its surmise relation (see Theorem 3.8.3). A discriminative and quasi ordinal space is a *(partially) ordinal space.* The surmise relation of such a space is a partial order. It is clear that a quasi ordinal space is finitary (cf. Definition 3.6.1).

3.8.2 Theorem. *Let $\mathcal{K}$ and $\mathcal{K}'$ be two quasi ordinal spaces on the same domain Q. Then,*

$$(\forall q, s \in Q : \quad \mathcal{K}_q \subseteq \mathcal{K}_s \Leftrightarrow \mathcal{K}'_q \subseteq \mathcal{K}'_s) \quad \Longleftrightarrow \quad \mathcal{K} = \mathcal{K}'. \tag{3.7}$$

PROOF. The implication from right to left is trivial. To establish the converse implication, suppose that $K \in \mathcal{K}$. For some state $K' \in \mathcal{K}'$, we then have

$$K \subseteq \bigcup_{q \in K} (\cap \mathcal{K}'_q) = K' . \tag{3.8}$$

We show that, in fact, $K = K'$. Take any $s \in K'$. There must be some $q \in K$ such that $s \in \cap\mathcal{K}'_q$. Thus, $\mathcal{K}'_q \subseteq \mathcal{K}'_s$, which implies $\mathcal{K}_q \subseteq \mathcal{K}_s$, by the l.h.s. of (3.7). We obtain $s \in \cap\mathcal{K}_q$, yielding $s \in K$. This gives $K' \subseteq K$, and by (3.8), $K = K'$. We conclude that $\mathcal{K} \subseteq \mathcal{K}'$, and by symmetry, $\mathcal{K} = \mathcal{K}'$. □

3.8.3 Theorem. (Birkhoff, 1937) *There exists a one-to-one correspondence between the collection of all quasi ordinal spaces $\mathcal{K}$ on a domain Q, and the collection of all quasi orders $\mathcal{Q}$ on Q. Such a correspondence is defined by the two equivalences*

$$p\mathcal{Q}q \iff (\forall K \in \mathcal{K} : \; q \in K \Rightarrow p \in K) \tag{3.9}$$
$$K \in \mathcal{K} \iff (\forall (p,q) \in \mathcal{Q} : \; q \in K \Rightarrow p \in K). \tag{3.10}$$

Under this correspondence, ordinal spaces are mapped onto partial orders.

Notice that Equation (3.9) can be written more compactly as

$$p\mathcal{Q}q \iff \mathcal{K}_p \supseteq \mathcal{K}_q. \tag{3.11}$$

Thus, by (3.5), the quasi order $\mathcal{Q}$ defined by (3.11) from the knowledge structure $\mathcal{K}$ is just the surmise relation of $\mathcal{K}$.

PROOF. Equation (3.9) clearly defines a quasi order on Q (cf. Theorem 3.7.2). Conversely, for any quasi order $\mathcal{Q}$ on Q, Equation (3.10) defines a family $\mathcal{K}$ of subsets of Q. We establish the relevant properties of this last family. First, the family $\mathcal{K}$ necessarily contains Q and also $\varnothing$, since the implication $q \in \varnothing \Rightarrow p \in \varnothing$ is vacuously true for any $(p,q) \in \mathcal{Q}$. Thus, $\mathcal{K}$ is a knowledge structure. We show that $\mathcal{K}$ is closed under intersection. Take any $K, K' \in \mathcal{K}$ and suppose that $p\mathcal{Q}q$, with $q \in K \cap K'$. We obtain $q \in K$, $q \in K'$, which by (3.10) implies that $p \in K$, $p \in K'$, yielding $p \in K \cap K'$. Thus, $K \cap K' \in \mathcal{K}$. Similarly, the intersection of any subfamily of $\mathcal{K}$ belongs to $\mathcal{K}$. The proof that $\mathcal{K}$ is closed under union is similar.

It remains to show that the two equivalences (3.9) and (3.10) define a bijection. We write $\mathfrak{K}^{\mathbf{so}}$ for the collection of all quasi ordinal spaces $\mathcal{K}$ on a domain Q, and $\mathfrak{R}^{\mathbf{o}}$ for the collection of all quasi orders $\mathcal{Q}$ on Q. The result obtains if the two mappings

$$f : \mathfrak{K}^{\mathbf{so}} \to \mathfrak{R}^{\mathbf{o}} : \mathcal{K} \mapsto f(\mathcal{K}) = \mathcal{Q},$$
$$g : \mathfrak{R}^{\mathbf{o}} \to \mathfrak{K}^{\mathbf{so}} : \mathcal{Q} \mapsto g(\mathcal{Q}) = \mathcal{K}$$

respectively defined by (3.9) and (3.10) are mutual inverses. By Equation (3.11) and Theorem 3.8.2, f is an injective function. Let $\mathcal{Q}$ be any quasi order on Q, with $\mathcal{K} = g(\mathcal{Q})$ and $f(\mathcal{K}) = \mathcal{Q}'$. Using (3.10), $p\mathcal{Q}q$ implies that for all $K \in \mathcal{K}$, $q \in K \Rightarrow p \in K$, yielding $p\mathcal{Q}'q$ by (3.9).

Moreover, if $p\mathcal{Q}'q$, we take $K = \{x \in Q \,|\, x\mathcal{Q}q\}$: since $K \in \mathcal{K}$ and $q \in K$, we have $p \in K$, hence $p\mathcal{Q}q$. Thus, $\mathcal{Q} = \mathcal{Q}'$. Hence, any quasi order $\mathcal{Q}$ is in the range of the function f. We conclude that f and g are mutually inverse functions.

The last assertion regarding ordinal spaces is obvious. □

3.8.4 Definition. Referring to the correspondence described in Theorem 3.8.3, we say that the quasi ordinal space $g(\mathcal{Q})$ is *derived* from the quasi order $\mathcal{Q}$, and similarly that the quasi order $f(\mathcal{K})$ is *derived* from the quasi ordinal knowledge structure $\mathcal{K}$.

For later reference, we point out that Equation (3.10) can be used with any relation $\mathcal{Q}$ to produce a knowledge space. The proof of the next theorem is left as Problem 7.

3.8.5 Theorem. *Let $\mathcal{Q}$ be any relation on a domain Q, and define a collection $\mathcal{K}$ of subsets of Q by the equivalence:*

$$K \in \mathcal{K} \quad \Longleftrightarrow \quad (\forall (p,q) \in \mathcal{Q} : \; q \in K \Rightarrow p \in K). \tag{3.10}$$

Then $\mathcal{K}$ is a quasi ordinal knowledge space on Q.

3.8.6 Definition. In the context of Theorem 3.8.5, we say that the quasi ordinal space $\mathcal{K}$ is *derived* from the relation $\mathcal{Q}$.

We leave the proof of the next result as Problem 15.

3.8.7 Theorem. *Any finite ordinal space is a learning space.*

3.9 Original Sources and Related Works

Most of our original paper (Doignon and Falmagne, 1985) was restricted to the finite case. Other relevant references are mentioned in Section 1.7. The results presented in this chapter for the infinite case were formulated in Doignon and Falmagne (1999), as for instance, those concerning the concept of a granular knowledge structure. Note, however, that a granular knowledge space is just the dual of a 'convex structure' in the sense of Van de Vel (1993).

A paper by Dowling (1993b) contains two algorithms. One constructs the base of a finite knowledge space; the other generates the space spanned by a finite family of sets. For the second task, we provide in 3.5.5 another algorithm which is similar in spirit but is much easier to grasp. It is also slightly more efficient on the average. A different algorithm, due to Ganter (1984, 1987, see also Ganter and Wille, 1996) in the framework of concept lattices, can also be used for the second task; it has the same overall theoretical efficiency but avoids storing the previously generated states.

In our terminology, Birkhoff's Theorem[4] concerns quasi ordinal spaces. A variant for knowledge spaces in general is given in Chapter 5. We mentioned in Example 3.3.2 a few mathematical examples of families of sets closed under intersection. A closure space (Definition 3.3.1) is often also called a 'convexity space.' The first term is used by Birkhoff (1967) and Buekenhout (1967), for example, while the second can be found in particular in Sierksma (1981). Birkhoff (1967) also refers to any family of subsets closed under intersection as a 'Moore family.' The excellent monograph of Van de Vel (1993) concerns 'convex structures' (also called 'aligned spaces' after Jamison-Waldner, 1982). These structures are dual to finitary knowledge spaces (cf. Definition 3.6.1). The word 'finitary' was used to qualify such closure spaces (which Buekenhout, 1967, calls 'espaces à fermeture finie'). It is motivated by the following result (often taken as a definition of 'finitary closure space'): *The closure space* $(Q, \mathcal{L})$ *is finitary if and only if the closure of any subset* A *of* Q *is the union of all closures of finite subsets of* A (see Problem 13). Chapter 8 extends the concept of a closure to the general context of a quasi order.

Problems

1. How many states are contained in the dual of the knowledge structure $\mathcal{H}$ of Example 2.1.3? Specify some of these states.
2. Prove that the collection $\mathcal{L}$ of states in Example 3.3.3 is closed under intersection. Explain how this result is related to Theorem 3.2.1.
3. Does Theorem 3.2.1 still hold if $\mathcal{R}$ is a relation on 2^Q? Provide a counterexample if your response is negative.
4. Show that Algorithm 3.5.1 also correctly builds the base when provided with a spanning family instead of the space itself.
5. Show that Algorithm 3.5.5 also correctly builds the space when provided with a spanning family instead of the base itself.
6. Prove Equation (3.5) in Definition 3.7.1.
7. Prove Theorem 3.8.5.
8. If a knowledge space is ordinal (resp. quasi ordinal), is it true that any of its projections and children are also ordinal (resp. quasi ordinal)?
9. Suppose that all the projections of a knowledge structure $\mathcal{K}$ are spaces, (resp. discriminative structures, closure spaces). Is it necessarily true, then, that $\mathcal{K}$ is itself a space (resp. a discriminative structure, a closure space)?

---

[4] Our Theorem 3.8.3.

10. For each of the following properties of a knowledge structure, check whether it implies the same property for the dual structure:
    (a) being a space;
    (b) being quasi ordinal;
    (c) being ordinal.

11. Prove Conditions (i), (ii) and (iii) in Theorem 3.3.4. Prove that the correspondence mentioned in the statement is 1-1.

12. Establish the assertions made regarding the execution time of Algorithm 3.5.5 in Remarks 3.5.8)(c).

13. *(Finitary closure spaces.)* As in Definitions 3.3.1 and 3.3.5, let $(Q, \mathcal{L})$ be a closure space, with A' denoting the closure of a subset A of Q. Dually to Definition 3.6.1, we say that $(Q, \mathcal{L})$ is $\cap$-*finitary* when the union of any chain of states is a state. Show that $(Q, \mathcal{L})$ is $\cap$-finitary if it satisfies the following condition: for any $p \in Q$ and $A \subseteq Q$, we have $p \in A'$ if and only if $p \in F'$ for some *finite* subset F of A. The converse also holds, but its proof may be more difficult. (In this connection, see for example: Cohn, 1965; Van de Vel, 1993).

14. (*Feasible Symbologies.*) Not all arbitrarily chosen set of symbols constitutes a symbology (or alphabet) that is appropriate for communication purposes. Conflicting considerations enter into the construction of an acceptable symbology S. On the one hand, any symbol in S must, in principle, be readily recognized as such, which means that these symbols must be easy to distinguish from other symbols available in some larger set. On the other hand, these symbols must also be discriminable from each other (see Jameson, 1992). For example, it is plausible that the set

$$\{\spadesuit, \heartsuit, \diamondsuit, \clubsuit, 0, 1, \dots, 9\} \tag{3.12}$$

would not be considered to form an appropriate symbology, while its two subsets $\{\spadesuit, \heartsuit, \diamondsuit, \clubsuit\}$ and $\{0, 1, \dots, 9\}$ would no doubt be suitable. From a formal viewpoint, this situation is similar to that of Example 3.3.3. Consider a set C of symbols forming the universe of discourse. That is, the symbols in C are the only ones under consideration. (The set specified in Equation (3.12) is an instance of such a set S.) It is conceivable that several subsets of C could form acceptable symbologies. Discuss this example in the style of Example 3.3.3. (Try to adapt Theorem 3.2.1.)

15. Prove Theorem 3.8.7.

4

Well-Graded Knowledge Structures

Wellgradedness was introduced in Definition 2.2.2 as a powerful property implied by the two axioms [L1] and [L2] defining learning spaces (cf. Section 2.2). As stated in Theorem 2.2.4, any well-graded knowledge space is in fact a learning space and conversely. In this chapter, we focus on the wellgradedness property per se. We define some new concepts, derive important consequences, and describe a variety of applications to topics quite different from education. The results of this chapter will have applications elsewhere in this book. For example, they will provide the combinatoric skeleton for the learning theories developed in Chapters 9 and 12 and for some of the assessment procedures described in Chapters 13 and 14. To avoid minor technicalities, we restrict consideration to discriminative structures.

4.1 Learning Paths, Gradations, and Fringes

The knowledge state of an individual may vary over time. For example, the following learning scheme is reasonable. A novice student is in the empty state and thus knows nothing at all. Then, one or a few items are mastered; next, another batch is absorbed, etc., up to the eventual mastery of the full domain of the knowledge structure. There may be many possible learning sequences, however. Forgetting may also take place. More generally, there may be many ways of traversing a knowledge structure, evolving at each step from one state to another closely resembling one, and various reasons for doing so.

4.1.1 Definition. A *learning path* in a knowledge structure $(Q, \mathcal{K})$ (finite or infinite) is a maximal chain $\mathcal{C}$ in the partially ordered set $(\mathcal{K}, \subseteq)$. According to the definition of 'chain' in 1.6.10, we have thus $C \subseteq C'$ or $C' \subseteq C$ for all $C, C' \in \mathcal{C}$. Saying that the chain $\mathcal{C}$ is maximal means that whenever $\mathcal{C} \subseteq \mathcal{C}'$ for some chain of states $\mathcal{C}'$, then $\mathcal{C} = \mathcal{C}'$. Thus, a maximal chain necessarily contains $\varnothing$ and Q.

J.-C. Falmagne, J.-P. Doignon, *Learning Spaces*,
DOI 10.1007/978-3-642-01039-2_4, © Springer-Verlag Berlin Heidelberg 2011

In some situations, the student could master the items one at a time. For example, in the case of a finite domain Q containing m elements, a learning path could take the form

$$\varnothing \subset \{q_1\} \subset \{q_1, q_2\} \subset \cdots \subset \{q_1, q_2, \ldots, q_m\} = Q \tag{4.1}$$

for some particular order q_1, q_2, ..., q_m of the elements of the domain Q. We call such a learning path a 'gradation' (cf. Definition 4.1.3). Note that a gradation in the sense of Equation (4.1) exists in a knowledge structure only if it is discriminative. Indeed, for any two items, there must be a state in the gradation of Equation (4.1) containing one item and not the other. Hence, these two items cannot be equally informative (cf. Definition 2.1.5). This means that each notion contains a single item. In other words, the knowledge structure is discriminative. On the other hand, a learning path in a discriminative structure is not necessarily a gradation. In fact, some discriminative structures have no gradations.

4.1.2 Example. Take a domain Q containing more than two elements, and let $\mathcal{F}$ be the family containing $\varnothing$ and Q, and all the subsets of Q containing exactly two elements. Then $\mathcal{F}$ is a discriminative knowledge structure, and all the learning paths are of the form: $\varnothing \subset \{q, r\} \subset Q$ with $q \neq r$. Note that Q may be infinite.

We introduce the basic tools of this chapter.

4.1.3 Definition. Let $(Q, \mathcal{K})$ be a finite knowledge structure. A learning path $\mathcal{C}$ in $\mathcal{K}$ is called a *gradation* if for any $K \in \mathcal{C} \setminus \{Q\}$, there exists $q \in Q \setminus K$ such that $K \cup \{q\} \in \mathcal{C}$. (Or equivalently: for any $K \in \mathcal{C} \setminus \{\varnothing\}$, there exists $q \in K$ such that $K \setminus \{q\} \in \mathcal{C}$.)

We recall (from 1.6.12) that d denotes the canonical distance between sets: $d(K, L) = |K \triangle L|$. We know from 2.2.2 that a tight path between two states K and L is a sequence

$$K_0 = K, K_1, \ldots, K_n = L \tag{4.2}$$

such that

$$d(K_i, K_{i+1}) = 1 \qquad (0 \leq i \leq n-1) \tag{4.3}$$

with

$$d(K, L) = n. \tag{4.4}$$

The sequence (4.2) is a *stepwise path* between the states K and L if it satisfies (4.3), but not necessarily (4.4).

We say that $(Q, \mathcal{K})$ is 1-*connected* if there is a stepwise path between any two of its (distinct) states. Remember from 2.2.2 that $(Q, \mathcal{K})$ is well-graded if there exists a tight path between any two of it states.

Even for a discriminative space, 1-connectedness is not equivalent to well-gradedness.

4.1.4 Example. Let $\mathcal{K}$ be the knowledge space with base

$$\{\{c\},\{a,b\},\{b,c\},\{c,d\},\{d,e\}\}.$$

As there is a stepwise path from any state to $\cup\mathcal{K} = \{a,b,c,d,e\}$, the space $\mathcal{K}$ is 1-connected. However, it is not well-graded: there is no tight path connecting the two states $\{a,b\}$ and $\{d,e\}$.

Theorem 4.1.7, the central result of this chapter, contains various characterizations of well-graded knowledge structures. It is formulated in terms of a few additional concepts which we illustrate by our next example.

4.1.5 Example. Consider the knowledge structure

$$\mathcal{H} = \{\varnothing,\{b\},\{e\},\{d,e\},\{a,b,c\},\{a,c,d\},\{a,b,c,d,\},U\}.$$

with domain $U = \{a,b,c,d,e\}$. There are (exactly) three states which are at distance 1 from the state $\{a,b,c,d\}$; they are:

$$\begin{aligned}\{a,c,d\} &= \{a,b,c,d\}\setminus\{b\},\\ \{a,b,c\} &= \{a,b,c,d\}\setminus\{d\},\\ U &= \{a,b,c,d\}\cup\{e\}.\end{aligned}$$

The first two of these states result from the removal of either item b or item d from the state $\{a,b,c,d\}$. We say that the set $\{b,d\}$ forms the 'inner fringe' of the state $\{a,b,c,d\}$. Similarly, the state U results from adding the item e to $\{a,b,c,d\}$. We say that the set $\{e\}$ forms the 'outer fringe' of $\{a,b,c,d\}$.

4.1.6 Definition. The *inner fringe* of a state K in a discriminative knowledge structure $(Q,\mathcal{K})$, is the subset of items

$$K^{\mathcal{I}} = \{q \in K \mid K\setminus\{q\} \in \mathcal{K}\}.$$

The *outer fringe* of a state K in such a structure is the subset

$$K^{\mathcal{O}} = \{q \in Q\setminus K \mid K\cup\{q\} \in \mathcal{K}\}.$$

The *fringe* of K is the union of the inner and outer fringes. We write

$$K^{\mathcal{F}} = K^{\mathcal{I}} \cup K^{\mathcal{O}}.$$

Let $\mathcal{N}(K,h)$ be the set of all states whose distance from K is at most h, thus:

$$\mathcal{N}(K,h) = \{L \in \mathcal{K} \mid d(K,L) \le h\}. \tag{4.5}$$

Then we also have $K^{\mathcal{F}} = \big(\cup\mathcal{N}(K,1)\big)\setminus\big(\cap\mathcal{N}(K,1)\big)$ (cf. Problem 14). In the sequel, we refer to $\mathcal{N}(K,h)$ as the *h-neighborhood* of K, or sometimes as the *ball of radius h centered* at the state K.[1]

[1] Obviously, these concepts would have to be redefined in terms of notions in the case of non discriminative structures.

We recall that the complement of a state K in a knowledge structure $(Q, \mathcal{K})$ is denoted by $\overline{K} = Q \setminus K$ (cf. 2.2.2).

4.1.7 Theorem. *For any family $\mathcal{K}$ of finite sets called states, the following five conditions are equivalent:*

(i) *$\mathcal{K}$ is well-graded;*
(ii) *for any two states K and L, there exists a stepwise path $K = K_0$, K_1, $\ldots$, $K_n = L$ satisfying*

$$K_j \cap L \subseteq K_{j+1} \subseteq K_j \cup L \qquad (0 \leq j \leq n-1); \qquad (4.6)$$

(iii) *for any two distinct states K, L, we have*

$$(K \bigtriangleup L) \cap K^{\mathcal{F}} \neq \varnothing; \qquad (4.7)$$

(iv) *any two states K and L in $\mathcal{K}$ which satisfy $K^{\mathcal{I}} \subseteq L$ and $K^{\mathcal{O}} \subseteq \overline{L}$ must be equal;*
(v) *any two states K and L in $\mathcal{K}$ which satisfy $K^{\mathcal{I}} \subseteq L$, $K^{\mathcal{O}} \subseteq \overline{L}$, $L^{\mathcal{I}} \subseteq K$, $L^{\mathcal{O}} \subseteq \overline{K}$ must be equal.*

Note that this result applies to uncountable families: take, for example, the family of all finite subsets of $\mathbb{R}$.

Proof. We prove (i) $\Rightarrow$ (ii) $\Rightarrow$ (iii) $\Rightarrow$ (iv) $\Rightarrow$ (v) $\Rightarrow$ (i).

(i) $\Rightarrow$ (ii). Take any two states K and L, with $d(K, L) = h$. By the definition of wellgradedness in 2.2.2, there exists a tight path from K to L. Any tight path is obviously a stepwise path. We leave to the reader to verify that any tight path connecting K and L satisfies the two inclusions properties in (4.6) (Problem 5).

(ii) $\Rightarrow$ (iii). Take any two states $K \neq L$, and let $(K_j)_{0 \leq j \leq n}$ be a stepwise path described in Condition (ii). Then K and K_1 differ by exactly one element q, and we have moreover $K \cap L \subseteq K_1 \subseteq K \cup L$. Either q belongs to K, or it belongs to L, but not both. Hence q belongs to $(K \bigtriangleup L) \cap K^{\mathcal{F}}$.

(iii) $\Rightarrow$ (iv). We proceed by contradiction. Let K and L be two distinct states satisfying $K^{\mathcal{I}} \subseteq L$ and $K^{\mathcal{O}} \subseteq \overline{L}$. Take any $q \in (K \bigtriangleup L) \cap K^{\mathcal{F}}$. If $q \in K$, then $q \in K^{\mathcal{I}} \subseteq L$, contradicting $q \in K \bigtriangleup L$. Hence $q \notin K$, but then $q \in L \cap K^{\mathcal{O}}$, and we obtain $q \in L$ and $q \in K^{\mathcal{O}} \subseteq \overline{L}$, a contradiction.

(iv) $\Rightarrow$ (v). Obvious.

(v) $\Rightarrow$ (i). Let K and L be two distinct states in $\mathcal{K}$ with $d(K, L) = h > 0$. We construct a tight path $(K_i)_{0 \leq j \leq h}$ with $K = K_0$ and $K_n = L$. Since $K \neq L$, Condition (v) implies that:

$$(K^{\mathcal{I}} \not\subseteq L) \text{ or } (K^{\mathcal{O}} \not\subseteq \overline{L}) \text{ or } (L^{\mathcal{I}} \not\subseteq K) \text{ or } (L^{\mathcal{O}} \not\subseteq \overline{K}).$$

So, there must be some element q satisfying

$$q \in (K^{\mathcal{I}} \setminus L) \cup (K^{\mathcal{O}} \cap L) \cup (L^{\mathcal{I}} \setminus K) \cup (L^{\mathcal{O}} \cap K).$$

If $q \in K^{\mathcal{I}} \setminus L$, we set $K_1 = K \setminus \{q\}$. In the other three cases, we set $K_1 = K \cup \{q\}$ or $K_{h-1} = L \setminus \{q\}$ or $K_{h-1} = L \cup \{q\}$. We obtain either $d(K_1, L) = h - 1$ (in the first two cases), or $d(K, K_{h-1}) = h - 1$ (in the last two cases). The result follows by induction. □

4.1.8 Remarks. a) The equivalence of (i) and (iv) in Theorem 4.1.7 has important applications in education. This result tells us that, in a well-graded structure $\mathcal{K}$, a state is fully specified by its two fringes in the sense that

$$\forall K, L \in \mathcal{K}: \quad (K^{\mathcal{I}} = L^{\mathcal{I}} \text{ and } K^{\mathcal{O}} = L^{\mathcal{O}}) \quad \Longleftrightarrow \quad K = L. \tag{4.8}$$

In practice, in an educational software such as ALEKS for example, learning spaces are used. Since learning spaces are well-graded by Theorem 2.2.4, the equivalence (4.8) applies to them. This equivalence means that at the end of an assessment, the state uncovered can be accurately described by two typically short lists: one containing all the items in the inner fringe of the student's state, and the other all the items in the outer fringe. The interest of such a representation is not only its economy. The two fringes have a pedagogical meaning for a student and her teacher. The inner fringe indicates the high points in a student's state. These are items that the student may have learned only recently; the mastery of such items may still be shaky. The outer fringe is even more useful: it contains the items that the student is ready to learn. The outer fringe is thus a window to further study.

b) With regard to the stepwise path $(K_j)_{0 \le j \le h}$ in Theorem 4.1.7(ii), we stress that K_{i+1} is derived either by removing from K_i an element in $K_i \setminus L$, or by adding to K_i an element in $L \setminus K_i$.

c) Note that a knowledge structure in which all the learning paths are gradations is not necessarily well-graded. As an example, consider the knowledge structure

$$\{\varnothing, \{a\}, \{c\}, \{a,b\}, \{b,c\}, \{a,b,d\}, \{b,c,d\}, \{a,b,c,d\}\}.$$

This structure has two learning paths, both of which are gradations, but is not well-graded: the two states $\{a,b\}$ and $\{b,c\}$ have the same inner fringe $\{b\}$ and outer fringe $\{d\}$ but are different, contradicting Condition (iv) of Theorem 4.1.7 and the equivalence (4.8). Moreover, there is no tight path connecting those states. We have in fact

$$(\{a,b\} \triangle \{b,c,\}) \cap \{a,b\}^{\mathcal{F}} = \{a,c\} \cap \{b,d\} = \varnothing.$$

We mention a consequence of Theorem 4.1.7 which specifies the meaning of Axiom [L1] from 2.2.1.

4.1.9 Theorem. *For any knowledge space $(Q, \mathcal{K})$, the following three conditions are equivalent:*

(i) $(Q, \mathcal{K})$ *is well-graded;*
(ii) $(Q, \mathcal{K})$ *is finite and all its learning paths are gradations;*
(iii) *Axiom* [L1] *holds.*

PROOF. (i) $\Rightarrow$ (ii). Remember that any well-graded knowledge structure is necessarily finite (cf. 2.2.2). Let $\mathcal{C}$ be any learning path, and take any K in $\mathcal{C} \setminus \{Q\}$. Denote by L the state that immediately follows K in $\mathcal{C}$. By Theorem 4.1.7, (i) $\Rightarrow$ (iii), there exists some q in $(K \triangle L) \cap K^{\mathcal{F}}$. Thus, $K + \{q\}$ is a state included in L. Since $\mathcal{C}$ is maximal, we must have $K + \{q\} = L \in \mathcal{C}$. This implies that $\mathcal{C}$ is a gradation.

(ii) $\Rightarrow$ (iii). We leave this part of the proof to the reader as Problem 6.

(iii) $\Rightarrow$ (i). This follows from Lemma 2.2.3. □

Two other characterizations of finite, discriminative well-graded knowledge spaces will be given in Chapter 11 (Theorems 11.5.3 and 11.5.4).

4.1.10 Theorem. *Any finite ordinal space $(Q, \mathcal{K})$ is well-graded.*

In Theorem 4.3.5, we extend this result to all ordinal spaces (on the basis of the concept of '∞-wellgradedness' for infinite structures defined in 4.3.3).

PROOF. In view of Theorem 2.2.4, we need only prove that Axiom [MA] holds, that is: any state K contains an item q such that $K \setminus \{q\} \in \mathcal{K}$. It suffices to take for q any item maximal in K for the partial order from which $\mathcal{K}$ derives. It is clear that $K \setminus \{q\}$ is then also a state. □

4.2 A Well-Graded Family of Relations: the Biorders*

An interesting example of a well-graded structure arises in the theory of order relations, in the guise of the 'biorders.' In fact, several well-known families of relations— regarded as sets of pairs—can be shown to be well-graded in the sense of Definition 2.2.2. We indulge in this detour into the theory of order relations to illustrate potential applications of our results beyond the main focus of this monograph. We restrict considerations to finite structures.

4.2.1 Definition. Let X and Y be two basic finite, nonempty sets, with Y not necessarily disjoint or distinct from X. Following our convention in 1.6.1, we abbreviate the pair $(x, y) \in X \times Y$ as xy. A relation R from X to Y, that is $R \subseteq X \times Y$, is called a *biorder* if for all $x, x' \in X$ and $y, y' \in Y$, we have

[BO] $\quad (xRy,\ \neg(x'Ry) \text{ and } x'Ry') \Rightarrow xRy'.$

Using the compact notation for the (relative) product introduced in 1.6.2, Condition [BO] can also be stated by the formula

[BO′] $\quad R\bar{R}^{-1}R \subseteq R.$

It is easy to check that the complement $\bar{R}$ of a biorder is itself a biorder. Accordingly, [BO′] is equivalent to

[BO″] $\quad \bar{R}R^{-1}\bar{R} \subseteq \bar{R}$.

The interest for biorders stems in part from their numerical representation. Ducamp and Falmagne (1969) have shown that for finite sets X and Y, Condition [BO] was necessary and sufficient to ensure the existence of two functions $f : X \to \mathbb{R}$ and $g : Y \to \mathbb{R}$ satisfying

$$xRy \quad \Longleftrightarrow \quad f(x) > g(y). \tag{4.9}$$

The term "biorder" was coined by Doignon, Ducamp, and Falmagne (1984) who extended this representation to infinite sets X and Y. The concept plays an important role in psychometrics, where X and Y represent, respectively, a set of subjects and a set of questions of a test of some ability. The notation xRy formalizes the fact that subject x has solved question y. The r.h.s. of the equivalence (4.9) is then interpreted as meaning: the ability $f(x)$ of subject x exceeds the difficulty $g(y)$ of question y. In this context, the relation R is coded as a rectangular 0-1 array and referred to as a *Guttman's scale* (from a famous paper of Guttman, 1944). Condition [BO] means that such a 0-1 array never contains a sub-array of the form shown in Table 4.1.

Table 4.1. Forbidden pattern in a 0-1 array representing a biorder.

	y	y'
x	1	0
x'	0	1

Condition [BO] enters as one of the defining conditions of other standard order relations encountered in measurement theory and utility theory such as the interval orders and the semiorders. The semiorders were introduced by Luce (1956) (see also Scott and Suppes, 1958). The interval orders are due to Fishburn (1970, 1985). For background and references, the reader is referred to the Sources section at the end of the chapter.

For the rest of this section, we consider the full family of all the biorders from X to Y as a family of subsets of a basic finite set $Q = X \times Y$. Thus, each of the biorders is regarded as a set of pairs. We shall prove the following result, which is due to Doignon and Falmagne (1997).

4.2.2 Theorem. *The family $\mathfrak{B}$ of all the biorders between two finite sets X and Y is a well-graded discriminative knowledge structure. Moreover, the inner and outer fringes (cf.* 4.1.6*) of any relation R in $\mathfrak{B}$ are specified by the two equations:*

$$R^{\mathfrak{I}} = R \setminus R\bar{R}^{-1}R, \tag{4.10}$$

$$R^{\mathfrak{O}} = \bar{R} \setminus \bar{R}R^{-1}\bar{R}. \tag{4.11}$$

Clearly, both $\varnothing$ and $X \times Y$ are biorders. Hence, $\mathfrak{B}$ is a knowledge structure, which is also discriminative because $\{xy\}$ is a biorder for all $x \in X$ and $y \in Y$. Checking that expressions for the inner and outer fringes are indeed those specified Equations (4.10) and (4.11) is easy, and is left to the reader (cf. Problem 7). To establish that the family $\mathfrak{B}$ of all biorders between two finite sets X and Y is well-graded, we shall establish Condition (iv) in Theorem 4.1.7. The proof given in 4.2.5 relies on some auxiliary results.

Notice that for any relation R, the products $R\bar{R}^{-1}$ and $\bar{R}^{-1}R$ are irreflexive relations. Moreover, if R is a biorder, then for any positive integer n, the nth power $(R\bar{R}^{-1})^n$ of the product $R\bar{R}^{-1}$ is also irreflexive. We shall use the following fact:

4.2.3 Lemma. *If R is a biorder from a finite set X to a finite set Y, then we necessarily have*

$$R = \bigcup_{k=0}^{\infty} (R\bar{R}^{-1})^k R = \bigcup_{k=0}^{\infty} \left(R^{\mathfrak{I}}(R^{\mathfrak{O}})^{-1}\right)^k R^{\mathfrak{I}}.$$

PROOF. We show that the following inclusions hold:

$$R \subseteq \cup_{k=0}^{\infty}(R\bar{R}^{-1})^k R \subseteq \cup_{k=0}^{\infty}\left(R^{\mathfrak{I}}(R^{\mathfrak{O}})^{-1}\right)^k R^{\mathfrak{I}} \subseteq R.$$

The first inclusion is obvious: take $k = 0$ and use a convention from 1.6.2: as $R\bar{R}^{-1}$ is a relation on X, we have $(R\bar{R}^{-1})^0$ equal to the identity relation on X. To establish the second inclusion, suppose that $xy \in (R\bar{R}^{-1})^k R$ for some $k \geq 0$. Because $(R\bar{R}^{-1})^n$ is irreflexive for any positive integer n and X is finite, we can assume without loss of generality that k is maximal. This implies that each of the $k+1$ factors R in the formula $(R\bar{R}^{-1})^k R$ can be replaced with $R^{\mathfrak{I}}$ while keeping xy in the full product. Indeed, if this were not the case, such a factor R could be replaced with $R\bar{R}^{-1}R$ and we would find $xy \in (R\bar{R}^{-1})^{k+1} R$, contradicting the maximality of k. The fact that each of the k factors $\bar{R}^{-1}$ in the formula $(R\bar{R}^{-1})^k R$ can be replaced with $(R^{\mathfrak{O}})^{-1}$ is proved by similar arguments. We conclude that the second inclusion holds.

The third inclusion results from the biorder inclusion $R\bar{R}^{-1}R \subseteq R$ together with $R^{\mathfrak{I}} \subseteq R$ and $R^{\mathfrak{O}} \subseteq \bar{R}$. □

4.2.4 Theorem. *Let R and S be two biorders from X to Y. Then*

$$(R^{\mathfrak{I}} \subseteq S \text{ and } R^{\mathfrak{O}} \subseteq \bar{S}) \implies R = S.$$

PROOF. The inclusion $R \subseteq S$ follows from

$$\begin{aligned} xy \in R \quad &\Rightarrow \quad x\left(R^{\mathfrak{I}}(R^{\mathfrak{O}})^{-1}\right)^k R^{\mathfrak{I}} y, && \text{for some } k \geq 0 \text{ (by Proposition 4.2.3)}\\ &\Rightarrow \quad x(S\bar{S}^{-1})^k S y && \text{(by hypothesis, } R^{\mathfrak{I}} \subseteq S \text{ and } R^{\mathfrak{O}} \subseteq \bar{S})\\ &\Rightarrow \quad xy \in S && \text{(by Proposition 4.2.3).} \end{aligned}$$

To prove the converse inclusion, notice that $\bar{R}$ and $\bar{S}$ are themselves biorders. Moreover, $(\bar{R})^{\mathfrak{I}} = R^{\mathfrak{O}}$ and $(\bar{R})^{\mathfrak{O}} = R^{\mathfrak{I}}$. This means that our hypothesis can be translated as $(\bar{R})^{\mathfrak{I}} \subseteq \bar{S}$ and $(\bar{R})^{\mathfrak{O}} \subseteq \overline{(\bar{S})}$. The argument used above thus gives $\bar{R} \subseteq \bar{S}$, that is $S \subseteq R$. □

4.2.5 Proof of Theorem 4.2.2. In view of our discussion after the statement, it only remains to show here that $\mathfrak{B}$ is well-graded. This results from Theorem 4.2.4 which establishes Condition (iv) of Theorem 4.1.7. □

4.2.6 Remarks. a) The family $\mathfrak{B}$ of biorders of Theorem 4.2.2 is neither a knowledge space nor a closure space (in the sense of Definitions 2.2.2 and 3.3.1). Indeed, with $a \neq b$ and $a' \neq b'$, each of the four relations $\{ab\}, \{a'b'\}, \{ab, a'b, a'b'\}, \{ab, ab', a'b'\}$ is a biorder from $\{a, b\}$ to $\{a', b'\}$, but

$$\{ab\} \cup \{a'b'\} = \{ab, a'b, a'b'\} \cap \{ab, ab', a'b'\} = \{ab, a'b'\} \notin \mathfrak{B}.$$

(In fact, $\{ab, a'b'\}$ is a case of the forbidden subrelation represented by the 0-1 array of Table 4.1.)

b) As mentioned earlier, similar result have been obtained regarding the wellgradedness of other families of order relations under a slightly more general definition of a knowledge structure which does not require that the domain of the knowledge structure be a state. Examples are the partial orders, the interval orders and the semiorders (Doignon and Falmagne, 1997, and see Problems 8-11).

4.3 Infinite Wellgradedness⋆

We now extend the concept of wellgradedness to the case where a state in a gradation may be obtained as the 'limit' of the states that it includes in this gradation. Thus, the concept of distance no longer applies. To simplify the exposition, we limit consideration to discriminative knowledge structures[2].

4.3.1 Definition. An ∞-*gradation* in a discriminative knowledge structure $(Q, \mathcal{K})$ is a learning path $\mathcal{C}$ such that for any $K \in \mathcal{C} \setminus \{\varnothing\}$, we have:

either $\quad K = K' \cup \{q\}$, for some $q \in K$ and $K' \in \mathcal{C} \setminus \{K\}$, $\quad$ (4.12)

or $\quad K = \bigcup \{L \in \mathcal{C} \mid L \subset K\}$. $\quad$ (4.13)

When the knowledge structure $(Q, \mathcal{K})$ is finite, the 'limit' situation described by Equation (4.13) does not occur and the ∞-gradations are just gradations in the sense of Definition 4.1.3.

[2] Extending the concepts and results to non necessarily discriminative structures is straightforward; cf. Problem 12.

4.3.2 Example. As in 3.4.3, let $\mathcal{O}$ be the knowledge space formed by the collection of all open subsets of $\mathbb{R}$. Then any two states O, O' in $\mathcal{O}$ with $O \subset O'$ belong to a ∞-gradation. In fact, any maximal chain $\mathcal{C}$ of states containing O and O' must be a ∞-gradation. (By Hausdorff maximality principle, cf. 1.6.10, there exists at least one such maximal chain.) Indeed, suppose that some K in $\mathcal{C}$ does not satisfy Equation (4.13). Pick some item q in $K \setminus \bigcup\{L \in \mathcal{C} \mid L \subset K\}$. Then $K \setminus \{q\} = K' \in \mathcal{C}$ since $\bigcup\{L \in \mathcal{C} \mid L \subset K\} \subset K' \subset K$, with $K = K' \cup \{q\}$, and K satisfies Equation (4.12).

Defining well-graded structures in the infinite case requires more powerful tools than the finite sequences used in 2.2.2. A suitable device is suggested by the formulation in Theorem 4.1.7 (iii) (cf. Remark 4.1.8(b)). We begin by generalizing the concept of a path connecting two states.

4.3.3 Definition. A family $\mathcal{D}$ of states in a knowledge structure $(Q, \mathcal{K})$ is a *bounded path* connecting a state K and a state L if it contains K and L and the following three conditions hold: for all distinct D and E in $\mathcal{D}$,

(1) $$K \cap L \subseteq D \subseteq K \cup L;$$

(2) either $D \setminus L \subseteq E \setminus L$ and $D \setminus K \supseteq E \setminus K$,

or $D \setminus L \supseteq E \setminus L$ and $D \setminus K \subseteq E \setminus K$;

(3) either (a) $\exists F \in \mathcal{D} \setminus \{D\}, \exists q \in D \setminus F : F \cup \{q\} = D$;

or (b) $$\begin{cases} D \setminus K = \cup\{G \setminus K \mid G \in \mathcal{D},\ G \setminus K \subset D \setminus K\}, \\ \text{and} \\ D \setminus L = \cup\{G \setminus L \mid G \in \mathcal{D},\ G \setminus L \subset D \setminus L\}. \end{cases}$$

The knowledge structure is ∞-*well-graded* if any two of its states are connected by a bounded path. When the knowledge structure is finite, this definition of wellgradedness becomes identical to that in 2.2.2: Case (3)(b) does not arise, and Theorem 4.1.7(ii) applies.

4.3.4 Example. Examples of bounded paths are easy to manufacture. In the knowledge space formed by the open subsets of $\mathbb{R}$ (cf. 4.3.2), consider the two states $]a, b[$ and $]c, d[$, with $a < c < b < d$. Define a bounded path $\mathcal{A}$ connecting these two states by

$$\mathcal{A} = \{]a, b[,\ ]c, d[\} \cup \{A(x) \mid x \in \mathbb{R}\}$$

containing all the open intervals

$$A(x) =]g(x)(c - a) + a,\ g(x)(d - b) + b[\,,$$

where $g : \mathbb{R} \to]0, 1[$ is a continuous, strictly increasing function, satisfying

$$\lim_{x \to -\infty} g(x) = 0, \qquad \lim_{x \to +\infty} g(x) = 1.$$

(Take for example $g(x) = (1 + e^{-x})^{-1}$.) The family $\mathcal{A}$ is a bounded path connecting $]a, b[$ and $]c, d[$. We have

$$a < g(x)(c - a) + a < c < b < g(x)(d - b) + b < d,$$

verifying Condition (1) in 4.3.3. Notice that we also have

$$\begin{aligned} A(x) \setminus]c, d[&=]g(x)(c - a) + a,\, c], \\ A(x) \setminus]a, b[&= [b,\, g(x)(d - b) + b[\,. \end{aligned}$$

This yields for any $x \leq y$

$$A(y) \setminus]c, d[\;\subseteq\; A(x) \setminus]c, d[\quad \text{and} \quad A(y) \setminus]a, b[\;\supseteq\; A(x) \setminus]a, b[\,,$$

which verifies Condition (2) in 4.3.3. Finally, we observe that

$$\begin{aligned} A(x) \setminus]a, b[&= \cup_{z<x} \big(A(z) \setminus]a, b[\big), \\ A(x) \setminus]c, d[&= \cup_{x<z} \big(A(z) \setminus]c, d[\big), \end{aligned}$$

establishing Condition (3).

As a partial extension of Theorem 4.1.9 in the case of possibly infinite knowledge spaces, we have:

4.3.5 Theorem. *For any discriminative knowledge space* $(Q, \mathcal{K})$, *the following two conditions are equivalent:*

(i) $(Q, \mathcal{K})$ *is* ∞*-well-graded;*
(ii) *all the learning paths in* $(Q, \mathcal{K})$ *are* ∞*-gradations.*

Moreover, Conditions (i) *and* (ii) *are implied by*

(iii) *for any two distinct states* K *and* L, *we have*

$$(K \triangle L) \cap K^{\mathcal{F}} \neq \varnothing.$$

Proof. (i) $\Rightarrow$ (ii). Let $\mathcal{C}$ be any learning path, and take any state K in $\mathcal{C} \setminus \{\varnothing\}$. Defining $U = \cup\{L \in \mathcal{C} \,|\, L \subset K\}$, assume that $U \neq K$. Notice that $U \in \mathcal{C}$ follows from the closure of $\mathcal{K}$ under union and the maximality of $\mathcal{C}$. By (i), there is a bounded path $\mathcal{D}$ from U to K. Since $U \subset K$, we must have $U \subseteq D \subseteq K$ for all $D \in \mathcal{D}$. Suppose that there is some D in $\mathcal{D}$ such that $U \subset D \subset K$. Then $\mathcal{C} \cup \{D\}$ is a chain, which is impossible because $\mathcal{C}$ is maximal. Thus $\mathcal{D} = \{U, K\}$. Since $\mathcal{D}$ is a bounded path, we derive the existence of $q \in K$ such that $U \cup \{q\} = K$. This proves that $\mathcal{C}$ is a ∞-gradation.

(ii) $\Rightarrow$ (i). Let $K, L \in \mathcal{K}$, thus also $K \cup L \in \mathcal{K}$. Take a learning path $\mathcal{C}_1$ containing K and $K \cup L$, and a learning path $\mathcal{C}_2$ containing L and $K \cup L$. Then

$$\{D \in \mathcal{C}_1 \,|\, K \subseteq D \subseteq K \cup L\} \cup \{E \in \mathcal{C}_2 \,|\, L \subseteq E \subseteq K \cup L\}$$

is a bounded path from K to L.

(iii) $\Rightarrow$ (ii). Assume again that $\mathcal{C}$ is a learning path, $K \in \mathcal{C} \setminus \{\varnothing\}$, and $U = \cup\{L \in \mathcal{C} \mid L \subset K\} \neq K$. Thus $U \in \mathcal{K}$ and $U \in \mathcal{C}$. By (iii), there is an item q in $(K \setminus U) \cap U^{\mathcal{F}}$. Then $U \cup \{q\}$ is a state that must be equal to K. This shows that $\mathcal{C}$ is a ∞-gradation. □

Example 4.3.2 shows that for discriminative spaces, Condition (iii) in Theorem 4.3.5 does not follow from Conditions (i) and (ii). (Because in this case the outer fringe of any open interval of $\mathbb{R}$ is empty.) There are also ordinal spaces that can be used as similar counter-examples; for instance, take the set $\mathbb{R}$ of real numbers with its usual order.

4.3.6 Theorem. *Any ordinal space $(Q, \mathcal{K})$ is ∞-well-graded.*

PROOF. We establish Condition (ii) in Theorem 4.3.5. Let K be a state of some learning path $\mathcal{C}$ in $(Q, \mathcal{K})$. Then $U = \cup\{L \in \mathcal{C} \mid L \subset K\}$ is a state of $\mathcal{C}$. It suffices to show that $K \setminus U$ contains at most one item. If p and q were two distinct items in $K \setminus U$, we could find a state M with either $p \in M$ and $q \notin M$, or $q \in M$ and $p \notin M$. A contradiction follows by considering the state $U \cup (M \cap K)$. □

4.4 Finite Learnability

4.4.1 Example. Consider the discriminative knowledge structure

$$\mathcal{J} = \{\varnothing, \{a\}, \{a,b,c\}, \{a,b,d\}, \{a,c,d\}, \{a,b,c,d,e\}\}. \tag{4.14}$$

Suppose that some individual in state $\{a\}$ wishes to acquire item d. Since there is no intermediate state between $\{a\}$ and $\{a,b,d\}$ or $\{a,c,d\}$, this can only be done by mastering simultaneously either b and d, or c and d.

This situation does not arise in the knowledge structure

$$\mathcal{J} \cup \{\{a,b\}\},$$

in which the individual may progress from state $\{a\}$ to a state containing d by steps involving one new item at a time. In learning spaces, which are finite structures, the issue is taken care of by Axiom [L1], which requires that the items be learnable one at a time. We consider a more general situation here. The next definition applies to any discriminative knowledge structure, finite or not. It allows for the possibility that the student must learn several items simultaneously. But the number of such items is finite and bounded.

4.4.2 Definition. A discriminative structure is *finitely learnable* if there is a positive integer l such that, for any state K and any item $q \notin K$, there exists a positive integer h and a chain of states $K = K_0 \subset K_1 \subset \cdots \subset K_h$ satisfying

(i) $q \in K_h$;
(ii) $d(K_i, K_{i+1}) \leq l$, for $0 \leq i \leq h-1$.

A finitely learnable knowledge structure $(Q, \mathcal{K})$ necessarily has a smallest number l satisfying these conditions, which is called the *learnstep number* of $(Q, \mathcal{K})$. We write then

$$\texttt{lst}\,(\mathcal{K}) = l.$$

With $\mathcal{J}$ defined by (4.14), we have, for example

$$\texttt{lst}\,(\mathcal{J} \cup \{\{a, b\}\}) = 2.$$

Indeed, from state $\{a, b, d\}$, item e can be mastered only by mastering simultaneously items c and e. Note that any well-graded structure has learning step number 1 (see Problem 16).

4.4.3 Remarks. a) Obviously, any finite knowledge structure is finitely learnable. However, some infinite knowledge structures are also finitely learnable. For example, for every infinite set Q, we clearly have $\texttt{lst}\,(2^Q) = 1$: for any set $K \subset Q$ and $q \in Q \setminus K$, we have $K \subset K_h = K \cup \{q\}$ with $h = l = 1$.

b) The ordinal space on $\mathbb{R}$ derived (in the sense of Definition 3.8.4) from the usual order of the reals is not finitely learnable.

c) A finite knowledge space $(Q, \mathcal{K})$ may satisfy $\texttt{lst}\,(\mathcal{K}) = 1$ without being well-graded. An example is the structure

$$\mathcal{K} = \{\varnothing, \{a\}, \{a, b\}, \{a, b, c\}, \{a, c, d\}, \{a, b, c, d\}\},$$

in which there is no tight path between $\{a\}$ and $\{a, c, d\}$.

4.5 Verifying Wellgradedness for a ∪-Closed Family

In Section 3.2, we introduced the concept of a relation $\mathcal{R}$ on $2^Q \setminus \{\varnothing\}$ capable of faithfully representing a particular knowledge space on a domain Q. In practice, such a relation $\mathcal{R}$ can be constructed by interviewing expert teachers, or by assessment statistics (cf. the equivalence discussed in Subsection 3.2.3 on page 45). Chapter 15 is devoted to the description of a relevant algorithm, named **QUERY**. An application of **QUERY** consists in a step-by-step construction of the relation $\mathcal{R}$, the final output delivering the knowledge space defined by $\mathcal{R}$ (see Theorem 3.2.1). Our goal, however, is to obtain a learning space, rather than just a knowledge space, and there is no guarantee that a knowledge space constructed by the above procedure is well-graded.

There are two possible methods for solving this problem. One is to modify **QUERY** so that only learning spaces are produced on each step. We analyze this possibility in Section 16.2, where an algorithm performing the required construction can be found.

A rather different method relies on first building a knowledge space $\mathcal{S}$, for example by a straightforward application of **QUERY**. If $\mathcal{S}$ is not well-graded, we then correct $\mathcal{S}$ by the addition of a minimal number of states ensuring wellgradedness (while preserving $\cup$-closure). Since any knowledge space is specified by its base, a typically much smaller set, it is natural to think of suitably expanding the base of $\mathcal{S}$. However, some theoretical investigations are needed to assess the value of this approach, and to extend it. Section 16.3 is devoted to this topic. Here, we consider the more general case in which the generating subfamily is not necessarily the base. In order to cover these more general results, we introduce a restricted concept of span (the latter was defined in 3.4.1).

4.5.1 Definition. The *span*† of a family of sets $\mathcal{G}$ is the collection $\mathcal{F}$ of all sets X such that $X = \cup\mathcal{H}$ for some **nonempty** $\mathcal{H} \subseteq \mathcal{G}$. We write then $\mathbb{S}^\dagger(\mathcal{G}) = \mathcal{F}$, and we say that $\mathcal{G}$ *spans*† $\mathcal{F}$.

Notice that, while $\mathbb{S}(\mathcal{G})$ is always a knowledge space, $\mathbb{S}^\dagger(\mathcal{G})$ is a knowledge space only if $\varnothing \in \mathcal{G}$.

The discussion at the beginning of this section motivates the four problems below for a family $\mathcal{G}$ of finite sets.

4.5.2 Four Problems.

A. Find necessary and sufficient conditions for $\mathcal{G}$ to be spanning† a well-graded, partially $\cup$-closed family of sets (see 2.2.6 for 'partially $\cup$-closed').

B. Find such conditions when the spanned family is a learning space. (These conditions should be simpler than in Problem A.)

C. Provide efficient algorithms for testing the conditions on a family $\mathcal{G}$ uncovered in A and B.

D. Supposing that some family $\mathcal{G}$ fails to satisfy the conditions in Problems A and B, provide algorithms for modifying $\mathcal{G}$ in some optimal sense to yield a family $\mathcal{G}'$ satisfying such conditions.

Problems A and B are solved in this section and Problems C and D in Section 16.3. Except for a couple of additional results in this section, we follow closely Eppstein, Falmagne, and Uzun (2009).

The following lemma allows us to infer the wellgradedness of a family from that of its base or of any of its spanning subfamilies.

4.5.3 Lemma. *The span*† *of a finite well-graded family is well-graded.*

PROOF. Let $\mathbb{S}^\dagger(\mathcal{G})$ be the span† of some finite wg-family $\mathcal{G}$. By the same argument as in Lemma 2.2.3, we only need to prove that there is a tight path from K to L for K, $L \in \mathbb{S}^\dagger(\mathcal{G})$ satisfying $K \subset L$. By definition of the span†, there exist nonempty $\mathcal{K}, \mathcal{L} \subseteq \mathcal{G}$ such that $K = \cup\mathcal{K}$ and $L = \cup\mathcal{L}$. First notice that $K \triangle L$ must be finite (so the distance between K and L is well defined). Indeed, $\mathcal{G}$ being finite, K and L are unions of finite numbers of elements of $\mathcal{G}$, and $\mathcal{G}$ being well-graded, the distance between any two of its members is finite. Second, take some K' in $\mathcal{K}$ and some L' in $\mathcal{L}$ such that $L' \setminus K \neq \varnothing$. There is some tight path from K' to L', say $K' = K'_0, K'_1, \ldots, K'_h = L'$. Let K'_j be the first element in the sequence that contains an element of $L \setminus K$. Then $K'_j \setminus K$ consists of one element lying in L. Setting $K_1 = K \cup K'_1$, we begin the construction of a tight path in $\mathbb{S}^\dagger(\mathcal{G})$ from K to L.

An induction completes the proof. □

It is easily verified that a similar result does not hold in general for the span of a well-graded family. Moreover, the finiteness assumption made in Lemma 4.5.3 cannot be dispensed with, as shown by the next example.

4.5.4 Example. Form a family $\mathcal{G}$ by taking the empty set and each one-element subset of $\mathbb{N}$. Both the span and the span† of $\mathcal{G}$ are equal to the collection of all subsets of $\mathbb{N}$. Notice that the family $\mathcal{G}$ is well-graded, while $\mathbb{S}^\dagger(\mathcal{G})$ is not.

Next we provide a characterization of families whose spans are well-graded. This is the first of the two main results in this section, and a solution to Problem B in 4.5.2.

4.5.5 Theorem. *Suppose that $\mathcal{G}$ is a finite family of sets and let $\mathbb{S}(\mathcal{G})$ be its span. Then the two following conditions are equivalent:*

(i) *$\mathbb{S}(\mathcal{G})$ is a well-graded family;*
(ii) *for each q in $\cup\mathcal{G}$ and each set G in $\mathcal{G}$ which is minimal for the property of containing q, the set $G \setminus \{q\}$ is the union of some subfamily of $\mathcal{G}$.*

Theorem 4.5.5 is in the spirit of a result of Koppen (1998). We give in Theorem 5.4.1 a version of Koppen's result in the infinite case[3]. Note that Theorem 4.5.5 is no longer true if we replace the span by the span† and assume that the family $\mathcal{G}$ does not contain the empty set (cf. our Counterexample 5.4.2).

Proof. (i) ⇒ (ii). Assume $\mathbb{S}(\mathcal{G})$ is well-graded, and let $q \in \cup\mathcal{G}$, $G \in \mathcal{G}$ be as in (ii). By assumption, there is a tight path $K_0, K_1, \ldots, K_h$ in $\mathbb{S}(\mathcal{G})$ from $\varnothing$ to G. Because of the minimality of G, we have $K_{h-1} = G \setminus \{q\}$. Thus $G \setminus \{q\}$ is a union of elements of $\mathcal{G}$.

(ii) ⇒ (i). In view of Lemma 2.2.3, we need only consider elements K and L in $\mathbb{S}(\mathcal{G})$ satisfying $K \subset L$, and prove that there is a tight path in $\mathbb{S}(\mathcal{G})$ from K to L. Pick q in $L \setminus K$. There exists at least one set G of $\mathcal{G}$ such that $q \in G \subseteq L$. By the finiteness of $\mathcal{G}$, we may assume that G is minimal with respect to those properties. Our assumption implies that $G \setminus \{q\}$ belongs to $\mathbb{S}(\mathcal{G})$. We have then either $K \subset K \cup G \subseteq L$ with $|K \cup G| = |K| + 1$ or $K \subset K \cup (G \setminus \{q\}) \subseteq L$. By induction, there exists a tight path from K to L. □

The base of a well-graded knowledge space need not be well-graded.

[3] Koppen's result is the equivalence (i) ⇔ (ii) of Theorem 5.4.1.

4.5.6 Example. The well-graded knowledge space

$$\mathcal{F} = \{\varnothing, \{a\}, \{b\}, \{c\}, \{a,b\}, \{a,c\}, \{b,c\}, \{c,d\}, \{a,b,c\}, \\ \{a,c,d\}, \{b,c,d\}, \{a,b,c,d\}, \{a,b,c,d,e\}\} \tag{4.15}$$

has the base $\{\{a\}, \{b\}, \{c\}, \{c,d\}, \{a,b,c,d,e\}\}$, which is not well-graded. Moreover, $\mathcal{F}$ has two different minimal well-graded subfamilies spanning $\mathcal{F}$:

$$\{\{a\}, \{b\}, \{c\}, \{a,b\}, \{a,c\}, \{b,c\}, \{c,d\}, \{a,b,c\}, \\ \{a,c,d\}, \{a,b,c,d\}, \{a,b,c,d,e\}\}, \tag{4.16}$$

$$\{\{a\}, \{b\}, \{c\}, \{a,b\}, \{a,c\}, \{b,c\}, \{c,d\}, \{a,b,c\}, \\ \{b,c,d\}, \{a,b,c,d\}, \{a,b,c,d,e\}\}. \tag{4.17}$$

Other minimal well-graded subfamilies spanning $\mathcal{F}$ are obtained by first adding $\varnothing$ to the base, and then other subsets as needed.

4.5.7 Example. The base of a knowledge space which is closed under intersection is not necessarily well-graded. Indeed, consider the knowledge space

$$\mathcal{F} = \{\varnothing, \{a\}, \{b\}, \{d\}, \{a,b\}, \{a,d\}, \{b,d\}, \{a,b,c\}, \{a,b,d\}, \\ \{a,b,c,d\}, \{a,b,c,d,e\}\},$$

whose base is $\{\{a\}, \{b\}, \{d\}, \{a,b,c\}, \{a,b,c,d,e\}\}$.

We turn to the second of the two main results of this section. As a consequence of Lemma 4.5.3, we derive a characterization of families whose spans† are well-graded.

4.5.8 Theorem. *Let $\mathcal{F}$ be a partially $\cup$-closed family spanned† by some finite family $\mathcal{G}$. Then $\mathcal{F}$ is a wg-family if and only if, for any two distinct sets G and H in $\mathcal{G}$, there is a tight path in $\mathcal{F}$ from G to $G \cup H$. If $\mathcal{G}$ contains the empty set, then $\mathcal{F}$ is well-graded if and only if there is a tight path in $\mathcal{F}$ from $\varnothing$ to H for any H in $\mathcal{G}$.*

This result provides another solution to Problems A and B in 4.5.2. The solution is not ideal, however, since it refers to a tight path in the spanned† family $\mathcal{F}$ and involves the state $G \cup H$ which does not necessarily belong to the spanning† family $\mathcal{G}$. (See the Open Problem 18.2.4 in Chapter 18 in this connection.)

Proof. As $\mathcal{F}$ is well-graded and contains $\mathcal{G}$, the necessity is clear for both statements. To establish sufficiency in the first statement, we notice that the argument in the proof of Lemma 2.2.3 also works for partially union-closed families. That is, we consider K, L in $\mathcal{F}$ and establish the existence of a tight path from K to L in $\mathcal{F}$ only under the assumption $K \subset L$. There exist subfamilies $\mathcal{K}$ and $\mathcal{L}$ of $\mathcal{G}$ such that $K = \cup\mathcal{K}$ and $L = \cup\mathcal{L}$. Take arbitrarily G in $\mathcal{K}$ and H in $\mathcal{L}$ with nevertheless $H \not\subseteq K$. By our assumption, there exists a tight path $G = G_0, G_1, \ldots, G_k = G \cup H$ in $\mathcal{F}$ from G to $G \cup H$. Let j be the smallest index such that $G_j \setminus K \neq \varnothing$. Then $G_j \setminus K = \{q\}$, for some q in $\cup \mathcal{G}$. We have thus $K \subset K \cup \{q\} \subseteq L$, with $K \cup \{q\} \in \mathcal{F}$. An induction completes the proof that $\mathcal{F}$ is well-graded.

We now show that if $\varnothing \in \mathcal{G}$ and $\mathcal{G}$ satisfies the latter condition of the statement, then there is a tight path in $\mathcal{F}$ from G to $G \cup H$ for any G and H in $\mathcal{G}$. Thus, the sufficiency of the second statement follows from that in the first statement. Indeed, let $H_0 = \varnothing, H_1, \ldots, H_h = H$ be a tight path in $\mathcal{F}$. It is easily seen that, after removal of identical terms if need be, the sequence $G \cup H_0 = G, G \cup H_1, \ldots, G \cup H_h = G \cup H$ is a tight path in $\mathcal{F}$ from G to $G \cup H$. □

4.6 Original Sources and Related Works

The concepts of learning paths and well-graded knowledge structures were introduced by Falmagne and Doignon (1988b) in the finite case. As mentioned earlier in this book (see the comments after Definition 2.2.1 on page 26), learning spaces are dual to the so-called '(intersection) antimatroids' or 'convex geometries' in the sense of Edelman and Jamison (1985). Specifically, a finite closure space is a convex geometry exactly when its dual $(Q, \mathcal{K})$ is a knowledge space in which all learning paths are gradations[4]. Definition 4.3.3 extends the concept of wellgradedness to the infinite case. This definition is a natural one in the context of knowledge structures in education, but would probably not be suitable for abstract convexity.

There are differences in the way we cast the concept of wellgradedness in this book as compared to Doignon and Falmagne (1999). The latter exposition took also care of the non-discriminative case, using notions (cf. Definition 2.1.5) while we use items here.

The application of the wellgradedness concept to families of relations, especially to biorders and semiorders, is taken from Doignon and Falmagne (1997) (see also Falmagne and Doignon, 1997). Ovchinnikov (1983) was a forerunner for the special case of partial orders. Biorders appeared under other names in the literature: Guttman scales (Guttman, 1944), Ferrers relations[5] (Riguet, 1951; Cogis, 1982), bi-quasi-series (Ducamp and Falmagne, 1969). Among the more recent papers, we mention Doignon, Ducamp, and Falmagne (1984) (where the term 'biorder' originated), and Doignon, Monjardet, Roubens, and Vincke (1986). Two important special cases of biorders are the semiorders[6] introduced by Luce (1956) (see also Scott and Suppes, 1958) and the interval

[4] Additional characterizations of finite, well-graded knowledge spaces can be derived from Edelman and Jamison (1985).

[5] From Norman Macleod Ferrers, a nineteen century British mathematician.

[6] We define those relations in Problem 9.

orders due to Fishburn (1970). For an introduction to these concepts and their applications in the social sciences, the reader is referred to Roberts (1979), Roubens and Vincke (1985), Suppes, Krantz, Luce, and Tversky (1989), or Pirlot and Vincke (1997). Purely mathematical expositions can be found in the monographs by Fishburn (1985) and Trotter (1992).

Section 4.5 follows (with some additions) Eppstein, Falmagne, and Uzun (2009), a paper motivated by the problems encountered in building a learning space empirically. Techniques for developing a knowledge space by questioning experts or by assessment statistics have been available for quite some time, based on theoretical results such as Theorem 3.2.1. However, as argued in Section 3.1, such a knowledge space is not necessarily the final step. It must still be tested for wellgradedness, and corrected in some optimal fashion if the test turns out to be negative. This leads to the four problems spelled in 4.5.2. The first two are solved in Section 4.5, and the last two in Section 16.3. The issues of constructing a knowledge space or a learning space in practice are considered again in Chapter 11 and, especially, Chapters 15 and 16.

Problems

1. Check whether the equivalence between the three statements in Theorem 4.1.9 still holds when the axiom of closure under union is replaced by Axiom [JS] of Problem 6 from Chapter 2. Prove your result.

2. Suppose that a knowledge structure is well-graded (resp. 1-connected, 1-learnable, i.e. has learnstep number equal to 1). Does that imply that the dual structure is also well-graded (resp. 1-connected, 1-learnable)?

3. Definition: We call property heritable when it necessarily holds for all the children of a knowledge structure which satisfies the property. Are wellgradedness, 1-connectedness, upgradability, and downgradability heritable properties?

4. Let $\mathcal{K}$ be a family of sets (infinite or not, containing only finite subsets or not) which is well-graded in the sense of Definition 2.2.2 (or 4.1.3). Show that $\mathcal{K}$ is discriminative if and only if $|\cap\mathcal{K}| \leq 1$. Is there a similar result for a family assumed to be ∞-well-graded (Definition 4.3.3)?

5. Prove the implication (i) $\Rightarrow$ (ii) in Theorem 4.1.7.

6. Prove the implication (ii) $\Rightarrow$ (iii) in Theorem 4.1.9.

7. Prove that the inner and outer fringes of a biorder R in the family of all the biorders between two finite sets X and Y are defined by the two equations (4.10) and (4.11).

8. Consider the collection $\mathcal{P}$ of all partial orders (cf. 1.6.1) on a finite set X, regarded as sets of pairs. Describe the inner and outer fringes of any partial order on X. Prove that $\mathcal{P}$ is well-graded.

9. A semiorder on a set X is an irreflexive biorder R between X and X satisfying the following additional condition:
[S] $RR\bar{R}^{-1} \subseteq R$.
Working with the collection $\mathcal{S}$ of all semiorders on a finite set X, regarded as sets of pairs, compute the inner and outer fringes of a given semiorder on X (cf. Doignon and Falmagne, 1997).

10. (Continuation. Difficult.) Prove that $\mathcal{S}$ is well-graded (see Doignon and Falmagne, 1997).

11. Would the results regarding wellgradedness in Problems 8 and 10 still hold if we drop the requirement that X is finite? Provide proofs of your responses.

12. Formulate the definitions and the results in the infinite case for non necessarily discriminative structures (cf. Section 4.3).

13. For the knowledge space $\mathcal{O}$ of Example 4.3.4, construct a bounded path connecting the two states $\{x \,|\, a < x < b \text{ or } c < x < d\}$ and $]e, f[$, with $a < e < b < f < c < d$.

14. Prove the following equality for the fringe of a state K of a knowledge structure $(Q, \mathcal{K})$ (cf. Definition 4.1.6):

$$K^{\mathcal{F}} = \big(\cup\mathcal{N}(K,1)\big) \setminus \big(\cap\mathcal{N}(K,1)\big).$$

15. Does a gradation $\mathcal{C}$ in a discriminative knowledge structure $(Q, \mathcal{K})$ necessarily satisfy the following property? For $K \in \mathcal{C} \setminus \{Q\}$, (i) or (ii) must hold, with:
(i) $K = K' \setminus \{q\}$, for some $K' \in \mathcal{C}$ and $q \in K'$;
(ii) $K = \cap\{L \in \mathcal{C} \,|\, K \subset L\}$.

16. Verify that the learnstep number of a well-graded knowledge structure is necessarily equal to 1.

5

Surmise Systems

When a knowledge structure is a quasi ordinal space, it can be faithfully represented by its surmise relation (cf. Theorem 3.8.3). In fact, as illustrated by Example 3.7.4, a finite ordinal space is completely recoverable from the Hasse diagram of the surmise relation. However, for knowledge structures in general, and even for knowledge spaces, the information provided by the surmise relation may be insufficient. In this chapter, we study the 'surmise system,' a concept generalizing that of a surmise relation, and allowing more than one possible learning 'foundation'[1] for an item[2]. One of the two main results of this chapter is Theorem 5.2.5 which establishes, in the style of Theorem 3.8.3 for quasi ordinal spaces, a one-to-one correspondence between knowledge spaces and surmise systems.

The surmise systems are closely related to the AND/OR graphs encountered in artificial intelligence. A section of this chapter is devoted to clarifying the relationship between the two concepts. This chapter also describes, in the form of Theorem 5.4.1, the relationship between well-graded knowledge spaces and a particular kind of surmise systems. This is the second main result of this chapter. Other highlights are: a generalization of the concept of a Hasse diagram, and a study of intractable 'cyclic' foundations which leads us to formulate conditions precluding such situations.

5.1 Basic Concepts

5.1.1 Example. Consider the knowledge structure

$$\begin{aligned}\mathcal{H} = \{\varnothing, \{a\}, \{b,d\}, \{a,b,c\}, \{b,c,e\}, \{a,b,d\},\\ \{a,b,c,d\}, \{a,b,c,e\}, \{b,c,d,e\}, \{a,b,c,d,e\}\}\end{aligned} \tag{5.1}$$

[1] We also use the terms 'clause for an item' or 'background of an item' as synonyms of 'foundation for an item.'

[2] The surmise relation only permits one foundation for any item q, which is formed by all the items preceding q in the surmise relation.

J.-C. Falmagne, J.-P. Doignon, *Learning Spaces*,
DOI 10.1007/978-3-642-01039-2_5, © Springer-Verlag Berlin Heidelberg 2011

on the domain $Q = \{a, b, c, d, e\}$. This knowledge structure is a discriminative knowledge space, with a surmise (or precedence) relation $\precsim$ represented in Figure 5.1 by its Hasse diagram (the relation $\precsim$ was defined in 3.7.1).

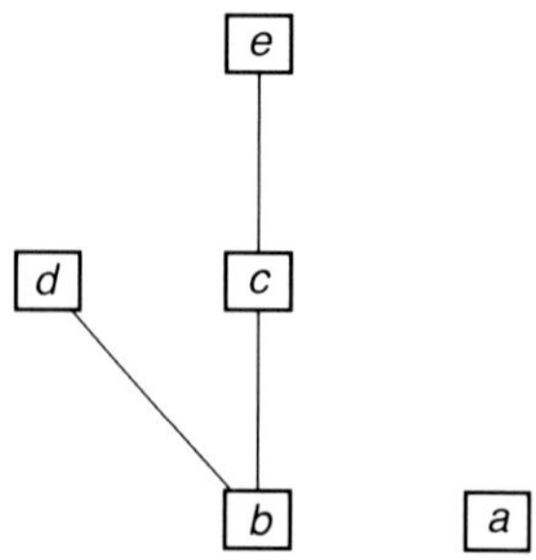

Figure 5.1. Hasse diagram of the surmise relation of the knowledge structure $\mathcal{H}$ specified by Equation (5.1).

Note that $\{q \in Q \,|\, q \precsim b\} = \{b\}$: there are no items in Q that must be mastered before b. This information, however, gives a distorted picture of the situation. Examining Equation (5.1) leads to the conclusion that b can be learned only if d was acquired simultaneously, or both a and c, or both c and e. Indeed,

$$\{b, d\}, \quad \{a, b, c\}, \quad \{b, c, e\}$$

are the three atoms at b, that is, the three minimal states of $\mathcal{H}$ containing b.

In a quasi ordinal space $\mathcal{K}$ with surmise relation $\precsim$, the situation is simpler. For any item q, there is always exactly one atom at q, namely $\cap\mathcal{K}_q$, which contains all the precedents of q in $\precsim$ (cf. Theorem 3.6.7 and Definition 3.8.1). In fact, considering the equivalence $r \precsim q \Leftrightarrow r \in \cap\mathcal{K}_q$, the full information concerning the quasi ordinal space is obtained by listing the unique atom $\cap\mathcal{K}_q$ at each question q. In the case of a knowledge space in general, however, there may be several atoms at a question, or possibly no atoms (as in Examples 3.4.3 or 3.4.9). This is illustrated by the knowledge space $\mathcal{H}$ of Equation (5.1), whose atoms at each item are listed in Table 5.1. The items b and c have three and two atoms, respectively.

Table 5.1. Items and their atoms in the knowledge space of Equation (5.1).

Items	Atoms
a	$\{a\}$
b	$\{b, d\}$, $\{a, b, c\}$, $\{b, c, e\}$
c	$\{a, b, c\}$, $\{b, c, e\}$
d	$\{b, d\}$
e	$\{b, c, e\}$

This example motivates the generalization of the surmise relation into a 'surmise function' which associates, to each item q of a domain Q, a collection of subsets of Q called the 'clauses' for q. Each of these clauses represents a possible 'foundation' for the mastery of item q. The concept of a surmise function is formalized by four conditions consistent with this interpretation. First, there must be at least one clause for each item. Second, we ask that each clause for an item contains this item. Third, each item in any clause C for an item has itself a clause included in C. (We give a pictorial illustration of this axiom in Figure 5.3). Finally, different clauses for the same item must be incomparable with respect to set inclusion.

We introduce the concepts of a surmise function and its clauses in the next definition. Note that we do not assume the existence of a knowledge structure. However, it turns out that any surmise function uniquely defines a granular knowledge space (cf. Theorem 5.2.5). (We recall from Definition 3.6.1 that a knowledge space $\mathcal{K}$ is granular when for each state K in $\mathcal{K}$ and each item q in K there exists an atom A at q with $q \in A \subseteq K$.)

5.1.2 Definition. Let Q be a nonempty set of items, and let σ be a function mapping Q into 2^{2^Q}. Thus, every value of σ is a family of subsets of Q. We say that σ is an *attribution (function)* on the set Q if this family is always nonempty, that is

(i) if $q \in Q$, then $\sigma(q) \neq \varnothing$.

For each $q \in Q$, any $C \in \sigma(q)$ is said to be a *clause for* q, or synonymously, a *foundation for* q (in σ). We formulate three additional conditions: for all $q, q' \in Q$, and $C, C' \subseteq Q$,

(ii) if $C \in \sigma(q)$, then $q \in C$;
(iii) if $q' \in C \in \sigma(q)$, then $C' \subseteq C$ for some $C' \in \sigma(q')$;
(iv) if $C, C' \in \sigma(q)$ and $C' \subseteq C$, then $C = C'$.

When all four conditions are satisfied, the pair (Q, σ) is a *surmise system* and the function σ is called a *surmise function* on Q. A surmise system (Q, σ) is *discriminative* if whenever $\sigma(q) = \sigma(q')$ for some $q, q' \in Q$, then $q = q'$. In such a case, the surmise function σ is also called *discriminative*.

5.1.3 Remarks. Condition (i) is reasonable in view of the intended meaning of the concept, but note that we can have $\sigma(q) = \{\varnothing\}$. (See Problem 1.) Condition (ii) is introduced for convenience and plays a minor role. Condition (iii) is natural if a clause for an item q is interpreted as a possible minimal foundation for the mastery of q: if q' is in a clause C for q, there must be a path to the mastery of q' within C, and so there must be a foundation for q' included in C. Condition (iv) ensures that the conceptual foundations formalized by the clauses are not redundant: suppose that C is a foundation of q, and C' is also a clause for q, with C' included in C; then C' must be equal to C, since, otherwise, C would not be a minimal foundation for q.

Surmise functions generalizes quasi orders. In particular, Conditions (ii) and (iii) correspond to reflexivity and transitivity, respectively. Actually, except for a trivial change in the encoding, any binary relation is a special case of an attribution.

5.1.4 Definition. Let $\mathcal{R}$ be any binary relation on a nonempty set Q. Define an attribution that has exactly one clause for each item q of Q by the equation:

$$\sigma(q) = \{\{r \in Q \,|\, r\mathcal{R}q\}\}.$$

We say then that $\mathcal{R}$ is *cast* as the attribution σ. Notice that σ and $\mathcal{R}$ contain exactly the same information. It is easily checked that

(i) $\mathcal{R}$ is reflexive iff σ satisfies Condition (ii) of a surmise function;
(ii) $\mathcal{R}$ is transitive iff σ satisfies Condition (iii) of a surmise function[3].

Thus, the collection of all surmise functions on Q encompasses the collection of all quasi orders on Q. Also, the collection of all attributions on Q encompasses the collection of all binary relations on Q.

5.1.5 Example. To get a pictorial display of an attribution in the case of a small finite set Q, we extend the standard conventions used for the graph of a binary relation. For instance, Figure 5.2 displays the graph of the attribution σ on $Q = \{a, b, c, d, e\}$ with

$$\sigma(a) = \{\{a, b, c\}, \{c, d\}\}, \qquad \sigma(b) = \{\{e\}\},$$
$$\sigma(c) = \{\{c\}\}, \qquad \sigma(d) = \{\{d\}\}, \qquad \sigma(e) = \{\{a, d\}, \{b\}\}.$$

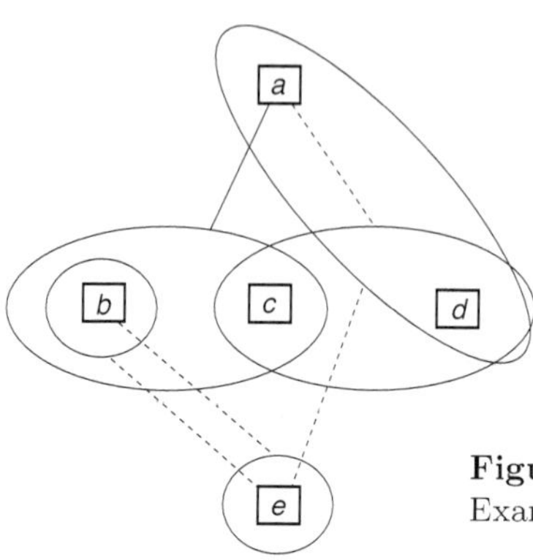

Figure 5.2. Graph of the attribution in Example 5.1.5.

The rules governing such a representation are as follows. Consider a clause C for some q in Q and suppose that C contains q and at least one item $q' \neq q$. Then C is represented by an ellipse surrounding all the elements of C **but** q, and linked to q by a solid segment. This is illustrated in Figure 5.2 by the

[3] Problem 3 asks the reader to verify the equivalence (ii).

ellipse surrounding the points b and c and linked to the point a by a solid segment. Indeed, $\{a, b, c\}$ is a clause for a in the attribution σ. If C is a clause for q which does not contain q, but contains some $q' \neq q$, the representation of that clause is the same, but the segment linking q to the ellipse is dashed. Four examples of such a linking are drawn in Figure 5.2. Finally, if a clause for q contains only q, then no ellipse is drawn, as in the case of the point c of the figure because we have $\sigma(c) = \{\{c\}\},$. Note that there is no dashed lines in the representation of a surmise function because of Condition (ii) in Definition 5.1.2 : for any item q, any clause C for q contains q.

Such figures may become very intricate[4]. On the other hand, a given partial order can be encoded, in a minimal efficient way, by its Hasse diagram. Surmise functions generalize quasi orders and thus partial orders. This evokes the potential concept of a 'Hasse system' capable of faithfully summarizing the information in a surmise system. Section 5.5 will be devoted to a precise definition of such a concept and some of its consequences.

Figure 5.3, a display of the kind just introduced, illustrates Condition (iii) in the definition of a surmise system (see 5.1.2).

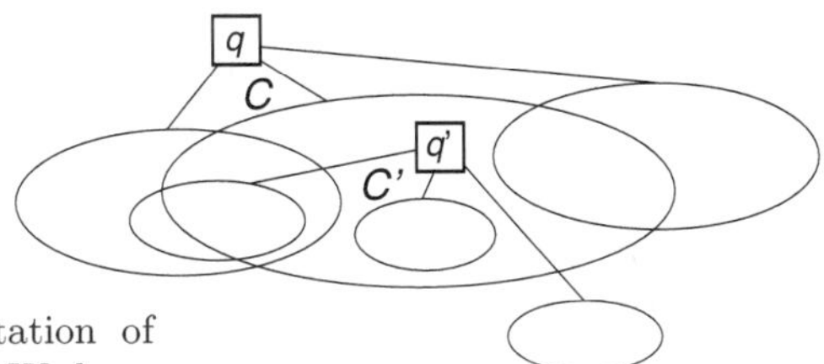

Figure 5.3. Pictorial representation of Condition (iii) in Definition 5.1.2. We have $q' \in C \in \sigma(q)$ and $C' \in \sigma(q')$ with $C' \subseteq C$.

5.2 Knowledge Spaces and Surmise Systems

The following four definitions and examples pave the way to a fundamental relationship between knowledge spaces and surmise systems that will be made precise in Theorem 5.2.5.

5.2.1 Definition. Let $(Q, \mathcal{K})$ be a granular knowledge structure (cf. 3.6.1). Accordingly, any item q has at least one atom. Let $\sigma : Q \to 2^{2^Q}$ be a function defined by the equivalence:

$$C \in \sigma(q) \quad \Longleftrightarrow \quad C \text{ is an atom at } q,$$

with $C \subseteq Q$ and $q \in Q$. It is easily seen that σ is a surmise function on Q. We shall say that σ is the surmise function on Q *derived* from $(Q, \mathcal{K})$.

[4] We invite the reader to graph the surmise function of the knowledge structure of Example 5.1.1, whose atoms are given in Table 5.1.

Note that if a granular knowledge structure $(Q, \mathcal{K})$ is closed under intersection, its derived surmise function σ has only one clause for each item. So, there exists a well-defined relation $\mathcal{R}$ on Q that is cast as σ in the sense of 5.1.4. If the knowledge structure $(Q, \mathcal{K})$ is a quasi ordinal space, this relation $\mathcal{R}$ is exactly the quasi order derived from $(Q, \mathcal{K})$ (see Definition 3.8.4).

5.2.2 Example. Applying the construction of Definition 5.2.1 to the knowledge structure $\mathcal{H}$ of Equation (5.1), we obtain from Table 5.1 the surmise function σ specified by:

$$\sigma(a) = \{\{a\}\}, \qquad \sigma(b) = \{\{b, d\}, \{a, b, c\}, \{b, c, e\}\},$$
$$\sigma(c) = \{\{a, b, c\}, \{b, c, e\}\}, \quad \sigma(d) = \{\{b, d\}\}, \quad \sigma(e) = \{\{b, c, e\}\}.$$

5.2.3 Definition. Any attribution σ on a set Q defines a knowledge space $(Q, \mathcal{K})$ by the equivalence

$$K \in \mathcal{K} \iff \forall q \in K, \exists C \in \sigma(q) : C \subseteq K. \tag{5.2}$$

The verification that $(Q, \mathcal{K})$ is indeed a knowledge space is left to the reader (Problem 2). In such a case, we say that the attribution σ *produces* the knowledge space $(Q, \mathcal{K})$, or equivalently, that $(Q, \mathcal{K})$ is *derived* from the attribution σ on Q.

When (Q, σ) is a surmise system, its derived knowledge structure is always a granular knowledge space. In particular, each clause for an item q is an atom at q in $\mathcal{K}$ (cf. Definition 5.1.2). Thus, the states in $\mathcal{K}$ are unions of clauses, and conversely, any union of clauses is a state.

This construction of $(Q, \mathcal{K})$ from (Q, σ) is a natural outcome of our interpretation of surmise systems: a set K of items forms a knowledge state when K includes, for each of its items q, a minimal foundation leading to q.

5.2.4 Example. With the surmise function obtained in Example 5.2.2, we verify that the set $K = \{a, b, c, e\}$ satisfies the r.h.s. of the equivalence (5.2) and so is a knowledge state. Indeed, remark that

$$\begin{aligned}
\{a\} &\in \sigma(a) &\quad \text{and} \quad && \{a\} &\subseteq \{a, b, c, e\},\\
\{a, b, c\} &\in \sigma(b) &\quad \text{and} \quad && \{a, b, c\} &\subseteq \{a, b, c, e\},\\
\{a, b, c\} &\in \sigma(c) &\quad \text{and} \quad && \{a, b, c\} &\subseteq \{a, b, c, e\},\\
\{b, c, e\} &\in \sigma(e) &\quad \text{and} \quad && \{b, c, e\} &\subseteq \{a, b, c, e\}.
\end{aligned}$$

(Notice that we could have used the clause $\{b, c, e\}$ for b.) On the other hand, the subset $\{a, c, d, e\}$ is not a state because it does not include a clause for e. The family $\mathcal{K}$ of all states is easily constructed and coincides with the original family $\mathcal{H}$ in Example 5.1.1.

In fact, the constructions of σ in Definition 5.2.1 and of $\mathcal{K}$ in Definition 5.2.3 are mutual inverses. We have the following general result.

5.2.5 Theorem. *There is a one-to-one correspondence between the collection of all granular knowledge spaces on a set Q and the collection of all surmise functions on Q. It is defined, for all granular knowledge space $\mathcal{K}$ and surmise function σ, by the following equivalence, where $S \subseteq Q$ and $q \in Q$:*

$$S \text{ is an atom at } q \text{ in } \mathcal{K} \iff S \text{ is a clause for } q \text{ in } \sigma. \tag{5.3}$$

Under this correspondence, the image of a discriminative, granular knowledge space is a discriminative surmise function.

PROOF. Let s be the function from the collection $\mathfrak{K}^{\mathbf{g}}$ of all granular knowledge spaces on a set Q to the collection $\mathfrak{F}^{\mathbf{s}}$ of all surmise functions on Q, defined as in Definition 5.2.1 by the equivalence (where $\mathcal{K} \in \mathfrak{K}^{\mathbf{g}}$ and $\sigma \in \mathfrak{F}^{\mathbf{s}}$)

$$s(\mathcal{K}) = \sigma \iff \forall q \in Q : \sigma(q) = \{S \in 2^Q \mid S \text{ is an atom at } q\}. \tag{5.4}$$

Suppose now that $\mathcal{K}$ and $\mathcal{K}'$ are two distinct granular knowledge spaces on Q, with $\sigma = s(\mathcal{K})$ and $\sigma' = s(\mathcal{K}')$. Then, $\mathcal{K}$ and $\mathcal{K}'$ must have different bases. In particular, there must be an item q such that the set of all atoms at q in $\mathcal{K}$ differs from the set of all atoms at q in $\mathcal{K}'$ (Theorems 3.6.6 and 3.4.8). We thus have $\sigma(q) \neq \sigma'(q)$, and therefore $\sigma \neq \sigma'$. We conclude that s is an injective mapping of $\mathfrak{K}^{\mathbf{g}}$ into $\mathfrak{F}^{\mathbf{s}}$. The mapping s is actually surjective onto $\mathfrak{F}^{\mathbf{s}}$. Indeed, Definition 5.1.2 implies that, for any σ in $\mathfrak{F}^{\mathbf{s}}$, a clause of σ cannot be a union of other clauses. Consequently, all the clauses of σ form the base of some granular knowledge space $\mathcal{K}$ with, automatically, $s(\mathcal{K}) = \sigma$.

The argument regarding the discriminative spaces and surmise functions is straightforward. □

Theorem 5.2.5 highlights an important consequence of the knowledges spaces provides some additional motivation for the choice of the knowledge space as one of our core concepts. In the case of a finite set Q, the axiom of closure under union selects, among all knowledge structures, the families of knowledge states that can be derived from clauses for the items.

According to Definitions 5.2.1 and 5.2.3, a granular space $(Q, \mathcal{K})$ and a surmise system (Q, σ) related as in Theorem 5.2.5 are said to be derived one from the other. This terminology is consonant with Definition 3.8.4 in the sense that the correspondence in Theorem 5.2.5 extends the correspondence between quasi ordinal spaces and quasi orders obtained in Birkhoff's Theorem 3.8.3.

5.3 AND/OR Graphs

We show here how attributions can be regarded as a formalization of a type of AND/OR graphs, and how these constructs generate knowledge spaces. The AND/OR graphs are used in the field of artificial intelligence, although often without a formal definition. They model the organization of a task into subtasks, for example in the resolution of a practical problem. Each of these subtasks might itself require the preliminary completion of one of some sets of other subtasks—or maybe no subtask at all. Here, the (sub)tasks are called 'OR-vertices.' A set of subtasks whose completion delivers the solution of another (sub)task is encoded as an 'AND-vertex.' An edge from an AND-vertex α to an OR-vertex a specifies that the combination α of subtasks gives a way to solve task a. An edge from an 'OR-vertex' b to an 'AND-vertex' α indicates that task b is involved in the combination α of tasks. To eliminate possible ambiguities, we will formulate as axioms some assumptions that are often left implicit elsewhere. The main difference between AND/OR graphs and surmise systems lies in the introduction of artificial AND-vertices representing combinations of subtasks; these 'AND-vertices' play the role of the clauses in the setting of attributions (cf. Definition 5.1.2).

The next definition is illustrated by Example 5.3.2 and Figure 5.4.

5.3.1 Definition. An *AND/OR graph* is a directed graph $G = (V, E)$, where the nonempty set V of vertices is the disjoint union of two subsets V_{AND} of *AND-vertices* and V_{OR} of *OR-vertices*. An element of the set E is a (directed) *edge*, that is an ordered pair of vertices. We also require that:

(i) either the initial vertex of an edge belongs to V_{AND}, and the terminal vertex belongs to V_{OR}, or vice versa;
(ii) each AND-vertex α belongs to exactly one edge (α, a), where $a \in V_{OR}$;
(iii) each OR-vertex a belongs to at least one edge (α, a), where $\alpha \in V_{AND}$.

The interpretation of these three conditions relies on the meaning of the vertices and edges discussed before Definition 5.3.1. Condition (i) essentially says that the edges admit exactly one of the two intended meanings. Condition (ii) requires that any combination of tasks relates to a definite task. Condition (iii) imposes that any task is accessible through some combination of (sub)tasks (including the empty combination).

5.3.2 Example. Suppose that $V_{AND} = \{\alpha_1, \alpha_2, \beta_1, \beta_2, \gamma_1, \gamma_2, \delta, \epsilon, \eta, \gamma, \vartheta\}$ and $V_{OR} = \{a, b, \dots, h\}$. An AND/OR graph $V = V_{OR} \cup V_{AND}$ with 21 edges is displayed in Figure 5.4. We use $\vee$ and $\wedge$ to mark the OR-vertex and the AND-vertex, respectively.

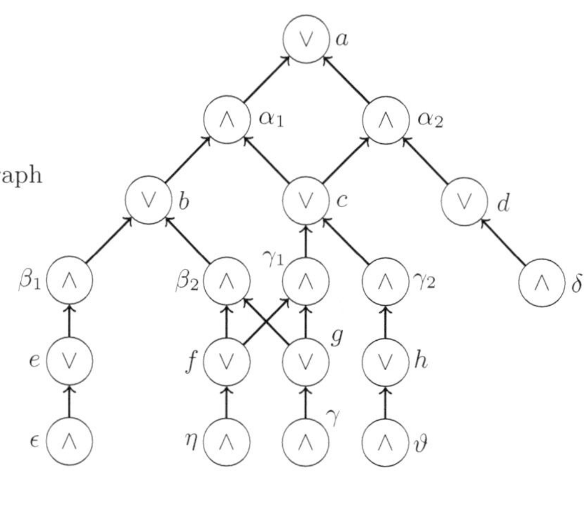

Figure 5.4. The AND/OR graph used in Example 5.3.2.

As mentioned at the beginning of this section, the usual interpretation of such a graph is in terms of tasks represented by OR-vertices. For instance, task a requires the previous completion of (sub)tasks b and c, or of (sub)tasks c and d. Each AND-vertex specifies a set of tasks whose overall completion allows to start the unique task to which it is linked.

The 'exactly one' in Condition (ii) from Definition 5.3.1 is replaced by some authors with 'at least one.' Our additional requirement is not a severe restriction, in the sense that it can be fulfilled after addition of AND-vertices without altering the intended meaning of the graph. Indeed, if some AND-vertex α belonged to several edges (α, b_1), (α, b_2), ..., (α, b_n), we could always replace α by clones, one for each b_i, with $i = 1, \ldots, n$. Such a cloning process would become necessary in Example 5.3.1 if the AND-vertices β_2 and γ_1 were collapsed.

Notice that our Definition 5.3.1 does not rule out 'intractable' situations involving 'cycles' of subtasks. An analysis of these concepts will be given in Section 5.6.

Each AND/OR graph generates a knowledge space on the set of its OR-vertices. This should be clear to the reader who has perceived the link between AND/OR graphs and attributions; Theorem 5.3.4 makes this link explicit. As the arguments establishing the result below are straightforward, we skip the proof.

5.3.3 Theorem. *Let $G = (V, E)$ be an AND/OR graph. A subset K of V_{OR} is said to be a state of G if it satisfies the condition: for any a in K, there exists an edge (α, a) with $\alpha \in V_{AND}$ such that $b \in K$ for each edge (b, α). Then the family of states is a knowledge space on V_{OR}.*

The next theorem shows how to transform an attribution into an AND/OR graph by taking the items as OR-vertices and the clauses as AND-vertices. Some care must be taken in a case where one set is a clause for several items. To obtain Condition (ii) in Definition 5.3.1 of an AND/OR graph, we transform a clause C for an item q into an AND-vertex (q, C).

5.3.4 Theorem. *The following two constructions are mutual inverses and establish a one-to-one correspondence between the collection of all attributions and the collection of all AND/OR graphs.*

Suppose that σ is an attribution on the set Q. Construct the corresponding AND/OR graph by setting $V_{OR} = Q$, $V_{AND} = \{(q, C) \,|\, q \in C \in \sigma(q)\}$, and defining the edges as follows. Declare an edge $((q, C), a)$, whenever $q = a \in Q$ and C is a clause for item a in σ. Declare an edge $(b, (q, C))$ whenever C is a clause for q in σ, and $b \in C$.

Conversely, if $G = (V, E)$ is an AND/OR graph, let $Q = V_{OR}$ and define an attribution σ on Q by the following statement. A subset C of Q is a clause for item q if there exists an edge (α, q) with α some AND-vertex such that $C = \{b \in V_{OR} \,|\, (b, \alpha) \text{ is an edge}\}$.

The two constructions described in the theorem indicate that attributions on Q and AND/OR graphs with $V_{OR} = Q$ are obvious rephrasings of one another, the items corresponding to OR-vertices, and the clauses (or rather the pairs (C, q), with C a clause for the item q) corresponding to the AND-vertices. We leave the verifications to the reader.

5.4 Surmise Functions and Wellgradedness

As in Theorem 5.2.5, we consider a surmise system (Q, σ) and its derived granular knowledge space $(Q, \mathcal{K})$. Thus, for $K \subseteq Q$,

$$K \in \mathcal{K} \iff \forall q \in K, \exists C \in \sigma(q) : C \subseteq K.$$

Remember from Theorem 4.1.9 that the space $(Q, \mathcal{K})$ is well-graded if and only if $(Q, \mathcal{K})$ is finite and all its learning paths are gradations. Theorem 4.3.5 partially extends this result in the infinite case, that is, for ∞-wellgradedness and ∞-gradations.

We now investigate how wellgradedness is reflected in the surmise system (Q, σ).

5.4.1 Theorem. *Suppose that $\mathcal{K}$ is a knowledge space having a base $\mathcal{B}$. (By Theorem 3.4.8, the base is thus formed by the collection of all the atoms.) Let σ be the surmise function of $\mathcal{K}$. Then, the following three conditions are equivalent:*

(i) *$\mathcal{K}$ is ∞-well-graded*[5]*;*
(ii) *any atom of $\mathcal{K}$ is an atom at only one item; in other terms, the family $\{\sigma(x) \,|\, x \in \cup\mathcal{F}\} \subseteq 2^{\mathcal{B}}$ is a partition of $\mathcal{B}$;*
(iii) *for any atom B at any item q, the set $B \setminus \{q\}$ is a state; in other terms, any clause for an item, minus that item, is a state.*

We give two proofs of this result. The first one only applies to the finite case and is intended for readers having skipped the starred Section 4.3 on ∞-wellgradedness in Chapter 4.

Proof for the finite case. (i) $\Rightarrow$ (ii). For any atom B of $\mathcal{K}$, there is a tight path in $\mathcal{K}$ from $\varnothing$ to B, say $\varnothing = K_0, K_1, \ldots, K_h = B$. Then $B \setminus K_{h-1}$ consists of a single item q, and $K_{h-1} = B \setminus \{q\}$ is a state. So, B can be an atom only at item q. It follows immediately that $\{\sigma(x) \,|\, x \in \cup\mathcal{F}\} \subseteq 2^{\mathcal{B}}$ is a partition of $\mathcal{B}$.

(ii) $\Rightarrow$ (iii). Let B be an atom at item q. Condition (ii) implies that, for any $r \in B \setminus \{q\}$, there is some clause $C(r)$ for r such that $r \in C(r) \subset B \setminus \{q\}$. Hence $B \setminus \{q\} = \cup_{r \in B \setminus \{q\}} C(r)$, and $B \setminus \{q\}$ is a state.

(iii) $\Rightarrow$ (i). This is a consequence of Theorem 4.5.5. □

[5] In the case of a finite domain Q, "∞-well-graded" may be replaced with "well-graded."

The second proof applies to both the finite and infinite cases.

PROOF*. (i) $\Rightarrow$ (ii). We proceed by contradiction. Let $(Q, \mathcal{K})$ be well-graded, with two distinct questions q and q' having a common clause C. Consider any learning path $\mathcal{L}$ containing the knowledge state C (such a learning path exists by the maximality principle). As C is a knowledge state minimal for $q \in C$, we cannot have $C = K \cup \{x\}$ for any state K in $\mathcal{L} \setminus \{C\}$ and item $x \neq q$. Similarly, as C is also minimal for $q' \in C$, we cannot have $C = K \cup \{q\}$ for any K in $\mathcal{L} \setminus \{C\}$. For the same reasons[6], we have

$$C \neq \bigcup \{L \in \mathcal{L} \,|\, L \subset C\}.$$

Hence the learning path $\mathcal{L}$ cannot be a gradation, in contradiction with the assumed ∞-wellgradedness.

(ii) $\Rightarrow$ (iii). Suppose that for some question q and some clause C for q, the subset $C \setminus \{q\}$ is not a state. Then there must be some q' in $C \setminus \{q\}$ such that no clause for q' is included in $C \setminus \{q\}$. On the other hand, there is some clause C' for q' included in the state C. Thus, $q \in C'$. As C is a state minimal for $q \in C$, we derive $C = C'$. We obtain in this way a common clause for q and q', the required contradiction.

(iii) $\Rightarrow$ (i). Assume Condition (iii), and also that the space $(Q, \mathcal{K})$ is not well-graded. Hence, by Theorem 4.3.5, there exists some learning path $\mathcal{L}$ which is not a gradation: in other terms, we can find some K in $\mathcal{L}$ satisfying both $K \neq \bigcup \{L \in \mathcal{L} \,|\, L \subset K\}$, and $K \setminus \{q\} \notin \mathcal{L}$ for each item $q \in K$. Define $K^\circ = \bigcup \{L \in \mathcal{L} \,|\, L \subset K\}$, noticing that $K^\circ \in \mathcal{L}$. There must exist some question r in $K \setminus K^\circ$ with a clause C for r included in K. Condition (iii) implies that $C \setminus \{r\} \in \mathcal{K}$. Setting $L = K^\circ \cup (C \setminus \{r\}$, we also have $L \in \mathcal{L}$. Because $K^\circ \subseteq L \subset K$, the definition of K° implies $K^\circ = L \setminus \{r\}$. Hence $L = K \setminus \{r\}$, contradicting our choice of K. □

Theorem 5.4.1 does not extend to partial knowledge spaces (in the sense of Definition 2.2.6). The term "spanned†" that we use below was introduced in Definition 4.5.1.

5.4.2 Counterexample. Consider the family $\mathcal{K}$ spanned† by the base

$$\mathcal{B} = \{\{a, b, c\}, \{b, d\}, \{c, d\}\}.$$

It is easily checked that $\mathcal{K}$ is discriminative and well-graded. However, neither (ii) nor (iii) in Theorem 5.4.1 are satisfied: the surmise function

$$\sigma(a) = \{\{a, b, c\}\}, \quad \sigma(b) = \{\{a, b, c\}, \{b, d\}\},$$
$$\sigma(c) = \{\{a, b, c\}, \{c, d\}\}, \quad \sigma(d) = \{\{b, d\}, \{c, d\}\}$$

does not define a partition of the base $\mathcal{B}$ since $\{c, d\}$ is an atom at both c and d and, because $\{c, d\} \setminus \{d\}$ is not a union of states of the base, it is not a state of $\mathcal{K}$.

[6] Cf. Section 4.3.1, Equation (4.12).

5.5 Hasse Systems

Let $\mathcal{R}$ be a relation on a set Q. If we cast $\mathcal{R}$ as an attribution function in the sense of Definition 5.1.4, then a subset K of Q is a state of $\mathcal{R}$ if it satisfies

$$\forall(p,q) \in \mathcal{R}: \ q \in K \Longrightarrow p \in K$$

(cf. Definition 5.2.3.) In the case of a relation $\mathcal{R}$ that is a quasi order, the states of $\mathcal{R}$ are exactly the states of the space derived from $\mathcal{R}$ in the sense of Definition 3.8.4. The Hasse diagram $\breve{P}$ of a partial order $\mathcal{P}$ on a finite set Q (cf. 1.6.8) has the same states as $\mathcal{P}$. Moreover, $\breve{\mathcal{P}}$ is the smallest relation having exactly those states (where 'smallest' means 'minimal for inclusion'). In that sense, $\breve{P}$ is a most economical summary of the partial order $\mathcal{P}$. The concern of this section is to develop a similar concept of 'most economical summary' for surmise systems (and thus also for AND/OR graphs, cf. Theorem 5.3.4). We shall define a 'Hasse system' as an 'economical' attribution, where the precise meaning of 'economical' relies on the comparison method for attributions which we introduce in the next definition.

5.5.1 Definition. We define the relation $\precsim$ on the collection $\mathfrak{F}$ of all attributions on a nonempty set Q by the equivalence

$$\sigma' \precsim \sigma \iff \forall q \in Q, \forall C \in \sigma(q), \exists C' \in \sigma'(q) : C' \subseteq C \qquad (\sigma, \sigma' \in \mathfrak{F}). \quad (5.5)$$

The relation $\precsim$ is always a quasi order, but not necessarily a partial order (however, see Problem 6). It will be referred to as the *attribution order* on $\mathfrak{F}$. Note that the restriction of $\precsim$ to the set of all relations on Q (cast as attributions) is the usual inclusion comparison of relations.

In general, there may be several attributions producing the states of a particular granular knowledge space $(Q, \mathcal{K})$ (in the sense of Definition 5.2.3). A natural condition on an attribution σ providing an economical description of $\mathcal{K}$ is that σ be a minimal element in the subset of all attributions producing $\mathcal{K}$ (where minimal refers to the attribution order $\precsim$). In the case of an infinite domain Q, the existence of at least one such minimal element is not ensured. In this connection, remember that for infinite partial orders, Hasse diagrams can be empty (take for example the usual linear order on the set of real numbers).

5.5.2 Example. The knowledge space $\mathcal{H}$ of Equation (5.1),

$$\mathcal{H} = \{\varnothing, \{a\}, \{b,d\}, \{a,b,c\}, \{b,c,e\}, \{a,b,d\}, \{a,b,c,d\}, \\ \{a,b,c,e\}, \{b,c,d,e\}, \{a,b,c,d,e\}\},$$

has a derived surmise function σ, which was described in Example 5.2.2 as

$$\sigma(a) = \{\{a\}\}, \qquad \sigma(b) = \{\{b,d\}, \{a,b,c\}, \{b,c,e\}\},$$
$$\sigma(c) = \{\{a,b,c\}, \{b,c,e\}\}, \quad \sigma(d) = \{\{b,d\}\}, \quad \sigma(e) = \{\{b,c,e\}\}.$$

This attribution σ is *not* minimal (for $\mathcal{H}$), since the following attribution ϵ satisfies $\epsilon \precsim \sigma$, but not $\sigma \precsim \epsilon$, while having the same knowledge states:

$$\epsilon(a) = \{\varnothing\}, \qquad \epsilon(b) = \big\{\{c\}, \{d\}\big\},$$
$$\epsilon(c) = \big\{\{a, b\}, \{b, e\}\big\}, \quad \epsilon(d) = \big\{\{b\}\big\}, \quad \epsilon(e) = \big\{\{c\}\big\}.$$

We leave to the reader to verify that ϵ is *not* a minimal attribution for $\mathcal{H}$. Any deletion of items from a clause would change the collection of states, but adding the clause $\{d, e\}$ to $\epsilon(c)$ gives a strictly 'smaller' attribution, in the sense of $\precsim$, which still produces $\mathcal{H}$. Notice that each state of $\mathcal{H}$ containing $\{c\} \cup \{d, e\}$ also contains another clause for c. Hence the addition of the clause $\{d, e\}$ for c would be superfluous in an 'economical' attribution that produces $\mathcal{H}$.

5.5.3 Example. The following knowledge space is ordinal

$$\mathcal{G} = \big\{\varnothing, \{a\}, \{a, b\}, \{a, c\}, \{a, b, c\}, \{a, b, c, d\}\big\},$$

since it is derived from a partial order cast as the surmise function δ with

$$\delta(a) = \big\{\{a\}\big\}, \qquad \delta(b) = \big\{\{b, a\}\big\},$$
$$\delta(c) = \big\{\{c, a\}\big\}, \qquad \delta(d) = \big\{\{d, a, b, c\}\big\}.$$

The space $\mathcal{G}$ is also derived from the attribution γ, with

$$\gamma(a) = \{\varnothing\}, \qquad \gamma(b) = \big\{\{a\}, \{c\}\big\},$$
$$\gamma(c) = \big\{\{a\}\big\}, \qquad \gamma(d) = \big\{\{b, c\}\big\}.$$

There is thus some minimal attribution μ for $\mathcal{G}$ with $\mu \precsim \gamma$ and $\{c\} \in \mu(b)$. (The construction of μ is not necessary for our argument). This is cumbersome in view of the extraneous clause $\{c\}$ in $\mu(b)$, which is not contained in the (unique) atom $\{a, b\}$ at b. Moreover, each state containing $\{b\} \cup \{c\}$ also contains the clause $\{a\}$ for b. The condition defined below rules out such extraneous clauses.

5.5.4 Definition. An attribution σ on the nonempty set Q is *tense* when for any item q and any clause C for q, there is some state K (in the knowledge space derived from σ in the sense of Definition 5.2.3) which contains q and includes C but no other clause for q.

Notice that any relation cast as an attribution is tense. Also, any surmise function is tense.

5.5.5 Theorem. *Any attribution σ on Q which is tense satisfies Condition* (iv) *in the definition of a surmise system (see* 5.1.2*), namely*

$$\forall q \in Q, \forall C, C' \in \sigma(q) : \; C \subseteq C' \implies C = C'.$$

PROOF. Assume $C, C' \in \sigma(q)$ with $C' \subset C$. Then any state containing q and C would also contains C', in contradiction with the tensity of σ. □

5.5.6 Theorem. *If an attribution is tense and produces a granular knowledge space $\mathcal{K}$, then any clause for any item q is contained in some atom of $\mathcal{K}$ at q.*

PROOF. Let σ be the attribution and suppose that $C \in \sigma(q)$. Select a state K containing $\{q\} \cup C$ but no clause for q distinct from C. By granularity, there exists some atom A at q with $A \subseteq K$. Since $A \in \mathcal{K}$, there is some clause D for q such that $D \subseteq A \subseteq K$. We must have $C = D$ in view of the choice of K. □

5.5.7 Remarks. a) Granularity is automatically fulfilled when Q or $\mathcal{K}$ is finite. Our guess is that the conclusion of Theorem 5.5.6 does not hold when granularity is not assumed. We leave this as one of the problems listed in Chapter 18 (see Open Problem 18.3.1). We shall return in Chapter 8 to attributions that produce granular knowledge spaces[7].

b) Notice also that the atom mentioned in Theorem 5.5.6 is not necessarily unique, as shown by the attribution ϵ in Example 5.5.2 with $q = b$.

Our preparation is now complete. Theorem 5.5.10 will show that the concept of a 'Hasse system' given in the next definition is a genuine generalization of the Hasse diagrams for partial orders.

5.5.8 Definition. A *Hasse system* for a granular knowledge space $(Q, \mathcal{K})$, or for its derived surmise system, is any attribution σ on Q which is minimal for the quasi order $\precsim$ defined by (5.5) in the set of all attributions that

(i) are tense;
(ii) produce $\mathcal{K}$.

In this situation, we also say that (Q, σ) is a *Hasse system* of $(Q, \mathcal{K})$.

In Problem 8, we ask the reader to verify whether the attribution ϵ in Example 5.5.2. is a Hasse system. Here is another example.

5.5.9 Example. Consider the attribution γ that produces the ordinal knowledge space $\mathcal{G}$ in Example 5.5.3. There is no Hasse system α for $\mathcal{G}$ with $\alpha \precsim \gamma$. (There cannot be one, because γ is not a tense attribution in view of its clause $\{c\}$ for b.) On the other hand, the Hasse diagram of the partial order δ cast as an attribution β leads to the Hasse system

$$\beta(a) = \{\varnothing\}, \quad \beta(b) = \{\{a\}\}, \quad \beta(c) = \{\{a\}\}, \quad \beta(d) = \{\{b, c\}\}.$$

Clearly, any finite knowledge space admits at least one Hasse system. Indeed, the number of tense attributions producing $(Q, \mathcal{K})$ is positive and finite. Thus one of them must be a minimal element with respect to the attribution order $\precsim$. Any such minimal element (Q, σ) is by definition a Hasse system.

[7] See Definition 8.5.2.

5.5.10 Theorem. *Any finite, ordinal knowledge space $(Q, \mathcal{K})$ admits exactly one Hasse system. For each item q in Q, this system has a unique clause which contains all the items covered by q in the partial order on Q derived from $\mathcal{K}$.*

Thus the Hasse diagram of the partial order on Q derived from $\mathcal{K}$ is cast as the unique Hasse system mentioned in the Theorem.

PROOF. By the arguments given before the statement of the theorem, we know there exists at least one Hasse system for $(Q, \mathcal{K})$. Let us denote by σ such a Hasse system. We have to show that σ has exactly one clause for each item q, and that this clause consists of all items covered by q in the partial order P on Q derived from $\mathcal{K}$. Take any clause C in $\sigma(q)$ (by Condition (i) in Definition 5.1.2 of an attribution, we have $\sigma(q) \neq \varnothing$). From Theorem 5.5.6, we know that xPq holds for all items x in C. The minimality of σ, together with Theorem 5.5.5, implies that $q \notin C$. Let K be the smallest state containing C. It is easily checked that $K \cup \{q\}$ is also a state. Hence each item y covered by q belongs to K. Moreover, such an item y must be in C (otherwise, there would be an element z of C satisfying $yPzPq$, and q would not cover y). We have thus proved that each clause for q contains all the elements covered by q. From minimality again, no other item can belong to this clause. □

5.5.11 Remarks. Had we not required tensity in the definition of a Hasse diagram, the last theorem would not hold (see Example 5.5.3). A quasi order has more than one Hasse system if it has at least three elements and a notion with more than one element. This is illustrated in the two examples below.

5.5.12 Example. Let $\{(b, a), (b, c), (c, b)\}$ be a relation on $Q = \{a, b, c\}$. This relation is cast as the attribution σ, with

$$\sigma(a) = \{\{b\}\}, \qquad \sigma(b) = \{\{c\}\}, \qquad \sigma(c) = \{\{b\}\}.$$

The derived knowledge space $\mathcal{K} = \{\varnothing, \{b, c\}, \{a, b, c\}\}$ is quasi ordinal. In fact, σ is a Hasse system of $\mathcal{K}$. It is not difficult to construct another one.

5.5.13 Example. Let $Q = \{a, b, c\}$ and let S be the relation consisting of the pairs (a, b), (b, c), and (c, a). The resulting quasi ordinal space $\mathcal{K}$ has only two states. It admits several Hasse systems, one of which is S cast as an attribution.

We do not know how to characterize efficiently the granular knowledge spaces that admit a unique Hasse system (see Open Problem 18.1.4).

5.6 Resolubility and Acyclicity

In our discussion of AND/OR graphs, we mentioned that any such graph has an interpretation as an organizing device for subtasks of a main task. We also indicated that Definition 5.3.1 does not preclude intractable situations involving cycles of subtasks. In this section, we consider constraints ruling out such cases. This leads us to introduce the concepts of 'resoluble' attributions. In the context of knowledge assessment and learning, the idea is that any item can be mastered via one or more learning tracks (that is, the prerequisites are not self-contradictory). This rules out intractable situations such as that of Example 5.5.13, in which each of items b and c is a prerequisite for the other.

A priori, two meanings can be given to 'resolubility': it can be either local (each individual question can be mastered), or global (a strategy for gradually mastering the whole structure can be designed). We begin by showing the equivalence of the two conceptions in the finite case.

Bear in mind that the notation $\mathcal{T}^{-1}(x)$ in Condition (ii) below stands for the set $\{y \in Q \mid y\mathcal{T}x\}$.

5.6.1 Theorem. *Consider the two following conditions for an attribution σ on a nonempty set Q:*

(i) *for each item q in Q, there exists some natural number k and some sequence of items $q_1, \ldots, q_k = q$ such that, for each i in $\{1, \ldots, k\}$,*

$$\exists C \in \sigma(q_i) : C \subseteq \{q_1, \ldots, q_i\};$$

(ii) *there exists a linear order $\mathcal{T}$ on Q satisfying, for each item q in Q:*
   (a) $\exists C \in \sigma(q) : C \subseteq \mathcal{T}^{-1}(q)$*;*
   (b) $\mathcal{T}^{-1}(q)$ *is finite.*

Then (ii) $\Rightarrow$ (i), *and if Q is finite,* (i) $\Rightarrow$ (ii).

In Condition (i), by allowing $i = 1$, we impose in particular that some clause for q_1 be included in $\{q_1\}$.

PROOF. (ii) $\Rightarrow$ (i). Let $\mathcal{T}$ be the linear order of Condition (ii), and pick any item $q \in Q$. The sequence $q_1, \ldots, q_k$ of Condition (i) is formed by the items that precede or equal q in the order $\mathcal{T}$.

(i) $\Rightarrow$ (ii), with Q finite. Consider all the subsets Y of Q that can be equipped with a linear order $\mathcal{T}$ in such a way that (a) and (b) in Condition (ii) are satisfied for all $q \in Y$. There exists at least one such subset Y, namely a set $\{q_1\}$ as in Condition (i). Now take some maximal subset among all these subsets, and call it again Y with $\mathcal{T}$ the linear order as above. We prove $Y = Q$. Suppose that there is some $q \in Q \setminus Y$, and take a sequence $q_1, \ldots, q_k$ as in Condition (i). There is a smallest index j such that $q_j \notin Y$. We may add q_j to Y and extend the linear order $\mathcal{T}$ to $Y \cup \{q_j\}$ by putting q_j after the elements of Y. The resulting linearly ordered set contradicts the maximality of Y. □

That (i) $\Rightarrow$ (ii) is not true in general is shown by the following example. Take an uncountable set Q with the trivial attribution σ defined by the equations $\sigma(q) = \{\varnothing\}$ for all $q \in Q$. Then Condition (i) is satisfied (even with $k = 1$), but not Condition (ii). Notice that a linearly ordered set $(Q, \mathcal{T})$, with $\mathcal{T}$ satisfying (b) in Theorem 5.6.1(ii), is necessarily isomorphic to a subset of the natural numbers with the usual order.

5.6.2 Definition. An attribution σ on the nonempty set Q is *resoluble* when it satisfies Condition (ii) in Theorem 5.6.1. The order $\mathcal{T}$ is called a *resolution order.*

5.6.3 Theorem. *Let σ be an attribution on a nonempty set Q, and let $\mathcal{K}$ be the knowledge space produced by σ. Then σ is resoluble if and only if $\mathcal{K}$ contains some chain $\mathcal{C}$ of states such that*

(i) $\varnothing \in \mathcal{C}$;
(ii) $\forall K \in \mathcal{C} \setminus \{Q\}, \exists q \in Q \setminus K : K \cup \{q\} \in \mathcal{C}$;
(iii) $\forall K \in \mathcal{C} : K$ *is finite;*
(iv) $\bigcup \mathcal{C} = Q$.

Proof. Let $\mathcal{T}$ be a resolution order for σ. The empty set plus all sets $\mathcal{T}^{-1}(q)$, for $q \in Q$, constitute a chain $\mathcal{C}$ satisfying Conditions (i) to (iv). Conversely, if we have a chain $\mathcal{C}$ satisfying (i) to (iv), then a resolution order $\mathcal{T}$ is defined by the equivalence

$$q \,\mathcal{T}\, r \quad \Longleftrightarrow \quad (\forall K \in \mathcal{C} : r \in K \Rightarrow q \in K) \qquad (q, r \in Q).$$

□

5.6.4 Corollary. *If two attributions produce the same knowledge space, then both are resoluble, or neither is.*

5.6.5 Definition. A knowledge space is *resoluble* when it is produced by at least one resoluble attribution.

The definitions of a resoluble attribution and of a resoluble knowledge space are not very appealing, because they involve an existential quantifier on linear orders. We now give other conditions for resolubility.

5.6.6 Theorem. *An attribution on a finite, nonempty set is resoluble if the derived knowledge space is well-graded.*

We leave the proof to the reader as Problem 10, in which we also ask to show that the converse is not true.

5.6.7 Definition. A relation $\mathcal{R}$ on a set Q is *acyclic* when there does not exist any finite sequence $x_1, x_2, \ldots, x_k$ of elements of Q such that $x_1 \mathcal{R} x_2$, $x_2 \mathcal{R} x_3, \ldots, x_{k-1} \mathcal{R} x_k$, $x_k \mathcal{R} x_1$ and $x_1 \neq x_k$. (Note that an acyclic relation may be reflexive.)

5.6.8 Theorem. *Any finite partially ordered set (cast as an attribution) is resoluble. More generally, a relation on a finite set is resoluble if and only if it is acyclic.*

PROOF. This follows from the existence of a linear order extending a given partial order (Szpilrajn's Theorem; see e.g. Szpilrajn, 1930; Trotter, 1992), and of a partial order extending a given acyclic relation. □

5.6.9 Example. The space $\mathcal{H}$ of Example 5.1.1 does not include any gradation, and by Theorem 5.6.3 is thus not resoluble. We recall the surmise or precedence relation $\precsim$ of $\mathcal{H}$, defined earlier by

$$r \precsim q \iff r \in \cap\mathcal{H}_q \qquad \text{(in Definition 3.7.1)},$$
$$r \precsim q \iff r \in \cap\sigma(q) \qquad \text{(in terms of the surmise function } \sigma).$$

The precedence relation $\precsim$ is represented in Figure 5.1 by its Hasse diagram. Notice that it is acyclic. The second characterization suggests another relation $\mathcal{R}$, defined by

$$r\mathcal{R}q \iff r \in \cup\sigma(q). \tag{5.6}$$

The relation $\mathcal{R}$ has many cycles. For instance, since $\{b, d\} \in \sigma(b) \cap \sigma(d)$, we have $b\mathcal{R}d\mathcal{R}b$.

5.6.10 Notation. For an attribution σ on the set Q, we define the relation $\mathcal{R}_\sigma$ on Q by the equivalence

$$q\mathcal{R}_\sigma q' \quad \iff \quad \exists C \in \sigma(q') : q \in C \qquad (q, q' \in Q).$$

We leave the proof of the next theorem as Problem 12.

5.6.11 Theorem. *Let σ be an attribution on a finite, nonempty set Q. Consider the following three conditions:*

(i) *the relation $\mathcal{R}_\sigma$ is acyclic;*
(ii) *the space $\mathcal{K}$ produced by σ is resoluble;*
(ii) *the precedence relation of $\mathcal{K}$ is acyclic.*

Then (i) $\Rightarrow$ (ii) $\Rightarrow$ (iii).

We shall run into acyclicity again in Chapter 8; see in particular Theorem 8.5.6, in which the following concept is used.

5.6.12 Definition. An attribution σ is *acyclic* when the relation $\mathcal{R}_\sigma$ is acyclic.

5.7 Original Sources and Related Works

The link between knowledge spaces and surmise systems was established in our original paper (Doignon and Falmagne, 1985) in the finite case. We have spelled out in Theorem 5.3.4 the close relationship existing between attributions and AND/OR graphs. For the latter concept, the reader may consult textbooks on artificial intelligence, such as Barr and Feigenbaum (1981) or Rich (1983), for example.

The correspondence between a knowledge space and its derived surmise system (Theorem 5.2.5) is rephrased as follows for simple closure spaces (cf. Definition 3.3.1). For a nonempty set Q, consider a mapping $\gamma : Q \to 2^{2^Q}$ with $\gamma(x)$ being a family of subsets of Q called 'semi-spaces' or 'co-points at x.' In the case of all convex subsets in real affine spaces, a semi-space would be a convex subset that is maximal for the property of not containing x. In the case of all vector subspaces, a semi-space at x would be any hyperplane avoiding x. We formulate four axioms on semi-spaces, where $x, x' \in Q$:

(i) $\gamma(x) \neq \varnothing$;
(ii) if $S \in \gamma(x)$, then $x \notin S$;
(iii) if $x' \notin S \in \gamma(x)$, then $S' \supseteq S$ for some $S' \in \gamma(x')$;
(iv) if $S, S' \in \gamma(x)$ and $S' \subseteq S$, then $S = S'$.

The attentive reader has surely noticed that each one of these four axioms is the dual of the corresponding condition of Definition 5.1.2 defining a surmise function. Functions γ on Q satisfying these four axioms are in a one-to-one correspondence with simple closure spaces on Q.

Theorem 5.2.5, which links knowledge spaces and surmise systems, can also be inferred from Flament (1976). A related result, in a different context and expressed in a very different language, can be found in Davey and Priestley (1990). Their Theorem 3.38 relates what, in our terminology, would be on the one hand finitary spaces (cf. Definition 3.6.1), and on the other hand a variant of surmise systems.

The translation of wellgradedness into a property of surmise systems was obtained in the finite case by Koppen (1989) (see also Koppen, 1998). (More precisely, Koppen uses Condition (ii) in Theorem 5.4.1.) A closely related work can be found in the context of 'convex geometries'; see for example Edelman and Jamison (1985) or Van de Vel (1993).

The definition of Hasse systems given in 5.5.8 relies on the tensity property, rather than on the Axiom [M] used by Doignon and Falmagne (1985). The new definition sharpens the focus on the minimality of an attribution. While acyclicity was already considered in our first paper on this subject, the concept of resolubility is new here.

Problems

1. In Definition 5.1.2, the first condition on a surmise system (Q, σ) is that $\sigma(q)$ is never identical to the empty subfamily of 2^{2^Q}. Show that removing this condition would correspond to dropping the requirement that $Q \in \mathcal{K}$ for a knowledge space $(Q, \mathcal{K})$. In other words, state and prove a result analogous to Theorem 5.2.5 for the modified concepts of 'pseudo' surmise system and 'pseudo' knowledge space.

2. Prove that any attribution defines a knowledge space; cf. Definition 5.2.3 and the equivalence (5.2).

3. Let σ be an attribution on a set Q, and let $\mathcal{R}$ be a relation on Q defined by $r\mathcal{R}q \Leftrightarrow r \in \cup\sigma(q)$. Suppose that $|\sigma(q)| = 1$ for all q in Q. Show that the relation $\mathcal{R}$ is transitive if and only if σ satisfies Condition (ii) of a surmise function (see Definition 5.1.2).

4. Describe the surmise system derived from the granular knowledge space $(Q, \mathcal{K})$ in each of the following cases:
   a) $Q = \{1, \ldots, 100\}$, and $\mathcal{K} = \{K \in 2^Q \,|\, |K| = 0 \text{ or } |K| \geq 50\}$;
   b) $Q = \{a, b, \ldots, z\}$, and $\mathcal{K} = \{K \in 2^Q \,|\, K = \varnothing \text{ or } a \in K\}$;
   c) $Q = \mathbb{R}^2$, and $\mathcal{K} = \{K \subseteq \mathbb{R}^2 \,|\, \mathbb{R}^2 \setminus K \text{ is an affine subspace}\}$ (an affine subspace is either the empty set, a subset formed with a single point, a (straight) line or the whole $\mathbb{R}^2$);
   d) $\mathcal{K} = \{\varnothing, Q\}$;
   e) Q is finite and $\mathcal{K}$ is a chain of subsets of Q.

5. Suppose $(Q, \mathcal{K})$ is the knowledge space derived from the surmise system (Q, σ). Give a necessary and sufficient condition, expressed in terms of clauses of σ, ensuring that $(Q, \mathcal{K})$ is discriminative (cf. the proof of Theorem 5.2.5).

6. Show that the relation $\precsim$ on the collection $\mathfrak{F}$ of all attribution functions on a set Q (cf. Definition 5.5.1) is a quasi order, but not necessarily a partial order. Show however that this relation, restricted to the collection $\mathfrak{F}^{\mathrm{s}}$ of all surmise functions on Q, is a partial order. Do you need all the axioms of a surmise function to prove this?

7. Let $Q = \{a, b, c, d\}$ and
$$\mathcal{K} = \{\varnothing, \{a\}, \{c\}, \{a, c\}, \{c, d\}, \{a, b, c\}, \{a, c, d\}, Q\}.$$
Is the knowledge structure $(Q, \mathcal{K})$ produced by some attribution? In the case of a positive response, find a surmise function producing $\mathcal{K}$; is this surmise function unique? Solve the same problem in the two cases:
   a) $(Q, \mathcal{K}')$, with $\mathcal{K}' = \mathcal{K} \cup \{\{a, b\}, \{b, c\}, \{b, c, d\}\}$;
   b) $(Q, \mathcal{K}'')$, with $\mathcal{K}'' = \mathcal{K} \cup \{\{b\}, \{b, c\}, \{b, c, d\}\}$.

8. Verify that the attribution ϵ in Example 5.5.2 is a Hasse system.

9. Describe the knowledge space produced by each of the attributions σ in the following cases:
   a) $Q = \{1, \ldots, 100\}$, and $\sigma(q) = \{\{q\}\}$;
   b) $Q = \mathbb{N}$, $\sigma(0) = \{\varnothing\}$, and $\sigma(q) = \{\{q-1\}\}$ for $q \geq 1$;
   c) Q is infinite and $\sigma(q)$ consists of all the infinite subsets of Q, plus $\varnothing$;
   d) $\sigma(q) = \{Q\}$.

10. Prove Theorem 5.6.6 and show that the converse does not hold.

11. Find all implications among the following conditions on a finite knowledge structure $(Q, \mathcal{K})$:
    a) $(Q, \mathcal{K})$ has learnstep number equal to 1 (cf. Definition 4.4.2);
    b) $(Q, \mathcal{K})$ is well-graded (cf. Definition 4.1.3);
    c) $(Q, \mathcal{K})$ is produced by an acyclic attribution (cf. Definition 5.6.2).

    Do the implications remain true for an infinite knowledge structure $(Q, \mathcal{K})$?

12. Prove Theorem 5.6.11. Are the converse implications also true? Does the answer change if we assume that the attribution σ is a surmise function?

6

Skill Maps, Labels and Filters

So far, cognitive interpretations of our mathematical concepts have been limited to the use of mildly evocative words such as 'knowledge state', 'learning path' or 'gradation.' This makes sense since, as suggested by our Examples in 1.4.1, 1.4.2 and 1.4.3, many of our results are potentially applicable to widely different fields. It must be realized, however, that our basic concepts are consistent with traditional explanatory features of psychometric theory, such as 'skills' or 'latent trait' (cf. Lord and Novick, 1974; Weiss, 1983; Wainer and Messick, 1983; Wainer, Dorans, Eignor, Flaugher, Green, Mislevy, Steinberg, and Thissen, 2000). Some possible relationships between knowledge states and skills, and other features of the items, are explored in this chapter.

6.1 Skills

Following Marshall (1981) and others (Falmagne, Koppen, Villano, Doignon, and Johannesen, 1990; Albert, Schrepp, and Held, 1992; Lukas and Albert, 1993), we assume the existence of some basic set S of 'skills.' These skills may consist in methods, algorithms or tricks which could in principle be identified. The idea is to associate with each question q in the domain, the skills in S which are useful or instrumental to solve this problem, and to deduce from this association what the knowledge states are. Our discussion will be illustrated by an example of a question which could be included in a test of proficiency in the UNIX operating system.

6.1.1 Example. Question a: *How many lines of the file* **lilac** *contain the word 'purple'? (Only one command line is allowed.)*

The subject tested must respond by entering a line of UNIX commands. This question can be solved by a variety of methods, three of which are listed below. For each method, we state the command line in typewriter style face, following the prompt '>.'

J.-C. Falmagne, J.-P. Doignon, *Learning Spaces*,
DOI 10.1007/978-3-642-01039-2_6, © Springer-Verlag Berlin Heidelberg 2011

(1) > `grep purple lilac | wc`
The system responds by listing three numbers; the first one is the response to the question. (The command '`grep`', followed by the two arguments '`purple`' and '`lilac`', extracts all the lines containing the word 'purple' from the file **lilac**; the 'pipe' command '|' directs this output to the '`wc`' (word count) command, which computes the number of lines, words and characters in this output.)

(2) > `cat lilac | grep purple | wc`
A less efficient solution achieving the same result. (The '`cat`' command requires a listing of the file **lilac**, which is unnecessary.)

(3) > `more lilac | grep purple | wc`
This is similar to the preceding solution.

Examining these three methods suggests several possible types of association between the skills and the questions, and corresponding ways of constructing the knowledge states consistent with those skills. A simple idea is to regard each one of the three methods as a skill. The complete set S of skills would contain those three skills and some others. The linkage between the questions and skills is then formalized by a function $\tau : Q \to 2^S$ associating to each question q a subset $\tau(q)$ of skills. In particular, we would have[1]:

$$\tau(a) = \{1, 2, 3\}.$$

Consider a subject endowed with a particular subset T of skills, containing some of the skills in $\tau(a)$ plus some other skills relevant to different questions; for example,

$$T = \{1, 2, s, s'\}.$$

This subject is able to solve Question a because $T \cap \tau(a) = \{1, 2\} \neq \varnothing$. In fact, the knowledge state K of this subject contains all those items that can be solved by at least one skill possessed by the subject; that is,

$$K = \{q \in Q \,|\, \tau(q) \cap T \neq \varnothing\}.$$

This linkage between skills and states is investigated in the next section, under the name 'disjunctive model.' We shall see that the knowledge structure induced by the disjunctive model is necessarily a knowledge space. This fact is established by Theorem 6.2.3.

We also briefly consider, for completeness, a model that we call 'conjunctive' and that is the dual of the disjunctive model. In the disjunctive model, only one of the skills assigned to an item q suffices to master that item. In the case of the conjunctive model, all the skills assigned to an item are required. Thus, K is a state if there is a set T of skills such that, for any item q, we have $q \in K$ exactly when $\tau(q) \subseteq T$ (rather than $\tau(q) \cap T \neq \varnothing$ as in the disjunctive

[1] There are many ways of solving Question a in the UNIX system. We only list three of them here to simplify our discussion.

model). The conjunctive model formalizes a situation in which, for any question q, there is a unique solution method represented by the set $\tau(q)$, which gathers all the skills required. The resulting knowledge structure is closed under intersection (cf. Theorem 6.4.3). We leave to the reader the analysis of a model producing a knowledge structure closed under both intersection and union (see Problem 1).

A different type of linkage between skills and states will also be discussed. The disjunctive and conjunctive models were obtained from a rather rudimentary analysis of Example 6.1.1, which regarded the three methods themselves as skills, even though several commands are required in each case. A more refined analysis would proceed by considering each command as a skill, including the 'pipe' command '|.'

The complete set S of skills would be of the form[2]

$$S = \{\mathtt{grep}, \mathtt{wc}, \mathtt{cat}, |, \mathtt{more}, s_1, \ldots, s_k\}$$

where, as before, $s_1, \ldots, s_k$ refer to skills relevant to the other questions in the domain under consideration. To solve Question a, a suitable subset of S may be used. For example, a subject equipped with the subset of skills

$$R = \{\mathtt{grep}, \mathtt{wc}, |, \mathtt{more}, s_1, s_2\}$$

would be able to solve Question a using either Method 1 or Method 3. Indeed, the two relevant sets of commands are included in the subject's set of skills R; we have

$$\{\mathtt{grep}, \mathtt{wc}, |\} \subseteq R,$$
$$\{\mathtt{more}, \mathtt{grep}, \mathtt{wc}, |\} \subseteq R.$$

This example is suggestive of a more complicated association between questions and skills. We shall postulate the existence of a function $\mu : Q \to 2^{2^S}$ associating to each question q the collection of all the subsets of skills corresponding to the possible solutions. In the case of question a, we have

$$\mu(a) = \{\{\mathtt{grep}, |, \mathtt{wc}\}, \{\mathtt{cat}, \mathtt{grep}, |, \mathtt{wc}\}, \{\mathtt{more}, \mathtt{grep}, |, \mathtt{wc}\}\}.$$

In general, a subject having some set R of skills is capable of solving some question q if there exists at least one C in $\mu(q)$ such that $C \subseteq R$. Each of the subsets C in $\mu(q)$ will be referred to as a 'competency for' q. This particular linkage between skills and states will be discussed under the name 'competency model.' We shall see that this model is consistent with general knowledge structures, that is not necessarily closed under union or intersection (cf. Theorem 6.5.3).

[2] It could be objected that this analysis omits the skill associated with the proper sequencing of the commands. However, it is reasonable to subsume this skill in the command |, whose sole purpose is to link two commands.

Example 6.1.1 may lead one to believe that the skills associated with a particular domain could always be identified easily. In fact, it is by no means obvious how such an identification might proceed in general. For most of this chapter, we shall leave the set of skills unspecified and regard S as an abstract set. Our focus will be the formal analysis of some possible linkages between questions, skills, and knowledge states, along the lines sketched above. Cognitive or educational interpretations of these skills will be postponed until the last section of this chapter, where we discuss a possible systematic labeling of the items, which could lead to an identification of the skills, and more generally to a description of the content of the knowledge states themselves.

6.2 Skill Maps: The Disjunctive Model

6.2.1 Definition. A *skill map* is a triple (Q, S, τ), where Q is a nonempty set of *items*, S is a nonempty set of *skills*, and τ is a mapping from Q to $2^S \setminus \{\varnothing\}$. When the sets Q and S are specified by the context, we shall sometimes refer to the function τ itself as the *skill map*. For any q in Q, the subset $\tau(q)$ of S will be referred to as the set of skills *assigned* to q (by the skill map τ).

Let (Q, S, τ) be a skill map and T a subset of S. We say that $K \subseteq Q$ is the knowledge state *delineated* by T *(via the disjunctive model)* if

$$K = \{q \in Q \,|\, \tau(q) \cap T \neq \varnothing\}.$$

Notice that the empty subset of skills delineates the empty knowledge state (because $\tau(q) \neq \varnothing$ for each item q), and that S delineates Q. The family of all knowledge states delineated by subsets of S is the knowledge structure *delineated* by the skill map (Q, S, τ) *(via the disjunctive model)*. When the term 'delineate' is used in the framework of a skill map without reference to any particular model, it must always understood with respect to the disjunctive model. Occasionally, when all ambiguities are removed by the context, the family of all states delineated by subsets of S will be referred to as the *delineated knowledge structure*.

6.2.2 Example. With $Q = \{a, b, c, d, e\}$ and $S = \{s, t, u, v\}$, we define the function $\tau : Q \to 2^S$ by

$$\tau(a) = \{t, u\}, \qquad \tau(b) = \{s, u, v\}, \qquad \tau(c) = \{t\},$$
$$\tau(d) = \{t, u\}, \qquad \tau(e) = \{u\}.$$

Thus (Q, S, τ) is a skill map. The knowledge state delineated by $T = \{s, t\}$ is $\{a, b, c, d\}$. On the other hand, $\{a, b, c\}$ is *not* a knowledge state, since it cannot be delineated by any subset R of S. Indeed, such a subset R would necessarily contain t (because of item c); thus, the knowledge state delineated by R would also contain d. The delineated knowledge structure is

$$\mathcal{K} = \{\varnothing, \{b\}, \{a, c, d\}, \{a, b, c, d\}, \{a, b, d, e\}, Q\}.$$

Notice that $\mathcal{K}$ is a knowledge space. This is not an accident, for we have the following result:

6.2.3 Theorem. *Any knowledge structure delineated (via the disjunctive model) by a skill map is a knowledge space. Conversely, any knowledge space is delineated by at least one skill map.*

PROOF. Assume that (Q, S, τ) is a skill map, and let $(K_i)_{i \in I}$ be some arbitrary subcollection of delineated states. If, for any $i \in I$, the state K_i is delineated by a subset T_i of S, it is easily checked that $\bigcup_{i \in I} K_i$ is delineated by $\bigcup_{i \in I} T_i$; that is, $\bigcup_{i \in I} K_i$ is also a state. Thus, the knowledge structure delineated by a skill map is always a space.

Conversely, let $(Q, \mathcal{K})$ be a knowledge space. We build a skill map by taking $S = \mathcal{K}$, and letting $\tau(q) = \mathcal{K}_q$ for any $q \in Q$. (The knowledge states containing q are thus exactly the skills assigned to q; notice that $\tau(q) \neq \varnothing$ follows from $q \in Q \in \mathcal{K}$). For $\mathcal{T} \subseteq S = \mathcal{K}$, we check that the state K delineated by $\mathcal{T}$ belongs to $\mathcal{K}$. Indeed, we have

$$\begin{aligned} K &= \{q \in Q \,|\, \tau(q) \cap \mathcal{T} \neq \varnothing\} \\ &= \{q \in Q \,|\, \mathcal{K}_q \cap \mathcal{T} \neq \varnothing\} \\ &= \{q \in Q \,|\, \exists K' \in \mathcal{K} : q \in K' \text{ and } K' \in \mathcal{T}\} \\ &= \{q \in Q \,|\, \exists K' \in \mathcal{T} : q \in K'\} \\ &= \cup \mathcal{T}, \end{aligned}$$

yielding $K \in \mathcal{K}$ since $\mathcal{K}$ is a space. Finally, we show that any state K from $\mathcal{K}$ is delineated by some subset of S, namely by the subset $\{K\}$. Denoting by L the state delineated by the subset $\{K\}$, we get

$$\begin{aligned} L &= \{q \in Q \,|\, \tau(q) \cap \{K\} \neq \varnothing\} \\ &= \{q \in Q \,|\, \mathcal{K}_q \cap \{K\} \neq \varnothing\} \\ &= \{q \in Q \,|\, K \in \mathcal{K}_q\} \\ &= K. \end{aligned}$$

We conclude that the space $\mathcal{K}$ is delineated by $(Q, \mathcal{K}, \tau)$. □

6.3 Minimal Skill Maps

In the last proof, we constructed, for any knowledge space, a specific skill map that delineates this space. It is tempting to regard such a representation as a possible explanation of the organization of the collection of states, in terms of the skills used to master the items. In science, an explanation of a phenomena is typically not unique, and there is a tendency to favor 'economical' ones. The material in this section is inspired by such considerations.

We begin by studying a situation in which two distinct skill maps only differ by a mere relabeling of the skills. In such a case, not surprisingly, we shall talk about 'isomorphic skill maps', and we shall sometimes say of such skill maps that they assign 'essentially the same skills' to any item q. We introduce this concept of isomorphism in the next definition.

6.3.1 Definition. Two skill maps (Q, S, τ) and (Q, S', τ') (thus with the same set Q of items) are said to be *isomorphic* if there exists a one-to-one mapping f from S onto S' that satisfies, for any $q \in Q$:

$$\tau'(q) = f(\tau(q)) = \{f(s) \,|\, s \in \tau(q)\}.$$

The function f is called an *isomorphism* between (Q, S, τ) and (Q, S', τ').

Definition 6.3.1 defines 'isomorphism for skill maps' with the same set of items. A more general situation is considered in Problem 2.

6.3.2 Example. Let $Q = \{a, b, c, d, e\}$ and $S' = \{1, 2, 3, 4\}$. Define the skill map $\tau' : Q \to 2^{S'}$ by

$$\tau'(a) = \{1, 4\}, \qquad \tau'(b) = \{2, 3, 4\}, \qquad \tau'(c) = \{1\}$$
$$\tau'(d) = \{1, 4\}, \qquad \tau'(e) = \{4\}.$$

The skill map (Q, S', τ') is isomorphic to the one given in Example 6.2.2: an isomorphism $f : S' \to S$ obtains by setting

$$f(1) = t, \qquad f(2) = s, \qquad f(3) = v, \qquad f(4) = u.$$

The following result is clear:

6.3.3 Theorem. *Two isomorphic skill maps (Q, S, τ) and (Q, S', τ') delineate exactly the same knowledge space on Q.*

6.3.4 Remark. Two skill maps may delineate the same knowledge space without being isomorphic. As an illustration, notice that deleting skill v from the set S in Example 6.2.2 and redefining τ by setting $\tau(b) = \{s, u\}$ yields the same delineated space $\mathcal{K}$. Skill v is thus superfluous for the delineation of $\mathcal{K}$. As recalled in the introduction to this section, it is a standard practice in science to search for parsimonious explanations of the phenomena under study. In our context, this translates into favoring skill maps with small, possibly minimal, sets of skills. Specifically, we shall call a skill map 'minimal' if the deletion of any of its skills modifies the delineated knowledge space. If this knowledge space is finite, a minimal skill map always exists, and has the smallest possible number of skills. (This assertion follows from Theorem 6.3.3.) In the infinite case, the situation is a bit more complicated because a minimal skill map does not necessarily exist. However, a skill map delineating the space and having a set of skills with minimum cardinality always exists (because the class of all cardinals is well-ordered, cf. Dugundji, 1966). It must be noted that such a skill map with minimum number of skills is not necessarily unique—even up to isomorphism (see Problem 10).

6.3.5 Example. Consider the family $\mathcal{O}$ of all open subsets of the set $\mathbb{R}$ of real numbers, and let $\mathcal{I}$ be any family of open intervals of $\mathbb{R}$ spanning $\mathcal{O}$. For $x \in \mathbb{R}$, set $\tau(x) = \mathcal{I}_x = \{I \in \mathcal{I} \,|\, x \in I\}$. Then the skill map $(\mathbb{R}, \mathcal{I}, \tau)$ delineates the space $(\mathbb{R}, \mathcal{O})$. Indeed, a subset T of $\mathcal{I}$ delineates $\{x \in \mathbb{R} \,|\, \mathcal{I}_x \cap T \neq \varnothing\} = \cup T$, and moreover an open subset O is delineated by $\{I \in \mathcal{I} \,|\, I \subseteq O\}$. It is well-known that there are countable families $\mathcal{I}$ satisfying the above conditions. Notice that such countable families will generate skill maps with minimum number of skills, that is, with a set of skills of minimum cardinality. Nevertheless, there is no minimal skill map. This can be proved directly, or inferred from Theorem 6.3.8.

With respect to uniqueness, minimal skill maps delineating a given knowledge space—if they exist—behave in a better way. In fact, any two of them are isomorphic. This will be shown in Theorem 6.3.8. This Theorem also provides a characterization of knowledge spaces having a base (in the sense of 3.4.1). Those knowledge spaces are exactly those that can be delineated by some minimal skill map.

6.3.6 Definition. The skill map (Q', S', τ') *prolongs* (resp. *strictly prolongs*) the skill map (Q, S, τ) if the following conditions hold:

(i) $Q' = Q$;
(ii) $S' \supseteq S$ (resp. $S' \supset S$);
(iii) $\tau(q) = \tau'(q) \cap S$, for all $q \in Q$.

A skill map (Q, S', τ') is *minimal* if there is no skill map delineating the same space while being strictly prolonged by (Q, S', τ').

6.3.7 Example. Deleting skill v in the skill map of Example 6.2.2, we now set $Q = \{a, b, c, d, e\}$, $S = \{s, t, u\}$, and

$$\tau(a) = \{t, u\}, \qquad \tau(b) = \{s, u\}, \qquad \tau(c) = \{t\},$$
$$\tau(d) = \{t, u\}, \qquad \tau(e) = \{u\}.$$

It can be checked that (Q, S, τ) is a minimal skill map.

6.3.8 Theorem. *A knowledge space is delineated by some minimal skill map if and only if it has a base. In such a case, the cardinality of the base is equal to that of the set of skills. Moreover, any two minimal skill maps delineating the same knowledge space are isomorphic. Also, any skill map (Q, S, τ) delineating a space $(Q, \mathcal{K})$ having a base prolongs a minimal skill map delineating the same space.*

PROOF. Consider any (not necessarily minimal) skill map (Q, S, τ), and denote by $(Q, \mathcal{K})$ the delineated knowledge space. For any $s \in S$, we write $K(s)$ for the state from $\mathcal{K}$ delineated by $\{s\}$. We have thus

$$q \in K(s) \quad \Longleftrightarrow \quad s \in \tau(q). \tag{6.1}$$

Take any state $K \in \mathcal{K}$, and consider a subset T of skills that delineates it. For any item q, we have

$$\begin{aligned} q \in K \quad &\iff \quad \tau(q) \cap T \neq \varnothing \\ &\iff \quad \exists s \in T : s \in \tau(q) \\ &\iff \quad \exists s \in T : q \in K(s) \qquad\qquad (\text{by } (6.1)) \\ &\iff \quad q \in \cup_{s \in T} K(s) \end{aligned}$$

yielding $K = \cup_{s \in T} K(s)$. Consequently, $\mathcal{A} = \{K(s) \,|\, s \in S\}$ spans $\mathcal{K}$. If we now assume that the skill map (Q, S, τ) is minimal, then the spanning family $\mathcal{A}$ must be a base. Indeed, if $\mathcal{A}$ is not a base, then some $K(s) \in \mathcal{A}$ can be expressed as a union of other members of $\mathcal{A}$. Deleting s from S would result in a skill map strictly prolonged by (Q, S, τ) and still delineating $(Q, \mathcal{K})$, contradicting the supposition that (Q, S, τ) is minimal. We conclude that any knowledge space delineated by a minimal skill map has a base. Moreover the cardinality of the base is equal to that of the set of skills. (When (Q, S, τ) is minimal, we have $|\mathcal{A}| = |S|$.)

Suppose now that the space $(Q, \mathcal{K})$ has a base $\mathcal{B}$. From Theorem 6.2.3, we know that $(Q, \mathcal{K})$ has at least one skill map, say (Q, S, τ). By Theorem 3.4.2, the base $\mathcal{B}$ of $(Q, \mathcal{K})$ must be included in any spanning subset of $\mathcal{K}$. We have thus, in particular, $\mathcal{B} \subseteq \mathcal{A} = \{K(s) \,|\, s \in S\}$, where again $K(s)$ is delineated by $\{s\}$. Defining $S' = \{s \in S \,|\, \exists B \in \mathcal{B} : K(s) = B\}$ and $\tau'(q) = \tau(q) \cap S'$, it is clear that (Q, S', τ') is a minimal skill map.

Notice that a minimal skill map (Q, S, τ) for a knowledge space with base $\mathcal{B}$ is isomorphic to the minimal skill map $(Q, \mathcal{B}, \psi)$ with $\psi(q) = \mathcal{B}_q$. The isomorphism is $s \mapsto K(s) \in \mathcal{B}$, with as above, $K(s)$ delineated by $\{s\}$. Two minimal skill maps are thus always isomorphic to each other.

Finally, let (Q, S, τ) be any skill map delineating a knowledge space $\mathcal{K}$ having a base $\mathcal{B}$. Defining $K(s)$, S' and τ' as before, we obtain a minimal skill map prolonged by (Q, S, τ). □

6.4 Skill Maps: The Conjunctive Model

In the conjunctive model, the knowledge structures that are delineated by skill maps are the simple closure spaces in the sense of Definition 3.3.1 (see Theorem 6.4.3 below). Since these structures are dual to the knowledge spaces delineated via the disjunctive model, we will not go into much detail.

6.4.1 Definition. Let (Q, S, τ) be a skill map and let T be a subset of S. The knowledge state K *delineated* by T *via the conjunctive model* is specified by

$$K = \{q \in Q \,|\, \tau(q) \subseteq T\}.$$

The resulting family of all knowledge states is the knowledge structure *delineated via the conjunctive model* by the skill map (Q, S, τ).

6.4.2 Example. As in 6.2.2, let $Q = \{a, b, c, d, e\}$, and $S = \{s, t, u, v\}$, with $\tau : Q \to S$ defined by

$$\tau(a) = \{t, u\}, \qquad \tau(b) = \{s, u, v\}, \qquad \tau(c) = \{t\},$$
$$\tau(d) = \{t, u\}, \qquad \tau(e) = \{u\}.$$

Then $T = \{t, u, v\}$ delineates the knowledge state $\{a, c, d, e\}$ via the conjunctive model. On the other hand, $\{a, b, c\}$ is *not* a knowledge state. Indeed if $\{a, b, c\}$ were a state delineated by some subset T of S, then T would include $\tau(a) = \{t, u\}$ and $\tau(b) = \{s, u, v\}$; thus, d and e would also belong to the delineated knowledge state. The knowledge structure delineated by the given skill map is

$$\mathcal{L} = \{\varnothing, \{c\}, \{e\}, \{b, e\}, \{a, c, d, e\}, Q\}.$$

Notice that $\mathcal{L}$ is a simple closure space (cf. 3.3.1). The dual knowledge structure $\overline{\mathcal{L}}$ coincides with the knowledge space $\mathcal{K}$ delineated by the same skill map via the disjunctive model; this space $\mathcal{K}$ was obtained in Example 6.2.2.

6.4.3 Theorem. *The knowledge structures delineated via the disjunctive and conjunctive model by the same skill map are dual one to the other. As a consequence, the knowledge structures delineated via the conjunctive model are exactly the simple closure spaces.*

The verification of these simple facts is left to the reader.

6.4.4 Remark. In the finite case, Theorems 6.2.3 and 6.4.3 are mere rephrasing of a known result on 'Galois lattices' of relations; for 'Galois lattice', see Chapter 8, especially Definition 8.3.10. We can reformulate a skill map (Q, S, τ), with Q and S finite, as a relation R between the sets Q and S: for $q \in Q$ and $s \in S$, we define

$$q\,R\,s \quad \Longleftrightarrow \quad s \notin \tau(q).$$

Then the knowledge state delineated by a subset T of S via the conjunctive model is the set

$$K = \{q \in Q \,|\, \forall s \in S \setminus T : qRs\}.$$

These sets K can be regarded as the elements of the 'Galois lattice' of the relation R. It is also well-known that any finite family of finite sets that is closed under intersection can be obtained as the elements of the 'Galois lattice' of some relation. Theorems 6.2.3 and 6.4.3 restate this result and extend it to infinite sets[3]. There is of course a direct analogue of Theorem 6.3.8 for families of sets closed under intersection.

---

[3] This extension is straightforward.

6.5 Skill Multimaps: The Competency Model

The two preceeding sections dealt with the delineation of knowledge structures closed under union or under intersection. We still need to discuss the general case. The delineation of arbitrary knowledge structures will be achieved by generalizing the concept of skill map. The intuition behind this generalization is natural enough. To each item q, we associate a collection $\mu(q)$ of subsets of skills. Any subset C of skills in $\mu(q)$ can be viewed as a method—called 'competency' in the next definition—for solving question q. Thus, possessing just one of these competencies is sufficient to solve question q.

6.5.1 Definition. By a *skill multimap*, we mean a triple $(Q, S; \mu)$, where Q is a nonempty set of *items*, S is a nonempty set of *skills*, and μ is a mapping that associates to any item q a nonempty family $\mu(q)$ of nonempty subsets of S. Thus, the mapping μ is from the set Q to the set $(2^{2^S \setminus \{\varnothing\}}) \setminus \{\varnothing\}$. We call *competency* for the item q any set belonging to $\mu(q)$.

A subset K of Q is said to be *delineated* by some subset T of skills if K contains all the items having at least one competency included in T; formally

$$q \in K \iff \exists C \in \mu(q) : C \subseteq T.$$

Taking $T = \varnothing$ and $T = S$, we see that $\varnothing$ is delineated by the empty set of skills, and that Q is delineated by S. The set $\mathcal{K}$ of all delineated subsets of Q thus forms a knowledge structure. We say then that the knowledge structure $(Q, \mathcal{K})$ is *delineated* by the skill multimap $(Q, S; \mu)$. This model is referred to as the *competency model*.

6.5.2 Example. With $Q = \{a, b, c, d\}$ and $S = \{s, t, u\}$, define the mapping $\mu : Q \to 2^S$ by listing the competencies for each item in Q:

$$\mu(a) = \{\{s,t\}, \{s,u\}\}, \qquad \mu(b) = \{\{u\}, \{s,u\}\},$$
$$\mu(c) = \{\{s\}, \{t\}, \{s,u\}\}, \qquad \mu(d) = \{\{t\}\}.$$

Applying Definition 6.5.1, we see that this skill multimap delineates the knowledge structure:

$$\mathcal{K} = \{\varnothing, \{b\}, \{c\}, \{c,d\}, \{a,b,c\}, \{a,c,d\}, \{b,c,d\}, Q\}.$$

Notice that $\mathcal{K}$ is neither closed under union nor under intersection.

6.5.3 Theorem. *Each knowledge structure is delineated by at least one skill multimap.*

PROOF. Let $(Q, \mathcal{K})$ be a knowledge structure. A skill multimap is defined by setting $S = \mathcal{K}$, and for $q \in Q$,

$$\mu(q) = \{\mathcal{K} \setminus \{M\} \mid M \in \mathcal{K}_q\}.$$

Thus, for each knowledge state M containing question q, we create the competency $\mathcal{K} \setminus \{M\}$ for q. Notice that $\mathcal{K} \setminus \{M\}$ is nonempty, because it has the empty subset of Q as a member. To show that $(Q, S; \mu)$ delineates $\mathcal{K}$, we apply Definition 6.5.1. For any $K \in \mathcal{K}$, we consider the subset $\mathcal{K} \setminus \{K\}$ of $\mathcal{K}$ and compute the state L that it delineates:

$$\begin{aligned} L &= \{q \in Q \mid \exists M \in \mathcal{K}_q : \mathcal{K} \setminus \{M\} \subseteq \mathcal{K} \setminus \{K\}\} \\ &= \{q \in Q \mid \exists M \in \mathcal{K}_q : M = K\} \\ &= \{q \in Q \mid K \in \mathcal{K}_q\} \\ &= K. \end{aligned}$$

Thus, each state in $\mathcal{K}$ is delineated by some subset of S.

Conversely, if $\mathcal{T} \subseteq S = \mathcal{K}$, the state L delineated by $\mathcal{T}$ is defined by

$$\begin{aligned} L &= \{q \in Q \mid \exists M \in \mathcal{K}_q : \mathcal{K} \setminus \{M\} \subseteq \mathcal{T}\} \\ &= \begin{cases} Q, & \text{when } \mathcal{T} = \mathcal{K}, \\ K, & \text{when } \mathcal{T} = \mathcal{K} \setminus \{K\} \text{ for some } K \in \mathcal{K}, \\ \varnothing, & \text{when } |\mathcal{K} \setminus \mathcal{T}| \geq 2, \end{cases} \end{aligned}$$

and we see that L belongs to $\mathcal{K}$. Thus $\mathcal{K}$ is indeed delineated by the skill multimap $(Q, S; \mu)$. □

We shall not pursue any further the study of the skill multimaps $(Q, S; \mu)$. As in the case of the simple skill map, we could investigate the existence and uniqueness of a minimum skill multimap for a given knowledge structure. Other variants of delineation are conceivable. For example, we could define a knowledge state as a subset K of Q consisting of all items q whose competencies all meet a particular subset of S (depending on K). These developments will be left to the interested reader.

6.6 Labels and Filters

On any question in a genuine domain of knowledge, such as arithmetic or grammar, there typically is a wealth of information which could have a bearing on the relevant skills and on the associated knowledge structure. This background information could also be used to paraphrase the knowledge state of a student in a description intended for a parent or for a teacher. Indeed, the complete list of items contained in the student knowledge state may have hundreds of items and may be hard to assimilate, even for an expert. A meaningful summary could be provided, which could rely on the information available on the items forming the knowledge state of the student. This summary might cover much more than the skills possessed (or lacked) by a student, and may include such features as a prediction of success in a future test, a recommendation of a course of study, or an assignment of some remedial work.

This section outlines a program of description (labeling) of the items and integration (filtering) of the corresponding background information contained in the knowledge states. We begin with some examples taken from the ALEKS system outlined in Section 1.3.

6.6.1 Examples of labels. Suppose that a large pool of questions has been selected, covering all the main concepts of the high school mathematics curriculum in some country. Detailed information concerning each of these items can be gathered under 'labels' such as

1. Descriptive name of the item.
2. Grade where the item is to be mastered.
3. Topic (section of a standard book) to which the item belongs.
4. Chapter (of a standard book) where the item is presented.
5. Division of the curriculum to which the item belongs.
6. Concepts and skills involved in the mastery of the item.
7. Type of the item (word problem, computation, reasoning, etc.).
8. Type of response required (word, sentence, formula).

Needless to say, the above list is only meant as an illustration. The actual list could be much longer, and would evolve from an extensive collaboration with experts in the field (in this case, experienced teachers). Two examples of items with their associated labels are given in Table 6.1.

Each of the items in the pool would be labeled in the same manner. The task is to develop a collection of computer routines permitting the analysis of knowledge states in terms of the labels. In other words, suppose that a particular knowledge state K has been diagnosed by some assessment routine like those described in Chapters 13 and 14. The labels associated with the items specifying that knowledge state will be passed through a collection of 'filters', resulting in a number of statements expressed in everyday language in terms of educational concepts[4].

6.6.2 The grade level reflected by the assessment. Suppose that, at the beginning of a school year, a teacher wishes to know which grade (in mathematics, say) is best suited for a student newly arrived from a foreign country. A knowledge assessment routine has been used, which has determined that the state of a student is K. A suitable collection of filters could be designed along the following lines. As before, we write Q for the domain. For each grade n, $1 \leq n \leq 12$, a filter computes the subset G_n of Q containing all the items to be mastered at that grade or earlier (label (2) in the list above).

If the educational system is sensible, we should have

$$G_1 \subset G_2 \subset \cdots \subset G_{12}.$$

[4] Note that one could also restrict this filtering to the items in the fringes of the uncovered state (cf. 4.1.6).

Table 6.1. Two examples of items and their associated lists of labels.

Specification List	Item
(1) Measure of missing angle in a triangle (2) 7 (3) Sum of angles in a plane triangle (4) Geometry of the triangle (5) Elementary euclidean geometry (6) Measure of an angle, sum of the angles in a triangle, addition, subtraction, deduction (7) Computation (8) Numerical	In a triangle ABC, the measure of angle A is $X°$ and the measure of angle B is $Y°$. What is the number of degrees in the measure of angle C?
(1) Addition and subtraction of 2-place decimal numbers with carry (2) 5 (3) Addition and subtraction of decimal numbers (4) Decimal numbers (5) Arithmetic (6) Addition, subtraction, decimals, carry, currency (7) Word problem and computation (8) Numerical	Mary bought two books costing \$$X$ and \$$Y$. She gave the clerk \$$Z$. What amount of change should she receive?

We may find

$$G_{n-1} \subset K \subset G_n \tag{6.2}$$

for some n, in which case the student could be assigned to grade $n-1$. However, this would not be the best solution when $G_n \setminus K$ is very small. We need more information. Moreover, we must provide for situations in which (6.2) does not hold for any n. Next, the filter calculates the standard distance $|K \bigtriangleup G_n|$ for all grades n, and retains the set

$$S(K) = \{n_j \,|\, |K \bigtriangleup G_{n_j}| \leq |K \bigtriangleup G_n|, 1 \leq n \leq 12\}. \tag{6.3}$$

Thus, $S(K)$ contains all the grades which minimize the distance to K. Suppose that $S(K)$ contains a single element n_j, and that we also have $G_{n_j} \subset K$. It would seem reasonable then to recommend the placement of the student in grade $n_j + 1$.

But $S(K)$ may very well contain more than one element. We still need more information. In particular, the content of K, with its strengths and weaknesses relative to its closest sets G_{n_j} must be summarized in some useful way. Without going into the technical details of such summary, we outline an example of a report that the system might produce at this juncture:

> *The closest match for Student X is grade 5. However, X would be an unusual student in that grade. Her knowledge of elementary geometry far exceeds that of a representative student in grade 5. For example, X is aware of the Pythagorean Theorem and capable of using it in applications. On the other hand, X has surprising weaknesses in arithmetic. For example, etc...*

Descriptions of this type would require the development of a varied collection of new filters, beyond those involved in the computation of $S(K)$ in Equation (6.3). Moreover, the system must have the capability of transforming, via a natural language generator, the output of such filters into grammatically correct statements in everyday language. We shall not pursue this discussion here. The point of this paragraph was to illustrate how the labeling of the items, vastly extending the concept of skills, could lead to refined descriptions of the knowledge states that could be useful for various purposes.

6.7 Original Sources and Related Works

Skill maps were not introduced initially in knowledge space theory. As indicated in the introductory paragraph of this chapter, we originally eschewed cognitive interpretations of our concepts since we believed that the overall machinery of knowledge spaces had potential use in a variety of empirical contexts quite different from mental testing. Nevertheless, traditional interpretations of tests results could not be ignored in view of the widespread use of such tests, especially in the US. In fact, inquiries were often raised about the possibility of 'explaining' these states from a small number of 'basic' aptitudes (see e.g. Albert, Schrepp, and Held, 1992; Lukas and Albert, 1993). A first pass at establishing such a linkage was made in Falmagne, Koppen, Villano, Doignon, and Johannesen (1990). Many of the details about skill maps were provided by Doignon (1994b). Further results on skill maps can also be found in Düntsch and Gediga (1995a). To this date, nothing is published concerning of labels and filters in practical applications of knowledge space theory.

Problems

1. For which type of relation $\mathcal{Q}$ on a set Q is it true that there exist some set S and some mapping $\tau : Q \to 2^S$ such that $q\mathcal{Q}r \iff \tau(q) \subseteq \tau(r)$?

2. Definition 6.3.1 of the isomorphism between skill maps was formulated for two maps defined on the same set of items. Drop this assumption and propose a more general type of isomorphic skill assignment. Show then that the knowledge spaces delineated according to the disjunctive model by two isomorphic skill assignments (in this new sense) are isomorphic. (The isomorphism of structures was introduced in Problem 14 of Chapter 2.)

3. Following up on Example 6.3.5, prove that no minimal skill map exists, without making reference to Theorem 6.3.8.

4. Verify that the skill map of Example 6.3.7 is minimal.

5. Give a proof of Theorem 6.4.3.

6. Under which condition on a skill multimap (Definition 6.5.1) is the delineated structure a knowledge space? Construct an example.

7. Solve a similar problem for the case of a knowledge structure closed under intersection.

8. Design an appropriate set of filters capable of listing all the items that a student in state K would not know, but would be ready to master.

9. Find a necessary and sufficient condition on a disjunctive model ensuring that the delineated knowledge space is discriminative.

10. Prove the assertion in the last sentence of Remark 6.3.4. (Hint: in Example 6.3.5, use two countable families $\mathcal{J}$ having different properties).

7

Entailments and the Maximal Mesh

In practice, how can we build a knowledge structure for a specific body of information? The first step is to select the items forming a domain Q. For real-life applications, we will typically assume this domain to be finite. The second step is then to construct a list of all the subsets of Q that are feasible knowledge states, in the sense that anyone of them could conceivably occur in the population of reference. To secure such a list, we could in principle rely on one or more experts in the particular body of information. However, if no assumption is made on the structure to be uncovered, the only exact method consists in the presentation of all subsets of Q to the expert, so that he can point out the states. As the number of subsets of Q grows exponentially with the size $|Q|$ of Q, this method becomes impractical even for relatively small sets Q. (For example: for just 20 items, there are $2^{20} - 2 = 1,048,574$ subsets that are potential states for the expert to consider).

Three complementary solutions were investigated for building knowledge structures in practical situations. The first one relies on supposing that the knowledge structure under consideration satisfies some conditions, the closure under union and/or intersection being prime examples. Such assumptions may result in a considerable reduction of the number of questions to be asked from an expert. An empirical example is discussed in Chapter 15 where it is shown that, at least for some empirical domains, a practical technique is feasible with 50 items[1]. The first part of the present chapter is devoted to some relevant theoretical results.

The second solution is also described in this chapter. The idea is to build a large knowledge structure by combining a number of small ones. Suppose that we have obtained—using experts and the method of the first solution, for example—all the structures on subdomains of at most seven items, say. These

[1] An even more convincing case is provided by the ALEKS system (see Section 1.3 and Chapter 17) which uses a knowledge structure with about 350 items in the beginning algebra curriculum. This structure has been built in part by a technique elaborating on the methods of this chapter (see also Chapters 15 and 16).

J.-C. Falmagne, J.-P. Doignon, *Learning Spaces*,
DOI 10.1007/978-3-642-01039-2_7, © Springer-Verlag Berlin Heidelberg 2011

structures on subdomains can be regarded as projections (in the sense of Definition 2.4.2) of some unknown structure on the full domain. They can then be combined into a global one on the whole domain Q. Here, "combining" means selecting in a sensible way, if possible, one structure on Q that is the parent structure of all the substructures on the subsets of seven items. Theoretical results will isolate the situations allowing for such a construction. Moreover, we study how properties of structures are preserved by the construction.

The third solution is based on collecting the responses of a large number of subjects to the items of the domain. By an appropriate statistical analysis of such data, the knowledge states can in principle be uncovered. This has been demonstrated by Villano (1991) for small domains, using real data. The technique is also applicable to large domains provided that the statistical analysis of the data has been preceded by a 'pruning down' of the collection of potential states by the methods of the first and/or second solutions (see Cosyn and Thiéry, 2000). This particular technique relies on a heavy use of some stochastic procedures for knowledge assessment (such as those presented in Chapters 13 and 14) and is described in Chapter 15.

7.1 Entailments

We begin by examining the case of a quasi ordinal knowledge space $(Q, \mathcal{K})$. From Birkhoff's Theorem 3.8.3, we know that this space is completely specified by its derived quasi order $\mathcal{Q}$, defined by

$$p\mathcal{Q}q \iff (\forall K \in \mathcal{K} : q \in K \Rightarrow p \in K),$$

where $p, q \in Q$ (cf. Definition 3.8.4). As a practical application, we may uncover a quasi ordinal space on a given domain by asking an (ideal) expert all queries of the form

[Q0] *Suppose a student has failed to solve item p. Do you believe this student will also fail item q? Answer this query under the assumption that chance factors, such as lucky guesses and careless errors, play no role in the student's performance.*

Assuming that the expert's responses[2] are consistent with the unknown, quasi ordinal space $(Q, \mathcal{K})$, we form a relation $\mathcal{Q}$ on Q by collecting all pairs (p, q) for which the expert gave a positive response to query [Q0]. The family $\mathcal{K}$ is then obtained by applying Theorem 3.8.3, since

$$K \in \mathcal{K} \iff \big(\forall (p,q) \in \mathcal{Q} : \ p \notin K \Rightarrow q \notin K\big).$$

[2] We recall that the responses to such queries as [Q0] (or [Q1] on the next page) can also be obtained by a different method, which relies on the statistical analysis of assessment data (see Remark 3.2.3).

If we drop the assumption that the unknown knowledge space $(Q, \mathcal{K})$ is quasi ordinal, the responses to all queries of the form [Q0] do not suffice to construct the space. As explained in 1.1.9, we consider in that case the more general type of query:

[Q1] *Suppose that a student has failed to solve items $p_1, \ldots, p_n$. Do you believe this student would also fail to solve item q? You may assume that chance factors, such as lucky guesses and careless errors, do not interfere in the student's performance.*

Such a query is summarized by the nonempty set $\{p_1, \ldots, p_n\}$ of items, paired with the single item q. Thus, all the positive responses to the queries form a relation $\mathcal{P}$ from 2^Q to Q. The expert is consistent with the (unknown) knowledge space $(Q, \mathcal{K})$ exactly when the following equivalence is satisfied for $A \in 2^Q \setminus \{\varnothing\}$ and $q \in Q$:

$$A \mathcal{P} q \iff (\forall K \in \mathcal{K} : A \cap K = \varnothing \Rightarrow q \notin K). \tag{7.1}$$

7.1.1 Example. For the knowledge space $(Q, \mathcal{K})$ defined by $Q = \{a, b, c\}$ and $\mathcal{K} = \{\varnothing, \{a, b\}, \{a, c\}, Q\}$, the queries (A, q) with $q \notin A$ which call for a positive response are listed below:

$$(\{a\}, b), \quad (\{a\}, c), \quad (\{a, b\}, c), \quad (\{a, c\}, b), \quad (\{b, c\}, a).$$

7.1.2 Example. With $k, m \in \mathbb{N}$ and $k < m$, consider the knowledge space $(Q, \mathcal{K})$, where Q has m elements, and $\mathcal{K}$ is the family of all subsets of Q having either 0 or at least k elements. For the corresponding relation $\mathcal{P}$, we have: for all $A \in 2^Q \setminus \{\varnothing\}$ and $q \in Q$,

$$A \mathcal{P} q \iff (q \in A \text{ or } |A| > m - k).$$

We return to the general situation. To design an efficient procedure for questioning the expert, we need to examine the relations $\mathcal{P}$ obtained through Equation (7.1) from all the knowledge spaces $\mathcal{K}$ on Q.

7.1.3 Theorem. *Let $(Q, \mathcal{K})$ be a knowledge structure, and suppose that $\mathcal{P}$ is the relation from $2^Q \setminus \{\varnothing\}$ to Q defined by Equation (7.1). Then, necessarily:*

(i) *$\mathcal{P}$ extends the reverse membership relation, that is: if $p \in A \subseteq Q$, then $A \mathcal{P} p$;*
(ii) *if A, $B \in 2^Q \setminus \{\varnothing\}$ and $p \in Q$, then $A \mathcal{P} b$ for all $b \in B$ and $B \mathcal{P} p$ imply $A \mathcal{P} p$.*

PROOF. Condition (i) is immediate. Suppose that A, B and p are as in Condition (ii) with $A \mathcal{P} b$ for all $b \in B$ and $B \mathcal{P} p$. We have to show that for all $K \in \mathcal{K}$, $A \cap K = \varnothing$ implies $p \notin K$. Take any $K \in \mathcal{K}$ with $A \cap K = \varnothing$. Thus, by Equation (7.1), we have $b \notin K$, for all $b \in B$. This means that $B \cap K = \varnothing$. Using (7.1) again and the fact that $B \mathcal{P} p$, we get $p \notin K$, which yields $A \mathcal{P} p$. □

The next Theorem shows that all relations from $2^Q \setminus \{\varnothing\}$ to Q satisfying Conditions (i) and (ii) are necessarily obtained, as in Theorem 7.1.3, from some knowledge space. Since these relations will play a fundamental role in the sequel, we give them a name.

7.1.4 Definition. An *entailment for* the nonempty domain Q (which may be infinite) is a relation $\mathcal{P}$ from $2^Q \setminus \{\varnothing\}$ to Q that satisfies Conditions (i) and (ii) in Theorem 7.1.3.

7.1.5 Theorem. *There is a one-to-one correspondence between the family of all knowledge spaces $\mathcal{K}$ on the same domain Q, and the family of all entailments $\mathcal{P}$ for Q. This correspondence is defined by the two equivalences*

$$A \mathcal{P} q \iff (\forall K \in \mathcal{K} : A \cap K = \varnothing \Rightarrow q \notin K), \tag{7.2}$$

$$K \in \mathcal{K} \iff \big(\forall (A, p) \in \mathcal{P} : A \cap K = \varnothing \Rightarrow p \notin K\big). \tag{7.3}$$

Proof. To each knowledge space $(Q, \mathcal{K})$, we associate the relation $\mathcal{P} = f(\mathcal{K})$ via Equation (7.2). The fact that $\mathcal{P}$ is an entailment follows from Theorem 7.1.3 and Definition 7.1.4. Conversely, let $\mathcal{P}$ be any entailment for Q. We define then a family $\mathcal{K} = g(\mathcal{P})$ of subsets of Q by Equation (7.3) and show that $\mathcal{K}$ is a space on Q. It is clear that $\varnothing$, $Q \in \mathcal{K}$. Suppose that $K_i \in \mathcal{K}$ for all i in some index set I. We have to show that $\cup_{i \in I} K_i \in \mathcal{K}$. Assume $A\mathcal{P}p$ and $A \cap (\cup_{i \in I} K_i) = \varnothing$. Then $A \cap K_i = \varnothing$ for all $i \in I$, thus $p \notin K_i$. It follows that $p \notin \cup_{i \in I} K_i$. Applying the equivalence (7.2), we obtain $\cup_{i \in I} K_i \in \mathcal{K}$.

We now show that f and g are mutual inverses. We proceed in two steps.

(1) We prove that $(g \circ f)(\mathcal{K}) = \mathcal{K}$. Let $\mathcal{K}$ be a space on Q and let $\mathcal{P} = f(\mathcal{K})$. Defining $\mathcal{L} = g(\mathcal{P})$, we show $\mathcal{L} = \mathcal{K}$. By definition:

$$L \in \mathcal{L} \iff (\forall A \in 2^Q \setminus \{\varnothing\}, p \in Q : (A\mathcal{P}p \text{ and } A \cap L = \varnothing) \Rightarrow p \notin L)\,.$$

Writing $A\mathcal{P}p$ in the r.h.s. explicitly in terms of $\mathcal{K}$ and omitting the quantifiers for A and p, we obtain

$$\begin{aligned} & L \in \mathcal{L} \\ \iff & \Big(((\forall K \in \mathcal{K} : A \cap K = \varnothing \Rightarrow p \notin K) \text{ and } A \cap L = \varnothing) \Rightarrow p \notin L\Big) \qquad (7.4) \\ \Longleftarrow & \; L \in \mathcal{K}. \end{aligned}$$

To prove the converse of the last implication, assume $L \in \mathcal{L}$ together with $L \notin \mathcal{K}$. Denote by L° the largest state contained in L. (Because $\mathcal{K}$ is a space, L° is well defined: it is equal to the union of all the states contained in L.) Since $L \notin \mathcal{K}$, there must exist some item p with $p \in L \setminus L^\circ$. Setting $A = Q \setminus L$, we have for any $K \in \mathcal{K}$:

$$\begin{aligned} A \cap K = \varnothing \quad & \Longrightarrow \quad K \subseteq L, \\ & \Longrightarrow \quad K \subseteq L^\circ, \\ & \Longrightarrow \quad p \notin K. \end{aligned}$$

As we also have $A \cap L = \varnothing$, the r.h.s. of (7.4) gives $p \notin L$, a contradiction. This completes the proof that $\mathcal{K} = \mathcal{L}$. We conclude that $(g \circ f)(\mathcal{K}) = \mathcal{K}$ for each space $\mathcal{K}$ on Q.

(2) We prove that $(f \circ g)(\mathcal{P}) = \mathcal{P}$. Take any entailment $\mathcal{P}$ for Q. With $\mathcal{K} = g(\mathcal{P})$ and $\mathcal{Q} = f(\mathcal{K})$, we show that $\mathcal{Q} = \mathcal{P}$. For $A \in 2^Q \setminus \{\varnothing\}$ and $p \in Q$, it is easily checked that

$$\begin{aligned} A\mathcal{Q}p \quad &\iff \quad (\forall K \in \mathcal{K} : A \cap K = \varnothing \Rightarrow p \notin K) \\ &\iff \quad \Big(\forall K \in 2^Q : ((\forall B \in 2^Q \setminus \{\varnothing\}, \forall q \in Q : (B\mathcal{P}q \text{ and} \\ &\qquad\qquad B \cap K = \varnothing) \Rightarrow q \notin K) \text{ and } A \cap K = \varnothing\big) \Rightarrow p \notin K\Big). \end{aligned}$$

Denoting by X the r.h.s. of the last equivalence, we clearly have $A\mathcal{P}p \Rightarrow X$. To prove that we also have $X \Rightarrow A\mathcal{P}p$, we proceed by contradiction. Suppose that X holds and that $A\mathcal{P}p$ is false. Set $K = \{q \in Q \mid \text{not } A\mathcal{P}q\}$. For any $B \in 2^Q \setminus \{\varnothing\}$ and $q \in Q$, we see that $B\mathcal{P}q$ together with $B \cap K = \varnothing$ implies $q \notin K$. Indeed, $B \cap K = \varnothing$ implies $A\mathcal{P}b$, $\forall b \in B$; as we also have $B\mathcal{P}q$, Condition (ii) of Definition 7.1.4 implies $A\mathcal{P}q$, thus $q \notin K$. Moreover, by Condition (i) of that definition, we have $A \cap K = \varnothing$. Since $p \in K$ by the definition of K, we have reached a contradiction with X. We have thus proved $\mathcal{Q} = \mathcal{P}$, that is $(f \circ g)(\mathcal{P}) = \mathcal{P}$. □

7.1.6 Definition. When an entailment $\mathcal{P}$ for Q and a knowledge space $\mathcal{K}$ on Q correspond to each other as in Equations (7.2) and (7.3) in Theorem 7.1.5, we say that they are *derived* from one another.

The correspondence obtained in Theorem 7.1.5 can be reformulated in an intuitive way. Starting from the space $(Q, \mathcal{K})$, it can be checked that $A\mathcal{P}q$ holds exactly when q does not belong to the largest state L_A disjoint from A. That is, for $A \in 2^Q \setminus \{\varnothing\}$ and $q \in Q$, Equation (7.2) is equivalent to

$$A \,\mathcal{P}\, q \quad \iff \quad q \notin L_A. \tag{7.5}$$

(The proof of the equivalence is left as part of Problem 1.)

In terms of the closure space dual to $\mathcal{K}$ (cf. Definition 3.3.1), $A\mathcal{P}p$ holds exactly when p belongs to the closure of A. On the other hand, for $K \in 2^Q$, Equation (7.3) is equivalent to

$$K \in \mathcal{K} \quad \iff \quad K = \{p \in Q \mid \text{not } (Q \setminus K)\mathcal{P}p\}, \tag{7.6}$$

(see Problem 2). This equivalence is rephrased as follows in terms of closed sets (that is, in terms of the complements of states): a subset F of Q is closed if and only if it contains all items p satisfying $F\mathcal{P}p$.

7.2 Entail Relations

Condition (ii) in Theorem 7.1.3 is the key requirement for entailments. It must be recognized as a disguised form of a transitivity condition for a relation. To see this, we associate to any relation $\mathcal{P}$ from $2^Q \setminus \{\varnothing\}$ to Q a relation $\mathcal{Q}$ on $2^Q \setminus \{\varnothing\}$ by defining

$$A \mathcal{Q} B \quad \Longleftrightarrow \quad (\forall b \in B : A \mathcal{P} b). \tag{7.7}$$

Condition (ii) in Theorem 7.1.3 for $\mathcal{P}$ can be restated in terms of $\mathcal{Q}$, yielding

$$(A \mathcal{Q} B \text{ and } B \mathcal{Q} \{p\}) \quad \Longrightarrow \quad A \mathcal{Q} \{p\}, \tag{7.8}$$

for $A, B \in 2^Q \setminus \{\varnothing\}$ and $p \in Q$. In the last formula, we can replace the singleton set $\{p\}$ with any subset C of Q. Thus, Equation (7.8) essentially states that $\mathcal{Q}$ is transitive. The following Theorem characterizes such relations $\mathcal{Q}$. (Note that Equation (7.7) implies $A \mathcal{P} b \Leftrightarrow A \mathcal{Q} \{b\}$.)

7.2.1 Theorem. *Equation* (7.7) *establishes a one-to-one correspondence between the family of all entailments $\mathcal{P}$ for Q and the family of all relations $\mathcal{Q}$ on $2^Q \setminus \{\varnothing\}$ that satisfy the three conditions*

(i) *$\mathcal{Q}$ extends the reverse inclusion, that is: for $A, B \in 2^Q \setminus \{\varnothing\}$, we have $A \mathcal{Q} B$ when $A \supseteq B$;*
(ii) *$\mathcal{Q}$ is a transitive relation;*
(iii) *if A, $B_i \in 2^Q \setminus \{\varnothing\}$ for i in some nonempty index set I, then $A \mathcal{P} B_i$ for all $i \in I$ implies $A \mathcal{P} (\cup_{i \in I} B_i)$.*

The proof is left to the reader, to whom we also leave to establish the following additional assertions (see Problems 3 and 4). For a relation $\mathcal{Q}$ on $2^Q \setminus \{\varnothing\}$ which is a transitive extension of reverse inclusion, it can be checked that Condition (iii) in the last Theorem is equivalent to any of the following two conditions:

(iv) for each $A \in 2^Q \setminus \{\varnothing\}$, there is a maximum subset B of Q such that $A \mathcal{Q} B$ (here, maximum means maximum for the inclusion);
(v) for all $A, B \in 2^Q \setminus \{\varnothing\}$, we have $A \mathcal{Q} B$ iff $A \mathcal{Q} \{b\}$ holds for each $b \in B$.

With a finite domain Q, Condition (iii) in Theorem 7.2.1 is also equivalent to

(vi) $A \mathcal{Q} B$ implies $A \mathcal{Q} (A \cup B)$ for all $A, B \in 2^Q \setminus \{\varnothing\}$.

7.2.2 Definition. A relation $\mathcal{Q}$ on $2^Q \setminus \{\varnothing\}$ satisfying Conditions (i), (ii) and (iii) in Theorem 7.2.1 is called an *entail relation for Q*.

7.3 Meshability of Knowledge Structures

We now turn to another approach for building knowledge structures. In many situations, the straightforward procedure for securing an entailment from a particular expert and building the associated space is not practical. No matter how competent the expert may be, his reliability in the course of many hours of questioning may not be perfect, and the resulting space may be partly erroneous[3]. Moreover, the domain may simply be so large that the number of questions required to obtain an entailment would be unacceptable. These objections call for other strategies.

We consider here the possibility of combining a number of small structures into a big one. These small structures may for example be obtained from several different experts, each of them being questioned for a short time on a small subset of the domain; or they may result from a statistical analysis of the responses from a large number of subjects, as in the work of Villano (1991). The origin of the small structures is not relevant here. We simply suppose that a number of projections of some unknown knowledge structure are available, and we consider ways of assembling these pieces into a coherent whole. Before getting into the theoretical background of such a construction, we need a further look at the notion of projection already encountered in Chapter 2. No finiteness assumption will be made, except when otherwise mentioned.

From Definition 2.4.2 and Theorem 2.4.8, we recall that the projection of a knowledge structure $(Q, \mathcal{K})$ on a nonempty subset A of Q is the knowledge structure $(A, \mathcal{H})$ characterized by:

$$\mathcal{H} = \{H \in 2^A \mid H = A \cap K \text{ for some } K \in \mathcal{K}\}. \tag{7.9}$$

We also say that the state $H = A \cap K$ of $\mathcal{H}$ is the trace of the knowledge state K on the subset A. The knowledge structure $\mathcal{H}$ is called a projection of the knowledge structure $(Q, \mathcal{K})$. The terms 'substructure' may be used as synonyms of 'projection.' Note that many of the properties of a knowledge structure are also automatically transferred to their projections, as for instance the property for a structure of being a learning space (cf. Theorem 2.4.8) or being discriminative, quasi ordinal, ordinal, well-graded, or 1-connected[4]. By contrast, none of these properties necessarily holds for the whole structure when it is valid for some of its projections. Positive results for some of these properties obtain when they hold for all the projections, however (cf. Problem 9 in Chapter 3).

We now turn to the combination of two structures on possibly overlapping sets into a structure on the union of these two sets.

---

[3] Some of the issues related to a possible unreliability of experts have been analyzed in detail by Cosyn and Thiéry (2000). Their results are reviewed in Chapter 15. We also recall our Remark 3.2.3 concerning the possibility of replacing a human expert by assessment statistics.

[4] In this regard, see Problem 8 in Chapter 2 and Problem 8 in Chapter 3.

7.3.1 Definition. The knowledge structure $(X, \mathcal{K})$ is called a *mesh* of the knowledge structures $(Y, \mathcal{F})$ and $(Z, \mathcal{G})$ if

(i) $X = Y \cup Z$;
(ii) $\mathcal{F}$ and $\mathcal{G}$ are the projections of $\mathcal{K}$ on X and Y, respectively.

As shown by the following examples, two knowledge structures may have more than one mesh, or no mesh at all. Two knowledge structures having a mesh are *meshable*; if this mesh is unique, they are *uniquely meshable*.

We exercise the concept of mesh on a few examples.

7.3.2 Example. The two knowledge structures (which are ordinal spaces)

$$\mathcal{F} = \big\{\varnothing, \{a\}, \{a, b\}\big\}, \qquad \mathcal{G} = \big\{\varnothing, \{c\}, \{c, d\}\big\}$$

admit the two meshes (which are also ordinal spaces)

$$\mathcal{K}_1 = \big\{\varnothing, \{a\}, \{a, b\}, \{a, b, c\}, \{a, b, c, d\}\big\},$$
$$\mathcal{K}_2 = \big\{\varnothing, \{a\}, \{a, b\}, \{a, c\}, \{a, b, c\}, \{a, b, c, d\}\big\}.$$

7.3.3 Example. Suppose that $(\{a, b, c, d\}, \mathcal{K})$ is a mesh of the two ordinal knowledge spaces

$$\mathcal{F} = \big\{\varnothing, \{a\}, \{a, b\}, \{a, b, c\}\big\}, \qquad \mathcal{G} = \{\varnothing, \{c\}, \{b, c\}, \{b, c, d\}\big\}.$$

Then $\mathcal{K}$ must contain a state K such that $K \cap \{b, c, d\} = \{c\} \in \mathcal{G}$. Hence either $K = \{a, c\}$ or $K = \{c\}$, and since $K \subseteq \{a, b, c\}$, either $\{a, c\}$ or $\{c\}$ must be a state of $\mathcal{F}$, which is not true. Thus, $\mathcal{F}$ and $\mathcal{G}$ are not meshable.

7.3.4 Example. The two knowledge structures

$$\mathcal{F} = \{\, \varnothing,\ \{a\},\ \{a, b\}\,\}, \qquad \mathcal{G} = \{\, \varnothing,\ \{b\},\ \{b, c\}\,\}$$

are uniquely meshable. Indeed, they have the unique mesh

$$\mathcal{K} = \{\, \varnothing,\ \{a\},\ \{a, b\},\ \{a, b, c\}\,\}.$$

(If a state contains c, it has to contain both of a and b; if it contains b, it has to contain a). Notice that, in this example, the mesh $\mathcal{K}$ does not include the union of the two component knowledge structures $\mathcal{F}$ and $\mathcal{G}$, since $\{b, c\} \notin \mathcal{K}$.

We shall first investigate conditions under which a mesh exists.

7.3.5 Definition. A knowledge structure $(Y, \mathcal{F})$ is *compatible with* a knowledge structure $(Z, \mathcal{G})$ if, for any $F \in \mathcal{F}$, the intersection $F \cap Z$ is the trace on Y of some state of $\mathcal{G}$. When two knowledge structures are compatible with each other, we shall simply say that they are *compatible*.

In other words, two knowledge structures $(Y, \mathcal{F})$ and $(Z, \mathcal{G})$ are compatible if and only if they induce the same projection on $Y \cap Z$.

7.3.6 Theorem. *Two knowledge structures are meshable if and only if they are compatible.*

PROOF. Let $(Y \cup Z, \mathcal{K})$ be a mesh of the two knowledge structures $(Y, \mathcal{F})$ and $(Z, \mathcal{G})$, and suppose that $F \in \mathcal{F}$. By the definition of a mesh, there is $K \in \mathcal{K}$ such that $K \cap Y = F$. Thus, $K \cap Z \in \mathcal{G}$ and $(K \cap Z) \cap Y = F \cap Z$. Hence $(Y, \mathcal{F})$ is compatible with $(Z, \mathcal{G})$. The other case follows by symmetry.

Conversely, suppose that $(Y, \mathcal{F})$ and $(Z, \mathcal{G})$ are compatible. Define

$$\mathcal{K} = \{K \in 2^{Y \cup Z} \mid K \cap Y \in \mathcal{F},\ K \cap Z \in \mathcal{G}\}. \tag{7.10}$$

It is clear that $(Y \cup Z, \mathcal{K})$ is a knowledge structure. For any $F \in \mathcal{F}$, we have $F \cap Z = G \cap Y$ for some $G \in \mathcal{G}$. Defining $K = F \cup G$, we obtain $K \cap Y = F$ and $K \cap Z = G$, yielding $K \in \mathcal{K}$. Thus $\mathcal{F}$ is included in the projection of $\mathcal{K}$ on Y. By the definition of $\mathcal{K}$, the reverse inclusion is trivial, so $\mathcal{F}$ is this projection. Again, the other case results from symmetry. We conclude that $\mathcal{K}$ is a mesh of $\mathcal{F}$ and $\mathcal{G}$. □

The construction of the mesh used in the above proof is of interest and deserves a separate investigation.

7.4 The Maximal Mesh

7.4.1 Definition. Let $(Y, \mathcal{F})$ and $(Z, \mathcal{G})$ be two compatible knowledge structures. The knowledge structure $(Y \cup Z, \mathcal{F} \star \mathcal{G})$ defined by the equation

$$\mathcal{F} \star \mathcal{G} = \{K \in 2^{Y \cup Z} \mid K \cap Y \in \mathcal{F},\ K \cap Z \in \mathcal{G}\}$$

is the *maximal mesh* of $\mathcal{F}$ and $\mathcal{G}$. Indeed, we have $\mathcal{K} \subseteq \mathcal{F} \star \mathcal{G}$ for any mesh $\mathcal{K}$ of $\mathcal{F}$ and $\mathcal{G}$. The operator $\star$ will be referred to as the *maximal meshing operator*. An equivalent definition of the maximal mesh is as follows:

$$\mathcal{F} \star \mathcal{G} = \{F \cup G \mid F \in \mathcal{F}, G \in \mathcal{G} \text{ and } F \cap Z = G \cap Y\}.$$

Obviously, we always have $\mathcal{F} \star \mathcal{G} = \mathcal{G} \star \mathcal{F}$. Notice in passing that, if $F \in \mathcal{F}$ and $F \subseteq Y \setminus Z$, then $F \in \mathcal{F} \star \mathcal{G}$. A corresponding property holds of course for the knowledge structure $\mathcal{G}$.

7.4.2 Example. The maximal mesh of the two ordinal knowledge spaces from Example 7.3.2 is the ordinal space

$$\mathcal{F} \star \mathcal{G} = \big\{\varnothing, \{a\}, \{c\}, \{a,b\}, \{a,c\}, \{c,d\}, \{a,b,c\}, \{a,c,d\}, \{a,b,c,d\}\big\}.$$

7.4.3 Theorem. *If $\mathcal{F}$ and $\mathcal{G}$ are compatible knowledge structures, then $\mathcal{F} \star \mathcal{G}$ is a space (respectively discriminative space) if and only if both $\mathcal{F}$ and $\mathcal{G}$ are spaces (respectively discriminative spaces).*

The proof is left to the reader as Problem 7.

If $\mathcal{F}$ and $\mathcal{G}$ are compatible knowledge structures and $\mathcal{F} \star \mathcal{G}$ is well-graded, then $\mathcal{F}$ and $\mathcal{G}$ are both well-graded. The maximal mesh of well-graded knowledge structures, or even learning spaces, is not necessarily well-graded, however. The counterexample below establishes this fact.

7.4.4 Example. Consider the two learning spaces

$$\mathcal{F} = \{\varnothing, \{a\}, \{b\}, \{a,b\}, \{a,c\}, \{b,c\}, \{a,b,c\}\}$$

and

$$\mathcal{G} = \{\varnothing, \{c\}, \{d\}, \{b,c\}, \{b,d\}, \{c,d\}, \{b,c,d\}\},$$

which are compatible. Their maximal mesh (necessarily a space)

$$\mathcal{F} \star \mathcal{G} = \{\varnothing, \{a\}, \{d\}, \{a,c\}, \{a,d\}, \{b,c\}, \{b,d\}, \{a,b,c\}, \\ \{b,c,d\}, \{a,b,d\}, \{a,c,d\}, \{a,b,c,d\}\}$$

is not well-graded since it contains $\{b,c\}$, but neither $\{b\}$ nor $\{c\}$.

7.4.5 Definition. A mesh $\mathcal{K}$ of two knowledge structures $\mathcal{F}$ and $\mathcal{G}$ is called *(union) inclusive* if $F \cup G \in \mathcal{K}$ for any $F \in \mathcal{F}$ and $G \in \mathcal{G}$.

7.4.6 Theorem. *Consider the following three conditions on two knowledge structures $(Y, \mathcal{F})$ and $(Z, \mathcal{G})$:*

(i) *$\mathcal{F}$ and $\mathcal{G}$ admit some inclusive mesh;*
(ii) *$\mathcal{F} \star \mathcal{G}$ is inclusive;*
(iii) *($\forall F \in \mathcal{F} : F \cap Z \in \mathcal{G}$) and ($\forall G \in \mathcal{G} : G \cap Y \in \mathcal{F}$).*

Then (i) $\Leftrightarrow$ (ii) $\Rightarrow$(iii). *Moreover, if $\mathcal{F}$ and $\mathcal{G}$ are spaces, then* (ii) $\Leftrightarrow$ (iii).

We leave the proof to the reader (as Problem 8). The following examples shows that in general, Condition (iii) does not imply Condition (ii).

7.4.7 Examples. a) Consider $\mathbb{R}^3$ and the two families $\mathcal{F}$ and $\mathcal{G}$, where $\mathcal{F}$ contains all the convex subsets of the plane $y = 0$, and $\mathcal{G}$ contains all the convex subsets of the plane $z = 0$. Let thus Y and Z denote the planes $y = 0$ and $z = 0$, respectively. Obviously, we do not in general have $F \cup G$ in $\mathcal{F} \star \mathcal{G}$ for any F in $\mathcal{F}$ and G in $\mathcal{G}$.

b) An example with a finite domain is easily constructed. Still in $\mathbb{R}^3$, take $Y = \{(0,0,1), (0,0,0), (1,0,0), (2,0,0)\}$ and $Z = \{(0,1,0), (0,0,0), (1,0,0), (2,0,0)\}$, with the states being the traces of the convex sets on Y and Z respectively. The maximal mesh $\mathcal{F} \star \mathcal{G}$ is not inclusive since $\{(0,0,1), (2,0,0)\} \in \mathcal{F}$ and $\{(0,0,0)\} \in \mathcal{G}$ but the union of these two states is not in $\mathcal{F} \star \mathcal{G}$.

7.4.8 Theorem. *If the maximal mesh $\mathcal{F} \star \mathcal{G}$ of two knowledge structures $\mathcal{F}$ and $\mathcal{G}$ is inclusive, then $\mathcal{F} \cup \mathcal{G} \subseteq \mathcal{F} \star \mathcal{G}$. When $\mathcal{F}$ and $\mathcal{G}$ are spaces, $\mathcal{F} \cup \mathcal{G} \subseteq \mathcal{F} \star \mathcal{G}$ implies that $\mathcal{F} \star \mathcal{G}$ is inclusive.*

Again, we omit the proof (see Problem 9). The examples in 7.4.7 proves that we cannot replace "spaces" by "structures" in Theorem 7.4.8.

7.4.9 Theorem. *If the maximal mesh of two finite, compatible, well-graded knowledge structures is inclusive, then it is necessarily well-graded.*

Example 7.3.4 shows that the inclusiveness condition is not necessary for two well-graded knowledge structures (or even spaces) to have a maximal mesh that is also well-graded.

PROOF. Let $(Y, \mathcal{F})$ and $(Z, \mathcal{G})$ be two well-graded knowledge structures, and suppose that $\mathcal{F} \star \mathcal{G}$ is inclusive. To prove that $\mathcal{F} \star \mathcal{G}$ is well-graded, we use Theorem 4.1.7(ii). Take any $K, K' \in \mathcal{F} \star \mathcal{G}$. As $K \cap Y$ and $K' \cap Y$ are two states of the well-graded knowledge structure $\mathcal{F}$, there exist a positive integer h and some sequence of states in $\mathcal{F}$

$$K \cap Y = Y_0,\, Y_1,\, \ldots, Y_h = K' \cap Y$$

such that for $i = 0, 1, \ldots, h-1$:

$$|Y_i \bigtriangleup Y_{i+1}| = 1 \quad \text{and} \quad Y_i \cap K' \subseteq Y_{i+1} \subseteq Y_i \cup K'.$$

Similarly, there exist a positive integer p and some sequence of states in $\mathcal{G}$

$$K \cap Z = Z_0,\, Z_1,\, \ldots,\, Z_p = K' \cap Z$$

such that for $j = 0, 1, \ldots, p-1$:

$$|Z_j \bigtriangleup Z_{j+1}| = 1 \quad \text{and} \quad Z_j \cap K' \subseteq Z_{j+1} \subseteq Z_j \cup K'.$$

We then form the sequence

$$\begin{aligned} X_0 = Y_0 \cup (K \cap Z), \quad X_1 = Y_1 \cup (K \cap Z), \quad \ldots, \\ X_h = Y_h \cup (K \cap Z) = (K' \cap Y) \cup Z_o, \\ X_{h+1} = (K' \cap Y) \cup Z_1, \quad X_{h+2} = (K' \cap Y) \cup Z_2, \quad \ldots, \\ X_{h+p} = (K' \cap Y) \cup Z_p. \end{aligned}$$

Clearly, $X_0 = K$ and $X_{h+p} = K'$. Since $\mathcal{F} \star \mathcal{G}$ is inclusive, we also have $X_k \in \mathcal{F} \star \mathcal{G}$, for $k = 0, 1, \ldots, h+p$. On the other hand, for $i = 0, 1, \ldots, h-1$:

$$\begin{aligned} X_i \cap K' &= (Y_i \cup (K \cap Z)) \cap K' \\ &\subseteq (Y_i \cap K') \cup (K \cap Z) \\ &\subseteq Y_{i+1} \cup (K \cap Z) \\ &= X_{i+1}, \end{aligned}$$

and

$$\begin{aligned} X_{i+1} &= Y_{i+1} \cup (K \cap Z) \\ &\subseteq Y_i \cup K' \cup (K \cap Z) \\ &= X_i \cup K'. \end{aligned}$$

In a similar way, one proves for $j = h, h+1, \ldots, h+p-1$:

$$X_j \cap K' \subseteq X_{j+1} \subseteq X_j \cup K'.$$

Finally, it is easy to show that $|X_i \triangle X_{i+1}|$ equals 0 or 1. Thus, after deletion of repeated subsets in the sequence X_i, we obtain a sequence as in Theorem 4.1.7(ii). □

We also indicate a simple result, which is very useful for the practical applications.

7.4.10 Theorem. *Suppose that* $(\mathcal{F}, \mathcal{G})$, $(\mathcal{F} \star \mathcal{G}, \mathcal{K})$, $(\mathcal{G}, \mathcal{K})$ *and* $(\mathcal{F}, \mathcal{G} \star \mathcal{K})$ *are four pairs of compatible knowledge structures. Then, necessarily*

$$(\mathcal{F} \star \mathcal{G}) \star \mathcal{K} \;=\; \mathcal{F} \star (\mathcal{G} \star \mathcal{K}).$$

PROOF. Let X, Y and Z be the domains of $\mathcal{K}$, $\mathcal{F}$ and $\mathcal{G}$, respectively. The result follows immediately from the following string of equivalences:

$$\begin{aligned}
& K \in (\mathcal{F} \star \mathcal{G}) \star \mathcal{K} \\
\iff\quad & K \cap (Y \cup Z) \in \mathcal{F} \star \mathcal{G} \quad \text{and} \quad K \cap X \in \mathcal{K} \\
\iff\quad & K \cap (Y \cup Z) \cap Y \in \mathcal{F} \quad \text{and} \quad K \cap (Y \cup Z) \cap Z \in \mathcal{G} \quad \text{and} \quad K \cap X \in \mathcal{K} \\
\iff\quad & K \cap Y \in \mathcal{F} \quad \text{and} \quad K \cap Z \in \mathcal{G} \text{ and } \quad K \cap X \in \mathcal{K}.
\end{aligned}$$

□

7.5 Original Sources and Related Works

Entail relations were independently investigated by Koppen and Doignon (1990) and Müller (1989, under the name of "implication relations"). Both sets of authors acknowledge an initial suggestion from Falmagne (see also Falmagne, Koppen, Villano, Doignon, and Johannesen, 1990). Müller obtains a version of Theorem 7.1.5 formulated in terms of implication relations, while our presentation follows Koppen and Doignon (1990). We learned from Bernard Monjardet that very similar results were obtained by Armstrong (1974) (cf. also Wild, 1994). Related algorithmic implementations will be discussed in Chapters 15 and 16 in the form of the `QUERY` routine.

For additional results on entail relations, see Dowling (1994) or Düntsch and Gediga (1995b). Another interesting question concerns the description of a knowledge space by a 'minimal' part of its entailment. It is investigated by Guigues and Duquenne (1986) in the framework of closure spaces and 'maximal informative implications' (see also Ganter, 1984).

The results on meshing are due to Falmagne and Doignon (1998). Extending the theory to the agregation of more than two structures, Heller and Repitsch (2008) point to further intricacies and establish many results to resolve them.

Problems

1. Prove the equivalence of the two Equations (7.2) and (7.5).

2. Prove the equivalence of the two Equations (7.3) and (7.6).

3. Prove Theorem 7.2.1.

4. For a relation $\mathcal{Q}$ on $2^Q \setminus \{\varnothing\}$ which is a transitive extension of reverse inclusion, show that Condition (iii) in Theorem 7.2.1 is equivalent to any of the following two conditions:
   (iv) for each $A \in 2^Q \setminus \{\varnothing\}$, there is a maximum subset B of Q such that $A\mathcal{Q}B$ (here, maximum means maximum for the inclusion);
   (v) for all $A, B \in 2^Q \setminus \{\varnothing\}$, we have $A\mathcal{Q}B$ iff $A\mathcal{Q}\{b\}$ for all $b \in B$,
   and in case the domain Q is finite, also to
   (vi) $A\mathcal{Q}B$ implies $A\mathcal{Q}(A \cup B)$ for all $A, B \in 2^Q \setminus \{\varnothing\}$.
   In general, does Condition (vi) implies Condition (v)?

5. Let $\mathcal{P}$ be an entailment for the domain Q, and let $\mathcal{K}$ be the derived knowledge space on Q. State and prove a necessary and sufficient condition on $\mathcal{P}$ for the
   (i) quasi ordinality of $\mathcal{K}$;
   (ii) wellgradedness of $\mathcal{K}$;
   (iii) granularity of $\mathcal{K}$.

6. Any knowledge space $\mathcal{K}$ on the finite domain Q is derived from exactly one surmise system σ on Q, and is also derived from exactly one entailment for Q. Make explicit the resulting one-to-one correspondence between surmise systems on Q and entailments for Q. Try to extend the result to the infinite case by considering granular knowledge spaces.

7. Show that the maximal mesh of two compatible knowledge spaces is again a space (cf. Theorem 7.4.3). If the two given spaces are (quasi) ordinal, is the maximal mesh also (quasi) ordinal?

8. Prove Theorem 7.4.6.

9. Prove Theorem 7.4.8.

10. Let $\mathcal{B}$ be the base of the maximal mesh $(X \cup Y, \mathcal{F} \star \mathcal{G})$ of two finite, compatible knowledge spaces $(X, \mathcal{F})$ and $(Y, \mathcal{G})$ with bases $\mathcal{C}$ and $\mathcal{D}$. Is there a simple construction of $\mathcal{B}$ from $\mathcal{C}$ and $\mathcal{D}$ (taking into account the intersection $Y \cap Z$) ?

11. From Theorems 7.1.5 and 7.2.1, there exists a one-to-one correspondence between the family of knowledge spaces $\mathcal{K}$ on Q and the family of entail relations $\mathcal{P}$ for Q. State explicitly when $\mathcal{K}$ and $\mathcal{P}$ correspond to each other. In particular, spell out the interpretation of $A\mathcal{P}B$, for A, $B \subseteq Q$, in terms of closed sets (i.e. complements of states in $\mathcal{K}$).

8

Galois Connections*

In various preceding chapters, a number of one-to-one correspondences were established between particular collections of mathematical structures. For instance, Birkhoff's Theorem 3.8.3 asserts the existence of a one-to-one correspondence between the collection of all quasi ordinal spaces on a domain Q and the collection of all quasi orders on Q. We will show here that all these correspondences derive from natural constructions. Each derivation will be obtained from the application of a general result about 'Galois connections.' A compendium of the notation for the various collections and the three 'Galois connections' of main interest to us is given at the end of the chapter, in Table 8.3 on page 148. We star the whole chapter because its content is more abstract than, and not essential to, the rest of this book.

8.1 Three Exemplary Correspondences

Table 8.1 summarizes three correspondences, gives references to relevant theorems, and recalls or introduces some notation.

Table 8.1. References, terminology, and notation for three one-to-one correspondences encountered earlier. Columns headings in the table are as follows:

1:	Theorem number
2 and 7:	Name of mathematical structure
3 and 6:	Typical symbol for this structure
4 and 5:	Notation for the collection of structures

1	2	3	4	5	6	7
3.8.3	quasi ordinal space	$\mathcal{K}$	$\mathfrak{K}^{\mathrm{so}}$	$\mathfrak{R}^{\mathrm{o}}$	$\mathcal{Q}$	quasi order
5.2.5	granular knowledge space	$\mathcal{K}$	$\mathfrak{K}^{\mathrm{sg}}$	$\mathfrak{F}^{\mathrm{s}}$	σ	surmise function
7.1.5	knowledge space	$\mathcal{K}$	$\mathfrak{K}^{\mathrm{s}}$	$\mathfrak{E}$	$\mathcal{P}$	entailment

J.-C. Falmagne, J.-P. Doignon, *Learning Spaces*,
DOI 10.1007/978-3-642-01039-2_8,

We assume throughout the chapter that the domain Q is a fixed nonempty set, which may be infinite. Line 1 of Table 8.1 refers to Birkhoff's Theorem. The existence of the one-to-one correspondences in this table proves that the corresponding collections have the same cardinality. A close examination of these correspondences reveals a more interesting situation. First, each one of the correspondences can be canonically derived from a construction relating two respectively larger collections of structures. Second, these larger collections and thus the original ones can be naturally (quasi) ordered, and the correspondence, as well as the constructions, are 'order reversing' between the (quasi) ordered sets.

8.1.1 Definition. Given two quasi ordered sets $(Y, \mathcal{U})$ and $(Z, \mathcal{V})$, a mapping $f : Y \to Z$ is *order reversing* when for all $x, y \in Y$,

$$x\mathcal{U}y \quad \Longrightarrow \quad f(x)\mathcal{V}^{-1}f(y).$$

The mapping f is an *anti-isomorphism* if it is bijective and satisfies the stronger condition

$$x\mathcal{U}y \quad \Longleftrightarrow \quad f(x)\mathcal{V}^{-1}f(y),$$

again for all $x, y \in Y$.

Taking the correspondence in the upper line of Table 8.1 as an example, the two larger collections are: on the one hand, the family of all knowledge structures on the fixed set Q, and on the other hand, the family of all relations on Q. Definition 3.7.1 associates to any knowledge structure $\mathcal{K}$ a particular relation, namely the surmise relation $\precsim$ (which happens to be a quasi order, cf. Theorem 3.7.2): we have, for $r, q \in Q$,

$$r \precsim q \quad \Longleftrightarrow \quad r \in \cap\mathcal{K}_q.$$

Conversely, Theorem 3.8.5 shows how to construct, for any given relation $\mathcal{Q}$ on Q, a derived knowledge structure $\mathcal{K}$ on Q (cf. Definition 3.8.6): a subset K of Q is a state of this structure when

$$\forall q \in K, \forall r \in Q : \ rRq \quad \Longrightarrow \quad r \in K.$$

It can be checked that both of the resulting mappings are inclusion reversing; moreover, they form a so-called 'Galois connection' in the sense defined below. As shown by Monjardet (1970), the one-to-one correspondence in the upper line of Table 8.1 consists of appropriate restrictions of these mappings.

8.2 Closure Operators and Galois Connections

A closure space is defined in 3.3.1 as a collection of subsets of a domain Q that is closed under intersection (and thus contains Q, as the intersection of the empty subcollection). Typical examples were given in 3.3.2, such as the Euclidean space $\mathbb{R}^3$ equipped with the family of all its affine subspaces, or with the family of all its convex subsets. Another example consists in the power set 2^E of a given set E, together will all knowledge spaces on E. It is indeed easily verified that the intersection $\cap_{i \in I} \mathcal{K}_i$ of any family $(\mathcal{K}_i)_{i \in I}$ of spaces on E is again a knowledge space on E, and also that 2^E is a knowledge space (see Problem 1).

For any closure space $(Q, \mathcal{L})$, we built in Theorem 3.3.4 a mapping $2^Q \to 2^Q : A \mapsto A'$, with A' the closure of A (cf. Definition 3.3.5). In the three examples just mentioned, we obtain the affine closure, the convex closure and the 'spatial closure[1]', respectively. In the third example, more explicitly, any knowledge structure $\mathcal{K}$ on a domain E admits a spatial closure, which is the smallest knowledge space on E containing $\mathcal{K}$, or in the terms of Definition 3.4.1, the space spanned by $\mathcal{K}$ (see Example 8.2.2(a) below).

These situations have in common that the domain of the 'closure operator' (such as the power set of $\mathbb{R}^3$, or the family of all knowledge structures on E) can be ordered by inclusion, and the resulting partial order is tightly intertwined with that operator. Given a closure space $(Q, \mathcal{L})$, we denote by $h(A) = A'$ the closure of the subset A of Q, and recall the fundamental properties of the 'closure operator' h (cf. Theorem 3.3.4): for all A, B in 2^Q,

1. $A \subseteq B$ implies $h(A) \subseteq h(B)$;
2. $A \subseteq h(A)$;
3. $h^2(A) = h(A)$;
4. $A \in \mathcal{L}$ iff $A = h(A)$.

We now consider a fairly abstract setting, taking any quasi ordered set $(X, \precsim)$ as the domain of the 'closure operator.'

8.2.1 Definition. Let $(X, \precsim)$ be a quasi ordered set, and let h be a mapping of X into itself. Then h is a *closure operator on* $(X, \precsim)$ if it satisfies the following three conditions: for all x, y in X,

(i) $x \precsim y$ implies $h(x) \precsim h(y)$;
(ii) $x \precsim h(x)$;
(iii) $h^2(x) = h(x)$.

Moreover, any x in X is *closed* when $h(x) = x$.

8.2.2 Examples. a) Let $\mathfrak{K}$ be the set of all knowledge structures on a set Q, with $\mathfrak{K}$ ordered by the inclusion relation. For any $\mathcal{K} \in \mathfrak{K}$, let $s(\mathcal{K})$ be the smallest space including $\mathcal{K}$, that is, the knowledge space spanned by $\mathcal{K}$. Then, s is a closure operator on $(\mathfrak{K}, \subseteq)$, and the closed elements are the spaces (Problem 1).

b) More generally, suppose that $(Q, \mathcal{L})$ is a closure space in the sense of Definition 3.3.1. For any $A \in 2^Q$, let $h(A) = A'$ be the smallest element of $\mathcal{L}$ including A. It is easily checked that the mapping $h : 2^Q \to 2^Q$ is a closure operator on $(2^Q, \subseteq)$, the closed elements being precisely the elements

[1] See Definition 8.5.4.

of $\mathcal{L}$ (cf. Problem 3). This example covers many fundamental structures in mathematics. A few of them are indicated below by the name given to the elements of $\mathcal{L}$ (together with the resulting closure operator):

— all the affine subsets of an affine space (affine closure);
— all the convex subsets of an affine space over an ordered skew field (convex closure);
— all the sublattices of a given lattice (generated sublattice);
— all the subgroups of a group (generated subgroup);
— all the closed sets in a topological space (topological closure);
— all the ideals of a ring (generated ideal).

Two further examples are contained in Definitions 8.4.1 and 8.5.4.

Our next definition extends to quasi orders a standard concept of ordered set theory (see e.g. Birkhoff, 1967). As an illustration, we recall the example mentioned in the second line of Table 8.1: take as a first quasi ordered set the family of all knowledge structures on a fixed domain Q, and as a second quasi ordered set the family of all relations on Q, both families being ordered by inclusion. Then, consider the mappings mentioned in the previous section.

8.2.3 Definition. Let $(Y, \mathcal{U})$ and $(Z, \mathcal{V})$ be two quasi ordered sets, and let $f : Y \to Z$ and $g : Z \to Y$ be any two mappings. The pair (f, g) is a *Galois connection between* $(Y, \mathcal{U})$ and $(Z, \mathcal{V})$ if the following six conditions hold: for all $y, y' \in Y$ and all $z, z' \in Z$,

(i) $y\mathcal{U}y'$ and $y'\mathcal{U}y$ imply $f(y) = f(y')$;
(ii) $z\mathcal{V}z'$ and $z'\mathcal{V}z$ imply $g(z) = g(z')$;
(iii) $y\mathcal{U}y'$ implies $f(y)\mathcal{V}^{-1}f(y')$;
(iv) $z\mathcal{V}z'$ implies $g(z)\mathcal{U}^{-1}g(z')$;
(v) $y\mathcal{U}(g \circ f)(y)$;
(vi) $z\mathcal{V}(f \circ g)(z)$.

The following facts will be useful, and are easily verified. We leave parts of the proof to the reader (see Problem 4.)

8.2.4 Theorem. *Let $(Y, \mathcal{U})$, $(Z, \mathcal{V})$, f and g be as in Definition 8.2.3. Then the following five properties hold:*

(i) *$g \circ f$ and $f \circ g$ are closure operators, respectively on $(Y, \mathcal{U})$ and $(Z, \mathcal{V})$;*
(ii) *there is at most one closed element in every equivalence class of the quasi ordered set $(Y, \mathcal{U})$ (resp. $(Z, \mathcal{V})$);*
(iii) *the set Y_0 of all the closed elements of Y (resp. Z_0, Z) is partially ordered by $\mathcal{U}_0 = \mathcal{U} \cap (Y_0 \times Y_0)$ (resp. $\mathcal{V}_0 = \mathcal{V} \cap (Z_0 \times Z_0)$);*
(iv) *if $z \in f(Y)$, there exists z_0 in Z_0 such that $z\mathcal{V}z_0$ and $z_0\mathcal{V}z$. Similarly, if $y \in g(Z)$, there exists y_0 in Y_0 with $y\mathcal{U}y_0$ and $y_0\mathcal{U}y$;*
(v) *the restriction f_0 of f to Y_0 is an anti-isomorphism between $(Y_0, \mathcal{U}_0)$ and $(Z_0, \mathcal{V}_0)$. Moreover $f_0^{-1} = g_0$, where g_0 is the restriction of g to Z_0.*

PROOF. We only prove parts (i), (ii) and (iii). In view of the symmetry of the statements, we only need to establish the facts concerning the quasi order $(Y, \mathcal{U})$ and the mapping $g \circ f$.

(i) We have to verify that, with $\mathcal{U} = \precsim$ and $g \circ f = h$, Conditions (i) to (iii) in Definition 8.2.1 are satisfied. Suppose that $x\mathcal{U}y$. Applying 8.2.3 (iii) and (iv) yields, successively, $f(x)\mathcal{V}^{-1}f(y)$ and $(g \circ f)(x)\mathcal{U}(g \circ f)(y)$, establishing 8.2.1(i). Up to a change of notation, Conditions 8.2.3(v) and 8.2.1(ii) are identical here. Finally, we have to show that, for all $x \in Y$, we have $h^2(x) = h(x)$, or more explicitly

$$g\big((f \circ g \circ f)(x)\big) = g\big(f(x)\big). \tag{8.1}$$

In view of 8.2.3(ii), Equation (8.1) holds if we have

$$(f \circ g \circ f)(x)\mathcal{V}f(x) \tag{8.2}$$

and

$$f(x)\mathcal{V}(f \circ g \circ f)(x). \tag{8.3}$$

Both of these formulas are true. By 8.2.3(v), we have $x\mathcal{U}(g \circ f)(x)$. Applying 8.2.3(iii) gives Equation (8.2). From 8.2.3(vi), we derive Equation (8.3). We conclude that $g \circ f$ is a closure operator on $(Y, \mathcal{U})$.

(ii) and (iii). Suppose that x and y are in the same equivalence class of $(Y, \mathcal{U})$. Thus, $x\mathcal{U}y$ and $y\mathcal{U}x$. By 8.2.3(i), this implies $f(x) = f(y)$. If both x and y are closed elements (for the closure operator $g \circ f$), we obtain

$$x = (g \circ f)(x) = (g \circ f)(y) = y.$$

This argument also shows that $\mathcal{U}_0 = \mathcal{U} \cap (Y_0 \times Y_0)$ is antisymmetric, and thus establishes (iii).

Parts (iv) and (v) are left to the reader (as Problem 4). □

When the quasi order $\mathcal{V}$ on Z happens to be a partial order, the first part of Condition (iv) in Theorem 8.2.4 can be reformulated as $f(Y) = Z_0$. The following example shows that this simplification does not hold in general.

8.2.5 Example. We build two weakly ordered sets $(Y, \mathcal{U})$ and $(Z, \mathcal{V})$, each having two classes, by setting

$$\begin{aligned} Y &= \{a, b\}, & x\mathcal{U}y &\Leftrightarrow (x = a \text{ or } y = b), \\ Z &= \{u, v, w\}, & z\mathcal{V}t &\Leftrightarrow (z = u \text{ or } t = v \text{ or } t = w). \end{aligned}$$

Define two mappings $f : Y \to Z$ and $g : Z \to Y$ by

$$\begin{gathered} f(a) = w, \qquad f(b) = v, \\ g(u) = g(v) = g(w) = b. \end{gathered}$$

The pair (f, g) is a Galois connection between $(Y, \mathcal{U})$ and $(Z, \mathcal{V})$. The only closed elements are b in Y, and v in Z. However $f(\{a, b\}) = \{v, w\} \neq \{v\}$.

8.3 Lattices and Galois Connections

When the quasi ordered sets between which a Galois connection is defined are 'lattices', the collections of closed elements are themselves 'lattices.' Before stating the relevant definition and results, we briefly present an important application of Theorem 8.2.4 in the field of ordinal data analysis. Let $\mathcal{R}$ be a relation from a set X to a set Y. A Galois connection (f, g) between $(2^X, \subseteq)$ and $(2^Y, \subseteq)$ will be built starting from $\mathcal{R}$. For $A \in 2^X$, we define

$$f(A) = \{y \in Y \mid \forall a \in A : a\mathcal{R}y\}, \tag{8.4}$$

and similarly for $B \in 2^Y$, we define

$$g(B) = \{x \in X \mid \forall b \in B : x\mathcal{R}b\}. \tag{8.5}$$

The next theorem makes use of the following concept.

8.3.1 Definition. A *maximal rectangle* of a relation $\mathcal{R}$ from X to Y is a pair of subsets A of X and B of Y that satisfies:

(i) for all $a \in A$, $b \in B$, we have $a\mathcal{R}b$;
(ii) for each x in $X \setminus A$ there is some b in B for which not $x\mathcal{R}b$;
(iii) for each y in $Y \setminus B$, there is some a in A for which not $a\mathcal{R}y$.

The term "maximal rectangle" is a natural one for a relation $\mathcal{R}$ between two finite sets when $\mathcal{R}$ is encoded into a 0-1 array; see the example below.

8.3.2 Example. Let $X = \{a, b, c, d, e\}$ and $Y = \{p, q, r, s\}$; a relation $\mathcal{R}$ from X to Y is specified by its 0-1 array in Table 8.2.

Table 8.2. The 0-1 array for the relation $\mathcal{R}$ in Example 8.3.2.

	p	q	r	s
a	1	1	0	1
b	1	0	1	1
c	1	0	0	0
d	0	1	0	1
e	0	0	1	1

For this particular relation $\mathcal{R}$, here are some maximal rectangles (A, B), with

$$\begin{array}{llll} & A = \{a, b, c\}, & & B = \{p\}, \\ \text{or} & A = \{b, e\}, & & B = \{r, s\}, \\ \text{or} & A = \{a, b, d, e\}, & & B = \{s\}, \\ \text{or} & A = \varnothing, & & B = \{p, q, r, s\}. \end{array}$$

All the maximal rectangles of $\mathcal{R}$ will be listed in Example 8.3.11.

8.3.3 Theorem. *Let $\mathcal{R}$ be a relation from X to Y. The pair (f, g) of mappings defined in Equations (8.4) and (8.5) form a Galois connection between the ordered sets $(2^X, \subseteq)$ and $(2^Y, \subseteq)$. The pairs (A, B) such that A is a closed set in 2^X and B is a closed set in 2^Y with $B = f(A)$, and thus also $A = g(B)$, are exactly the maximal rectangles of $\mathcal{R}$.*

PROOF. The first two requirements in Definition 8.2.3 for a Galois connection are automatically satisfied, since the domains of f and of g are ordered. If A_1, $A_2 \in 2^X$, then $A_1 \subseteq A_2$ implies $f(A_1) \supseteq f(A_2)$ because of the quantification in Equation (8.4); this establishes Condition (iii) in Definition 8.2.3, and Condition (iv) is similarly derived from Equation (8.5). Condition (v) here means: $A \subseteq g(f(A))$ for all $A \in 2^X$. This is a consequence of the definition of f and g, and so is Condition (vi). Finally, proving the assertion that pairs of related closed sets coincide with the maximal rectangles is easy and left to the reader. □

8.3.4 Definition. The Galois connection built in Theorem 8.3.3 is called the *Galois connection of* the relation $\mathcal{R}$.

As for any Galois connection, the two ordered collections of closed sets are anti-isomorphic (cf. Theorem 8.2.4(v)). In the situation of Theorem 8.3.3, they are moreover 'lattices' (see below). After recalling some terminology, we will derive this assertion from a general result on Galois connections between lattices.

8.3.5 Definition. An ordered set $(X, \mathcal{P})$ is a *lattice* if any two of its elements x, y admit a 'greatest lower bound' and a ' least upper bound.' The *greatest lower bound* of x and y is the element $x \wedge y$ in X satisfying $(x \wedge y)\mathcal{P}x$, $(x \wedge y)\mathcal{P}y$ and for all $l \in X$, $(l\mathcal{P}x$ and $l\mathcal{P}y)$ implies $l\mathcal{P}(x \wedge y)$. Similarly, the *least upper bound* of x and y is the element $x \vee y$ in X such that $x\mathcal{P}(x \vee y)$, $y\mathcal{P}(x \vee y)$, and for all $u \in X$, $(x\mathcal{P}u$ and $y\mathcal{P}u)$ implies $(x \vee y)\mathcal{P}u$.

Many examples of lattices appear as particular cases of the fairly general situation in the example below.

8.3.6 Example. Let $(Q, \mathcal{L})$ be a closure space; then $(\mathcal{L}, \subseteq)$ is a lattice in which, for $x, y \in \mathcal{L}$, we have $x \wedge y = x \cap y$ and $x \vee y = h(x \cup y)$ (with as usual $h(z)$ denoting the closure of z). This example is generalized to any closure operator in the next theorem.

8.3.7 Theorem. *Suppose that h is a closure operator on a lattice $(X, \precsim)$. The collection X_0 of all closed elements is itself a lattice for the induced order $\precsim_0 = \precsim \cap (X_0 \times X_0)$. For x, $y \in X_0$, the least upper bound $x \vee_0 y$ in $(X_0, \precsim_0)$ is equal to $h(x \vee y)$, where $x \vee y$ denotes the least upper bound in $(X, \precsim)$; on the other hand, the greatest lower bound of x and y in $(X_0, \precsim_0)$ and in $(X, \precsim)$ coincide.*

PROOF. We will use the axioms for a closure operator h without mentioning them explicitly. Let x, y be two elements closed for h, that is $h(x) = x$, and $h(y) = y$. As $x \precsim x \vee y$, we get $x = h(x) \precsim h(x \vee y)$, and similarly $y = h(y) \precsim h(x \vee y)$. Now if $z \in X_0$ satisfies $x \precsim_0 z$ and $y \precsim_0 z$, we infer $x \vee y \precsim z$, so also $h(x \vee y) \precsim_0 h(z) = z$. In all, this shows that $h(x \vee y)$ is the greatest lower bound in $(X_0, \precsim_0)$ of x and y.

Now, $x \wedge y \precsim x$ implies $h(x \wedge y) \precsim_0 h(x) = x$; as we have similarly $h(x \wedge y) \precsim_0 y$, we infer $h(x \wedge y) \precsim_0 x \wedge y$. Since we always have $x \wedge y \precsim h(x \wedge y)$, we conclude that $x \wedge y = h(x \wedge y)$, and hence $x \wedge y$ is a closed element, thus also the greatest lower bound of x and y in $(X_0, \precsim_0)$. □

8.3.8 Corollary. *Let (f, g) be a Galois connection between the lattice $(Y, \mathcal{U})$ and the quasi ordered set $(Z, \mathcal{V})$. Then the two collections Y_0 and Z_0 of closed elements in respectively Y and Z are anti-isomorphic lattices for the induced orders $\mathcal{U}_0 = \mathcal{U} \cap (Y_0 \times Y_0)$ and $\mathcal{V}_0 = \mathcal{V} \cap (Z_0 \times Z_0)$. For x, $y \in Y_0$, the least upper bound $x \vee_0 y$ is equal to $(g \circ f)(x \vee y)$, and the greatest lower bound is $x \wedge y$, where $\vee$ and $\wedge$ indicate that bounds are taken in Y.*

PROOF. As $g \circ f$ is a closure operator on the set $g(Y)$ of closed elements in X, the result is a straightforward consequence of the previous theorem. □

Clearly, there is a similar statement for the case in which $(Z, \mathcal{V})$ is a lattice. We now turn back to the Galois connection of a relation.

8.3.9 Theorem. *Let $\mathcal{R}$ be a relation from a set X to a set Y. The anti-isomorphic ordered sets of closed elements of the Galois connection of $\mathcal{R}$ are lattices.*

PROOF. This is a direct application of Corollary 8.3.8, since both $(2^X, \subseteq)$ and $(2^Y, \subseteq)$ are lattices. □

The last result has important applications in a class of situations where the data can be represented by a relation between two sets. Without going into much detail, we point out that the lattice of closed elements in 2^X obtained in Theorem 8.3.9 admits a description in other, maybe more appealing, terms. Its elements can be identified with the maximal rectangles of the relation $\mathcal{R}$, one of these rectangles, say (A, B), being smaller (in the lattice order) than another, say (C, D), iff $A \subseteq C$ iff $B \supseteq D$. For a particular case, notice that $\mathcal{R}$ is a biorder (in the sense of Definition 4.2.1) iff its lattice is a chain.

8.3.10 Definition. Let $\mathcal{R}$ be a relation from X to Y. The lattice of closed elements in 2^X of the Galois connection of $\mathcal{R}$ is the *Galois lattice* or *concept lattice* of $\mathcal{R}$.

The first term appears in Birkhoff (1967), Matalon (1965), and Barbut and Monjardet (1970), while the second was popularized by the Darmstadt school, see Ganter and Wille (1996).

8.3.11 Example. Going back to the relation $\mathcal{R}$ in Example 8.3.2, we display in Figure 8.1 the Hasse diagram of the Galois lattice of $\mathcal{R}$. Each node designates a maximal rectangle (A, B), specified by listing the elements in the subset A of $X = \{a, b, c, d, e\}$ and the elements in the subset B of $Y = \{p, q, r, s\}$, with the $-$ sign signaling the absence of such elements.

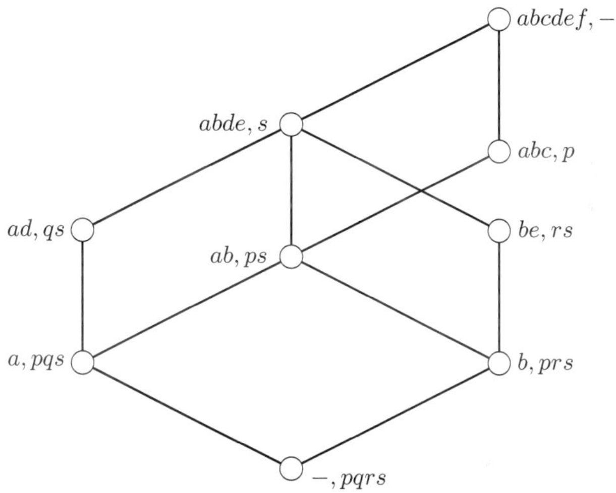

Figure 8.1. The Galois lattice of relation $\mathcal{R}$ in Example 8.3.11.

8.4 Knowledge Structures and Binary Relations

As will be shown in Corollary 8.4.3, Birkhoff's Theorem 3.8.3 arises as a case of Theorem 8.2.4 in which the sets Y and Z are, respectively, the set $\mathfrak{K}$ of all knowledge structures on a domain Q, and the set $\mathfrak{R}$ of all binary relations on Q. This formulation, which is due to Monjardet (1970), clarifies the special role played by the quasi orders, and the exact correspondence between the concepts involved.

8.4.1 Definition. Consider the set $\mathfrak{R}$ of all binary relations on a set Q, ordered by inclusion. Let $\mathcal{R} \mapsto t(\mathcal{R})$ be the mapping of $\mathfrak{R}$ to itself, defined by

$$t(\mathcal{R}) = \bigcup_{k=0}^{\infty} \mathcal{R}^k.$$

As definined in 1.6.4, $t(\mathcal{R})$ is the (reflexo-)transitive closure of $\mathcal{R}$. It can be verified that t is a closure operator on $(\mathfrak{R}, \subseteq)$ (Problem 6). The closed elements are the quasi orders. The mapping t will be called the *transitive closure operator on* $(\mathfrak{R}, \subseteq)$.

As before, we denote by $\mathfrak{K}$ the set of all knowledge structures on a set Q. We associate to any $\mathcal{K} \in \mathfrak{K}$ the smallest quasi ordinal space $u(\mathcal{K})$ including $\mathcal{K}$. The closure operator u will be referred to as the *quasi ordinal closure operator on* $(\mathfrak{K}, \subseteq)$. The closed elements are the quasi ordinal spaces. (See Problem 6).

8.4.2 Theorem. *Consider the set $\mathfrak{K}$ of all knowledge structures on a nonempty set Q, and the set $\mathfrak{R}$ of all binary relations on Q, both being ordered by inclusion. Let $\mathcal{K} \mapsto r(\mathcal{K})$ be a mapping from $\mathfrak{K}$ to $\mathfrak{R}$ defined by*

$$p\, r(\mathcal{K})\, q \quad \Longleftrightarrow \quad \mathcal{K}_p \supseteq \mathcal{K}_q \tag{8.6}$$

where $p, q \in Q$. For $\mathcal{R} \in \mathfrak{R}$, let $k(\mathcal{R})$ be the knowledge structure on Q defined from $\mathcal{R}$ by

$$K \in k(\mathcal{R}) \quad \Longleftrightarrow \quad (\forall (p, q) \in \mathcal{R} : \ q \in K \Rightarrow p \in K). \tag{8.7}$$

Thus, $\mathcal{R} \mapsto k(\mathcal{R})$ is a mapping from $\mathfrak{R}$ to $\mathfrak{K}$. Then, the pair (r, k) is a Galois connection between $(\mathfrak{K}, \subseteq)$ and $(\mathfrak{R}, \subseteq)$. Moreover, $k \circ r$ is the quasi ordinal closure operator on $(\mathfrak{K}, \subseteq)$ and $r \circ k$ is the transitive closure operator on $(\mathfrak{R}, \subseteq)$. The closed elements are respectively the quasi ordinal spaces in $\mathfrak{K}$, and the quasi orders in $\mathfrak{R}$.

The proof of this Theorem is given in 8.4.4. In this framework, the corollary below is a minor improvement of Birkhoff's Theorem 3.8.3. It shows that the one-to-one correspondence is actually an anti-isomorphism.

8.4.3 Corollary. *Let (r, k) be the Galois connection of Theorem 8.4.2. Then, the restriction r_0 of r to the set $\mathfrak{K}^{\mathbf{so}}$ of all quasi ordinal spaces on Q is an anti-isomorphism from the lattice $(\mathfrak{K}^{\mathbf{so}}, \subseteq)$ onto the lattice $(\mathfrak{R}^{\mathbf{o}}, \subseteq)$ of all quasi orders on Q. The inverse mapping r_0^{-1} is the restriction of k to $\mathfrak{R}^{\mathbf{o}}$. Moreover, the image $r_0(\mathcal{K})$ of any ordinal knowledge space $\mathcal{K}$ is a partial order.*

This immediately results from Theorems 8.4.2 and 8.2.4(v), together with Corollary 8.3.8 (notice that $(\mathfrak{K}, \subseteq)$ and $(\mathfrak{R}, \subseteq)$ are lattices).

8.4.4 Proof of Theorem 8.4.2. We first establish that (r, k) is a Galois connection. Because $(\mathfrak{K}, \subseteq)$ and $(\mathfrak{R}, \subseteq)$ are ordered sets, Conditions (i) and (ii) in Definition 8.2.3 are trivially true. Conditions (iii)-(vi) correspond to Conditions (a) to (d) below:

(a) $\quad \mathcal{K} \subseteq \mathcal{K}' \quad \Longrightarrow \quad (\forall p, q \in Q : \ \mathcal{K}'_p \supseteq \mathcal{K}'_q \ \Rightarrow \ \mathcal{K}_p \supseteq \mathcal{K}_q);$

(b) $\quad \mathcal{R} \subseteq \mathcal{R}' \quad \Longrightarrow \quad (\forall S \subseteq Q : \ S \in k(\mathcal{R}') \ \Rightarrow \ S \in k(\mathcal{R}));$

(c) $\quad \mathcal{K} \subseteq (k \circ r)(\mathcal{K});$

(d) $\quad \mathcal{R} \subseteq (r \circ k)(\mathcal{R}).$

We prove these four conditions.

(a) Take any $p, q \in Q$ and suppose that $K \in \mathcal{K}_q$, with $\mathcal{K} \subseteq \mathcal{K}'$ and $\mathcal{K}'_p \supseteq \mathcal{K}'_q$. Successively, $K \in \mathcal{K}'$, $K \in \mathcal{K}'_q$ (since $q \in K$), $K \in \mathcal{K}'_p$, $p \in K$, yielding $K \in \mathcal{K}_p$ (since $K \in \mathcal{K}$).

(b) Suppose that $S \in k(\mathcal{R}')$, with $\mathcal{R} \subseteq \mathcal{R}'$. By Equation (8.7), S is a state of $k(\mathcal{R}')$ if and only if whenever $p\mathcal{R}'q$ then $q \in S \Rightarrow p \in S$. We must show that S is also a state of $k(\mathcal{R})$. Take any $p, q \in Q$ and suppose that $p\mathcal{R}q$; thus $p\mathcal{R}'q$, which implies $q \in S \Rightarrow p \in S$ (since S is a state of $k(\mathcal{R}')$). Applying Equation (8.7), we obtain $S \in k(\mathcal{R})$.

(c) Successively,

$$\begin{aligned} K \in \mathcal{K} &\implies \forall p, q \in Q : \; (\mathcal{K}_p \supseteq \mathcal{K}_q,\; q \in K) \Rightarrow p \in K \\ &\iff \forall p, q \in Q : \; \big(p\, r(\mathcal{K})\, q,\; q \in K\big) \Rightarrow p \in K && \text{[by (8.6)]} \\ &\iff K \in (k \circ r)(\mathcal{K}). && \text{[by (8.7)]} \end{aligned}$$

(d) For all $p, q \in Q$,

$$\begin{aligned} p\mathcal{R}q &\implies \forall K \in 2^Q : \; \big(K \in k(\mathcal{R}),\; q \in K\big) \Rightarrow p \in K && \text{[by (8.7)]} \\ &\iff \forall K \in 2^Q : \; K \in \big(k(\mathcal{R})\big)_q \Rightarrow K \in \big(k(\mathcal{R})\big)_p \\ &\iff \big(k(\mathcal{R})\big)_p \supseteq \big(k(\mathcal{R})\big)_q \\ &\iff p\,(r \circ k)(\mathcal{R})\, q. && \text{[by (8.6)]} \end{aligned}$$

Since (r, k) is a Galois connection, by Theorem 8.2.4(i), $k \circ r$ and $r \circ k$ are closure operators on $(\mathfrak{K}, \subseteq)$ and $(\mathfrak{R}, \subseteq)$, respectively. The following two conditions derive from Definition 8.2.1(i):

(e) $\mathcal{K} \subseteq \mathcal{K}' \Rightarrow (k \circ r)(\mathcal{K}) \subseteq (k \circ r)(\mathcal{K}')$;

(f) $\mathcal{R} \subseteq \mathcal{R}' \Rightarrow (r \circ k)(\mathcal{R}) \subseteq (r \circ k)(\mathcal{R}')$.

Now, from Equation (8.6), it is clear that, for any knowledge structure $\mathcal{K}$, $r(\mathcal{K})$ is a quasi order on Q. By Equation (8.7), $k(\mathcal{R})$ is a quasi ordinal space for any relation $\mathcal{R}$ on Q. In particular, $(k \circ r)(\mathcal{K})$ is a quasi ordinal space on Q. Moreover, it is the smallest quasi ordinal space including $\mathcal{K}$. Indeed, for any quasi ordinal space $\mathcal{K}'$ on Q, it is easily seen that $(k \circ r)(\mathcal{K}') = \mathcal{K}'$. Hence, if $\mathcal{K}'$ includes $\mathcal{K}$, Condition (e) yields

$$(k \circ r)(\mathcal{K}) \subseteq (k \circ r)(\mathcal{K}') = \mathcal{K}'.$$

Thus, $k \circ r$ is the quasi ordinal closure on $(\mathfrak{K}, \subseteq)$.

We turn to the closure operator $r \circ k$. By Condition (d) and Equation (8.6), $(r \circ k)(\mathcal{R})$ is a transitive relation including $\mathcal{R}$. To prove that $r \circ k$ is the transitive closure operator on $(\mathfrak{R}, \subseteq)$, we have to show that, for any $\mathcal{R} \in \mathfrak{R}$ and any quasi order $\mathcal{R}'$ including $\mathcal{R}$, we have

$$(r \circ k)(\mathcal{R}) \subseteq \mathcal{R}'.$$

If $\mathcal{R}'$ is a quasi order, then $(r \circ k)(\mathcal{R}') = \mathcal{R}'$ (as can be checked easily). Thus, $\mathcal{R} \subseteq \mathcal{R}'$ implies

$$\begin{aligned}(r \circ k)(\mathcal{R}) &\subseteq (r \circ k)(\mathcal{R}') && \text{[by Condition (f)]}\\ &= \mathcal{R}'.\end{aligned}$$

Finally, the fact that $\mathfrak{R}^{\mathbf{o}}$ and $\mathfrak{K}^{\mathbf{so}}$ are the closed elements of $\mathfrak{R}$ and $\mathfrak{K}$, respectively, results from Theorem 8.2.4(iv). □

We now rephrase Definitions 3.8.6 and 3.7.1.

8.4.5 Definition. Referring to the mappings described in Theorem 8.4.2, we say that the quasi ordinal space $k(\mathcal{R})$ is *derived* from the relation $\mathcal{R}$, and similarly that the quasi order $r(\mathcal{K})$ is *derived* from the knowledge structure $\mathcal{K}$. Notice that $r(\mathcal{K})$ is the surmise relation (or precedence relation) of $\mathcal{K}$.

8.5 Granular Knowledge Structures and Granular Attributions

In Theorem 3.8.3, quasi ordinal knowledge spaces were put in a one-to-one correspondence with quasi orders. We just showed that this correspondence could be derived from a Galois connection (cf. Corollary 8.4.3). We discuss another Galois connection here, from which we will obtain a different proof of the one-to-one correspondence already established in Chapter 5 between the collection $\mathfrak{K}^{\mathbf{sg}}$ of all granular knowledge spaces on a set Q and the collection $\mathfrak{F}^{\mathbf{s}}$ of all surmise functions on Q (cf. Theorem 5.2.5).

The starting point is the construction of Definition 5.2.1. For any granular knowledge structure $(Q, \mathcal{K})$, we defined there a derived surmise function σ on the set Q by setting, for any q in Q,

$$C \in \sigma(q) \quad \Longleftrightarrow \quad C \text{ is an atom at } q.$$

On the other hand, according to Definition 5.2.3, each attribution σ on the nonempty set Q produces a derived collection $\mathcal{K}$ of knowledge states on Q. This collection $\mathcal{K}$ consists of all subsets K of Q satisfying

$$\forall q \in K, \exists C \in \sigma(q) : \; C \subseteq K.$$

However, the resulting knowledge space is not necessarily granular.

8.5.1 Example. Let Q be an infinite set of items, and let σ be the attribution mapping each item to the collection of all infinite subsets of Q. (Thus, the attribution σ is constant.) The knowledge space derived from σ consists of all infinite subsets of Q, plus the empty set. This space has no atom whatsoever.

8.5.2 Definition. An attribution is *granular* when the knowledge space it produces is granular. We denote by $\mathfrak{F}^{\mathbf{g}}$ the set of all granular attributions on the nonempty set Q.

We do not have a direct characterization of granular attributions (see the Open Problem 18.3.2). Such a characterization would be useful for the next Galois connection, which is between the collection $\mathfrak{K}^{\mathbf{g}}$ of all granular knowledge structures on Q and the collection $\mathfrak{F}^{\mathbf{g}}$ of all granular attributions on Q. The definition of a Galois connection (cf. 8.2.3) requires that these sets be first equipped with a quasi order. Here, $\mathfrak{K}^{\mathbf{g}}$ will be taken with the inclusion relation. On $\mathfrak{F}^{\mathbf{g}}$, we take a relation $\precsim$ similar to the one introduced in Definition 5.5.1, with for σ, σ' in $\mathfrak{F}^{\mathbf{g}}$:

$$\sigma' \precsim \sigma \quad \iff \quad (\forall q \in Q, \forall C \in \sigma(q), \exists C' \in \sigma'(q) : \; C' \subseteq C).$$

8.5.3 Theorem. *With $\mathfrak{K}^{\mathbf{g}}$ denoting the collection of all granular knowledge structures on some nonempty domain Q, and $\mathfrak{F}^{\mathbf{g}}$ the collection of all granular attributions on Q, consider the two mappings $a : \mathfrak{K}^{\mathbf{g}} \to \mathfrak{F}^{\mathbf{g}}$ and $g : \mathfrak{F}^{\mathbf{g}} \to \mathfrak{K}^{\mathbf{g}}$ defined as follows. The image $a(\mathcal{K})$ of a granular knowledge structure $\mathcal{K}$ is the attribution that associates to any question q the set of all atoms at q in $\mathcal{K}$. For any granular attribution σ, the image $g(\sigma)$ is the knowledge structure consisting of all subsets K of Q such that*

$$\forall q \in K, \, \exists C \in \sigma(q) : \; C \subseteq K.$$

Then the pair (a, g) of mappings forms a Galois connection between the quasi ordered sets $(\mathfrak{K}^{\mathbf{g}}, \subseteq)$ and $(\mathfrak{F}^{\mathbf{g}}, \precsim)$. The closed elements of this Galois connection are respectively in $\mathfrak{K}^{\mathbf{g}}$ the granular knowledge spaces, and in $\mathfrak{F}^{\mathbf{g}}$ the surmise functions. Moreover, the Galois connection induces between the two sets of closed elements the one-to-one correspondence obtained in Theorem 5.2.5.

Proof. First notice that for $\mathcal{K} \in \mathfrak{K}^{\mathbf{g}}$, we have $a(\mathcal{K}) \in \mathfrak{F}^{\mathbf{g}}$ (see Problem 9). All six conditions in Definition 8.2.3 for a Galois connection are easily established (Problem 10). It is also straightforward to check that all the closed elements in $\mathfrak{K}^{\mathbf{g}}$ are spaces, and that all the closed elements in $\mathfrak{F}^{\mathbf{g}}$ are surmise functions. To show that any granular knowledge space $\mathcal{K}$ is a closed element in $\mathfrak{K}^{\mathbf{g}}$, we have to show that $(g \circ a)(\mathcal{K}) = \mathcal{K}$. The inclusion $(g \circ a)(\mathcal{K}) \supseteq \mathcal{K}$ is true because (a, g) is a Galois connection. For the reverse inclusion, notice that if $K \in (g \circ a)(\mathcal{K})$, then K is a union of clauses of $a(\mathcal{K})$, thus a union of elements (in fact, atoms) of $\mathcal{K}$. As $\mathcal{K}$ is a space, it must contain K.

We now prove that $(a \circ g)(\sigma) = \sigma$ for any surmise function σ. If $q \in Q$, any σ-clause C for q is a state in $g(\sigma)$; moreover, there can be no element K' of $g(\sigma)$ such that $q \in K' \subset C$ (since no σ-clause for q could be included in K'). Thus $C \in \big((a \circ g)(\sigma)\big)(q)$. Conversely, if $C \in \big((a \circ g)(\sigma)\big)(q)$, then C is an element of $g(\sigma)$ which is minimal for the property $q \in C$. We leave to the reader to verify that $C \in \sigma(q)$.

The proof of the last sentence of the statement is also left to the reader.

□

8.5.4 Definition. In the notation of Theorem 8.5.3, $(g \circ a)(\mathcal{K})$ is the *spatial closure* of the granular knowledge structure $\mathcal{K}$, while $(a \circ g)(\sigma)$ is the *surmise closure* of the granular attribution σ. Note that $(g \circ a)(\mathcal{K})$ coincides with the space spanned by $\mathcal{K}$.

For relations cast as (necessarily granular) attributions (cf. Definition 5.1.4), it is easy to check that the surmise closure of a relation is exactly the transitive closure of this relation. We now state two properties of the spatial closure with respect to resolubility or acyclicity (in the sense of Definitions 5.6.2 and 5.6.12). The proof of the next theorem is left as Problem 11.

8.5.5 Theorem. *A granular attribution is resoluble if and only if its surmise closure is resoluble.*

8.5.6 Theorem. *If σ is a granular, acyclic attribution on a nonempty, finite set Q, then its surmise closure $(a \circ g)(\sigma)$ is also acyclic.*

PROOF. Setting $\tau = (a \circ g)(\sigma)$, we assume that the relation $\mathcal{R}_\sigma$ (cf. 5.6.10) is acyclic and we prove by contradiction that $\mathcal{R}_\tau$ is also acyclic. If $x_1, \dots, x_k$ is a cycle for $\mathcal{R}_\tau$, there is (by definition of $\mathcal{R}_\tau$) a clause C_i in $\tau(x_{i+1})$ that contains x_i (for a cyclic index i with $i = 1, \dots, k$). The thesis will result from the existence, for each value of i, of items $y_1^i, \dots, y_{\ell_i}^i$ such that

$$\begin{array}{llll} x_1 \mathcal{R}_\sigma y_1^1, & y_1^1 \mathcal{R}_\sigma y_2^1, & \dots, & y_{\ell_1}^1 \mathcal{R}_\sigma x_2, \\ x_2 \mathcal{R}_\sigma y_1^2, & y_1^2 \mathcal{R}_\sigma y_2^2, & \dots, & y_{\ell_2}^2 \mathcal{R}_\sigma x_3, \\ & \dots, & & \\ x_k \mathcal{R}_\sigma y_1^k, & y_1^k \mathcal{R}_\sigma y_2^k, & \dots, & y_{\ell_n}^n \mathcal{R}_\sigma x_1 \end{array}$$

(because we have here a cycle of $\mathcal{R}_\sigma$ in contradiction with our assumption).

To construct the finite sequence $y_1^i, \dots, y_{\ell_i}^i$, we first define a mapping η on a certain subset D of C_i. By the definition of $\tau = (a \circ g)(\sigma)$, the clause C_i in $\tau(x_{i+1})$ is a minimal element in $g(\sigma)$ among the elements containing x_{i+1}. In particular, $C_i \setminus \{x_i\}$ is not a state of σ. There must be some item y in $C_i \setminus \{x_i\}$ such that no clause in $\sigma(y)$ is included in $C_i \setminus \{x_i\}$. On the other hand, C_i being an element of $g(\sigma)$, there is some clause C_1^i in $\sigma(y)$ included in C_i. Thus $x_i \in C_1^i$. We set $\eta(y) = x_i$. If $y = x_{i+1}$, then the construction of η is completed, with $D = \{x_i, x_{i+1}\}$. If $y \neq x_{i+1}$, we initialize D to $\{x_i, y\}$. Then $C_i \setminus D$ is not a state of σ, but contains x_{i+1}. Again since $C_i \in g(\sigma)$ but $C_i \setminus D \notin g(\sigma)$, there is some item y' such that y or x_i belongs to a clause in $\sigma(y')$. We add y' to D, and set $\eta(y')$ equal to y or x_i accordingly. The same construction for an increasing set D is repeated until D contains x_{i+1} (which must happen at some time since C_i is finite). Clearly, there exists then a sequence $z_1 = \eta(x_{i+1})$, $z_2 = \eta(z_1)$, $\dots$, $z_\ell = \eta(z_{\ell-1})$ with moreover $\eta(z_\ell) = x_i$. We set $y_j^i = z_{\ell-j+1}$ for $j = 1, \dots, \ell$.

The construction of the finite sequence $y_1^i, \dots, y_{\ell_i}^i$ will be carried out for each value of the cyclic index i. □

The next example shows that the converse of Theorem 8.5.6 does not hold.

8.5.7 Example. Define an attribution σ on $Q = \{a, b, c, d\}$ by

$$\sigma(a) = \{\{a\}\}, \qquad \sigma(b) = \{\{a, b\}, \{b, d\}\},$$
$$\sigma(c) = \{\{a, c\}\}, \qquad \sigma(d) = \{Q\}.$$

The relation $\mathcal{R}_\sigma$ is **not** acyclic (since b, d is a cycle), and so σ is not an acyclic attribution. The surmise closure τ of σ is acyclic, however. It is given by

$$\tau(a) = \{\{a\}\}, \qquad \tau(b) = \{\{a, b\}\},$$
$$\tau(c) = \{\{a, c\}\}, \qquad \tau(d) = \{Q\}.$$

The same example shows that the relation $\mathcal{R}_\tau$ attached to the surmise closure τ of an attribution σ can differ from the transitive closure of the relation $\mathcal{R}_\sigma$.

8.6 Knowledge Structures and Associations

Theorem 7.1.5 describes a one-to-one correspondence α between the collection of all knowledge spaces on Q and the collection of all entailments for Q. Another one-to-one correspondence links the latter collection with that of entail relations for Q (Theorem 7.2.1 and Definition 7.2.2). So, composing these two correspondences, we also have a one-to-one correspondence β between the collection of all knowledge spaces on Q and the collection of all entail relations for Q. Each of the correspondences α and β can be produced from a specific Galois connection. We deal below with the case of α, leaving the case of β to the reader as Problem 12.

8.6.1 Definition. Let Q be a nonempty set. Define as follows a mapping v from the collection $\mathfrak{K}$ of all knowledge structures on Q to the collection $\mathfrak{E}$ of all relations from 2^Q to Q, where for $\mathcal{K} \in \mathfrak{K}$:

$$v(\mathcal{K}) = \mathcal{P} \iff$$
$$\big(\forall A \in 2^Q, \forall q \in Q: \; A\mathcal{P}q \;\Leftrightarrow\; (\forall K \in \mathcal{K}: \; q \in K \Rightarrow K \cap A \neq \varnothing)\big). \quad (8.8)$$

The relations from 2^Q to Q will be called *association relations*, or in short *associations*. Notice that if $\mathcal{K}$ is a knowledge space, then $v(\mathcal{K})$ is exactly its derived entailment (cf. Definition 7.1.6). When $\mathcal{K}$ is a general knowledge structure, we also say that $v(\mathcal{K})$ is *derived from* $\mathcal{K}$.

Define then a mapping $\ell : \mathfrak{E} \to \mathfrak{K}$ by setting for $\mathcal{P} \in \mathfrak{E}$:

$$\ell(\mathcal{P}) = \{K \in 2^Q \,|\, \forall r \in Q, \forall B \subseteq Q : (B\mathcal{P}r \text{ and } r \in K) \Rightarrow B \cap K \neq \varnothing\}. \quad (8.9)$$

The knowledge structure $\ell(\mathcal{P})$ is said to be *derived from* $\mathcal{P}$. A similar construction was encountered in Equation (7.3) of Theorem 7.1.5.

8.6.2 Theorem. *Let Q be a nonempty set. The mappings $v : \mathfrak{K} \to \mathfrak{E}$ and $\ell : \mathfrak{E} \to \mathfrak{K}$ form a Galois connection if both $\mathfrak{K}$ and $\mathfrak{E}$ are ordered by inclusion. The closed elements in $\mathfrak{K}$ form the lattice $(\mathfrak{K}^{\mathbf{s}}, \subseteq)$ of all knowledge spaces on Q; the closed elements in $\mathfrak{E}$ form the lattice $(\mathfrak{E}^{\mathbf{e}}, \subseteq)$ of all entailments. The anti-isomorphisms induced by v and ℓ between these two lattices provide the one-to-one correspondence in Theorem 7.1.5.*

PROOF. Conditions (i) and (ii) in the Definition 8.2.4 of a Galois connection are automatically satisfied because the inclusion relation, either on $\mathfrak{K}$ or on $\mathfrak{E}$, is a partial order. The other conditions are easily derived from the definitions of v and ℓ in Equations 8.8 and (8.9).

To show that the closed elements in $\mathfrak{K}$ constitute the collection $\mathfrak{K}^{\mathbf{s}}$ of all knowledge spaces on Q, it suffices to establish $\ell(\mathfrak{E}) = \mathfrak{K}^{\mathbf{s}}$. The inclusion $\ell(\mathfrak{E}) \subseteq \mathfrak{K}^{\mathbf{s}}$ is easily obtained. The opposite inclusion follows from the fact that $\mathcal{K} = (\ell \circ v)(\mathcal{K})$ for any space $\mathcal{K}$ on Q; notice that $\mathcal{K} \subseteq (\ell \circ v)(\mathcal{K})$ holds because (v, ℓ) is a Galois connection, while $(\ell \circ v)(\mathcal{K}) \subseteq \mathcal{K}$ is proved as follows. If $L \in (\ell \circ v)(\mathcal{K}) \setminus \mathcal{K}$, let K be the largest state of the space $\mathcal{K}$ included in L. Picking r in $L \setminus K$ and setting $B = Q \setminus L$, we get both $B\, v(\mathcal{K})\, r$ and $r \in L$, contradicting $L \in (\ell \circ v)(\mathcal{K})$.

Similarly, to prove that the closed elements in $\mathfrak{E}$ constitute the collection $\mathfrak{E}^{\mathbf{e}}$ of all entailments, it suffices to show that $v(\mathfrak{K}) = \mathfrak{E}^{\mathbf{e}}$. We leave this to the reader.

Finally, as $(\mathfrak{K}^{\mathbf{s}}, \subseteq)$ clearly is a lattice, we may apply Corollary 8.3.8. □

All the Galois connections introduced in the chapter for the various structures at the focus of this monograph are recorded in Table 8.3.

Table 8.3. The three Galois connections subsuming the one-to-one correspondences recalled in Table 8.1: names and notation for the domains and the closed sets. Columns headings are as follows:

1:	Theorem number
2 and 5:	Name of mathematical structure
3 and 4:	Notation for the collection

1	2	3	4	5
8.4.2	knowledge structure	$\mathfrak{K}$	$\mathfrak{R}$	relation
	quasi ordinal space	$\mathfrak{K}^{\mathbf{so}}$	$\mathfrak{R}^{\mathbf{o}}$	quasi order
8.5.3	granular knowledge structure	$\mathfrak{K}^{\mathbf{g}}$	$\mathfrak{F}^{\mathbf{g}}$	granular attribution
	granular knowledge space	$\mathfrak{K}^{\mathbf{sg}}$	$\mathfrak{F}^{\mathbf{s}}$	surmise function
8.6.2	knowledge structure	$\mathfrak{K}$	$\mathfrak{F}$	association
	knowledge space	$\mathfrak{K}^{\mathbf{s}}$	$\mathfrak{E}^{\mathbf{e}}$	entailment

8.7 Original Sources and Related Works

For background on Galois connections, we refer the reader to Birkhoff (1967) or Barbut and Monjardet (1970), for instance. The concept of a Galois connection is closely related to that of a 'residuated mapping', see e.g. Blyth and Janowitz (1972). Definition 8.2.3 slightly extends the concept of a Galois connection, by allowing quasi ordered sets instead of ordered sets. This construction comes from Doignon and Falmagne (1985)[2].

The idea of deriving Birkhoff's Theorem from a Galois connection is due to Monjardet (1970) (cf. our Corollary 8.4.3). The other applications given here of the theory of Galois connections, namely Theorems 8.5.3 and 8.6.2, come respectively from Doignon and Falmagne (1985, in the finite case), and Koppen and Doignon (1990). We have mentioned in passing the field of concept lattices, or Galois lattices. On this subject, we refer the reader to Matalon (1965) or Barbut and Monjardet (1970) and especially to Ganter and Wille (1996). Rusch and Wille (1996) have pointed out several relationships between the study of concept lattices and our investigation of knowledge spaces. Another link was mentioned in the Sources section of Chapter 1, namely Dowling (1993b) and Ganter (1984, 1987, see also Ganter, 1991). Algorithms can be reformulated in order to perform each other's task (see Problem 8).

Problems

1. Let $\mathfrak{K}$ be the set of all knowledge structures on a set Q. Prove that any intersection of knowledge spaces in $\mathfrak{K}$ is a knowledge space, and that the mapping s of Example 8.2.2(a) is a closure operator on $(\mathfrak{K}, \subseteq)$, the knowledge spaces being the closed elements.

2. If a knowledge structure is closed under intersection, is the same true for its spatial closure?

3. Check that the mapping $h : 2^Q \to 2^Q : A \mapsto h(A) = A'$ of Example 8.2.2(b) is a closure operator, with the elements of $\mathcal{L}$ being the closed elements.

4. Complete the proof of Theorem 8.2.4.

5. Does the Galois lattice of a relation determine this relation? Consider the case in which the elements of the lattice are marked as in Figure 8.1.

6. Prove that the function t of Definition 8.4.1 is a closure operator, with the quasi orders on Q forming the closed elements. Prove that u is also a closure operator, with the quasi ordinal spaces on Q being the closed elements.

[2] Note that a mistake in Proposition 2.7(iv) in this paper was pointed out by J. Heller.

7. Construct the Galois lattice of the following relations:
   a) the identity relation from a set to the same set;
   b) the membership relation from a set Q to its power set 2^Q;
   c) the relation defined by the table below.

Table 8.4. The 0-1 array for the relation in Problem 7 (c).

	p	q	r	s
a	1	1	0	1
b	1	0	1	1
c	1	0	0	0
d	1	1	0	1

8. Design an algorithm to compute the Galois lattice of a relation between two finite sets. Show that such an algorithm can be directly derived from Algorithm 3.5.5 (which constructs a knowledge space from any of its spanning families).

9. Let a be the mapping defined in Theorem 8.5.3. Show that for any granular knowledge structure $\mathcal{K}$, the image $a(\mathcal{K})$ is a granular attribution.

10. Complete the proof of Theorem 8.5.3.

11. Prove Theorem 8.5.5.

12. Establish a Galois connection between knowledge structures and entail relations from which follows the one-to-one correspondence β mentioned before Definition 8.6.1.

13. Show that mappings $f : Y \to Z$ and $g : Z \to Y$ between ordered sets $(Y, \mathcal{U})$ and $(Z, \mathcal{V})$ form a Galois connection iff for all y in Y and z in Z:
$$y\mathcal{U}g(z) \quad \Longleftrightarrow \quad z\mathcal{V}f(y)$$
(O. Schmidt, see Birkhoff, 1967, p. 124). Give a similar result for the general case in which $(Y, \mathcal{U})$ and $(Z, \mathcal{V})$ are quasi ordered sets. Use this other characterization of Galois connections to work out different proofs of results of this chapter.

14. A lattice $(L, \precsim)$ is *complete* if any family (possibly infinite) of elements of L admit a greatest lower bound and a least upper bound (define these terms). Does Theorem 8.3.7 extend to complete lattices? Find examples of complete lattices among the collections of objects studied in this monograph.

9

Descriptive and Assessment Languages*

How can we economically describe a state in a knowledge structure? The question is inescapable because, as pointed out earlier, realistic states will typically be quite large. In such cases, it is impractical to describe a state by giving the full list of items that it contains. It is also unnecessary: because of the redundancy in many real-life knowledge structures[1], a state will often be characterizable by a relatively small set of features. This idea is not new. In Chapter 4, we proved that any state in a well-graded knowledge structure could be fully described by simply listing its inner and outer fringes (cf. Theorem 4.1.7 and Remark 4.1.8(a)). Here, we consider this issue more systematically. This chapter is somewhat eccentric to the rest of this book and can be skipped without harm at first reading. We begin by illustrating the main ideas in the context of a simple example encountered earlier.

9.1 Languages and Decision Trees

9.1.1 Example. Consider the knowledge structure

$$\begin{aligned}\mathcal{G} = \big\{&\varnothing, \{a\}, \{b,d\}, \{a,b,c\}, \{b,c,e\}, \{a,b,d\},\\ &\{a,b,c,d\}, \{a,b,c,e\}, \{b,c,d,e\}, \{a,b,c,d,e\}\big\}\end{aligned} \tag{9.1}$$

on the domain $Q = \{a,b,c,d,e\}$. It was already used in Example 5.1.1 and is in fact a discriminative knowledge space. The state $\{a,b,c,e\}$ is the only state of $\mathcal{G}$ containing a, e and not d. It can be characterized by stating that $\{a,b,c,e\}$ is a particular state K of $\mathcal{G}$ satisfying

$$a \in K, \quad d \notin K \quad \text{and} \quad e \in K. \tag{9.2}$$

[1] For example, the domain of the learning space for beginning algebra used by the ALEKS system and discussed in Chapter 17 contains around 300 items, while the number of knowledge states does not exceed a few millions, which is a minute fraction of 2^{300}, the number of subsets of a set of size 300.

J.-C. Falmagne, J.-P. Doignon, *Learning Spaces*,
DOI 10.1007/978-3-642-01039-2_9, © Springer-Verlag Berlin Heidelberg 2011

Similarly, the state $M = \{b, c, d, e\}$ is specified by the statement

$$a \notin M \text{ and } d, e \in M. \tag{9.3}$$

We now adopt a compact notation. We shall represent (9.2) and (9.3) by the strings $a\bar{d}e$ and $\bar{a}de$, respectively. Extending this notation, we can, for example, represent all the states of $\mathcal{G}$ by the strings listed in Table 9.1.

Table 9.1. The states of $\mathcal{G}$ and their representing strings.

States	Strings
$\{a, b, c, d, e\}$	ade
$\{a, b, c, d\}$	$acd\bar{e}$
$\{a, b, c, e\}$	$a\bar{d}e$
$\{a, b, c\}$	$ac\bar{d}\bar{e}$
$\{a, b, d\}$	$a\bar{c}d$
$\{a\}$	$a\bar{b}$
$\{b, c, d, e\}$	$\bar{a}de$
$\{b, c, e\}$	$\bar{a}\bar{d}e$
$\{b, d\}$	$\bar{a}\bar{c}d$
$\varnothing$	$\bar{a}\bar{b}$

Such strings are referred to as 'words.' The set of words

$$G_1 = \{\, ade,\ acd\bar{e},\ a\bar{d}e,\ ac\bar{d}\bar{e},\ a\bar{c}d,\ a\bar{b},\ \bar{a}de,\ \bar{a}\bar{d}e,\ \bar{a}\bar{c}d,\ \bar{a}\bar{b} \,\}$$

is called a 'descriptive language' (for $\mathcal{G}$). Some descriptive languages are of particular interest in the framework of this monograph because they symbolize sequential procedures for recognizing the states. Such languages can in principle be used in assessing the knowledge states of individuals. For example, the language

$$G_2 = \{\, acde,\ acd\bar{e},\ ac\bar{d}e,\ ac\bar{d}\bar{e},\ a\bar{c}d,\ a\bar{c}\bar{d},\ \bar{a}ed,\ \bar{a}e\bar{d},\ \bar{a}\bar{e}b,\ \bar{a}\bar{e}\bar{b} \,\}$$

also specifies the states of $\mathcal{G}$, but with words satisfying certain rules. Notice that every word of G_2 begins with a or $\bar{a}$. Also, in any word beginning with a, the next symbol is either c or $\bar{c}$. Similarly, if a word begins with $\bar{a}$, the next symbol is either e or $\bar{e}$, etc. This illustrates a general pattern represented in Figure 9.1 in the form of a decision tree. The words of G_2 are read from the tree by following the branches from left to right. Each leaf corresponds to a word specifying a state. Such a tree represents a possible knowledge assessment procedure.

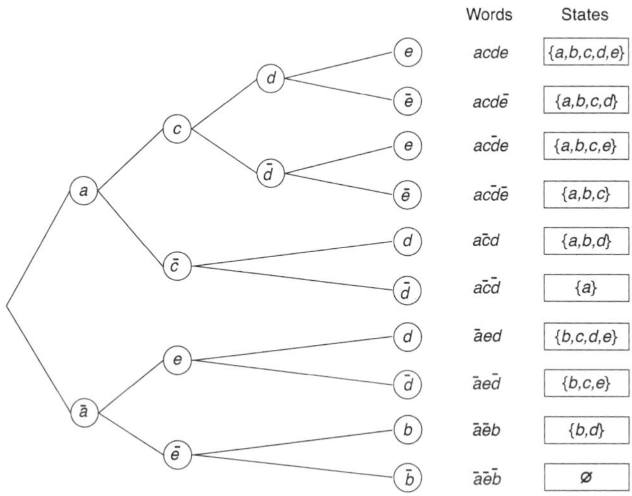

Figure 9.1. Sequential decision tree corresponding to the language G_2 for the knowledge structure $\mathcal{G}$.

9.1.2 Definition. We recall from 2.1.4 that if $\mathcal{K}$ is a knowledge structure and q is any item, then $\mathcal{K}_q = \{K \in \mathcal{K} \,|\, q \in K\}$. We similarly define

$$\mathcal{K}_{\bar{q}} = \{K \in \mathcal{K} \,|\, q \notin K\}. \tag{9.4}$$

We have thus $\mathcal{K} = \mathcal{K}_q \cup \mathcal{K}_{\bar{q}}$.

Beginning at the extreme left node of the tree in Figure 9.1, one could first check whether the knowledge state of a subject contains a, by proposing an instance of that item to the subject. Suppose that the subject solves a. We already know then that the subject's state is in

$$\mathcal{G}_a = \big\{\{a\}, \{a,b,c\}, \{a,b,d\}, \{a,b,c,d\}, \{a,b,c,e\}, \{a,b,c,d,e\}\big\}.$$

The next item proposed is c. Suppose that c is not solved, but that the subject then solves d. The complete questioning sequence in this case corresponds to the word $a\bar{c}d$ which identifies the state $\{a,b,d\}$ because

$$\mathcal{G}_a \cap \mathcal{G}_{\bar{c}} \cap \mathcal{G}_d = \big\{\{a,b,d\}\big\}.$$

A language which is representable by a decision tree, such as G_2, will be called an 'assessment language.' (We give an exact definition in 9.2.3.)

There are two difficulties with this deterministic approach to the assessment of knowledge in real-life cases. One is that human behavior in testing situations is often unreliable: subjects may temporarily forget the solution of a problem that they know well, or make careless errors in responding; they could also in some situations guess the correct response to a problem not already mastered. This feature suggests that a straightforward deterministic decision tree is inadequate as a knowledge assessment device.

The other difficulty is that we evidently wish to minimize (in some sense) the number of questions asked in an assessment, which raises the issue of 'optimality' of a decision tree. At least two parameters could be minimized. The first one is the largest number of questions to be asked for specifying any state. In Figure 9.1, this *worst case number* or *depth of the tree*, as it is also called, equals 4 (and is attained for four states). A second parameter often used is the average number of questions, which is here $3.4 = (4 \times 4 + 6 \times 3)/10$. An *optimal* tree is one that minimizes either the depth or the average number of internal nodes, depending on the context. The design of an optimal decision tree for the task at hand is known from theoretical computer science to be a hard problem—where 'hard' is meant in its technical sense, namely the corresponding decision problem is 'NP-complete.' We refer the reader to the standard text of Garey and Johnson (1979) for a good introduction to complexity theory; for the particular problem at hand here, our reference is Hyafill and Rivest (1976) (see in particular the proof of the main result). We shall not discuss here the various results obtained regarding the construction of optimal decision trees. The reason is that in our setting, deterministic procedures (encoded as binary decision trees) rely on the untenable assumptions that student's answers truly reflect the student's knowledge state.

In Chapters 13 and 14, we describe probabilistic assessment procedures which are more robust than those based on decision trees, and are capable of uncovering the state even when the subject's behavior is somewhat erratic.

There are nevertheless some theoretical questions worth studying in a deterministic framework, and which are also relevant to knowledge assessment. For instance, imagine that we observe a teacher conducting an oral examination of students. Idealizing the situation, suppose that this observation is taking place over a long period, and that we manage to collect all the sequences of questions asked by the teacher. Would we then be able to infer the knowledge structure relied upon by the teacher? We assume here that all potential sequences are revealed during the observation, and that the teacher is correct in assuming that the students' responses reflect their true knowledge state. An example of the results presented in this chapter is as follows: if a knowledge structure is known to be ordinal (in the sense of 3.8.1), then it can be uncovered from any of its assessment languages (cf. Corollary 9.3.6). In general, an arbitrary structure cannot be reconstructed on the basis of a single assessment language. Suppose however that we have observed many teachers for a long time, and that all the assessment languages have been observed. The corresponding result is that any knowledge structure can be uncovered from the set of all its feasible assessment languages (cf. 9.4.2).

Our presentation of these results relies on a basic terminology concerning 'words', 'languages', and related concepts which we introduce in the next section. This terminology is consistent with that of Formal Language Theory (as in Rozenberg and Salomaa, 1997, for example)[2].

[2] Some of the terminology is also used in Chapter 10.

9.2 Terminology

9.2.1 Definition. We start from a finite set Q which we call *alphabet*. Any element in Q is called a *positive literal*. The *negative literals* are the elements of Q marked with an overbar. Thus, for any $q \in Q$, we have the two literals q and $\bar{q}$. A *string* over the alphabet Q is any finite sequence $\alpha_1, \alpha_2, \ldots, \alpha_n$ of literals written as $\alpha_1\alpha_2\ldots\alpha_n$. Denoting by Σ the set of all literals, we equip the set Σ^* of all strings with the associative concatenation operation

$$(\rho, \rho') \longmapsto \rho\rho' \in \Sigma^* \qquad\qquad (\rho, \rho' \in \Sigma^*)$$

and hence get a semigroup. The neutral element is the empty string, which we denote by 1. Any subset L of Σ^* is called a *language* over Q. An element ω of a language L is called a *word* of that language.

A string ρ of Σ^* is a *prefix* (resp. *suffix*) of the language L if $\rho\rho'$ (resp. $\rho'\rho$) is a word of L for some string ρ' of Σ^*. A prefix (resp. suffix) ρ is *proper* if there exists a nonempty string π such that $\rho\pi$ (resp. $\pi\rho$) is a word. For any string ρ and any language L, we denote by ρL the language containing all words of the form $\rho\omega$, for $\omega \in L$.

Strings, words, prefixes and suffixes are *positive* (resp. *negative*) when they are formed with positive (resp. negative) literals only. If $\rho = \alpha_1\alpha_2\ldots\alpha_n$ is a string over the alphabet Q, we set $\bar{\rho} = \bar{\alpha}_1\bar{\alpha}_2\ldots\bar{\alpha}_n$, with the convention $\bar{\bar{\alpha}} = \alpha$ for any literal α.

9.2.2 Example. For $Q = \{a, b, c, d, e\}$, we have ten literals. Consider the language consisting of all words of length at most 2. This language consists of $1 + 10 + 10^2$ words. Every word is both a prefix and a suffix. However, there are only $1 + 10$ proper prefixes, and the same number of proper suffixes. The language has $1 + 5 + 5^2$ positive words (coinciding with the positive prefixes) and $1 + 5$ positive, proper prefixes.

We recall from Definition 3.3.1 that a collection $\mathcal{K}$ on a domain Q is a family of subsets of Q. We also write $(Q, \mathcal{K})$ to denote the collection and call states the elements of $\mathcal{K}$. Thus, a knowledge structure $(Q, \mathcal{K})$ is a collection $\mathcal{K}$ which contains both $\varnothing$ and Q. Notice that a collection may be empty. Generalizing Definition 2.4.2, we call *projection* of a collection $(Q, \mathcal{K})$ on a subset A of Q the family $\mathcal{K}_{|A} = \{K \cap A \mid K \in \mathcal{K}\}$. The meaning of "trace" and of $\mathcal{K}_q$ and $\mathcal{K}_{\bar{q}}$ are the same for collections and for structures (cf. Definitions 2.1.4 and 9.1.2).

We assume in this chapter that the domain Q is finite.

9.2.3 Definition. An *assessment language* for the collection $(Q, \mathcal{K})$ is a language L over the alphabet Q that is empty if $|\mathcal{K}| = 0$, has only the word 1 if $|\mathcal{K}| = 1$, and otherwise satisfies $L = qL_1 \cup \bar{q}L_2$, for some q in Q, where

[A1] L_1 is an assessment language for the projection $(\mathcal{K}_q)_{|Q\setminus\{q\}}$;
[A2] L_2 is an assessment language for the collection $\mathcal{K}_{\bar{q}}$ with domain $Q\setminus\{q\}$.

It is easily verified that the words of an assessment language L for $\mathcal{K}$ are in a one-to-one correspondence with the states of $\mathcal{K}$ (see Problem 2). Figure 9.1 lists the words of an assessment language for the knowledge structure of Equation (9.1).

We can also characterize assessment languages nonrecursively, by specific properties. The following concept is instrumental in that respect.

9.2.4 Definition. A *binary classification language* over a finite alphabet Q is a language L which satisfies the two following conditions:

[B1] a literal may not appear more than once in a word;
[B2] if π is a proper prefix of L, then there exist exactly two prefixes of the form $\pi\alpha$ and $\pi\beta$, where α and β are literals; moreover $\bar{\alpha} = \beta$.

Condition [B1] implies that L does not contain any word of the form $\pi x \rho x \sigma$ or $\pi x \rho \bar{x} \sigma$ with x a literal and π, ρ, σ strings. Consequently, a binary classification language L is finite. Notice that the empty language and the language consisting of the single word 1 are both binary classification languages: they trivially satisfy Conditions [B1] and [B2].

9.2.5 Theorem. *For any proper prefix ρ of a nonempty binary classification language L, there exist a unique positive suffix ν and a unique negative suffix μ such that $\rho\nu$, $\rho\mu \in L$. In particular, L has exactly one positive and one negative word.*

The proof is left as Problem 4. We now show that the conditions in Definition 9.2.4 provide a nonrecursive characterization of assessment languages.

9.2.6 Theorem. *Any assessment language is a binary classification language. Conversely, any binary classification language is an assessment language for some collection.*

PROOF. It is easily shown that an assessment language L for the collection $\mathcal{K}$ on Q is a binary classification language over the alphabet Q. Conversely, suppose that L is a binary classification language over Q that contains more than one word. Then the word 1 is a proper prefix, and by Condition [B2] of Definition 9.2.4, there is a letter q such that all words are of the form $q\pi$ or $\bar{q}\sigma$ for various strings π and σ. Notice that $L_1 = \{\pi \,|\, q\pi \in L\}$ is again a binary classification language, this time over the alphabet $Q \setminus \{q\}$. Proceeding by induction, we infer the existence of a collection $\mathcal{K}_1$ on $Q \setminus \{q\}$ for which L_1 is an assessment language. Similarly, $L_2 = \{\pi \,|\, \bar{q}\pi \in L\}$ is an assessment language for some collection $\mathcal{K}_2$ on $Q \setminus \{q\}$. It is easily verified that L is an assessment language for the collection $\{K \cup \{q\} \,|\, K \in \mathcal{K}_1\} \cup \mathcal{K}_2$. □

A class of languages less restrictive than the assessment languages, or equivalently the binary classification languages, will be used in the next section. We proceed to define it.

9.2.7 Definition. Let $\mathcal{K}$ be a collection on Q and let L be a language over Q. We write $\alpha \vdash \omega$ to mean that α is a literal of the word ω. A word ω *describes* a state K if K is the only state that satisfies

$$\text{for any } x \in Q: \ (x \vdash \omega \Rightarrow x \in K) \ \text{ and } \ (\bar{x} \vdash \omega \Rightarrow x \notin K).$$

The language L is a *descriptive language* for the collection $\mathcal{K}$ when the two following conditions hold:

[D1] any word of L describes a unique state in K;
[D2] any state in $\mathcal{K}$ is described by at least one word of L.

We may also say for short that *L describes $\mathcal{K}$*.

Every assessment language is a descriptive language. While the words of an assessment language for a collection $\mathcal{K}$ are in a one-to-one correspondence with the states of $\mathcal{K}$, we only have a surjective mapping from a descriptive language for $\mathcal{K}$ onto $\mathcal{K}$.

We denote by $\mathrm{ASL}(\mathcal{K})$ and $\mathrm{DEL}(\mathcal{K})$ the collections of all the assessment languages and all the descriptive languages for $\mathcal{K}$, respectively. We also write $(Q, \overline{\mathcal{K}})$ for the *dual* collection of a collection $(Q, \mathcal{K})$, with $\overline{\mathcal{K}} = \{Q \setminus K \mid K \in \mathcal{K}\}$. (This extends the notation and terminology introduced in Definition 2.2.2 in the case of a structure.) For any language L we define the corresponding language $\bar{L} = \{\alpha \mid \bar{\alpha} \in L\}$. (Remember our convention that $\bar{\bar{\alpha}} = \alpha$ for any literal α.) In Problem 5, we ask the reader to establish the two equivalences

$$\begin{aligned} L \in \mathrm{ASL}(\mathcal{K}) \quad &\Longleftrightarrow \quad \bar{L} \in \mathrm{ASL}(\overline{\mathcal{K}}), \\ L \in \mathrm{DEL}(\mathcal{K}) \quad &\Longleftrightarrow \quad \bar{L} \in \mathrm{DEL}(\overline{\mathcal{K}}). \end{aligned}$$

9.3 Recovering Ordinal Knowledge Structures

We prove here that any (partially) ordinal space $(Q, \mathcal{K})$ can be recovered from any of its descriptive languages, and thus also from any of its assessment languages. In this section, we will denote by $\mathcal{P}$ a partial order on Q from which $\mathcal{K}$ is derived (in the sense of Definition 3.8.4), and by $\mathcal{H}$ the covering relation or Hasse diagram of $\mathcal{P}$ (cf. 1.6.8). Maximality and minimality of elements of Q are understood with respect to $\mathcal{P}$. We will use $\mathcal{K}(q) = \cap \mathcal{K}_q$ to denote the smallest state that contains item q. (In the language of Definition 3.4.5, $\mathcal{K}(q)$ is thus an atom at q, which is unique in this case.) Finally, L will stand for a descriptive language for $\mathcal{K}$.

9.3.1 Lemma. *The following two statements are true for any word ω of L describing the state K from $\mathcal{K}$:*

(i) *if x is a maximal element of K, then $x \vdash \omega$;*
(ii) *if y is a minimal element of $Q \setminus K$, then $\bar{y} \vdash \omega$.*

PROOF. If x is maximal in K, then $K \setminus \{x\}$ is a state of $\mathcal{K}$. As ω distinguishes between K and $K \setminus \{x\}$, we conclude that $x \vdash \omega$. Since $K \cup \{y\}$ is also a state, the second assertion follows from a similar argument. □

9.3.2 Corollary. *For each q in Q, the language L uses both literals q and $\bar{q}$.*

PROOF. The smallest state $\mathcal{K}(q)$ containing q has q as a maximal element. Moreover, $\mathcal{K}(q) \setminus \{q\}$ is a state whose complement has q as a minimal element. □

9.3.3 Theorem. *Define a relation $\mathcal{S}$ on Q by declaring $q\mathcal{S}r$ to hold when the two following conditions are satisfied:*

(i) *there exists some word ω of L such that $q \vdash \omega$ and $\bar{r} \vdash \omega$;*
(ii) *there is no word ρ of L such that both $\bar{q} \vdash \rho$ and $r \vdash \rho$.*

Let $\widehat{\mathcal{P}}$ denote the strict partial order obtained from $\mathcal{P}$ by deleting all loops. Then, we necessarily have

$$\mathcal{H} \subseteq \mathcal{S} \subseteq \widehat{\mathcal{P}}. \tag{9.5}$$

PROOF. Assume $q\mathcal{H}r$ and set $K = \mathcal{K}(r) \setminus \{r\}$. Then K is a state having q as a maximal element; also, r is a minimal element in $Q \setminus K$. By Lemma 9.3.1, any word ω describing K makes Condition (i) true. Moreover, Condition (ii) also holds since $q\mathcal{H}r$ implies that any state containing r also contains q. This establishes $\mathcal{H} \subseteq \mathcal{S}$.

Assume now $q\mathcal{S}r$. Take any state K described by the word ω whose existence is asserted in Condition (i). As $q \in K$ and $r \notin K$, we have $q \neq r$ and $r\mathcal{P}q$ cannot hold. It only remains to show that q and r are comparable with respect to $\mathcal{P}$. If this were not the case, $(\mathcal{K}(q) \setminus \{q\}) \cup \mathcal{K}(r)$ would be a state K' having r as a maximal element; moreover, q would be minimal in $Q \setminus K'$. Any word describing K' would then contradict Condition (ii). □

Each of the two inclusions in Equation 9.5 can be either strict or an equality, as shown by the following two examples.

9.3.4 Example. Consider the set $Q = \{a, b, c\}$ equipped with the alphabetical order $\mathcal{P}$. Figure 9.2 specifies an assessment language for the corresponding ordinal structure $\mathcal{K}$; here we have $\mathcal{H} \subset \mathcal{S} = \widehat{\mathcal{P}}$.

9.3.5 Example. Figure 9.3 describes another assessment language for the same ordinal knowledge structure as in previous example. Here, we have $\mathcal{H} = \mathcal{S} \subset \widehat{\mathcal{P}}$.

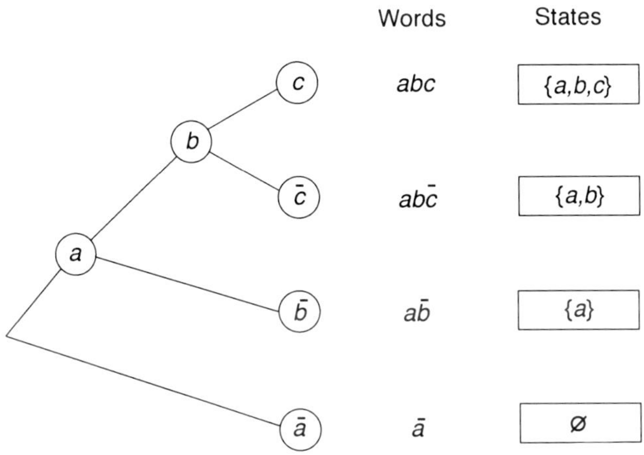

Figure 9.2. Decision tree, words and states from Example 9.3.4.

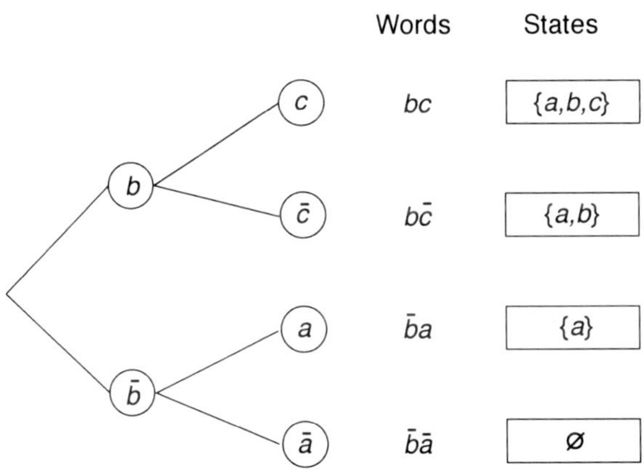

Figure 9.3. Decision tree, words and states from Example 9.3.5.

9.3.6 Corollary. *If a finite knowledge structure is known to be ordinal, it can be recovered from any of its descriptive languages.*

Corollary 9.3.6 is the main result of this section. It readily follows from Theorem 9.3.3: $\mathcal{P}$ is always the transitive closure of $\mathcal{S}$.

9.3.7 Remarks. A language can describe two distinct knowledge structures only one of which is ordinal. Such a case is described by Figure 9.4.

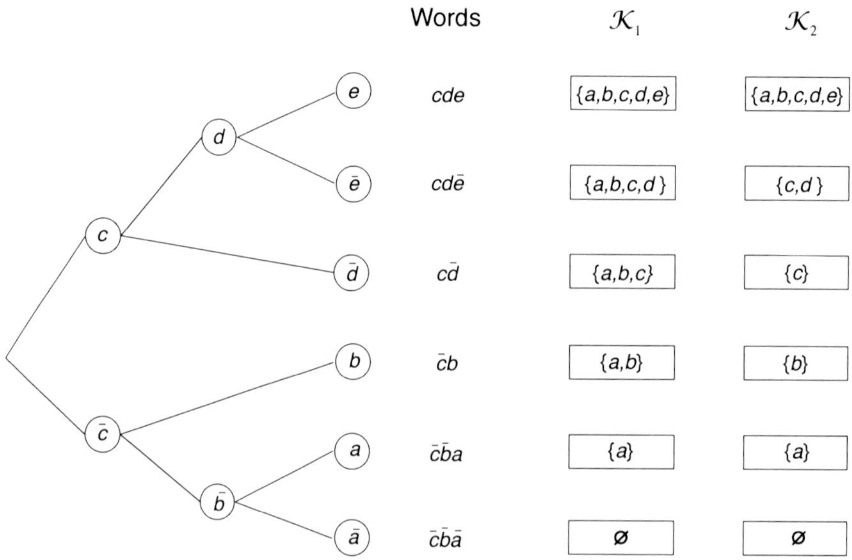

Figure 9.4. Decision tree, words and states from two distinct knowledge structures, each one described by the same language (cf. Remark 9.3.7). Only $\mathcal{K}_1$ is ordinal.

9.4 Recovering Knowledge Structures

As the example in Figure 9.4 indicates, a knowledge structure cannot be recovered from just one of its assessment languages (except if it is known to be ordinal, cf. Corollary 9.3.6). We will prove by induction on the number of items that any structure $\mathcal{K}$ can be recovered from the complete collection $\mathrm{ASL}(\mathcal{K})$ of all its assessment languages.

9.4.1 Lemma. *Suppose that $A \subseteq Q$ and let L_A be an assessment language for the projection $\mathcal{K}_{|A}$. There exists an assessment language L for $\mathcal{K}$ such that*

(i) *in each word of L, the letters from A precede the letters from $Q \setminus A$;*
(ii) *truncating the words from L to their literals from A give all the words in L_A (possibly with repetitions).*

Leaving the proof as Problem 7, we derive from 9.4.1 the main result of this section.

9.4.2 Theorem. *The two following statements are equivalent for any two knowledge structures* $(Q, \mathcal{K})$ *and* $(Q', \mathcal{K}')$*:*

(i) $ASL(\mathcal{K}) = ASL(\mathcal{K}')$;
(ii) $\mathcal{K} = \mathcal{K}'$.

Proof. (i) $\Rightarrow$ (ii). Suppose that Condition (i) holds. If some item q belongs to Q but not to Q', we can use it as the root of a decision tree for $\mathcal{K}$. So, the corresponding assessment language cannot belong to $\text{ASL}(\mathcal{K}')$. This contradiction establishes $Q \subseteq Q'$, and by symmetry $Q' = Q$.

We now prove by induction on $|Q|$ that the negation of Condition (ii) leads to a contradiction. As Condition (ii) does not hold, select in $\mathcal{K} \triangle \mathcal{K}'$ a maximal element, say K in $\mathcal{K} \setminus \mathcal{K}'$. Since $Q = Q'$, we must have $K \neq Q$. From Lemma 9.4.1, we have $\text{ASL}(\mathcal{K}_{|K}) = \text{ASL}(\mathcal{K}'_{|K})$, and thus also by the induction hypothesis $\mathcal{K}_{|K} = \mathcal{K}'_{|K}$. Using again Lemma 9.4.1, we construct an assessment language L for $\mathcal{K}$ by taking letters from K systematically before letters from $Q \setminus K$. The word ω in L describing K must then be of the form

$$k_1 k_2 \dots k_m \bar{y}_1 \bar{y}_2 \dots \bar{y}_n$$

with $k_i \in K$ and $y_j \in Q \setminus K$. By our assumption (i), L is also an assessment language for $\mathcal{K}'$. The word ω describes some state K' in $\mathcal{K}'$. In view of the prefix $k_1 k_2 \dots k_m$ and of $\mathcal{K}_{|K} = \mathcal{K}'_{|K}$, we get $K \subseteq K'$. Then from the maximality of K we derive $K' \in \mathcal{K}$. Hence the word ω of L describes two distinct states of $\mathcal{K}$, namely K and K', a contradiction.

(ii) $\Rightarrow$ (i). This is trivial. □

9.4.3 Corollary. *The two following statements are equivalent:*

(i) $DEL(\mathcal{K}) = DEL(\mathcal{K}')$;
(ii) $\mathcal{K} = \mathcal{K}'$.

9.5 Original Sources and Related Works

The chapter closely follows Degreef, Doignon, Ducamp, and Falmagne (1986). This paper contains some additional, open questions about how languages allow to recover knowledge structures. We formulate one of these questions as Open Problem 18.1.2 in Chapter 18.

Problems

1. Verify all the numbers in Example 9.2.2.
2. Prove that for any collection $\mathcal{K}$ and any assessment language L for $\mathcal{K}$, there is a one-to-one correspondence between the states of $\mathcal{K}$ and the words of L (cf. Definition 9.2.3).
3. Describe an optimal decision tree for the ordinal knowledge structure derived from a linear order.
4. Give a proof of Theorem 9.2.5.
5. Give a proof of the following two equivalences (see after Definition 9.2.7)
$$\begin{aligned} L \in \mathrm{ASL}(\mathcal{K}) \quad &\Longleftrightarrow \quad \bar{L} \in \mathrm{ASL}(\overline{\mathcal{K}}), \\ L \in \mathrm{DEL}(\mathcal{K}) \quad &\Longleftrightarrow \quad \bar{L} \in \mathrm{DEL}(\overline{\mathcal{K}}). \end{aligned}$$
6. Prove that if L is a descriptive language for $\mathcal{K}$, then any language formed by arbitrarily changing the orders of the literals in the words of L is also a descriptive language for $\mathcal{K}$.
7. Prove Lemma 9.4.1.

10

Learning Spaces and Media

A 'medium' is a collection of transformations on a set of states, specified by two constraining axioms. The term "medium" stems from the original intuition suggesting such a structure, which is that of a system exposed to a bombardment of bits of information, each of which is capable of modifying its state in a minute way (Falmagne, 1997). The system could be, for example, an individual subjected to a barrage of messages from the media—that is, the press in all its incarnations—regarding the candidates in an election (see Regenwetter, Falmagne, and Grofman, 1999, for a good example). An account of such an application is Falmagne, Hsu, Leite, and Regenwetter (2007).

Media generalize learning spaces in the sense that any learning space can be represented as a particular kind of medium, called an 'oriented rooted medium.' The converse does not hold, however. For instance, a medium may be uncountable, while a learning space is necessarily finite. The link between media and learning spaces lies in the wellgradedness property of the latter. Indeed, it turns out that the states of a medium have a natural representation as the sets of a well-graded family. The transformations consist then in adding or removing some element from a set, thereby forming another state. The distinction between adding an element to a set (a positive transformation) and removing an element from a set (a negative transformation) makes the medium oriented.

More specifically, we will show that discriminative, well-graded families of sets and oriented media are cryptomorphic, in the sense that they can represent one another faithfully. The technical statement is Theorem 10.4.11.

A detailed presentation of media theory can be found in the monograph by Eppstein, Falmagne, and Ovchinnikov (2008), published under that title. We only cover the essentials in this chapter, which focusses of the relationship between well-graded collections (in particular, learning spaces) and media.

Examples of media abound and may differ widely. We describe some of them in our first section. We then give the basic definitions and the two axioms defining a medium. In the following sections, we derive some of the main results and we characterize the media representing learning spaces.

J.-C. Falmagne, J.-P. Doignon, *Learning Spaces*,
DOI 10.1007/978-3-642-01039-2_10, © Springer-Verlag Berlin Heidelberg 2011

10.1 Main Concepts of Media Theory

10.1.1 Examples. a) The family $\mathcal{P}$ of all partial orders on a finite set X. Each partial order is a state, and a transformation consists in adding or removing a pair xy from a partial order, whenever such a transformation results in another partial order in the family $\mathcal{P}$; otherwise, the transformation returns the same partial order. Thus, the transformations come in pairs, one of which can undo the action of the other. As seen in Chapter 4 (Remark 4.2.6(b) and Problem 8), the family $\mathcal{P}$ is well-graded. This implies that, for any two states P and R in $\mathcal{P}$, there is a minimal sequence of transformations converting P into R. There may be more than one sequence of such transformations, however. The medium is the pair $(\mathcal{P}, \mathcal{T})$ where $\mathcal{T}$ is the family of all the transformations.

b) The family $\mathcal{F}$ of all finite subsets of $\mathbb{R}$. This example is similar to the previous one: the sets in $\mathcal{F}$ are the states, a transformation consists in adding (or removing) a single element to (or from) a set in $\mathcal{F}$, and the family $\mathcal{F}$ is well-graded. The transformations are always effective here[1], and while the family $\mathcal{P}$ of the above example is finite, $\mathcal{F}$ is uncountable. However, the symmetric difference distance (cf. 1.6.12) between two finite sets is finite. This implies that, for any two distinct states S and T in $\mathcal{F}$, there is a finite, minimal sequence of transformations producing S from T.

Note that in this example and in the previous one, the medium is equipped with an 'orientation' in the sense that there is a natural partition of the collection of transformations into the two classes respectively gathering the 'addition' and 'removal' of elements to and from the sets. This feature does not apply to the next two examples.

c) The family $\mathcal{L}$ of all linear orders on a finite set. This family is not well-graded[2]. The states are the linear orders and a transformation consists in removing a pair xy of contiguous elements in a linear order, replacing it by the opposite pair yx. We denote by τ_{yx} such a transformation. In the special case of the collection of all the linear orders on the set $\{1, 2, 3, 4\}$, we thus have for example, with obvious notation,

$$4312 \xmapsto{\tau_{13}} 4132$$
$$2134 \xmapsto{\tau_{31}} 2314$$
$$3142 \xmapsto{\tau_{24}} 3124\,.$$

The graph of such a medium is represented in Figure 10.1, which is reproduced (with permission and with some added features pictured in red) from Eppstein et al. (2008, Figure 1.4). In the relevant literature, such a graph is sometimes referred to as a *permutohedron* (cf. Bowman, 1972; Gaiha and Gupta, 1977; Le Conte de Poly-Barbut, 1990; Ziegler, 1995). Note that we omit the loops.

---

[1] Except in the case of the empty set, from which no element can be removed.

[2] Rather, it is '2-graded'; cf. Definition 10.3.6.

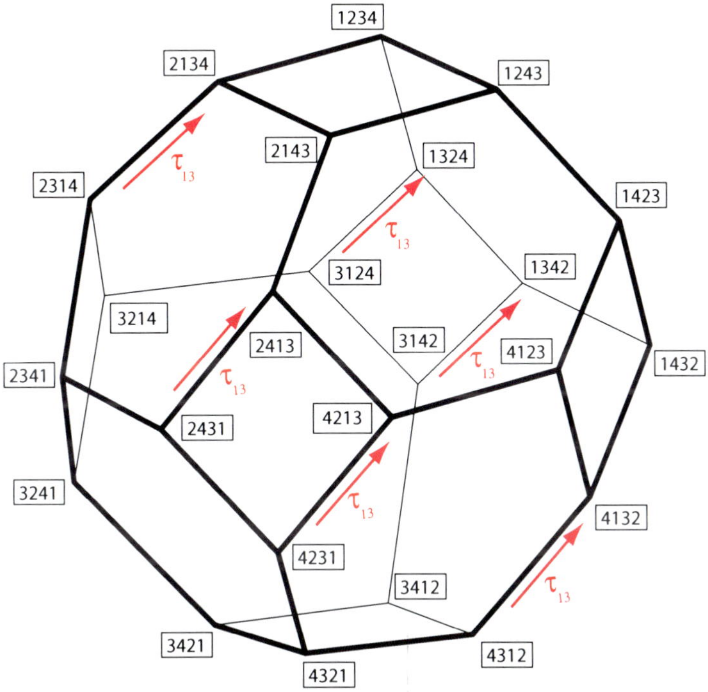

Figure 10.1. Permutohedron of $\{1, 2, 3, 4\}$. Graph of the medium of the set of linear orders on $\{1, 2, 3, 4\}$. Pictured in red, a representation of the transformation τ_{13}. Notice the location of these six arcs in the graph.

d) Let $\mathcal{H}$ be an arrangement of hyperplanes in $\mathbb{R}^n$, that is, a finite collection of hyperplanes in $\mathbb{R}^n$. The set $\mathbb{R}^n \setminus \cup\mathcal{H}$ is the union of open, convex polyhedral regions of $\mathbb{R}^n$ bounded by the hyperplanes. An example of a line arrangement in $\mathbb{R}^2$, with five lines, is pictured by Figure 10.2. Each of these polyhedral regions is a state of the medium, and a transformation consists in crossing a single hyperplane in $\mathcal{H}$. More precisely, to each hyperplane H in $\mathcal{H}$ correspond two ordered pairs (H_-, H_+) and (H_+, H_-) of half spaces bounded by H. These ordered pairs define two transformations τ_{H_-} and τ_{H_+}, where τ_{H_-} transforms a state in H_- bounded by H into an adjacent state in H_+, and τ_{H_+} has the opposite effect. Note that if some polyhedral region $X \subset H_-$ is not bounded by H, then the application of either τ_{H_+} or τ_{H_-} to X has no effect. Thus, while τ_{H_-} is capable of undoing the effect of τ_{H_+} and vice versa, the transformations are not mutual inverses. Indeed, if a region A satisfies $\tau_{H_+}(A) = B$ with $A \neq B$, then we also have $\tau_{H_+}(B) = B$ and the mapping τ_{H_+} cannot be reversed.

For other examples see Eppstein et al. (2008) (and Problem 1).

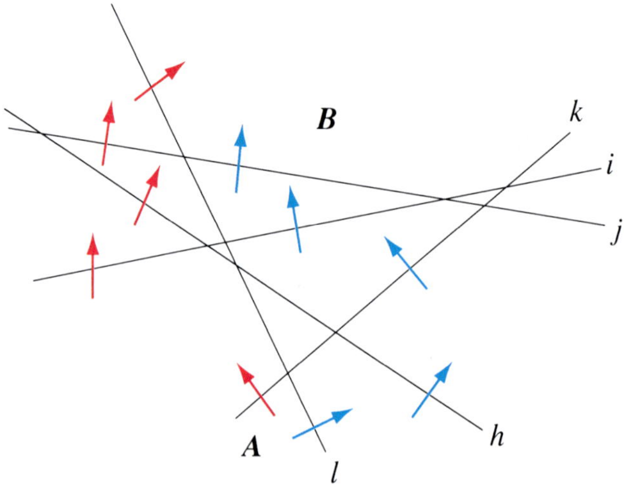

Figure 10.2. A line arrangement in the case of five straight lines l, h, i, j, and k in $\mathbb{R}^2$ delimiting sixteen states with ten pairs of transformations. Two direct paths from state A to state B cross these five lines in two different orders: $lhkij$ (pictured in blue) and $kihjl$ (in red).

10.1.2 Definition. Let $\mathcal{S}$ be a set of *states.* Let $\tau : S \mapsto S\tau$ be a function mapping $\mathcal{S}$ into itself. We thus write $S\tau = \tau(S)$, and

$$S\tau_1\tau_2 \cdots \tau_n = \tau_n(\cdots \tau_2(\tau_1(S)) \cdots)$$

for the function composition. Let $\mathcal{T}$ be a set of such functions on $\mathcal{S}$. The pair $(\mathcal{S}, \mathcal{T})$ is called a *token system* if it satisfies the following three conditions:

1. $|\mathcal{S}| \geq 2$;
2. $\mathcal{T} \neq \varnothing$;
3. the identity τ_0 on $\mathcal{S}$ does not belong to $\mathcal{T}$.

The functions in $\mathcal{T}$ are called *tokens.* So, the identity on $\mathcal{S}$ is not a token.

A state V is *adjacent* to a state S if $S \neq V$ and $S\tau = V$ for some token τ in $\mathcal{T}$. A token $\tilde{\tau}$ is a *reverse* of a token τ if for any two adjacent states S and V, the following equivalence holds:

$$S\tau = V \iff V\tilde{\tau} = S. \tag{10.1}$$

It is easily verified that a token has at most one reverse. If the reverse $\tilde{\tau}$ of τ exists, then τ and $\tilde{\tau}$ are mutual reverses. We thus have $\tilde{\tilde{\tau}} = \tau$, and adjacency is a symmetric relation on $\mathcal{S}$ (Problem 3).

The next definition introduces a convenient language. The key concept is that of a 'message' which is a composition of tokens permitting to transform a state into a not necessarily adjacent one.

10.1.3 Definition. A *message* in a token system $(\mathcal{S}, \mathcal{T})$ is a (possibly empty) string of tokens from $\mathcal{T}$. A nonempty message $\boldsymbol{m} = \tau_1 \dots \tau_n$ defines a function $S \mapsto S\tau_1 \cdots \tau_n$ mapping $\mathcal{S}$ into itself. Note that, by abuse of notation, we also write then $\boldsymbol{m} = \tau_1 \cdots \tau_n$ for the corresponding function composition. When $S\boldsymbol{m} = V$ for some states S, V and some message $\boldsymbol{m}$, we say that $\boldsymbol{m}$ *produces* V from S. The *content* of a message $\boldsymbol{m} = \tau_1 \cdots \tau_n$ is the set $\mathcal{C}(\boldsymbol{m}) = \{\tau_1, \dots, \tau_n\}$ of its tokens. We write $\ell(\boldsymbol{m}) = n$ to denote the *length* of the message $\boldsymbol{m}$. (We thus have $|\mathcal{C}(\boldsymbol{m})| \leq \ell(\boldsymbol{m})$.) A message $\boldsymbol{m}$ is *effective* (resp. *ineffective*) for a state S if $S\boldsymbol{m} \neq S$ (resp. $S\boldsymbol{m}{=}S$) for the corresponding function $S \mapsto S\boldsymbol{m}$. A message $\boldsymbol{m} = \tau_1 \cdots \tau_n$ is *stepwise effective* for S if $S\tau_1 \cdots \tau_k \neq S\tau_0 \cdots \tau_{k-1}$, $1 \leq k \leq n$. A message which is both stepwise effective and ineffective for some state is called a *return message* or, more briefy, a *return* (for that state).

We say that a message $\boldsymbol{m} = \tau_1 \cdots \tau_n$ is *inconsistent* if it contains both a token and its reverse, that is, if $\tau_j = \tilde{\tau}_i$ for some distinct indices i and j; otherwise, it is called *consistent*. A message consisting of a single token is thus consistent by default. Two messages **m** and **n** are *jointly consistent* if **mn** (or, equivalently, **nm**) is consistent. A consistent message, with no token occurring more than once, which is stepwise effective for some state S is said to be *concise* (for S). A message $\boldsymbol{m} = \tau_1 \cdots \tau_n$ is *vacuous* if its set of indices $\{1, \dots, n\}$ can be partitioned into pairs $\{i, j\}$, such that τ_i and τ_j are mutual reverses. The *reverse* of a message $\boldsymbol{m} = \tau_1 \cdots \tau_n$ is defined by $\widetilde{\boldsymbol{m}} = \tilde{\tau}_n \cdots \tilde{\tau}_1$.

For convenience, we may sometimes say that "$\boldsymbol{m}$ is the *empty message*" to mean that $\boldsymbol{m} = \tau_0$, the identity on $\mathcal{S}$ (even though the identity is not a token). In all such cases, $\boldsymbol{m}$ is a place holder symbol that can be deleted, as in: 'let $\boldsymbol{mn}$ be a message in which $\boldsymbol{m}$ is either a concise message or is empty (that is $\boldsymbol{mn} = \boldsymbol{n}$)'.

10.1.4 Axioms for media. A token system $(\mathcal{S}, \mathcal{T})$ is called a *medium (on $\mathcal{S}$)* if the following two axioms are satisfied.

[Ma] For any two distinct states S, V in $\mathcal{S}$, there is a concise message producing V from S.

[Mb] Any return message is vacuous.

A medium $(\mathcal{S}, \mathcal{T})$ is *finite* if $\mathcal{S}$ is a finite set. We leave to the reader to show that these two axioms are independent (Problem 4).

Except when stated otherwise, all statements from now on concern a medium $(\mathcal{S}, \mathcal{T})$. Thus, Axioms [Ma] and [Mb] are implicitly assumed to hold.

10.2 Some Basic Lemmas

We omit the proofs of some straightforward facts, such as those gathered in the next lemma (see Problem 5).

10.2.1 Lemma. (i) *Each token has a unique reverse.*

(ii) *If a message* $\boldsymbol{m}$ *is stepwise effective for* S*, then* $S\boldsymbol{m} = V$ *implies* $V\widetilde{\boldsymbol{m}} = S$.

(iii) $\tau \in \mathcal{C}(\boldsymbol{m})$ *if and only if* $\tilde{\tau} \in \mathcal{C}(\widetilde{\boldsymbol{m}})$.

(iv) *If* $\boldsymbol{m}$ *is consistent, so is* $\widetilde{\boldsymbol{m}}$.

10.2.2 Lemma. *Suppose that* $T\boldsymbol{n} = V\boldsymbol{m}$*, with* $\boldsymbol{m}$ *and* $\boldsymbol{n}$ *consistent, stepwise effective messages for the states* T *and* V*, respectively, and with* T *not necessarily distinct from* V*. Then,* $\boldsymbol{n}$ *and* $\boldsymbol{m}$ *are jointly consistent.*

PROOF. If $T \neq V$, we know from Axiom [Ma] that there is a concise message $\boldsymbol{w}$ producing T from V. Thus $\boldsymbol{n}\widetilde{\boldsymbol{m}}\boldsymbol{w}$ is a return for T, which must be vacuous by [Mb]. If $\boldsymbol{nm}$ is not consistent, there is some token $\tau \in \mathcal{C}(\boldsymbol{n}) \cap \mathcal{C}(\widetilde{\boldsymbol{m}})$. But as $\boldsymbol{n}\widetilde{\boldsymbol{m}}\boldsymbol{w}$ is vacuous and each of $\boldsymbol{n}$ and $\boldsymbol{m}$ is consistent, the token $\tilde{\tau}$ must appear at least twice in $\boldsymbol{w}$, contradicting the conciseness of $\boldsymbol{w}$.

If $T = V$, the message $\boldsymbol{n}\widetilde{\boldsymbol{m}}$ is a return for T which, by [Mb], must be vacuous. Suppose that $\boldsymbol{nm}$ is inconsistent. There is then some token τ that occurs in $\boldsymbol{n}$ and in $\widetilde{\boldsymbol{m}}$. Since $\boldsymbol{n}\widetilde{\boldsymbol{m}}$ is vacuous, the token $\tilde{\tau}$ must occur in $\boldsymbol{n}\widetilde{\boldsymbol{m}}$. This is impossible since $\boldsymbol{n}$ and $\widetilde{\boldsymbol{m}}$ are consistent. □

10.2.3 Lemma. (i) *No token is identical to its own reverse.*

(ii) *Any consistent message which is stepwise effective for some state is concise.*

(iii) *For any two adjacent states* S *and* V*, there is exactly one token producing* V *from* S.

(iv) *Let* $\boldsymbol{m}$ *be a message that is concise for some state, then*

$$\ell(\boldsymbol{m}) = |\mathcal{C}(\boldsymbol{m})|\,, \tag{10.2}$$

and

$$\mathcal{C}(\boldsymbol{m}) \cap \mathcal{C}(\widetilde{\boldsymbol{m}}) = \varnothing. \tag{10.3}$$

(v) *No token* τ *can be a bijection. Moreover, if* $S\tau = V$ *with* S, V *two distinct states, then* $V\tau = V$.

(vi) *Suppose that* $\boldsymbol{m}$ *and* $\boldsymbol{n}$ *are stepwise effective for* S *and* V*, respectively, with* $S\boldsymbol{m} = V$ *and* $V\boldsymbol{n} = W$*. Then* $\boldsymbol{mn}$ *is stepwise effective for* S*, with* $S\boldsymbol{mn} = W$.

(vii) *Any vacuous message which is stepwise effective for some state is a return message for that state*[3].

[3] Statement (vii) is thus a partial converse of Axiom [Mb].

The proofs of (i), (iii), (iv) and (vi) are short and left as Problem 6.

Proof. (ii) Suppose that $\boldsymbol{m}$ is a consistent, stepwise effective message producing V from S, and that some token τ occurs at least twice in $\boldsymbol{m}$. We thus have $S\boldsymbol{m} = \boldsymbol{n}_1\tau\boldsymbol{n}_2\tau\boldsymbol{n}_3 = V$ for some consistent stepwise effective messages $\boldsymbol{n}_1\tau$ and $\boldsymbol{n}_2\tau\boldsymbol{n}_3$. One or more of $\boldsymbol{n}_1$, $\boldsymbol{n}_2$ and $\boldsymbol{n}_3$ may of course be empty. On the other hand, τ may occur more than twice in $\boldsymbol{m}$. Without loss of generality, we may suppose that τ occurs exactly twice in $\boldsymbol{n}_1\tau\boldsymbol{n}_2\tau$. We thus have $S\boldsymbol{n}_1\tau\boldsymbol{n}_2\tau = V'$, for some state V'. We conclude that the messages $\boldsymbol{n}_1\tau$ and $\tilde{\tau}\tilde{\boldsymbol{n}}_2$ are consistent, stepwise effective messages producing the state $W = S\boldsymbol{n}_1\tau = V'\tilde{\tau}\tilde{\boldsymbol{n}}_2$ from S and V', respectively. These two messages are not jointly consistent, contradicting Lemma 10.2.2.

(v) Suppose that $S\tau = V$ for some token τ and two distinct states S and V. If $V\tau = W \neq V$ for some state W, then $V = S\tau = W\tilde{\tau}$, a contradiction of Lemma 10.2.2, because, by definition, τ is a consistent message. Hence, $S\tau = V\tau = V$, and so τ is not a bijection.

(vii) Let $\boldsymbol{m}$ be a vacuous message which is stepwise effective for some state S, with $S\boldsymbol{m} = V$. If $S \neq V$, then Axiom [Ma] implies that there is a concise message $\boldsymbol{n}$ producing S from V. Thus, $\boldsymbol{mn}$ is a return for S, which must be vacuous by [Mb]. Since $\boldsymbol{m}$ is vacuous, $\boldsymbol{n}$ must be vacuous. This cannot be true because $\boldsymbol{n}$ is concise. □

In the next section, we describe the states of a medium in terms of the consistent messages producing them. This is the first step toward representing the states of a medium by the sets forming some well-graded family.

10.3 The Content of a State

10.3.1 Definition. For any state S in a medium $(\mathcal{S}, \mathcal{T})$, we define the *(token) content* of S as the set $\widehat{S}$ of all the tokens contained in at least one concise message producing S. Formally, we thus have

$$\widehat{S} = \bigcup\{\mathcal{C}(\boldsymbol{m}) \,|\, \boldsymbol{m} \text{ is a concise message producing } S\}.$$

We refer to the family $\widehat{\mathcal{S}}$ of all the contents of the states in $\mathcal{S}$ as the *content family* of the medium $(\mathcal{S}, \mathcal{T})$.

10.3.2 Example. In the medium formed by the collection of the 4! linear orders on the set $\{1, 2, 3, 4\}$ which is represented by its permutohedron in Figure 10.1, the content of the state 2134 is the set of transformations

$$\widehat{2134} = \{\tau_{21}, \tau_{23}, \tau_{24}, \tau_{13}, \tau_{14}, \tau_{34}\},$$

in which each token corresponds to one of the ordered pairs contained in the linear order $2 < 1 < 3 < 4$. The content of the state 2143, which is adjacent to 2134 in the graph of Figure 10.1, is almost the same, namely

$$\widehat{2143} = \{\tau_{21}, \tau_{23}, \tau_{24}, \tau_{13}, \tau_{14}, \tau_{43}\},$$

that is, the token τ_{34} of $\widehat{2134}$ has been substituted by its reverse τ_{43}. For these two adjacent states, we thus have

$$d(\widehat{2134}, \widehat{2143}) = 2$$

where d stands as usual for the symmetric difference distance between sets (cf. 1.6.12). Notice also that $|\widehat{2134}| = 6$, which is half of the total number of tokens in this example (that is, $12 = 4 \cdot 3$).

We omit the proof of the next theorem, which generalizes these observations (see Problem 7).

10.3.3 Theorem. *For any state S and any token τ, we have either $\tau \in \widehat{S}$ or $\tilde{\tau} \in \widehat{S}$ (but not both). This implies that $|\widehat{S}| = |\widehat{V}|$ for any two states S and V. Moreover, if the states S and T are adjacent, then $d(\hat{S}, \hat{T}) = 2$. Finally, if the set of states $\mathcal{S}$ is finite, then $|\widehat{S}| = |\mathcal{T}|/2$ for any $S \in \mathcal{S}$.*

The following result is in the same vein but considers states which are not necessarily adjacent.

10.3.4 Theorem. *If $S\boldsymbol{m} = V$ for some nonempty, concise message $\boldsymbol{m}$ (thus $S \neq V$), then $\widehat{V} \setminus \widehat{S} = \mathcal{C}(\boldsymbol{m})$, and so $\widehat{V} \triangle \widehat{S} = \mathcal{C}(\boldsymbol{m}) + \mathcal{C}(\widetilde{\boldsymbol{m}}) \neq \varnothing$.*

PROOF. Since $\widehat{V}$ contains all the tokens from concise messages producing V, we necessarily have $\mathcal{C}(\boldsymbol{m}) \subseteq \widehat{V}$. As we also have $V\widetilde{\boldsymbol{m}} = S$, the same argument yields $\mathcal{C}(\widetilde{\boldsymbol{m}}) \subseteq \widehat{S}$. By Theorem 10.3.3, $\widehat{S}$ cannot contain both a token and its reverse; so $\mathcal{C}(\boldsymbol{m}) \subseteq \widehat{V} \setminus \widehat{S}$.

Turning to the converse inclusion, suppose that $\tau \in \widehat{V} \setminus \widehat{S}$ for some token τ. Thus, τ occurs in some concise message producing V. Without loss of generality, we may assume that $W\tau\boldsymbol{n} = V$ for some state W, with $\tau\boldsymbol{n}$ concise. Suppose that $W \neq S$ and let $\boldsymbol{q}$ be a concise message producing S from W. As the message $\boldsymbol{m}\tilde{\boldsymbol{n}}\tilde{\tau}\boldsymbol{q}$ is a return for S, it must be vacuous by [Mb]. Thus, we must have

$$\tau \in \mathcal{C}(\boldsymbol{m}) \cup \mathcal{C}(\tilde{\boldsymbol{n}}) \cup \mathcal{C}(\boldsymbol{q}).$$

We cannot have either $\tau \in \mathcal{C}(\boldsymbol{q})$ (because this would imply $\tau \in \widehat{S}$), or $\tau \in \mathcal{C}(\tilde{\boldsymbol{n}})$ (because this would yield $\tau, \tilde{\tau} \in \mathcal{C}(\tau\boldsymbol{n})$, with $\tau\boldsymbol{n}$ concise, a contradiction). We conclude that $\tau \in \mathcal{C}(\boldsymbol{m})$, and so $\hat{V} \setminus \hat{S} \subseteq \mathcal{C}(\boldsymbol{m})$. The case $W = S$ is straightforward.

The last equation of the theorem results from (10.3) in Lemma 10.2.3(iv). □

It is now easily shown that, in a medium, the content of a state defines that state (Problem 9).

10.3.5 Theorem. *For any two states S and V, we have*

$$S = V \iff \widehat{S} = \widehat{V}. \tag{10.4}$$

The statement of the last result of this section requires a definition.

10.3.6 Definition. A family of sets $\mathcal{F}$ is *2-graded* if, for any two distinct sets S and V in $\mathcal{F}$, there is a positive integer n and a sequence

$$S_0 = S, S_1, \ldots, S_n = V$$

of sets in $\mathcal{F}$ such that, for $1 \leq i \leq n$, we have $d(\hat{S}_{i-1}, \hat{S}_i) = 2$, and moreover $d(\hat{S}, \hat{V}) = 2n$.

10.3.7 Theorem. *The content family of a medium is 2-graded.*

PROOF. For any two distinct states S and V in a medium, there exists by [Ma] a concise message $\boldsymbol{m} = \tau_1\tau_2 \cdots \tau_n$ producing V from S. We write $S_0 = S$ and $S_i = S_{i-1}\tau_i$ for $1 \leq i \leq n$. Note that $S_{i-1} \neq S_i$ because $\boldsymbol{m}$ is stepwise effective. Thus, S_{i-1} and S_i are adjacent and so, by Theorem 10.3.3, $d(\hat{S}_{i-1}, \hat{S}_i) = 2$ for $1 \leq i \leq n$. The fact that $d(\hat{S}, \hat{V}) = 2n$ is an immediate consequence of the property $\widehat{S} \triangle \widehat{V} = \mathcal{C}(\boldsymbol{m}) + \mathcal{C}(\widetilde{\boldsymbol{m}}) \neq \varnothing$ of Theorem 10.3.4. □

The main goal of this chapter is the representation of a learning space by a particular kind of medium, with the states of the learning space corresponding to those of the medium. As a learning space is well-graded, Theorem 10.3.7 suggests a possible device. The tokens of a medium come in pairs of mutual reverses. So, we could arbitrarily choose one token in each pair to represent the addition of an item to a state (of the learning space to be constructed), and its reverse to the opposite operation.

With this idea in mind, we reconsider our introductory Example 10.1.1 (a), which dealt with the medium of all the partial orders on a finite set.

10.3.8 Example. Let $\mathcal{P}_3$ be the collection of the partial orders on $\{a, b, c\}$, including the partial order ι consisting only of loops. We take $\mathcal{P}_3$ to be the set of states of a token system, with the transformations consisting in adding/removing an ordered pair of distinct elements in $\{a, b, c, d\}$ to/from the partial orders in $\mathcal{P}_3$ (whenever possible, that is, whenever producing a partial order). We thus have for any partial order $P \in \mathcal{P}_3$ and any distinct $x, y \in \{a, b, c\}$,

$$P\tau_{xy} = \begin{cases} P \cup \{xy\} & \text{if } xy \notin P \text{ and } P \cup \{xy\} \in \mathcal{P}_3, \\ P & \text{otherwise;} \end{cases} \tag{10.5}$$

$$P\tilde{\tau}_{xy} = \begin{cases} P \setminus \{xy\} & \text{if } xy \in P \text{ and } P \setminus \{xy\} \in \mathcal{P}_3, \\ P & \text{otherwise.} \end{cases} \tag{10.6}$$

We write $\mathcal{T}_3$ for the set of all such transformations on $\mathcal{P}_3$. The pair $(\mathcal{P}_3, \mathcal{T}_3)$ is a medium (cf. Problem 10). Our goal is to examine the contents of the states in $\mathcal{P}_3$. The digraph of this medium is drawn in Figure 10.3, with the centrifugal arcs pictured by the red arrows. They represent the addition of a pair to a partial order. The transformations corresponding to each arc are indicated, also in red.

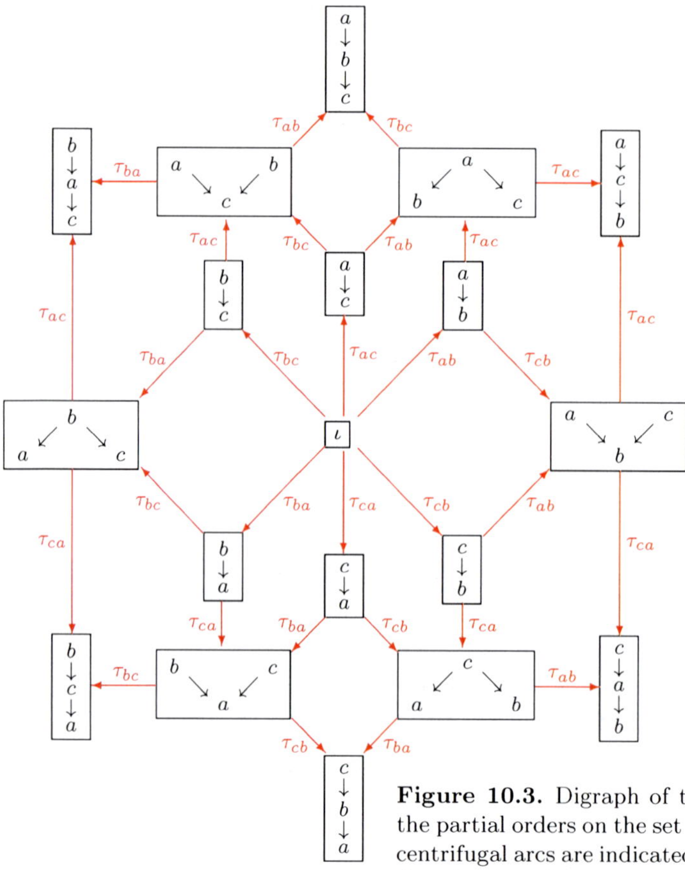

Figure 10.3. Digraph of the medium of all the partial orders on the set $\{a, b, c\}$. Only the centrifugal arcs are indicated. Each represents the addition of a pair to a partial order.

For the purpose of this example, anticipating on definitions to come, let us call such transformations 'positive'. In the same vein, the 'positive content' of a state is the subset of positive transformations in the content of that state.

Figure 10.4 displays a digraph having as nodes the contents of the states (with positive content easily identifiable), and arcs derived in a natural way from the positive tokens. This digraph is isomorphic to the digraph of Figure 10.3.

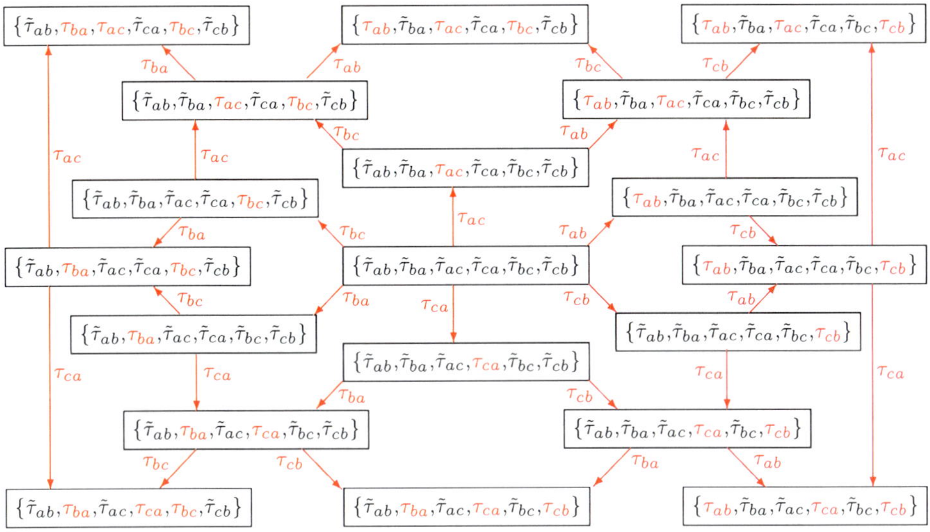

Figure 10.4. Digraph of state contents for the medium of all the partial orders on the set $\{a, b, c\}$. This digraph is isomorphic to that of Figure 10.3. Here, the vertices are marked by the contents of the states. A token printed in red correspond to the addition of a pair to a partial order (thus, a positive transformation). The linear orders are represented in the top and the bottom rows.

The collection of the positive contents of states in this example is easily seen to be well-graded, considering: (i) the wellgradednes of the family $\mathcal{P}_3$; and (ii) the isomorphism between the digraphs represented in Figures 10.3 and 10.4. It also contains the empty set, because the positive content of the partial order ι is empty. However, this collection does not contain the union of all the positive contents, that is, the state $\{\tau_{ab}, \tau_{ba}, \tau_{ac}, \tau_{ca}, \tau_{bc}, \tau_{cb}\}$. Thus, it is not closed under union and so is not a learning space.

This example prompts the following observation. The set of tokens of certain media can be partitioned into two classes, respectively called 'positive' and 'negative', in such a way that a token is positive if and only if its reverse is negative. Thus, the two classes are of equal size. The next section presents the relevant theory, starting with the construction of an 'oriented' medium for any discriminative, well-graded family. By the end of the section, a stronger result establishes a natural correspondence between the collection of discriminative, well-graded families and the collection of 'oriented' media (Theorem 10.4.11). Section 10.5 will then derive similar results for learning spaces and a specific subclass of the 'oriented' media.

10.4 Oriented Media

The medium constructed from a family of partial orders in Example 10.3.8 was endowed with a natural 'orientation' dictated by the nature of the tokens. There are many such examples of media arising from well-graded families of relations[4]. We have two results generalizing these examples. The first one below essentially says that a medium can be built from any discriminative well-graded family. The second one (Theorem 10.4.3) states that the resulting medium is 'oriented', in the sense of the next Definition 10.4.2.

Notice that a well-graded family $\mathcal{K}$ is discriminative if and only if $|\cap\mathcal{K}| \leq 1$ (see Problem 4 in Chapter 4).

10.4.1 Theorem. *Let $\mathcal{K}$ be a discriminative and well-graded family of sets, with $|\mathcal{K}| \geq 2$. Set $X = \cup\mathcal{K}\setminus\cap\mathcal{K}$, $\mathcal{S} = \mathcal{K}$ and let $\mathcal{T}$ consist of all transformations τ_q and $\tau_{\tilde{q}}$ of $\mathcal{S}$, for all $q \in X$, where, for all $K \in \mathcal{S}$:*

$$K\tau_q = \begin{cases} K \cup \{q\} & \text{if } q \notin K \text{ and } K \cup \{q\} \in \mathcal{K}, \\ K & \text{otherwise}, \end{cases} \tag{10.7}$$

$$K\tau_{\tilde{q}} = \begin{cases} K \setminus \{q\} & \text{if } q \in K \text{ and } K \setminus \{q\} \in \mathcal{K}, \\ K & \text{otherwise}. \end{cases} \tag{10.8}$$

Then $(\mathcal{S}, \mathcal{T})$ is a medium in which τ_q and $\tau_{\tilde{q}}$ are mutual reverses; so, $\tau_{\tilde{q}} = \tilde{\tau}_q$.

PROOF. By definition, all τ_q and $\tau_{\tilde{q}}$ are transformations on $\mathcal{S}$. We prove that none of them is the identity. Take any q in X. There thus exist sets K and L in $\mathcal{K}$ such that $q \in K$ and $q \notin L$. By assumption, there is a tight path $K = K_0, K_1, \ldots, K_h = L$. Let i be the smallest index such that $q \notin K_i$. Then $K_i\tau_q = K_{i-1}$ and $K_{i-1}\tau_{\tilde{q}} = K_i$. Thus $(\mathcal{S}, \mathcal{T})$ is a token system. Let us now show that $(\mathcal{S}, \mathcal{T})$ satisfies the axioms for a medium (Definition 10.1.4). If K, L are distinct states in $\mathcal{S}$, there is by our assumption a tight path $K = K_0$, $K_1, \ldots, K_h = L$. If $K_i \Delta K_{i-1} = \{q\}$, we have $K_i\tau_q = K_{i-1}$ or $K_i\tau_{\tilde{q}} = K_{i-1}$ (according to $q \notin K_i$ or $q \in K_i$). Consequently, there is a concise message from K to L, and Axiom [Ma] thus holds. To prove Axiom [Mb], notice that any return message to some state K must add and delete the same number of times a given element of X. □

The next definition specifies the concept of an orientation for media in general.

10.4.2 Definition. An *orientation* of a medium $\mathcal{M} = (\mathcal{S}, \mathcal{T})$ is a partition of its set of tokens $\mathcal{T}$ into two classes $\mathcal{T}^+$ and $\mathcal{T}^-$ respectively called *positive* and *negative* such that, for any $\tau \in \mathcal{T}$, we have

$$\tau \in \mathcal{T}^+ \iff \tilde{\tau} \in \mathcal{T}^- .$$

[4] Some of them are dealt with in Problems 11, 12 and 13.

The medium $\mathcal{M}$ is said to be *oriented* by the partition $\{\mathcal{T}^+, \mathcal{T}^-\}$. The tokens belonging to the class $\mathcal{T}^+$ (resp. $\mathcal{T}^-$) are then called *positive* (resp. *negative*).

An oriented medium $\mathcal{M} = (\mathcal{S}, \mathcal{T})$ will be implicitly taken to have its orientation denoted by $\{\mathcal{T}^+, \mathcal{T}^-\}$. The *positive* (resp. *negative*) *content* of a state S is the set $\widehat{S}^+ = \widehat{S} \cap \mathcal{T}^+$ (resp. $\widehat{S}^- = \widehat{S} \cap \mathcal{T}^-$) of the positive (resp. negative) tokens in $\widehat{S}$. The two families

$$\widehat{\mathcal{S}}^+ = \{\widehat{S}^+ \mid S \in \mathcal{S}\} \quad \text{and} \quad \widehat{\mathcal{S}}^- = \{\widehat{S}^- \mid S \in \mathcal{S}\}. \tag{10.9}$$

are called the *positive content family* and *negative content family* of $\mathcal{M}$, respectively. A message containing only positive (resp. negative) tokens is called *positive* (resp. *negative*).

Note that any medium can be oriented, and that a finite medium $(\mathcal{S}, \mathcal{T})$ can be given $2^{|\mathcal{T}|/2}$ different orientations (Problem 15).

We now show that the medium built from any discriminative well-graded family in Theorem 10.4.1 is endowed with a natural orientation. We keep the notation of that theorem.

10.4.3 Theorem. *Let $\mathcal{K}$ be a discriminative and well-graded family of sets, with $|\mathcal{K}| \geq 2$. The medium $(\mathcal{S}, \mathcal{T})$ constructed in Theorem 10.4.1 is oriented by the following partition:*

$$\mathcal{T}^+ = \{\tau_q \in \mathcal{T} \mid q \in X\}, \qquad \mathcal{T}^- = \{\tilde{\tau}_q \in \mathcal{T} \mid q \in X\}. \tag{10.10}$$

PROOF. That $\{\mathcal{T}^+, \mathcal{T}^-\}$ is a partition of $\mathcal{T}$ is clear, as well as the other requirements in Definition 10.4.2. □

We now turn to the converse construction which involves the manufacture of a well-graded family from an oriented medium. As a consequence of Theorems 10.3.3 and 10.3.5, we have:

10.4.4 Theorem. *The following equivalence holds for any two states S and V of an oriented medium $(\mathcal{S}, \mathcal{T})$:*

$$S = V \iff \widehat{S}^+ = \widehat{V}^+.$$

By symmetry, a similar equivalence also holds for the negative contents.

PROOF. The necessity is immediate. For the sufficiency, suppose that both $\widehat{S}^+ = \widehat{V}^+$ and $\widehat{S}^- = \widehat{V}^-$ hold. Thus, we must have

$$\widehat{S} = \widehat{S}^+ + \widehat{S}^- = \widehat{V}^+ + \widehat{V}^- = \widehat{V},$$

and so $S = V$ by Theorem 10.3.5. It suffices to prove that $\widehat{S}^+ = \widehat{V}^+$ implies $\widehat{S}^- = \widehat{V}^-$. Suppose that $\widehat{S}^+ = \widehat{V}^+$. We have successively, for any $\tau \in \mathcal{T}$,

$$\begin{aligned} \tau \in \widehat{S}^- &\iff \tilde{\tau} \in \mathcal{T}^+ \setminus \widehat{S}^+ && \text{(by Theorem 10.3.3)} \\ &\iff \tilde{\tau} \in \mathcal{T}^+ \setminus \widehat{V}^+ && \text{(because } \widehat{S}^+ = \widehat{V}^+\text{)} \\ &\iff \tau \in \widehat{V}^- && \text{(by Theorem 10.3.3).} \end{aligned}$$

□

10.4.5 Theorem. *If S and V are two distinct states in an oriented medium, with $S\boldsymbol{m} = V$ for some positive concise message $\boldsymbol{m}$, then $\widehat{V}^+ = \widehat{S}^+ + \mathcal{C}(\boldsymbol{m})$.*

The proof is left as Problem 16. In some situations, a natural orientation is dictated by the the specific structure of a medium.

One organizing principle for such an orientation is described in the next definition.

10.4.6 Definition. A state R in an oriented medium is a *root* if any concise message producing any other state S from R is positive. An oriented medium having a root is said to be *rooted*. By abuse of language, we may say that the orientation of a medium is *rooted* (at the state R) if the medium is rooted for that orientation and R is the root.

10.4.7 Theorem. *In an oriented medium, a state R is a root if and only if $\widehat{R}^+ = \varnothing$. An oriented medium has at most one root and does not necessarily have one.*

Proof. (Sufficiency.) Let R be a state with an empty positive content, that is $\widehat{R}^+ = \varnothing$. Suppose that $\boldsymbol{m}$ is a concise message such that $R\boldsymbol{m} = S$; this implies $S\widetilde{\boldsymbol{m}} = R$. By Theorem 10.3.4, we have $\widehat{R} \setminus \widehat{S} = \mathcal{C}(\widetilde{\boldsymbol{m}})$. So, if $\widetilde{\boldsymbol{m}}$ contains a positive token, then $\widehat{R}^+ \neq \varnothing$. Thus, $\widetilde{\boldsymbol{m}}$ is negative, and $\boldsymbol{m}$ is positive. Because this holds for any message $\boldsymbol{m}$ effective for R, the state R must be a root.

(Necessity.) If R is a root, then by Theorem 10.4.5 and the definition of a root, we have $\widehat{R}^+ \subset \widehat{S}^+$ for any state $S \neq R$. Suppose that $\widehat{R}^+$ contains some positive token τ. Then τ belongs to the contents of all the states. This implies that $\tilde{\tau}$ is not effective for any state; so, $\tilde{\tau}$ is the identity function τ_0, which is not a token by Definition 10.1.2. We conclude that $\widehat{R}^+$ must be empty.

Since by Theorem 10.3.5 a state is defined by its content, an oriented medium can have at most one root. We leave it to the reader to construct an oriented medium without a root (Problem 17). □

The two results in the next theorem are straightforward. We omit the short proofs. (Problem 18).

10.4.8 Theorem. *The following two statements are true for any medium.*

(i) *Any state R can be made a root by defining a suitable orientation.*

(ii) *There exists an orientation ensuring that the positive contents of all the states are finite sets. In particular, the states of any rooted medium have finite positive contents.*

10.4.9 Theorem. *The family $\widehat{\mathcal{S}}^+$ of all positive contents in an oriented medium $(\mathcal{S}, \mathcal{T})$ is well-graded and satisfies both $\cup\widehat{\mathcal{S}}^+ = \mathcal{T}^+$ and $\cap\widehat{\mathcal{S}}^+ = \varnothing$. Thus, $\widehat{\mathcal{S}}^+$ is also discriminative.*

Proof. Take any two distinct $\widehat{S}^+$, $\widehat{V}^+ \in \mathcal{S}^+$. From Theorem 10.3.7, we know that the content family of $(\mathcal{S}, \mathcal{T})$ is 2-graded. As in the proof of the latter theorem, if $\tau_1 \dots \tau_n$ is a concise message producing V from S, with

$$S = S_0,\ S_0\tau_1 = S_1,\ S_1\tau_2 = S_2,\ \dots,\ S_{n-1}\tau_n = S_n = V, \qquad (10.11)$$

then

$$\widehat{S}_i \setminus \widehat{S}_{i-1} = \{\tau_i\} \text{ and } \widehat{S}_{i-1} \setminus \widehat{S}_i = \{\tilde{\tau}_i\} \qquad \text{for } 1 \le i \le n. \qquad (10.12)$$

Moreover, we must have $d(\widehat{S}, \widehat{V}) = 2n$. The sequence (10.11) induces the corresponding sequence $\widehat{S}^+ = \widehat{S}_0^+, \widehat{S}_1^+, \dots, \widehat{S}_n^+ = \widehat{V}^+$. By Theorem 10.3.3, exactly one of the τ_i and $\tilde{\tau}_i$ in (10.12) is positive. We thus have

$$\text{either} \quad \widehat{S}_i^+ \setminus \widehat{S}_{i-1}^+ = \{\tau_i\} \subseteq \mathcal{T}^+ \quad \text{or} \quad \widehat{S}_{i-1}^+ \setminus \widehat{S}_i^+ = \{\tilde{\tau}_i\} \subseteq \mathcal{T}^+,$$

and so $d(\widehat{S}_i^+, \widehat{S}_{i+1}^+) = 1$, for $1 \le i \le n$, with $d(\widehat{S}^+, \widehat{V}^+) = d(\widehat{S}, \widehat{V})/2 = n$. Hence $\mathcal{S}^+$ is well-graded.

For any $\tau^+ \in \mathcal{T}^+$, there are distinct S and T in $\mathcal{S}$ such that $S\tau^+ = T$; so $\tau^+ \in \widehat{T}$, yielding $\cup\widehat{\mathcal{S}}^+ = \mathcal{T}^+$. The fact that the family $\widehat{\mathcal{S}}^+$ satisfies $\cap\widehat{\mathcal{S}}^+ = \varnothing$ results easily from the definition of the positive content of a state. Finally, we leave to the reader to establish that $\widehat{\mathcal{S}}^+$ is discriminative (see Problem 19). □

We turn to one of the two main results of this chapter. It formalizes how discriminative well-graded families and oriented media are cryptomorphic structures. The statement of this result relies on some simple concepts introduced in the next definition[5].

10.4.10 Definition. Two families of sets, $\mathcal{K}$ and $\mathcal{L}$, are *isomorphic*, which is denoted by $\mathcal{K} \sim \mathcal{L}$, if there exists a bijective mapping

$$a:\ (\cup\mathcal{K}) \setminus (\cap\mathcal{K}) \to (\cup\mathcal{L}) \setminus (\cap\mathcal{L})$$

satisfying

$$a(\{K \setminus \cap\mathcal{K} \,|\, K \in \mathcal{K}\}) \;=\; \{L \setminus \cap\mathcal{L} \,|\, L \in \mathcal{L}\}.$$

Two token systems $(\mathcal{S}, \mathcal{T})$ and $(\mathcal{U}, \mathcal{V})$ are *isomorphic*, which we denoted by $(\mathcal{S}, \mathcal{T}) \sim (\mathcal{U}, \mathcal{V})$, if there exist two bijective mappings

$$b:\ \mathcal{S} \to \mathcal{U} \qquad \text{and} \qquad c:\ \mathcal{T} \to \mathcal{V}$$

such that for all S, T in $\mathcal{S}$ and τ in $\mathcal{T}$

$$b(S)c(\tau) \;=\; b(T) \quad \Longleftrightarrow \quad S\tau \;=\; T.$$

Finally, two oriented mediums $(\mathcal{S}, \mathcal{T})$ and $(\mathcal{U}, \mathcal{V})$ are *sign-isomorphic*, which we also denote by $(\mathcal{S}, \mathcal{T}) \sim (\mathcal{U}, \mathcal{V})$, if there exist two bijective mappings

[5] These concepts are not used elsewhere in this book.

$$b : \mathcal{S} \to \mathcal{U} \qquad \text{and} \qquad c : \mathcal{T} \to \mathcal{V}$$

such that for all S, T in $\mathcal{S}$ and τ in $\mathcal{T}$

$$b(S)c(\tau) = b(T) \iff S\tau = T \qquad \text{and} \qquad c(\mathcal{T}^+) = \mathcal{V}^+. \tag{10.13}$$

10.4.11 Theorem. *For any discriminative, well-graded family $\mathcal{K}$ of sets, denote by $s(\mathcal{K})$ the oriented medium constructed in Theorems* 10.4.1 *and* 10.4.3. *For any oriented medium $(\mathcal{S}, \mathcal{T})$, denote by $f(\mathcal{S}, \mathcal{T})$ the discriminative, well-graded family formed by all its positive contents (cf. Theorem* 10.4.9*). We then have*

$$\mathcal{K} \sim (f \circ s)(\mathcal{K}) \qquad \textit{and} \qquad \mathcal{S} \sim (s \circ f)(\mathcal{S}). \tag{10.14}$$

PROOF. Given $\mathcal{K}$, select some $q \in \cup\mathcal{K} \setminus \cap\mathcal{K}$ and some $K \in \mathcal{K}$. Note that K stands for both an element of $\mathcal{K}$ and an element of $s(\mathcal{K})$, since these two sets are equal. We have $q \in K$ if and only if $\tau_q \in \widehat{K}^+$. With $(\mathcal{S}, \mathcal{T}) = s(\mathcal{K})$, the positive tokens of the oriented medium $(\mathcal{S}, \mathcal{T})$ are of the form τ_q, for q in $\cup\mathcal{K} \setminus \cap\mathcal{K}$. Also, $f(\mathcal{S}, \mathcal{T})$ is the family $\widehat{\mathcal{S}}^+$ of all positive contents of $(\mathcal{S}, \mathcal{T})$, and any positive token τ is an element of $\cup\widehat{\mathcal{S}}^+$; moreover, we have $\cap\widehat{\mathcal{S}}^+ = \varnothing$ by Theorem 10.4.9. The first isomorphism formula in (10.14) derives from the mapping

$$(\cup\mathcal{K}) \setminus (\cap\mathcal{K}) \to \widehat{\mathcal{S}}^+ : q \mapsto \tau_q.$$

Let $(\mathcal{S}, \mathcal{T})$ be an oriented medium, and let $(\mathcal{S}', \mathcal{T}')$ denote $(s \circ f)(\mathcal{S}, \mathcal{T})$. An element K of the set family $\mathcal{K} = f(\mathcal{S}, \mathcal{T})$ is exactly the positive content of some state of the medium $(\mathcal{S}, \mathcal{T})$. Each state in the medium $s(\mathcal{K}) = (\mathcal{S}', \mathcal{T}')$ is an element of $\mathcal{K}$, so is a positive content $\widehat{S}^+$ of some state S in the given medium $(\mathcal{S}, \mathcal{T})$. By Theorem 10.4.9, S is fully determined by $\widehat{S}^+$. We may thus set $b(S) = K$, which gives a bijective mapping $b : \mathcal{S} \to \mathcal{S}'$. Now, what about the tokens of $(\mathcal{S}', \mathcal{T}')$? According to Equations (10.7) and (10.8), a token τ of $s(\mathcal{K}) = (\mathcal{S}', \mathcal{T}')$ is either of the form τ_q or $\tau_{\bar{q}} = \tilde{\tau}_q$, where $q \in (\cup\mathcal{K}) \setminus (\cap\mathcal{K})$. Theorem 10.4.9 gives us here $\cup\mathcal{K} = \cup\widehat{\mathcal{S}}^+ = \mathcal{T}^+$ and $\cap\mathcal{K} = \cap\widehat{\mathcal{S}}^+ = \varnothing$. Then, any $q \in \cup\mathcal{K} = \mathcal{T}^+$ equals some positive token $\tau(q)$ of the given medium $(\mathcal{S}, \mathcal{T})$, and conversely. This yields the bijective mapping $c : \mathcal{T} \to \mathcal{T}'$ sending $\tau(q)$ to τ_q, and $\widetilde{\tau(q)}$ to $\tau_{\bar{q}}$. We leave to the reader to verify that the mappings b and c just constructed satisfy Equation (10.13). □

10.4.12 Remarks. a) The mappings s and f introduced in Theorem 10.4.11 make explicit the cryptomorphism between discriminative, well-graded families of sets on the one hand, and oriented media on the other hand.

b) In Theorem 10.4.11, we need oriented media rather than just media for the following reason. Let $(\mathcal{S}, \mathcal{T})$ be an oriented medium, and denote by $\mathcal{K} = f(\mathcal{S}, \mathcal{T})$ the corresponding well-graded family with empty intersection (with f as in Theorem 10.4.11). Suppose we change the orientation of $(\mathcal{S}, \mathcal{T})$

by moving all tokens in some subset $\mathcal{Y}$ of $\mathcal{T}^+$ to $\mathcal{T}^-$, and necessarily all tokens $\tilde{\tau}$ with $\tau \in \mathcal{Y}$ to $\mathcal{T}^+$. The effect on $\mathcal{K}$ is as follows, with $Y = \mathcal{Y}$ this time a subset of $\cup\mathcal{K}$: any set K in $\mathcal{K}$ gets replaced with the set $K \triangle Y$. Thus, for any element q in Y and any K in $\mathcal{K}$, membership of q to K is exchanged with non-membership. Say that $\mathcal{K}$ and the resulting family of sets are *linked* (whatever the subset Y of $\cup\mathcal{K}$ is). Hence, a (non oriented) medium $(\mathcal{S}, \mathcal{T})$ corresponds to a whole class of set families, which are two by two linked.

By combining previous theorems, we derive the converse of Theorem 10.4.1 and so establish the close connection between the notion of a medium and that of a well-graded family of sets. Notice the word "finite" in the statement.

10.4.13 Theorem. *Any medium is isomorphic to the medium constructed as in Theorem* 10.4.1 *from some discriminative, well-graded family of finite sets which contains the empty set.*

PROOF. By Theorem 10.4.8, any medium $(\mathcal{S}, \mathcal{T})$ admits an orientation with a root. For such an orientation, all the positive contents are finite and moreover the positive content of the root is empty. Remember from Theorem 10.4.9 that the family $\mathcal{K}$ of all the positive contents is a discriminative, well-graded family. Now, Theorem 10.4.11 ensures that the resulting oriented medium $(\mathcal{S}, \mathcal{T})$ is sign-isomorphic to the oriented medium constructed from the family $\mathcal{K}$. This concludes the proof: the given medium $(\mathcal{S}, \mathcal{T})$ is then isomorphic to the medium constructed from the discriminative, well-graded family $\mathcal{K}$ with $\varnothing \in \mathcal{K}$. □

10.5 Learning Spaces and Closed, Rooted Media

Learning spaces are, in particular, discriminative and well-graded families. In view of Theorem 10.4.11, they are thus set in a one-to-one correspondence with certain oriented media. We will characterize the latter by two additional properties, namely being 'closed' and having a 'root' (see Theorem 10.5.13).

10.5.1 Definition. An oriented medium $(\mathcal{S}, \mathcal{T})$ is *closed* if for any state S and any two distinct positive[6] tokens τ, τ' both effective for S, we have

$$(S\tau = V,\, S\tau' = W) \quad \Longrightarrow \quad V\tau' = W\tau. \tag{10.15}$$

Clearly, a medium can be closed under one orientation without being closed under some other orientation.

[6] Obviously, a corresponding concept of closure for negative tokens can be defined similarly.

10.5.2 Theorem. *In an oriented medium* $(\mathcal{S}, \mathcal{T})$, *the two conditions below are equivalent:*

(i) $(\mathcal{S}, \mathcal{T})$ *is closed.*

(ii) *Let* $\boldsymbol{m} = \tau_1 \dots \tau_n$ *be any positive concise message from some state* S, *with* $S_0 = S$, *and* $S_i = S_{i-1}\tau_i$ *for* $1 \leq i \leq n$. *If a positive token* $\tau \notin \mathcal{C}(\boldsymbol{m})$ *is effective for some state* S_i, $0 \leq i < n$, *it is also effective for any state* S_j, $i < j \leq n$.

We leave the short proof of this theorem as Problem 22. Condition (ii) of this theorem is conceptually germane to Axiom [L2] of a learning space and is a pointer to Theorem 10.5.12, the second main result of this chapter. This theorem states that the family $\widehat{\mathcal{S}}^+$ of positive contents of a finite, closed, rooted medium $(\mathcal{S}, \mathcal{T})$ is a learning space on the set $\mathcal{T}^+$ of positive tokens. We already know by Theorem 10.4.9 that such a family $\widehat{\mathcal{S}}^+$ is well-graded, with $\mathcal{T}^+ = \cup\widehat{\mathcal{S}}^+$. It remains to show that it contains the empty set and that it is closed under union. The next lemma is our first step.

10.5.3 Lemma. *Let* $\boldsymbol{n} = \boldsymbol{m}\tau_1\tau_2\boldsymbol{p}$ *be a concise message from some state* S *in a closed medium, with* τ_1 *negative,* τ_2 *positive, and both* $\boldsymbol{m}$ *and* $\boldsymbol{p}$ *possibly empty. Then* $S\boldsymbol{m}\tau_1\tau_2\boldsymbol{p} = S\boldsymbol{m}\tau_2\tau_1\boldsymbol{p}$.

In other words, two adjacent tokens in a concise message, with the first one negative and the second one positive, can be transposed without changing the state produced.

PROOF. Let $\boldsymbol{n}$ be as in the theorem and suppose that $T = S\boldsymbol{m}\tau_1$. Then, there must be two distinct states W and W' such that $T\tilde{\tau}_1 = W = S\boldsymbol{m}$ and $T\tau_2 = W'$. Since both $\tilde{\tau}_1$ and τ_2 are positive and the medium is closed, we get $W\tau_2 = W'\tilde{\tau}_1$, and thus also, successively

$$S\boldsymbol{m}\tau_2\tau_1 = W\tau_2\tau_1 = W'\tilde{\tau}_1\tau_1 = W' = T\tau_2 = S\boldsymbol{m}\tau_1\tau_2. \qquad \square$$

10.5.4 Definition. Suppose that $\boldsymbol{n} = \boldsymbol{m}\boldsymbol{p}\boldsymbol{m}'$, with $\boldsymbol{m}$ and $\boldsymbol{m}'$ two possibly ineffective messages, and $\boldsymbol{p}$ an effective one. Then $\boldsymbol{p}$ is a *segment* of $\boldsymbol{n}$. If $\boldsymbol{m}$ is empty, then (as $\boldsymbol{m}$ can be omitted) $\boldsymbol{p}$ is an *initial* segment or *prefix* of $\boldsymbol{n}$. Similarly, if $\boldsymbol{m}'$ is empty, then $\boldsymbol{p}$ is a *terminal* segment or *suffix* of $\boldsymbol{n}$. With respect to some orientation, a segment is said to be *positive* (resp. *negative*) if it contains only positive (resp. negative) tokens.

10.5.5 Definition. In an oriented medium, a concise message $\boldsymbol{m}$ producing a state V from a state S is called *canonical* if it satisfies one of the following three cases:

(i) $\boldsymbol{m}$ is positive;
(ii) $\boldsymbol{m}$ is negative;
(iii) $\boldsymbol{m} = \boldsymbol{n}\boldsymbol{n}'$ with $\boldsymbol{n}$ a positive prefix and $\boldsymbol{n}'$ a negative suffix.

In Case (iii) the canonical message $\boldsymbol{m} = \boldsymbol{n}\boldsymbol{n}'$ is said to be *mixed*.

10.5.6 Theorem. *For any two distinct states S and V in a closed medium, there is a canonical message producing V from S.*

PROOF. We know from Axiom [Ma] that $S\boldsymbol{p} = V$ for some concise message $\boldsymbol{p} = \tau_1 \ldots \tau_n$. Suppose that $\boldsymbol{p}$ is not canonical. Then there must be an index i such that τ_i is negative and τ_{i+1} positive. By Lemma 10.5.3, the tokens τ_i and τ_{i+1} can be transposed without changing the state produced. The result follows by induction. □

10.5.7 Lemma. *Suppose that a state V in an oriented medium is produced from a state S by a mixed canonical message $\boldsymbol{m} = \boldsymbol{n}\boldsymbol{n}'$, with $S\boldsymbol{n} = T$, $\boldsymbol{n}$ a positive prefix of $\boldsymbol{m}$, and $\boldsymbol{n}'$ a negative prefix of $\boldsymbol{m}$. We then have $\widehat{S}^+ \cup \widehat{V}^+ = \widehat{T}^+$.*

PROOF. Theorem 10.4.5 implies that $\widehat{S}^+ \subset \widehat{T}^+$. We have also $\widehat{V}^+ \subset \widehat{T}^+$ since $\widetilde{\boldsymbol{n}}'$ is a positive message which is concise for V and produces T from V. This yields $\widehat{S}^+ \cup \widehat{V}^+ \subseteq \widehat{T}^+$. We get

$$\widehat{T}^+ \setminus (\widehat{S}^+ \cup \widehat{V}^+) = (\widehat{T}^+ \setminus \widehat{S}^+) \cap (\widehat{T}^+ \setminus \widehat{V}^+) = \mathcal{C}(\boldsymbol{n}) \cap \mathcal{C}(\widetilde{\boldsymbol{n}}') = \varnothing,$$

the second equation holding in view of Theorem 10.4.5, and the last one because $\boldsymbol{n}\boldsymbol{n}'$ is concise for S. Thus, $\widehat{T}^+ \subseteq \widehat{S}^+ \cup \widehat{V}^+$, yielding the result. □

10.5.8 Theorem. *For any two states S and V in a closed medium $(\mathcal{S}, \mathcal{T})$, there is a unique state T whose positive content is the union of the positive contents of S and V. So, the family $\widehat{\mathcal{S}}^+$ of all the positive contents is closed under finite unions, and the family $\widehat{\mathcal{S}}^-$ of all the negative contents is closed under finite intersections.*

PROOF. By Theorem 10.5.6, there is a canonical message $\boldsymbol{p}$ producing V from S. Suppose that $\boldsymbol{p}$ is positive, then clearly $\widehat{S}^+ \subset \widehat{V}^+$, and so $\widehat{S}^+ \cup \widehat{V}^+ = \widehat{V}^+ \in \widehat{\mathcal{T}}^+$, yielding $T = V$. If $\boldsymbol{p}$ is negative, then $\widetilde{\boldsymbol{p}}$ is positive and produces S from V, with a similar result. The case where $\boldsymbol{p}$ is a mixed canonical message is an immediate consequence of Lemma 10.5.7. The set T is unique because any state is defined by its positive content (Theorem 10.4.4). The statement concerning the negative contents follows by duality. □

The next theorem and definition complete our preparation.

10.5.9 Theorem. *Any finite closed medium $(\mathcal{S}, \mathcal{T})$ has a unique state Λ which is produced only by positive messages. Accordingly, the content of Λ is confounded with its positive content. We thus have $\widehat{\Lambda} = \widehat{\Lambda^+} = \widehat{\mathcal{T}^+}$.*

PROOF. By finiteness and Theorem 10.5.8, the union of all the positive contents is the positive content of some state, which we denote by Λ. Theorem 10.4.9 implies $\widehat{\Lambda^+} = \widehat{\mathcal{T}^+}$; so, by Theorem 10.3.3, we get $\widehat{\Lambda} = \widehat{\Lambda^+}$. Finally, any state S with $\widehat{S} = \widehat{S^+}$ must be equal to Λ, because then $\widehat{S} = \widehat{S^+} = \widehat{\mathcal{T}^+}$, and so $\widehat{S} = \widehat{\Lambda}$ from which follows $S = \Lambda$ (Theorem 10.3.5). □

10.5.10 Remarks. a) Without the finiteness assumption, Theorem 10.5.9 no longer holds. As a counterexample, take the family of all finite subsets of $\mathbb{R}$, including the empty set. As this family is well-graded and discriminative, Theorem 10.4.3 can be used to built an oriented medium. This medium has no state produced only by positive messages. Note that this infinite oriented medium has a root.

b) The converse of Theorem 10.5.9 does not hold. An infinite medium with a state produced only by positive messages is obtained by exchanging the positive and the negative tokens in the counter-example specified in (a) (Λ is the empty set). This infinite oriented medium has no root.

10.5.11 Definition. A state S in an oriented medium $\mathcal{M} = (\mathcal{S}, \mathcal{T})$ is called the *apex* of $\mathcal{M}$ if $\widehat{S} = \mathcal{T}^+$.

Thus, the state Λ introduced in Theorem 10.5.9 is the apex of the finite closed medium. By Theorem 10.3.5, any oriented medium can have at most one apex. The infinite, closed medium of Remark 10.5.10 (a) has no apex.

10.5.12 Theorem. *The two following propositions are equivalent for an oriented medium $\mathcal{M} = (\mathcal{S}, \mathcal{T})$:*

(i) *$\mathcal{M}$ is a finite, closed and rooted medium.*
(ii) *The positive content family $\widehat{\mathcal{S}}^+$ of $\mathcal{M}$ is a learning space.*

PROOF. (i) $\Rightarrow$ (ii). We have to prove that both $\mathcal{T}^+$ and $\varnothing$ are in $\widehat{\mathcal{S}}^+$, and that Axioms [L1] and [L2] of a learning space are satisfied. By Theorem 10.5.9, the apex Λ of the medium $\mathcal{M}$ satisfies $\widehat{\Lambda} = \mathcal{T}^+ \in \widehat{\mathcal{S}}^+$. Since $\mathcal{M}$ is rooted, $\varnothing \in \widehat{\mathcal{S}}^+$ results from Theorem 10.4.7. From Theorems 10.4.9 and 10.5.8, we know that $\widehat{\mathcal{S}}^+$ is a wg-family that is $\cup$-closed. Accordingly, by Theorem 2.2.4, the collection $\widehat{\mathcal{S}}^+$ of positive contents must be a learning space.

(ii) $\Rightarrow$ (i). By Theorem 10.4.4, any state in an oriented medium is defined by its positive content. Observe that if the positive tokens τ and μ are effective for some state S of $\mathcal{M}$, then

$$\widehat{(S\tau)}^+ = \widehat{S}^+ + \{\tau\} \qquad \text{and} \qquad \widehat{(S\mu)}^+ = \widehat{S}^+ + \{\mu\}.$$

Because $\widehat{\mathcal{S}}^+$ is $\cup$-closed, the set $\widehat{V}^+ = \widehat{S}^+ + \{\tau\} + \{\mu\}$ is the positive content of some state V, with necessarily $V = S\tau\mu = S\mu\tau$. So, $\mathcal{M}$ is closed. We have $\varnothing \in \widehat{S}^+$ by definition of a learning space as a particular kind of knowledge structure. Thus $\varnothing$ is the positive content of some state R that is a root of $\mathcal{M}$, by Theorem 10.4.7. $\square$

We close the chapter with a theorem essentially stating that learning spaces are cryptomorphic to finite, closed and rooted media. The functions s and f are defined as in Theorem 10.4.11.

10.5.13 Theorem. *Suppose that $\mathcal{K}$ is a discriminative, well-graded family of sets and let $(\mathcal{S}, \mathcal{T})$ be the corresponding oriented medium, that is $s(\mathcal{K}) = (\mathcal{S}, \mathcal{T})$ and $f(\mathcal{S}, \mathcal{T}) = \mathcal{K}$. Then $\mathcal{K}$ is a learning space if and only if $(\mathcal{S}, \mathcal{T})$ is a finite, closed and rooted medium.*

The proof is left as Problem 23.

10.6 Original Sources and Related Works

The concept of a medium in the sense of this chapter originated with a paper of Falmagne (1997), whose motivation combined two quite different endeavors. One was the search for an algebraic generalization of the property of wellgradedness of some families of relations, such as the family of all partial orders on a finite set or the family of all finite subsets of $\mathbb{R}$. The other was quite different. It was empirical and prompted by the wish to model the behavior of a potential voter in an election. The tokens in such a model represent the many types of informations received by the voter in the course of time, each of which may be tiny, unobservable, and yet potentially capable of altering the voters opinions on the competing candidates in a minute way. The term 'medium' originated from that example.

David Eppstein and Sergei Ovchinnikov soon expressed interest in the topic, and the first paper was followed by several others (see in particular Ovchinnikov and Dukhovny, 2000; Eppstein and Falmagne, 2002; Falmagne and Ovchinnikov, 2002; Eppstein, 2005, 2007; Falmagne and Ovchinnikov, 2009).

The mathematical concept of a medium closely relates to (at least) two other ones. First, as noted by Ovchinnikov and Dukhovny (2000), the connection between well-graded families and media is mutual. Second, finite media are also cryptomorphically equivalent to 'partial cubes' (where a 'partial cube' is an 'isometric' subgraph of a hypercube graph; we refer the reader to Imrich and Klavžar, 2000, for the terminology, and a thorough exposition of the subject). Media are the topic of a recent book by Eppstein, Falmagne, and Ovchinnikov (2008).

Problems

1. Find other examples of media in combinatorics or in real life. (Think of games, for example.)
2. Verify that the collection $\mathcal{L}$ of all the linear orders on a finite set can be cast as a medium. Which axiom may be violated if $\mathcal{L}$ is infinite?
3. Verify that any token in a token system has at most one reverse, and that if the reverse $\tilde{\tau}$ of some token τ exists, then $\tilde{\tilde{\tau}} = \tau$. Construct an example in which some token in a token system has no reverse.

4. Manufacture two examples establishing the independence of Axioms [Ma] and [Mb].

5. (Lemma 10.2.1.) Prove that, in a medium: (i) each token has a unique reverse; (ii) if $\boldsymbol{m}$ is stepwise effective for S, then $S\boldsymbol{m} = V$ implies $V\widetilde{\boldsymbol{m}} = S$; (iii) $\tau \in \mathcal{C}(\boldsymbol{m})$ if and only if $\tilde{\tau} \in \mathcal{C}(\widetilde{\boldsymbol{m}})$; (iv) if $\boldsymbol{m}$ is consistent, so is $\widetilde{\boldsymbol{m}}$.

6. Prove the following facts stated in Lemma 10.2.3. (i) No token is identical to its own reverse. (iii) For any two adjacent states S and V, there is exactly one token producing V from S. (iv) Let $\boldsymbol{m}$ be a message that is concise for some state, then the following two equalities hold

$$\ell(\boldsymbol{m}) = |\mathcal{C}(\boldsymbol{m})| \qquad \text{and} \qquad \mathcal{C}(\boldsymbol{m}) \cap \mathcal{C}(\widetilde{\boldsymbol{m}}) = \varnothing.$$

(vi) Suppose that $\boldsymbol{m}$ and $\boldsymbol{n}$ are stepwise effective for S and V, respectively, with $S\boldsymbol{m} = V$ and $V\boldsymbol{n} = W$. Then $\boldsymbol{mn}$ is stepwise effective for S, with $S\boldsymbol{mn} = W$.

7. Prove Theorem 10.3.3.

8. Is it true that, in a medium, the content of any state is necessarily finite? Prove the last statement or give a counterexample.

9. Prove that, in a medium, a state is defined by its content (Theorem 10.3.5).

10. Verify that the pair $(\mathcal{P}_3, \mathcal{T}_3)$ defined in Example 10.3.8 satisfies Axioms [Ma] and [Mb]. Does the result extend to the family of partial orders on n elements, for all n in $\mathbb{N}$?

11. Verify that the family of all semiorders on a finite set can be represented as a medium.

12. (Continuation.) Define the positive content family of a the collection all semiorders on a finite set in the style used for the partial orders in Example 10.3.8 for the partial orders. (Thus, the positive tokens are those representing the addition of pairs to semiorders.) Is such a family always a learning space?

13. (Continuation.) How about the collection of almost connected orders or ac-orders on a finite set? Here, a relation R on a set $\mathcal{X}$ is an ac-order if it is asymmetric and 2-connected, that is, R satisfies the condition $R^2\bar{R}^{-1} \subseteq R$ (cf. for example Doble et al., 2001, and many other references to this concept there).

14. Verify that the family of positive contents in the graph of Figure 10.4 satisfies Axioms [Ma] and [Mb].

15. Prove that: (i) any medium can be oriented, and (ii) that a finite medium $(\mathcal{S}, \mathcal{T})$ can be given $2^{|\mathcal{T}|/2}$ orientations.

16. Prove Theorem 10.4.5.

17. Is it true that, for any medium, an orientation can always be defined so that the oriented medium has no root?

18. Prove the following two facts. (i) Any state in a medium can be made a root by defining a suitable orientation. (ii) There always exists some orientation ensuring that the positive contents of all the states are finite sets. In particular, the states of any rooted medium have finite positive contents. (Theorem 10.4.8).

19. Complete the proof of Theorem 10.4.9 by showing that $\widehat{\mathcal{S}}^{+}$ is discriminative.

20. Suppose that the positive class of an oriented medium is a learning space. What can you say about the negative class?

21. Let $\mathcal{F}$ be a well-graded family of sets. Construct the family $\mathcal{F}^{\cup}$ of all the sets which are unions of sets in $\mathcal{F}$. Under which necessary and sufficient conditions on $\mathcal{F}$ can a medium be obtained from $\mathcal{F}^{\cup}$ by defining the tokens along the lines of (10.5) and (10.6)?

22. Prove Theorem 10.5.2.

23. Prove Theorem 10.5.13.

11

Probabilistic Knowledge Structures

The concept of a knowledge structure is a deterministic one. As such, it does not provide realistic predictions of subjects' responses to the problems of a test. There are two ways in which probabilities must enter in a realistic model. For one, the knowledge states will certainly occur with different frequencies in the population of reference. It is thus reasonable to postulate the existence of a probability distribution on the collection of states. For another, a subject's knowledge state does not necessarily specify the observed responses. A subject having mastered an item may be careless in responding, and make an error. Also, in some situations, a subject may be able to guess the correct response to a question not yet mastered. In general, it makes sense to introduce conditional probabilities of responses, given the states. A number of simple probabilistic models will be described in this chapter. They will be used to illustrate how probabilistic concepts can be introduced within knowledge space theory. These models will also provide a precise context for the discussion of some technical issues related to parameter estimation and statistical testing. The material in this chapter must be regarded as a preparation for the stochastic theories discussed in Chapters 12, 13 and 14.

11.1 Basic Concepts and Examples

11.1.1 Example. As an illustration, we consider the knowledge structure

$$\begin{aligned}\mathcal{H} = \big\{\varnothing, \{a\}, \{b\}, \{a,b\}, \{a,b,c\}, \{a,b,d\},\\ \{a,b,c,d\}, \{a,b,c,e\}, \{a,b,c,d,e\}\big\}\end{aligned} \tag{11.1}$$

with domain $Q = \{a,b,c,d,e\}$ (see Figure 11.1). This example will be referred to many times in this chapter, under the name *standard example*. The knowledge structure $\mathcal{H}$ is actually an ordinal knowledge space (in the sense of Definition 3.8.1), with nine states. We suppose that any subject sampled from a population of reference will necessarily be in one of these nine states.

J.-C. Falmagne, J.-P. Doignon, *Learning Spaces*,
DOI 10.1007/978-3-642-01039-2_11, © Springer-Verlag Berlin Heidelberg 2011

More specifically, we assume that to each knowledge state $K \in \mathcal{H}$ is attached a probability $p(K)$ measuring the likelihood that a sampled subject is in that state. We thus enlarge our theoretical framework by a probability distribution p on the family of all knowledge states. In practice, the parameters $p(K)$ must be estimated from the assessment data.

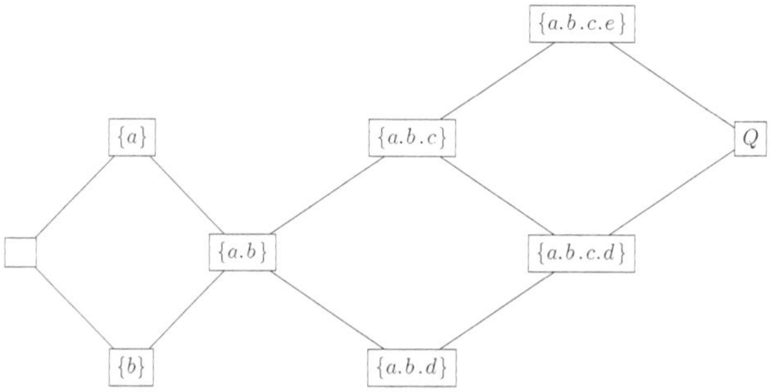

Figure 11.1. Inclusion graph of the knowledge space $\mathcal{H}$ of Equation (11.1).

Notice that the states may not be directly observable. If careless errors or lucky guesses are made, all kinds of 'response patterns' may be arise from the states in $\mathcal{H}$. A convenient coding will be adopted for these response patterns. Suppose that a subject has correctly solved questions c and d, and failed to solve a, b and e. We shall denote such a result by the subset $\{c, d\}$ of Q. In general, we shall represent a response pattern by the subset R of Q containing all the questions correctly solved by the subject. There are thus $2^{|Q|}$ possible response patterns.

For any $R \subseteq Q$ and $K \in \mathcal{H}$, we denote by $r(R, K)$ the conditional probability of response pattern R, given state K. For example, $r(\{c, d\}, \{a, b, c\})$ denotes the probability that a subject in state $\{a, b, c\}$ responds correctly to questions c and d, and fails to solve a, b, and e. For simplicity, let us suppose in this example that a subject never responds correctly to a question not in his state. Writing $\rho(R)$ for the probability of response pattern R, we obtain for instance[1]

$$\rho(\{c, d\}) = r(\{c, d\}, \{a, b, c, d\})\, p\{a, b, c, d\} + r(\{c, d\}, Q)\ p(Q). \tag{11.2}$$

Indeed, the only two states capable of generating the response pattern $\{c, d\}$ are the states including it, namely, $\{a, b, c, d\}$ and Q. We shall also assume that, given the state, the response to the questions are independent. Writing β_q for the probability of an incorrect response to a question q in the subject's state, and using this independence assumption, we get

$$r(\{c, d\}, \{a, b, c, d\}) = \beta_a\beta_b(1 - \beta_c)(1 - \beta_d), \tag{11.3}$$
$$r(\{c, d\}, Q) = \beta_a\beta_b\beta_e(1 - \beta_c)(1 - \beta_d), \tag{11.4}$$

[1] Note that we simplify the writing of the probabilities and use $p\{\ldots\}$ for $p(\{\ldots\})$.

and thus, from Equation (11.2),

$$\rho(\{c,d\}) = \beta_a\beta_b(1-\beta_c)(1-\beta_d)p\{a,b,c,d\} + \beta_a\beta_b\beta_e(1-\beta_c)(1-\beta_d)p(Q).$$

In psychometrics, the type of assumption exemplified by Equations (11.3) and (11.4) is often referred to as a 'local independence' condition. We follow this usage in the definition below which generalizes the example and includes the case in which a correct response to some question q may be elicited by a state not containing q (thus, in the framework of the theory, a lucky guess).

11.1.2 Definition. A *(finite, partial) probabilistic knowledge structure* is a triple $(Q, \mathcal{K}, p)$ in which

(i) $(Q, \mathcal{K})$ is a finite partial knowledge structure[2] with $|Q| = m$ and $|\mathcal{K}| = n$;
(ii) the mapping $p : \mathcal{K} \to [0,1] : K \mapsto p(K)$ is a probability distribution on $\mathcal{K}$; thus, for any $K \in \mathcal{K}$, we have $p(K) \geq 0$, and moreover, $\sum_{K\in\mathcal{K}} p(K) = 1$.

A *response function* for a probabilistic knowledge structure $(Q, \mathcal{K}, p)$ is a function $r : (R, K) \mapsto r(R, K)$, defined for all $R \subseteq Q$ and $K \in \mathcal{K}$, and specifying the probability of the *response pattern* R for a subject in state K. Thus, for any $R \in 2^Q$ and $K \in \mathcal{K}$, we have $r(R, K) \geq 0$; moreover, $\sum_{R\subseteq Q} r(R, K) = 1$. A quadruple $(Q, \mathcal{K}, p, r)$, in which $(Q, \mathcal{K}, p)$ is a probabilistic knowledge structure and r its response function will be referred to as a *basic probabilistic model*. This name is justified by the prominent place of this model in this book.

Writing $R \mapsto \rho(R)$ for the resulting probability distribution on the set of all response patterns, we obtain for any $R \subseteq Q$

$$\rho(R) = \sum_{K\in\mathcal{K}} r(R, K)p(K). \tag{11.5}$$

The response function r satisfies *local independence* if for each $q \in Q$, there are two constants $\beta_q, \eta_q \in [0, 1[$, respectively called *(careless) error probability* and *guessing probability* at q, such that, for all $R \subseteq Q$ and $K \in \mathcal{K}$, we have

$$r(R,K) = \Bigg(\prod_{q\in K\setminus R} \beta_q\Bigg)\Bigg(\prod_{q\in K\cap R} (1-\beta_q)\Bigg)\Bigg(\prod_{q\in R\setminus K} \eta_q\Bigg)\Bigg(\prod_{q\in \overline{R\cup K}} (1-\eta_q)\Bigg) \tag{11.6}$$

(with $\overline{R \cup K} = Q \setminus (R \cup K)$ in the last factor). The basic probabilistic model satisfying local independence will be called the *basic local independence model*. These concepts are fundamental in the sense that all the probabilistic models discussed in this book (in Chapters 12, 13 and 14) will satisfy Equation (11.5), and most of them also Equation (11.6) (possibly with $\eta_q = 0$ for all items q; see Remark 11.1.3 (a)). In some models, however, the existence of the probability distribution p on the collection of states $\mathcal{K}$ will not be an axiom.

[2] We recall that family of sets $\mathcal{K}$ is a partial knowledge structure if $\cup\mathcal{K} \in \mathcal{K}$ (cf. Definition 2.2.6).

Instead, it will appear as a consequence of more or less elaborate assumptions regarding the learning process by which a student moves around in the knowledge structure, gradually evolving from the state $\varnothing$ of total ignorance to the state Q of full mastery of the material.

11.1.3 Remarks. a) In the case of the knowledge space $\mathcal{H}$ of Equation (11.1), there would thus be $8 + 2 \cdot 5 = 18$ parameters to be estimated from the data: $8 = 9 - 1$ independent probabilities for the states, and two parameters β_q and η_q for each of the five questions in Q. This number of parameters will seem unduly large compared to the 31 degrees of freedom in the data ($31 = 32 - 1$ independent response frequencies of response patterns). There are two ways in which this number of parameters may be reduced. The questions of the test may be designed in such a manner that the probability of finding the correct response by chance or by a sophisticated guess is zero. We thus have $\eta_q = 0$ for all $q \in Q$. (We have in fact made that assumption in our discussion of Example 11.1.1.) More importantly, the number of independent state probabilities can also be reduced. Later in this chapter, and also in Chapter 12, we investigate a number of learning mechanisms purporting to explain how a subject may evolve from the state $\varnothing$ to the state Q. Typically, such mechanisms set constraints on the state probabilities, effectively reducing the number of parameters. Finally, we point out that this example, which was chosen for expository reasons, is not representative of most empirical situations in that the number of questions is unrealistically small. It is our experience that the ratio of the number of states to the number of possible response patterns decreases dramatically as the number of questions increases (cf. Chapter 15 and, especially, Villano, 1991). The ratio of the number of parameters to the number of response patterns will decrease accordingly.

b) The local independence assumption Equation (11.6) may be criticized on the grounds that the response probabilities (represented by the parameters β_q and η_q) are attached to the questions, and do not vary with the knowledge state of the subject. For instance, in Example 11.1.1, the probability β_c of a careless error to question c is the same in all four states: $\{a, b, c\}$, $\{a, b, c, d\}$, $\{a, b, c, e\}$ and Q. However, weakening this assumption would result in a costly increase of the already substantial number of parameters in the model. More to the point, the probabilistic models described here are intended for situations in which the core of the model, that is, the knowledge structure with the probability distribution on the collection of states, gives an essentially accurate picture of the data. The local independence assumption for careless errors is in keeping with this view: if item q belongs to the student's state, the probability of a careless error should be the same, no matter what that state is. As for the lucky guesses, their probabilities can be negligibly small if the items are framed appropriately. This requirement obviously disqualifies any multiple choice items. Note that the local independence asssumption is central to the so called *Latent Structure Analysis* (see Lazarsfeld and Henry, 1968, which is the main source for such models).

11.2 An Empirical Application

The basic local independence model has been successfully applied in a number of situations (see Falmagne et al., 1990; Villano, 1991; Lakshminarayan, 1995; Lakshminarayan and Gilson, 1998; Taagepera, Potter, Miller, and Lakshminarayan, 1997). For convenience of exposition, we have chosen to present here a simple application to some fictitious data, which will be used repeatedly in this chapter. Lakshminarayan's experiment is described in Chapter 12.

11.2.1 The Data. We consider a situation in which 1,000 subjects have responded to the five questions of the domain $Q = \{a, b, c, d, e\}$ of Example 11.1.1. The hypothetical frequencies of the 32 response patterns are contained in Table 11.1. We write $N(R)$ for the frequency of the response pattern R. In the example given in Table 11.1, we thus have $N(\varnothing) = 80$, $N(\{a\}) = 92$, and so forth. We also denote by

$$N = \sum_{R \subseteq Q} N(R), \tag{11.7}$$

the number of subjects in the experiment (thus $N = 1000$ in this case). As in earlier chapters, we set $|Q| = m$ (so $m = 5$ here). The basic local independence model (cf. Definition 11.1.2) depends upon three kinds of parameters: the state probabilities $p(K)$, and the response probabilities β_q and η_q entering in Equation (11.6). We denote by θ the vector of all these parameters. In the case of Example 11.1.1, we set

$$\theta = (p(\varnothing),\, p\{a\},\, \ldots,\, p(Q),\, \beta_a,\, \ldots,\, \beta_e,\, \eta_a,\, \ldots,\, \eta_e).$$

Table 11.1. Frequencies of the response patterns for 1,000 fictitious subjects.

R	Obs.	R	Obs.	R	Obs.	R	Obs.
$\varnothing$	80	*ad*	18	*abc*	89	*bde*	2
a	92	*ae*	10	*abd*	89	*cde*	2
b	89	*bc*	18	*abe*	19	*abcd*	73
c	3	*bd*	20	*acd*	16	*abce*	82
d	2	*be*	4	*ace*	16	*abde*	19
e	1	*cd*	2	*ade*	3	*acde*	15
ab	89	*ce*	2	*bcd*	18	*bcde*	15
ac	16	*de*	3	*bce*	16	*Q*	77

This parameter θ is a point in a *parameter space* Θ containing all possible vectors of parameter values. The probability $\rho(R)$ of a response pattern R depends upon θ, and we shall make that dependency explicit by writing

$$\rho_\theta(R) = \rho(R). \tag{11.8}$$

A key component in the application of the model is the estimation of the parameters, that is, the choice of a point of Θ yielding the best fit. We shall briefly review two closely related standard techniques for estimating the parameters and testing the model.

11.2.2 The chi-square statistic. Among the most widely used indices for evaluating the goodness-of-fit of a model to data is the *chi-square statistic*, defined in this case as the random variable

$$\mathrm{CHI}(\theta; D, N) = \sum_{R \subseteq Q} \frac{\left(N(R) - \rho_\theta(R)N\right)^2}{\rho_\theta(R)N}, \tag{11.9}$$

in which θ is the vector of parameters, D symbolizes the data (that is, the vector of observed frequencies $N(R)$) and N and $\rho_\theta(R)$ are as in Equations (11.7) and (11.8). Thus, the chi-square statistic is a weighted sum of all the square deviations between the observed frequencies $N(R)$ and the predicted frequencies $\rho_\theta(R)N$. The weights $\left(\rho_\theta(R)N\right)^{-1}$ are normalizing factors ensuring that CHI enjoys important asymptotic properties. Namely, if the model is correct, then the minimum of $\mathrm{CHI}(\theta; D, N)$ for θ varying in Θ converges (as N converges to ∞) to a chi-square random variable. This holds under fairly general conditions on the smoothness of ρ_θ as a function of θ, and under the assumption that the response patterns given by different subjects are independent. Specifically, for N large and considering D as a random vector, the random variable $\min_{\theta \in \Theta} \mathrm{CHI}(\theta; D, N)$ is approximately distributed as a chi-square random variable with

$$v = (2^m - 1) - (n - 1 + 2m) \tag{11.10}$$

degrees of freedom (remember that $|Q| = m$ and $|\mathcal{K}| = n$). Denoting by $\tilde{\theta}$ a value of θ that gives a global minimum, and by χ_v^2 a chi-square random variable with v degrees of freedom, we recap this result by the formula

$$\min_{\theta \in \Theta} \mathrm{CHI}(\theta; D, N) \;=\; \mathrm{CHI}(\tilde{\theta}; D, N) \overset{d}{\approx} \chi_v^2, \qquad \text{(for } N \text{ large)} \tag{11.11}$$

where $\overset{d}{\approx}$ means *approximately distributed as*. The r.h.s. of Equation (11.10) computes the difference between the number of degrees of freedom in the data, and the number of parameters estimated by the minimization. In the case of Example 11.1.1, we would have $v = (2^5 - 1) - (9 - 1 + 2 \cdot 5) = 13$. The convergence in Equation (11.11) is fast. A rule of thumb for the approximation in (11.11) to be acceptable is that, for every $R \subseteq Q$, the expected number of response patterns equal to R is greater than 5, i.e.

$$\rho_{\tilde{\theta}}(R)N > 5. \tag{11.12}$$

The result $c = \mathrm{CHI}(\tilde{\theta}; D, N)$ of such a computation for some observed data D is then compared with a standard table of representative values of the chi-square random variable with the appropriate degrees of freedom (which is v

in our case). The model is accepted if c lies in the bulk of the distribution, and rejected if c exceeds some critical value.

11.2.3 Decision procedures. Typical decision procedures are:

1. **Strong rejection:** $\mathbb{P}(\chi_v^2 > c) \leq .01$;
2. **Rejection:** $\mathbb{P}(\chi_v^2 > c) \leq .05$;
3. **Acceptance:** $\mathbb{P}(\chi_v^2 > c) > .05$.

A value c leading to a rejection (resp. strong rejection) will be called *significant* (resp. *strongly significant*). In practice, a significant or even strongly significant value of a chi-square statistic needs not always lead to a rejection of the model. This is especially true if the model is providing a simple, detailed description of complex data, and no sensible alternative model is yet available.

11.2.4 Remarks. a) For large, sparse data tables (as is Table 11.1), the criterion specified by (11.12) is sometimes judged to be too conservative, and a minimum expected cell size equal to 1 may be appropriate (see Feinberg, 1981). This less demanding criterion will be used in all the analyses reported in this chapter. When one fears that this criterion might fail, a simple solution is to group the cells with low values. Suppose, for instance, that only three cells have low frequencies, corresponding to the response patterns R, R', R''. These three cells would be grouped into one. In the chi-square statistic, the three terms corresponding to the response patterns R, R', R'' would be replaced by the single term

$$\frac{\Big(N(R) + N(R') + N(R'') - (\rho_{\dot\theta}(R) + \rho_{\dot\theta}(R') + \rho_{\dot\theta}(R''))N\Big)^2}{(\rho_{\dot\theta}(R) + \rho_{\dot\theta}(R') + \rho_{\dot\theta}(R''))N}.$$

This grouping would result in a loss of two degrees of freedom for the chi-square random variable in the decision procedure. The change of notation from $\tilde\theta$ to $\dot\theta$ for the estimate of θ in the last equation is meant to suggest that a different global minimum may be obtained after grouping. Obviously, such a grouping must proceed with caution, so that no bias ensue. Grouping the cells *a posteriori*, on the basis of their low expected frequencies, would be poor methodology.

b) It sometimes happens that the observed value c of the chi-square statistic lies in the other tail of the distribution, e.g. $\mathbb{P}(\chi_v^2 < c) < .05$. Such a result would suggest that the variability in the data is smaller than should be expected, which is usually indicative of an error (in the computation of the chi-square statistic, the number of degrees of freedom, or some other aspect of the procedure or even the experimental paradigm).

c) The minimum $\mathrm{CHI}(\tilde{\theta}; D, N)$ in Equation (11.9), often called the *chi-square statistic*, cannot usually be obtained by analytical methods. (Setting equal to zero the derivatives in (11.9) with respect to each of the parameters results in a non linear system.) Brute force search of the parameter space is possible. A modification, due to Brent (1973), of a conjugate gradient search method of Powell (1964) was used for our optimization problems. The actual program was `PRAXIS` (Brent, 1973; Gegenfurtner, 1992)[3]. It should be noted that one application of `PRAXIS` (or for that matter any other optimizing program of that kind) does not guarantee the achievement of a global minimum. Typically, the procedure is repeated many times, with varying the starting values of the parameters, until the researcher is reasonably sure that a global minimum has been reached.

d) All the statistical results and techniques used in this book are well-known, and we shall rarely give supporting references in the text. Listed in increasing order of difficulty, some standard references are Fraser (1958), Lindgren (1968), Brunk (1973), Cramér (1963), Lehman (1959). The last two are classics of the literature.

11.2.5 Results. The parameter values obtained from the minimization of the chi-square statistic for the data in Table 11.1 are given in Table 11.2. The value of the chi-square statistic $\min_{\theta \in \Theta} \mathrm{CHI}(\theta; D, N)$ was 14.7 for $31 - 18 = 13$ degrees of freedom. According to the decision procedures listed in 11.2.3, the model should be accepted, pending further results. Some comments on the values obtained for the parameters will be found in the next section.

Table 11.2. Estimated values of the parameters obtained from the minimization of the chi-square statistic of Equations (11.10)-(11.12). The value of the chi-square statistic was 14.7 for $31 - 18 = 13$ degrees of freedom.

Response Probabilities	
$\beta_a = .17$	$\eta_a = .00$
$\beta_b = .17$	$\eta_b = .09$
$\beta_c = .20$	$\eta_c = .00$
$\beta_d = .46$	$\eta_d = .00$
$\beta_e = .20$	$\eta_e = .03$

State Probabilities	
$p(\varnothing) = .05$	$p\{a,b,c\} = .04$
$p\{a\} = .11$	$p\{a,b,d\} = .19$
$p\{b\} = .08$	$p\{a,b,c,d\} = .19$
$p\{a,b\} = .00$	$p\{a,b,c,e\} = .03$
$p\{a,b,c,d,e\} = .31$	

[3] We are grateful to Michel Regenwetter and Yung-Fong Hsu for all the computations reported in this chapter.

11.3 The Likelihood Ratio Procedure

Another time-honored method, which will also be used in this book, is the so-called *likelihood ratio procedure.*

11.3.1 Maximum likelihood estimates. In estimating the parameters of a model, it makes good intuitive and theoretical sense to choose those parameter values which render the observed data most likely. Such estimates are called *maximum likelihood estimates.* In the case of the empirical example discussed in the previous section, the likelihood of the data (for a given parameter point θ) is obtained from the *likelihood function*

$$\prod_{R\subseteq Q} \rho_\theta(R)^{N(R)}. \tag{11.13}$$

This computation relies on the reasonable assumption that the response patterns given by different subjects are independent. In practice, maximum likelihood estimates of the parameter in the vector θ are obtained by maximizing the logarithm (in base 10 for example) of the likelihood function from Equation (11.13). Let us denote by $\hat{\theta}$ a value of θ maximizing $\sum_{R\subseteq Q} N(R)\log\rho_\theta(R)$, that is

$$\max_{\theta\in\Theta} \sum_{R\subseteq Q} N(R)\log\rho_\theta(R) = \sum_{R\subseteq Q} N(R)\log\rho_{\hat{\theta}}(R).$$

It turns out that, in most cases of interest, the maximum likelihood estimates are asymptotically (thus, for N converging to ∞) the same as those obtained from the minimization of the chi-square statistic. This means that, if the parameter values in the vector $\hat{\theta}$ are maximum likelihood estimates, then for large N, the chi-square statistic

$$\mathrm{CHI}(\hat{\theta}; D, N) = \sum_{R\subseteq Q} \frac{\left(N(R) - \rho_{\hat{\theta}}(R)N\right)^2}{\rho_{\hat{\theta}}(R)N}, \tag{11.14}$$

is approximately distributed as a chi-square random variable with

$$v = 2^m - 1 - (n-1) - 2m$$

degrees of freedom (compare with Equations (11.9) and (11.11)).

11.3.2 The likelihood ratio test. A statistical test of the model can be obtained from the fact that, for large N,

$$-2\log\frac{\prod_{R\subseteq Q} \rho_{\hat{\theta}}(R)^{N(R)}}{\prod_{R\subseteq Q} \left(\frac{N(R)}{N}\right)^{N(R)}} \quad \overset{d}{\approx} \quad \chi^2_v, \tag{11.15}$$

in which v and χ_v^2 have the same meaning as in Equations (11.10) and (11.11). This result holds if ρ is a smooth enough function of the variables in θ, which is the case in all the models considered here. The decision procedures are as in 11.2.3. The statistical test associated with Equation (11.15) is called a *likelihood ratio test*. The l.h.s. of (11.15) is referred to as a *log-likelihood ratio statistic*. The concordance between the likelihood ratio and the chi-square tests illustrated by Equations (11.12) and (11.11) hold in general. This fact is sometimes expressed by the statement that *the likelihood ratio test and the chi-square test are asymptotically equivalent.*

Maximum likelihood estimates of the parameters have been computed for the data of Table 11.1. The results are given in Table 11.3.

Table 11.3. Maximum likelihood estimates of the parameters of the basic local independence model defined in 11.1.2. The value of the log-likelihood ratio statistic was 12.6, for 31-18=13 degrees of freedom.

Response Probabilities	
$\beta_a = .16$	$\eta_a = .04$
$\beta_b = .16$	$\eta_b = .10$
$\beta_c = .19$	$\eta_c = .00$
$\beta_d = .29$	$\eta_d = .00$
$\beta_e = .14$	$\eta_e = .02$

State Probabilities	
$p(\varnothing) = .05$	$p\{a,b,c\} = .08$
$p\{a\} = .10$	$p\{a,b,d\} = .15$
$p\{b\} = .08$	$p\{a,b,c,d\} = .16$
$p\{a,b\} = .04$	$p\{a,b,c,e\} = .10$
$p\{a,b,c,d,e\} = .21$	

The value of the log-likelihood ratio statistic was 12.6 for $31 - 18 = 13$ degrees of freedom. As in the case of the chi-square test (and not surprisingly, in view of the asymptotic equivalence of the two tests), the data supports the model. A comparison between Tables 11.2 and 11.3 indicates a reasonably good agreement between the estimates of the parameters. In the rest of this book, we shall systematically use likelihood ratio tests. Notice that the estimated values of the η_i parameters in Table 11.3 are quite small, suggesting that the true value of these parameters may be zero. We show in 11.3.6 how to test this hypothesis in the framework of likelihood ratio procedures.

11.3.3 Remarks. a) If the statistical analyses reported above were based on real data, the high values obtained for some of the β_q parameters would be a cause of concern. In the case of Item d, for instance, it would be difficult to explain a situation in which a question has been fully mastered, but is nevertheless answered incorrectly in 46% of the cases (according to Table 11.2). However, the data are artificial and the analysis only intended to illustrate some statistical techniques.

b) The estimates obtained for the parameters were actually more unstable than what is suggested by a comparison between the values in Tables 11.2 and

11.3. For some starting values of the parameters used by the PRAXIS routine, the estimated values were sometimes quite different from those in the Tables (for a chi-square value not very different from those reported). This is not atypical for such search procedures. It usually suggests an overabundance of parameters. In such cases, an equally good fit may be obtained by varying only a subset of those parameters. An obvious candidate is the special case of the basic local independence model in which all the η_q values are kept fixed, equal to 0.

c) An examination of Equation (11.15) indicates that the log-likelihood ratio statistic is based on a comparison of two models, one represented in the denominator, and the other in the numerator. The denominator of (11.15) computes the likelihood of the data for a very general multinomial model assuming only that there is some probability $F(R)$ associated with each response pattern $R \subseteq Q$. It is well-known that the maximum likelihood estimates of the multinomial parameters $F(R)$ is their relative frequency: we have $\widehat{F}(R) = N(R)/N$. These maximum likelihood estimates are used in the computation of the likelihood in the denominator of (11.15). In the numerator of (11.15), we have a similar expression for the basic local independence model, involving also maximum likelihood estimates.

The approximation in Equation (11.15) is based on a general result which is stated informally below, leaving aside some technical conditions.

11.3.4 Theorem. *Let Ω be an s-dimensional subset of $\mathbb{R}^s$. Suppose that $f(\omega; D, N)$ is the likelihood of the data D according to some model, where $\omega \in \Omega$ represents a vector of s independent parameters of the model, N represents the number of observations, and f is a smooth function of ω. Let Ω' be a u-dimensional subset of Ω with $0 < u < s$. If the vector ω_0 of the true values of the parameters lies in Ω', then, under fairly general conditions on the sets Ω and Ω', we have, for large N,*

$$-2 \log \frac{\sup_{\omega \in \Omega'} f(\omega; D, N)}{\sup_{\omega \in \Omega} f(\omega; D, N)} \overset{d}{\approx} \chi^2_{s-u}. \tag{11.16}$$

11.3.5 Remark. The subset Ω' in Equation (11.16) specifies a submodel or a special case of the model represented in the denominator. The likelihood ratio statistic in (11.16) yields a test of the hypothesis $\omega_0 \in \Omega'$ against the general model that $\omega_0 \in \Omega$. The number of degrees of freedom in the chi-square is the difference between the number of estimated parameters in the denominator and in the numerator. In the case of the likelihood ratio test of the basic local independence model, we have $\Omega = \Theta =]0,1[^s$ with $s = 2^m - 1$, and $\Omega' = \Theta'$ is a u-dimensional surface in Θ, with $u = n - 1 + 2m < 2^m - 1$, specified by Equations (11.5) and (11.6). The importance of this theorem from the standpoint of applications is that it justifies a nested sequence of statistical tests of increasing specificity, corresponding to a chain of subsets $\Omega \supset \Omega' \supset \Omega'' \supset \cdots$ of decreasing dimensionality.

11.3.6 Application. We illustrate the use of this theorem to test, for the data of Table 11.1, the hypothesis that $\eta_q = 0$ in the framework of the basic local independence model. This hypothesis specifies the subset Θ'' of Θ' defined by

$$\Theta'' = \{\theta \in \Theta' \mid \forall q \in Q : \eta_q = 0\},$$

where θ symbolizes the vectors

$$\theta = (p(\varnothing), p\{a\}, \ldots, p(Q), \beta_a, \ldots, \beta_e, \eta_a, \ldots, \eta_e).$$

In other words, the model in which all the η_i's are equal to zero is now regarded as a submodel of the basic local independence model. Using Theorem 11.3.4, we obtain the statistic

$$-2\log \frac{\max_{\theta\in\Theta''} \prod_{R\subseteq Q} \rho_\theta(R)^{N(R)}}{\max_{\theta\in\Theta'} \prod_{R\subseteq Q} \rho_\theta(R)^{N(R)}} \overset{d}{\to}\approx \chi^2_5, \qquad (11.17)$$

with $5 = 18 - 13$. The denominator of this log-likelihood ratio statistic is the numerator of Equation (11.15). The value obtained for this chi-square was 1.6, which is nonsignificant (in the terms of 11.2.3). Accordingly, we temporarily retain, for these data, the special case $\eta_q = 0$ for all $q \in \{a, b, \ldots, e\}$. In the sequel, this model will be referred to as the *basic local independence model with no guessing*.

11.4 Learning Models

The number of knowledge states of actual knowledge structures tend to be quite large. For example, in an experiment reviewed in Chapter 17, the number of states is on the order of several million (for about 262 items). This presents a problem for practical applications of the basic local independence model, because it means that a prohibitively large number of parameters—e.g. the probabilities of all these states in the relevant population—may have to be estimated from the empirical frequencies of the response patterns. Even with substantial data sets, reliable estimates may be hard to obtain[4]. One possible solution to this difficulty is to set constraints on the state probabilities, effectively reducing the number of independent quantities involved.

A natural idea is to postulate some learning mechanism describing the successive transitions of a student, over time, from the state $\varnothing$ of complete ignorance to the state Q of full mastery of the material. Several examples of models will be described. All are based on the following basic principle:

The probability that, at the time of the test, a subject is in a state K of the structure is expressed as the probability that this subject

(i) *has successively mastered all the items in the state K, and*
(ii) *has failed to master any item immediately accessible from K.*

[4] However, see Villano (1991).

Here, assuming that the structure is discriminative (cf. Definition 2.1.5), we consider an item q to be immediately accessible, i.e. learnable, from the state K if $K + \{q\}$ is again a state. In other words, q belongs to the outer fringe of the state K, namely $K^{\mathcal{O}} = \{q \in Q \setminus K \,|\, K \cup \{q\} \in \mathcal{K}\}$ (cf. Definition 4.1.6).

11.4.1 A simple learning model. The model considered in this section makes strong independence assumptions regarding the learning process. Consider a discriminative knowledge structure $(Q, \mathcal{K})$. For each item q in Q, we introduce one parameter g_q, with $0 \leq g_q \leq 1$, intended to measure the probability that q is mastered. To define a probability distribution p on $\mathcal{K}$, we suppose that, for each state K, all the events in the two classes: '*mastering any item* $q \in K$' and '*failing to master any item* $q \in K^{\mathcal{O}}$' are (conditionally) independent in the sense of the formula:

$$p(K) = \prod_{q \in K} g_q \prod_{q' \in K^{\mathcal{O}}} (1 - g_{q'}). \tag{11.18}$$

This formula applies to all states K by virtue of the convention that a product of zero terms equals 1. When p is a probability distribution on $\mathcal{K}$, and a response function r is provided, the quadruple $(Q, \mathcal{K}, p, r)$ is called a *simple learning model*. This model specifies the state probabilities in terms of only $m = |Q|$ parameters, regardless of the number of states. In the case of our standard example (Figure 11.1), the probabilities of the states are given by the formulas in Table 11.4.

States	Probabilities
$\varnothing$	$(1 - g_a)(1 - g_b)$
$\{a\}$	$g_a(1 - g_b)$
$\{b\}$	$g_b(1 - g_a)$
$\{a, b\}$	$g_a g_b(1 - g_c)(1 - g_d)$
$\{a, b, c\}$	$g_a g_b g_c(1 - g_d)(1 - g_e)$
$\{a, b, d\}$	$g_a g_b g_d(1 - g_c)$
$\{a, b, c, d\}$	$g_a g_b g_c g_d(1 - g_e)$
$\{a, b, c, e\}$	$g_a g_b g_c g_e(1 - g_d)$
$\{a, b, c, d, e\}$	$g_a g_b g_c g_d g_e.$

Table 11.4. Probabilities of the states in the structure $\mathcal{H}$ of Equation (11.1) and Figure 11.1 for the simple learning model.

The probabilities of the nine states are thus expressed in terms of 5 parameters. It is easy to verify that these particular probabilities add up to one. In general, however, the quantities $p(K)$ defined by Equation (11.18) do not necessarily specify a probability distribution. We give two examples in the next section.

11.4.2 Test of the simple learning model. This model was tested on the data of Table 11.1. We made the assumption that the response function r was specified by the local independence condition with no guessing. The resulting model is then a special case of the basic local independence model with no guessing, which has already been tested. Accordingly, it makes sense to use a likelihood ratio procedure.

A likelihood ratio test was performed, which yielded a value 15.5 for the log-likelihood ratio statistic. The number of degrees of freedom is $3 = 13 - 10$. Indeed, the basic local independence model with no guessing has 13 parameters, while the simple learning model (with the same local independence assumptions) has 10 parameters, and no grouping of the response patterns was necessary. This value of the chi-square statistic is very significant, leading to a strong rejection of the model.

11.4.3 Remark. This model was introduced to illustrate, in a simple case, how assumptions about the learning process could dramatically reduce the number of parameters. Objections can certainly be made to the hypothesis that the probability of mastering an item q does not depend on the current state K of the subject, provided that the item is learnable from that state, that is, provided that $q \in K^{\mathcal{O}}$. This seems very strong. The independence assumptions are also difficult to accept. This model can be elaborated by assuming that the probability of mastering an item may depend upon past events, for example upon the last item learned. We shall not develop this idea here[5].

A different kind of model is considered later in this chapter, in which a knowledge structure is regarded as the state space of a Markov chain describing the learning process. Before discussing such a model, we return to a question left pending concerning the conditions under which the simple learning model can be defined. We shall ask: under which conditions do the quantities defined by Equation (11.18) add up to one?

11.5 A Combinatorial Result

The simple learning model introduced in 11.4.1 specifies the values $p(K)$ in terms of the parameters g_q, for all K in $\mathcal{K}$. However, the following two examples show that the real-valued mapping p defined by Equation (11.18) does not always provide a probability distribution on $\mathcal{K}$. This prompts searching for conditions on a knowledge structure that guarantee that the simple learning model is applicable. In other words, we look for conditions under which Equation (11.18) defines a genuine probability distribution. Interestingly, the concept of a learning space will be crucial.

[5] However, see Problem 3.

11.5.1 Example. Let $\mathcal{G} = \{\varnothing, \{a\}, \{b\}, \{c\}, \{a,b\}, \{a,c\}, \{a,b,c\}\}$ and define $g_q = \frac{1}{2}$ for all $q \in \{a,b,c\}$. Equation (11.18) gives

$$p(\varnothing) = p\{a\} = p\{a,b\} = p\{a,c\} = p\{a,b,c\} = \frac{1}{8},$$
$$p\{b\} = p\{c\} = \frac{1}{4},$$

yielding $\sum_{K\in\mathcal{G}} p(K) = \frac{9}{8}$. This knowledge structure is well-graded, but is not a knowledge space: we have $\{b\},\{c\} \in \mathcal{G}$, but $\{b,c\} \notin \mathcal{G}$.

11.5.2 Example. Let $\mathcal{H} = \{\varnothing, \{c\}, \{a,b\}, \{b,c\}, \{a,b,c\}\}$. Then according to Equation (11.18),

$$\sum_{K\in\mathcal{H}} p(K) = (1-g_c) + g_c(1-g_b) + g_a g_b(1-g_c) + g_b g_c(1-g_a) + g_a g_b g_c$$
$$= 1 + g_a g_b (1-g_c),$$

which is equal to 1 only for some special values of g_a, g_b, and g_c. It is easily checked that $\mathcal{H}$ is a knowledge space which is not well-graded.

We state a preparatory result.

11.5.3 Theorem. *Let $(Q,\mathcal{K})$ be a finite discriminative knowledge structure. Then $(Q,\mathcal{K})$ is a learning space if and only if it satisfies the following condition:*

[U] *Each set $A \subseteq Q$ includes at most one state K such that $K^{\mathcal{O}} \cap A = \varnothing$.*

PROOF. Suppose that Condition [U] holds. We first show that $(Q,\mathcal{K})$ is a knowledge space, and then that it is well-graded. If L, M are two states in $\mathcal{K}$, we prove $L \cup M \in \mathcal{K}$ by recurrence on $|L \triangle M|$. The conclusion clearly holds if $|L \triangle M| = 0$, so let us assume $|L \triangle M| > 0$. Condition [U] applied to $A = L \cup M$ entails that not both of the distinct states L and M can play the role of K. Thus $L^{\mathcal{O}} \cap A \neq \varnothing$ or $M^{\mathcal{O}} \cap A \neq \varnothing$ (or both). This implies the existence of some item q with

$$q \in M \setminus L \quad \text{and} \quad L \cup \{q\} \in \mathcal{K}$$

or

$$q \in L \setminus M \quad \text{and} \quad M \cup \{q\} \in \mathcal{K}.$$

In the first case, we get by recurrence $(L \cup \{q\}) \cup M \in \mathcal{K}$, that is $L \cup M \in \mathcal{K}$. The conclusion the second case follows by symmetry. Thus Condition [U] implies that $(Q,\mathcal{K})$ is a space.

Condition [U] also implies that $(Q,\mathcal{K})$ is well-graded. Indeed, if this were not true, we would derive by Lemma 2.2.3 the existence of two states L and M with $L \subset M$, and such that $|M \setminus L| \geq 2$ with no state N satisfying

$L \subset N \subset M$. Taking $A = M$, we derive the required contradiction since both L and M can play the role of K in Condition [U].

We conclude that $(Q, \mathcal{K})$ is a well-graded knowledge space. By Theorem 2.2.4, it is thus a learning space.

Conversely, suppose that $(Q, \mathcal{K})$ is a learning space. For any subset A of Q, the only state K as in Condition [U] is the largest state included in A (that is, the union of all the states included in A). $\square$

11.5.4 Theorem. *When $(Q, \mathcal{K})$ is a learning space space, the real-valued mapping p defined by Equation (11.18) specifies a probability distribution on $\mathcal{K}$ for any mapping g from $\mathcal{K}$ to $[0,1]$. As a partial converse, if $(Q, \mathcal{K})$ is any finite discriminative knowledge structure, g is a mapping from Q to $]0,1[$, and the mapping p defined from Equation (11.18) is a probability distribution, then $(Q, \mathcal{K})$ is a learning space.*

PROOF.[6] Suppose that $(Q, \mathcal{K})$ is a learning space. Thus, by Theorem 11.5.3, Condition [U] is satisfied. Let g be any function from Q to $[0,1]$. We define a function $h : 2^Q \to [0,1]$ by

$$h(A) = \prod_{q \in A} g_q \prod_{q \in Q \setminus A} (1 - g_q).$$

With $p : \mathcal{K} \to [0,1]$ from Equation (11.18), using the mapping h just defined, we get

$$\begin{aligned} p(K) &= \prod_{q \in K} g_q \prod_{q \in K^{\mho}} (1 - g_q) \\ &= \left(\prod_{q \in K} g_q \prod_{q \in K^{\mho}} (1 - g_q) \right) \times \prod_{q \in Q \setminus (K \cup K^{\mho})} (g_q + (1 - g_q)) \\ &= \sum_{A \in 2^Q,\, A \supseteq K,\, A \cap K^{\mho} = \varnothing} h(A). \end{aligned}$$

(The last equality follows by distributing the rightmost product). Thus

$$\sum_{K \in \mathcal{K}} p(K) = \sum_{K \in \mathcal{K}} \left(\sum_{A \in 2^Q,\, A \supseteq K,\, A \cap K^{\mho} = \varnothing} h(A) \right). \tag{11.19}$$

On the other hand,

$$1 = \prod_{q \in Q} (g_q + (1 - g_q)) = \sum_{A \in 2^Q} h(A). \tag{11.20}$$

[6] Except for the argument that Condition [U] implies wellgradedness, we owe this proof to an anonymous referee of Falmagne (1994).

By Condition [U], each subset A of Q includes exactly one state K with $A \cap K^{\mathcal{O}} = \varnothing$. We conclude that each term $h(A)$, for $A \in 2^Q$, appears exactly once in the right-hand side of Equation (11.19). Hence, by comparing (11.19) and (11.20), we derive $\sum_{K \in \mathcal{K}} p(K) = 1$. Moreover, the converse also holds since g takes its values in $]0, 1[$, (so $h(A) > 0$ for all subsets A). □

11.6 Markov Chain Models

The models described in this section also explain the state probabilities of the basic probabilistic model by a learning process. However, these models differ from the simple learning model of Section 11.4.1 in that the time is introduced explicitly, taking discrete values[7] $n = 1, 2, \ldots$ The response function r is specified by the parameters β_q and η_q as in the local independence Equation (11.6). We assume that the knowledge structure is well-graded[8]. All these models are formulated in the framework of Markov chain theory[9].

11.6.1 Markov chain Model 1. We assume that learning takes place in discrete steps. On any given step, at most one item can be mastered. In the case of our standard example in Figure 11.1, a transition between state $\varnothing$ to state $\{a\}$ or to state $\{b\}$ may occur on step one or later, if neither of the two states $\{a\}$ or $\{b\}$ have been achieved. The probabilities of such transitions are specified by some parameters g_a and g_b. We assume that these probabilities do not depend upon past events. Thus, the probability of a transition from state K to state $K + \{q\}$, with q in the outer fringe of K, is equal to g_q. A sample of subjects tested at a given time is assumed to have accomplished some number n of steps. This number is the same for all subjects and is a parameter, which has to be estimated from the data. If n is large, the probabilities of states containing many items will be large.

This model is a Markov chain having as a state space the knowledge structure $\mathcal{K}$ (so, the states of the Markov chain coincide with the knowledge states). The transition probabilities, contained in a matrix $M = (m_{KL})$ with m_{KL} specifying the probability of a transition from K to L, are defined by

$$m_{KL} = \begin{cases} g_q & \text{if } L = K + \{q\}, \text{ with } q \in K^{\mathcal{O}}, \\ 1 - \sum_{q \in K^{\mathcal{O}}} g_q & \text{if } L = K, \\ 0 & \text{otherwise.} \end{cases} \tag{11.21}$$

A case of such a matrix is given in Table 11.5 for our standard example of the knowledge structure $\mathcal{H}$ defined in (11.1.1). For simplicity, we represent

[7] Note that, in this section, n does not denote the number $|\mathcal{K}|$ of states.

[8] It is thus finite and discriminative; cf. 2.2.2.

[9] For an introduction, see Feller (1968), Kemeny and Snell (1960), Parzen (1994), or Shyryayev (1960).

any knowledge state by a string listing its elements. Also, since the matrix is quite large, we adopt the abbreviations

$$\overline{g}_q = 1 - g_q \quad \text{and} \quad \overline{g}_{qr} = 1 - g_q - g_r \qquad (\text{for } q, r \in \{a, b, c, d, e\}). \tag{11.22}$$

The process begins with a vector ν_0 specifying the initial probabilities of the states. (Since the states of the Markov chains are confounded with the knowledge states, no clash of terminology can arise.) The probabilities after one step are thus given by the row vector

$$\nu_1 = \nu_0 \, M.$$

Table 11.5. Transition matrix M of the Markov Model 1, for our standard example.

	$\varnothing$	a	b	ab	abc	abd	$abcd$	$abce$	Q
$\varnothing$	$\overline{g}_{ab}$	g_a	g_b	0	0	0	0	0	0
a	0	$\overline{g}_b$	0	g_b	0	0	0	0	0
b	0	0	$\overline{g}_a$	g_a	0	0	0	0	0
ab	0	0	0	$\overline{g}_{cd}$	g_c	g_d	0	0	0
abc	0	0	0	0	$\overline{g}_{de}$	0	g_d	g_e	0
abd	0	0	0	0	0	$\overline{g}_c$	g_c	0	0
$abcd$	0	0	0	0	0	0	$\overline{g}_e$	0	g_e
$abce$	0	0	0	0	0	0	0	$\overline{g}_d$	g_d
Q	0	0	0	0	0	0	0	0	1

It may often be sensible to suppose that the subjects start the learning process in state $\varnothing$. In this case, the initial probability vector takes the form

$$\nu_0 = (\underbrace{1, 0, \ldots, 0}_{|\mathcal{K}| \text{ terms}}).$$

If this assumption is used in the case of our example, the probabilities of the states after the first and the second step would then be

$$\begin{aligned} \nu_1 &= (\overline{g}_{ab},\, g_a,\, g_b,\, 0,\, 0,\, 0,\, 0,\, 0,\, 0), \\ \nu_2 &= (\overline{g}_{ab}^2,\, \overline{g}_{ab} g_a + g_a \overline{g}_b,\, \overline{g}_{ab} g_b + g_b \overline{g}_a,\, 2 g_a g_b,\, 0,\, 0,\, 0,\, 0,\, 0), \end{aligned}$$

that is, the first row of $M^1 = M$ and M^2 respectively. In general, the state probabilities after state n are obtained from

$$\nu_n = \nu_0 \, M^n.$$

Writing $\nu_n(K)$ for the probability of state K at step n, we obtain, for the probability $\rho_n(R)$ of a response pattern R at step n, the prediction

$$\rho_n(R) = \sum_{K \in \mathcal{H}} r(R, K)\nu_n(K). \tag{11.23}$$

Using standard concepts of the theory of Markov chains[10], it is easy to show that if $g_q > 0$ for all q, then

$$\lim_{n\to\infty} \nu_n(K) = \begin{cases} 1 & \text{if } K = Q \\ 0 & \text{otherwise.} \end{cases}$$

Notice that the number of parameters, including the parameters β_q and η_q entering in the specification of the function r in Equation (11.6) and the step number n representing the time of the test, cannot exceed $3m + 1$, with $m = |Q|$. We shall not develop this model any further here, and no empirical application will be presented.

11.6.2 Remarks. a) This model is vulnerable to the same kind of criticisms as those addressed to the simple learning model. In particular, the probability of learning a new item q depends upon past events only in a trivial way: this probability is equal to g_q if q is learnable from the subject's current state K, that is, if $q \in K^{\mathcal{O}}$, and is equal to 0 otherwise. However, a straightforward adaptation of this model is available, which is outlined in our next subsection (Markov Chain Model 2).

b) Another objection lies in the implicit assumption that all the subjects had the same amount of learning, represented by the step number n in Equation (11.23). The only difference between the subjects' states lies in the chance factors associated with the transition parameters g_q. This model could be generalized by postulating the existence of a probability distribution on the set of positive integers representing the learning step. For instance, we could assume that the learning step is distributed as a negative binomial. These developments will not be pursued here (but see Problem 6).

c) An essential difference between this model and the Simple Learning Model should not be overlooked. Markov Model 1 is capable of predicting the results of data obtained from a sample of subjects having been tested several times. Suppose, for example, that the same test has been given to a sample of subject before and after a special training period. A pair (R, R') of response patterns is thus available for each subject. The required prediction concerns the probabilities $\rho_{n,n+j}(R, R')$ of observing response pattern R at step n and response pattern R' at step $n+j$ (with n and j being parameters). Using standard techniques of Markov chains theory, these predictions could be derived from Markov Model 1 (and also from Markov Model 2 below). The number of parameters would not exceed $3m+2$: we would have the same

[10] Some basic terminology of Markov chains is recalled in Section 14.4.

parameters as in the case of a single test, plus one, namely the positive integer j entering in the specification of the step number $n + j$ of the second test. This parameter could be used as a measure of the efficiency of the training period. Such predictions could not be derived from the Simple Learning Model without further elaboration (cf. Problem 7).

11.6.3 Markov Chain Model 2. The main concepts are as in Markov Chain Model 1, except that the last item learned affects the probabilities of learning a new one. We also have a Markov chain, but we have to keep track of the last item learned. In other words, except for the empty set, the states of the Markov chain take the form of a pair (K, q), with q in the inner fringe $K^{\mathcal{I}} = \{s \in Q \,|\, K \setminus \{s\} \in \mathcal{K}\}$ of K (cf. Definition 4.1.6).

To avoid confusion, we shall refer to the states of the Markov chain as *m-states*. The probability of a transition from m-state (K, q), with $q \in K^{\mathcal{I}}$, to some m-state $(K \cup \{s\}, s)$ (this corresponds to a transition from the knowledge state K to a knowledge state $K + \{s\}$) depends only on q and s. We denote it as g_{qs}. For every m-state (K, q), the probabilities g_{qs} satisfy the constraint $\sum_{s \in K^{\mathcal{O}}} g_{qs} \leq 1$.

We also set the probability of remaining in m-state (K, q) equal to $1 - \sum_{s \in K^{\mathcal{O}}} g_{qs}$. Hence the probabilities of a transition from m-state (K, q) to any other m-state (K', s), with $K' \neq K + \{s\}$, are equal to zero. Needless to say, the empty set is also a m-state. The probability of a transition from that state to state $(\varnothing, q)$ is denoted by g_{0q}. All the other details are as in Model 1. We leave further developments as Problem 15.

11.7 Probabilistic Projections

In Section 2.4, we defined a projection of a knowledge structure $(Q, \mathcal{K})$ as a knowledge structure $(Q', \mathcal{K}_{|Q'})$ whose states are the traces of the states of $\mathcal{K}$ on some nonempty set $Q' \subset Q$, thus

$$\mathcal{K}_{|Q'} = \{L \in 2^{Q'} \,|\, \exists K \in \mathcal{K} : L = K \cap Q'\}.$$

The motivation for this concept in the context of this chapter is that some empirical knowledge structure may be so large that a direct statistical study aimed at estimating the state probabilities may not be feasible, even in the framework of learning models such as those that we just described.

In such a case, partial information on these probabilities can be obtained from studying projections of the given knowledge structure. Consider, for example, a knowledge structure with a domain containing 300 problems of high school mathematics covering say, beginning algebra[11]. The relevant knowledge structure may contain several millions knowledge states. The empirical

[11] Three hundred or so items is realistic for beginning algebra (see Chapter 17).

analysis of such a structure encounters two kinds of difficulties. For one, it may not be practical to propose a 300 problem test to a large enough number of students. For the other, even if such a test is given, the analysis of the data in terms of probabilistic knowledge structures may be very difficult, in particular because of the large number of parameters to be estimated. However, the analysis of a number of shorter tests made of problems taken from the same domain may be manageable, and would reveal useful information on the full structure.

Notice that the probabilities of the knowledge states of the parent structure have a natural importation into any projection. This is illustrated in the example below.

11.7.1 Example. Suppose that the states of the knowledge structure

$$\mathcal{H} = \{\varnothing, \{a\}, \{b\}, \{a,b\}, \{a,b,c\}, \{a,b,d\}, \{a,b,c,d\}, \\ \{a,b,c,e\}, \{a,b,c,d,e\}\}$$

of our standard example occur in a population of reference with the probabilities listed below:

$$\begin{aligned} p(\varnothing) &= .04 & p\{a,b\} &= .12 & p\{a,b,c,d\} &= .13 \\ p\{a\} &= .10 & p\{a,b,c\} &= .11 & p\{a,b,c,e\} &= .18 \\ p\{b\} &= .06 & p\{a,b,d\} &= .07 & p\{a,b,c,d,e\} &= .19 \end{aligned} \tag{11.24}$$

If a test consisting only of questions a, d and e is considered, the knowledge state $\{a,d\}$ of $\mathcal{H}' = \mathcal{H}_{|\{a,d,e\}}$ will occur in the population with a probability

$$p'\{a,d\} = p\{a,b,d\} + p\{a,b,c,d\} = .07 + .13 = .20.$$

Indeed, any student in state $\{a,b,d\}$ or $\{a,b,c,d\}$ of the structure $\mathcal{H}$ will appear to be in the state $\{a,d\}$ of the structure $\mathcal{H}'$ if only questions a, d and e are considered.

More generally, the probability $p'(J)$ of any state J of $\mathcal{H}'$ is the sum of the probabilities $p(K)$ of all the states K of $\mathcal{H}$ having their trace on $\{a,d,e\}$ equal to J. Therefore, all the state probabilities p' are as follows:

$$\begin{aligned} p'(\varnothing) = .10, \qquad p'\{a\} = .33, \qquad p'\{a,d\} = .20, \\ p'\{a,e\} = .18, \qquad p'\{a,d,e\} = .19. \end{aligned} \tag{11.25}$$

Definition 11.7.3 generalizes this example. We first recall some concepts and notation from an earlier definition (cf. 2.4.1).

11.7.2 Definition. Let $(Q, \mathcal{K})$ be a knowledge structure, and let $\mathcal{K}' = \mathcal{K}_{|Q'}$ be the projection of $\mathcal{K}$ on some proper subset $Q' \subset Q$. Notice that the mapping $K \mapsto K \cap Q' = J$ from $\mathcal{K}$ onto the projection $\mathcal{K}_{|Q'}$ defines an equivalence relation on the parent structure $\mathcal{K}$, with equivalence classes[12]

$$[K/Q'] = \{K' \in \mathcal{K} \mid K \cap Q' = K' \cap Q'\}.$$

For any $J \in \mathcal{K}_{|Q'}$ we write $J^\diamond = \{K \in \mathcal{K} \mid K \cap Q' = J\}$. The family $J^\diamond \subseteq \mathcal{K}$ is called the *parent family* of J, and we have $\cup_{J \in \mathcal{K}'} J^\diamond = \mathcal{K}$. The ambiguity that may arise from this compact notation when more than one induced family is considered will always be eliminated by the context.

11.7.3 Definition. Suppose that $(Q, \mathcal{K}, p)$ is a probabilistic knowledge structure (cf. 11.1.2), with $\mathcal{K}'$ and Q' as in Definition 11.7.2. In this case, the triple $(Q', \mathcal{K}', p')$ is called the *(probabilistic) projection induced* by Q' if for all $J \in \mathcal{K}'$ we have

$$p'(J) = \sum_{K \in J^\diamond} p(K). \tag{11.26}$$

As the parent families $J^\diamond$ are equivalence classes of a partition of $\mathcal{K}$, we have

$$\sum_{J \in \mathcal{K}'} p'(J) = \sum_{J \in \mathcal{K}'} \sum_{K \in J^\diamond} p(K) = \sum_{K \in \mathcal{K}} p(K) = 1.$$

From the standpoint of applications, the inverse case is the important one: the state probabilities of a projection are known, and one wishes to make some inferences on the state probabilities of the parent structure.

11.7.4 Example. Let $\mathcal{H}$ and $\mathcal{H}'$ be as Example 11.7.1, and suppose that only the state probabilities $p'(J)$, for $J \in \mathcal{H}'$, are available, their values being those given in Equation (11.25). For instance, the only thing we know about the state probabilities $p(\varnothing)$ and $p\{b\}$ is that they have to satisfy $p(\varnothing) + p\{b\} = p'(\varnothing) = .10$. In this case, it may seem reasonable to split the mass of the state $\varnothing$ of $\mathcal{H}'$ into two equal parts, and to set $p(\varnothing) = p\{b\} = p'(\varnothing)/2 = .05$.

In general, the idea is to assign, for each $J \in \mathcal{H}'$, the same probability to each state of the equivalence class $J^\diamond$. The resulting probabilities for all the states of $\mathcal{H}$ are as follows

$$\begin{aligned} p(\varnothing) &= .05 & p\{a,b\} &= .11 & p\{a,b,c,d\} &= .10 \\ p\{a\} &= .11 & p\{a,b,c\} &= .11 & p\{a,b,c,e\} &= .18 \\ p\{b\} &= .05 & p\{a,b,d\} &= .10 & p\{a,b,c,d,e\} &= .19 \end{aligned}$$

11.7.5 Definition. Let $\mathcal{K}' = \mathcal{K}_{|Q'}$ be the projection induced by a proper subset Q' of Q, and suppose that $(Q', \mathcal{K}', p')$ is a probabilistic knowledge structure. Then $(Q, \mathcal{K}, p)$ is a *uniform extension* of $(Q', \mathcal{K}', p')$ to $(Q, \mathcal{K})$ if for all $K \in \mathcal{K}$

$$p(K) = \frac{p'(K \cap Q')}{|(K \cap Q')^\diamond|}. \tag{11.27}$$

---

[12] The shorter notation $[K]$ was used in Definition 2.4.1 for these equivalence classes. The more explicit $[K/Q']$ is needed here because several projections are in play.

11.8 Nomenclatures and Classifications

We can also think of combining the information from several probabilistic projections of a given knowledge structure on different subsets of its domain in order to manufacture the probabilistic parent structure. This section sketches the mechanism of such a construction.

11.8.1 Definition. Let $Q_1, \ldots, Q_k$ be nonempty subsets of the domain of a knowledge structure $(Q, \mathcal{K})$. If $\cup_{i=1}^k Q_i = Q$, then the collection of projections $\mathcal{K}|_{Q_i}$, $i = 1, \ldots k$ is a *nomenclature* of $(Q, \mathcal{K})$. If $\{Q_1, \ldots, Q_k\}$ is a partition of Q, then the nomenclature is called a *classification*.

A simple result in this connection is as follows.

11.8.2 Theorem. *Let $\{Q_1, \ldots, Q_k\}$ be a finite collection of subsets of the domain Q of a knowledge structure $\mathcal{K}$. Then, for any state K, we have*

$$[K/(Q_1 \cup \cdots \cup Q_k)] = [K/Q_1] \cap \cdots \cap [K/Q_k]. \tag{11.28}$$

Moreover, if the collection of projections $\mathcal{K}|_{Q_i}$ is a nomenclature, then

$$\{K\} = [K/Q_1] \cap \cdots \cap [K/Q_k]. \tag{11.29}$$

PROOF. Denoting by $\wedge$ the logical conjunction, it is easily verified that

$$J \cap (Q_1 \cup \cdots \cup Q_k) = K \cap (Q_1 \cup \cdots \cup Q_k) \quad \Longleftrightarrow \quad \wedge_{i=1}^k (J \cap Q_i = K \cap Q_i).$$

This implies Equation (11.28). The special case follows immediately. □

11.8.3 Remark. A special case of Theorem 11.8.2 arises when each of the sets Q_i contains a single element. We have then, using the notation $\mathcal{K}_q$ (resp. $\mathcal{K}_{\bar{q}}$) for the collection of all states containing (resp. not containing) q (cf. 2.1.4, resp. 9.1.2),

$$[K/\{q\}] = \begin{cases} \mathcal{K}_q & \text{if } q \in K, \\ \mathcal{K}_{\bar{q}} & \text{if } q \notin K. \end{cases}$$

11.9 Independent Projections

Combining the state probabilities of several projections to construct a probability distribution for the parent structure is feasible in certain cases; for instance, if the projections can be considered as 'independent' in the sense of the next definition.

11.9.1 Definition. Let $(Q, \mathcal{K}, p)$ be a probabilistic knowledge structure (cf. 11.1.2), and let $\mathbb{P}$ be the induced probability measure on the power set of $\mathcal{K}$. (Thus, for any $\mathcal{F} \subseteq \mathcal{K}$, $\mathbb{P}(\mathcal{F}) = \sum_{K \in \mathcal{F}} p(K)$.) Take two proper subsets $Q', Q'' \subset Q$, and consider the probabilistic projections $(Q', \mathcal{K}', p')$ and $(Q'', \mathcal{K}'', p'')$ (in the sense of Definition 11.7.3). Two states $J \in \mathcal{K}'$, $L \in \mathcal{K}''$ are called *independent* if the events $J^\diamond = \{K \in \mathcal{K} \,|\, K \cap Q' = J\}$ and $L^\diamond = \{K \in \mathcal{K} \,|\, K \cap Q'' = L\}$ are independent in the probability space $(\mathcal{K}, 2^{\mathcal{K}}, \mathbb{P})$, that is if

$$\mathbb{P}(J^\diamond \cap L^\diamond) = \mathbb{P}(J^\diamond) \cdot \mathbb{P}(L^\diamond) = p'(J) \cdot p''(L).$$

The projections $(Q', \mathcal{K}', p')$ and $(Q'', \mathcal{K}'', p'')$ are *independent* if any state $K \in \mathcal{K}$ has independent traces on Q' and Q''.

These concepts extend in a natural manner to the case of an arbitrary (finite) number of structures. Suppose that $\Upsilon = (Q_i, \mathcal{K}_i, p_i)_{1 \leq i \leq k}$ is a collection of probabilistic projections of a probabilistic knowledge structure $(Q, \mathcal{K}, p)$. A collection of traces $J_i \in \mathcal{K}_i$, $i = 1, \ldots, k$ is *independent* if

$$\mathbb{P}(\bigcap_{i=1}^{k} J_i{}^\diamond) = \prod_{i=1}^{k} \mathbb{P}(J_i{}^\diamond) = \prod_{i=1}^{k} p_i(J_i).$$

(It is not sufficient to require that the traces J_i are pairwise independent; see Problem 13.) The collection Υ is said to be *independent* if any state K of the parent structure $\mathcal{K}$ has an independent collection of traces $K \cap Q_i$. If, in addition, Υ is a nomenclature (cf. 11.8.1), then it is called an *independent representation* of $(Q, \mathcal{K}, p)$. In this case, we must have for any $K \in \mathcal{K}$, using Equation (11.29) in Theorem 11.8.2,

$$\mathbb{P}(\{K\}) = \mathbb{P}(\cap_{i=1}^{k} [K/Q_i]) = \prod_{i=1}^{k} \mathbb{P}([K/Q_i]). \tag{11.30}$$

11.9.2 Examples. a) With $\mathcal{H}$ and $\mathcal{H}'$ as in Example 11.7.1, consider the projection

$$\mathcal{H}'' = \{\,\varnothing,\ \{b\},\ \{b,c\},\ \{b,c,e\}\,\},$$

induced by $\mathcal{H}$ on $\{b, c, e\}$. Using Equation (11.26) we obtain the state probabilities of $\mathcal{H}''$ from those of $\mathcal{H}$ given in Equation (11.24):

$$p''(\varnothing) = .14, \quad p''\{b\} = .25, \quad p''\{b,c\} = .24, \quad p''\{b,c,e\} = .37.$$

(Notice in passing that $\{\mathcal{H}', \mathcal{H}''\}$ form a nomenclature of $\mathcal{H}$.) Any state H of $\mathcal{H}$ belongs to a class of the partition generated by intersecting the classes of the two partitions associated with $\mathcal{H}'$ and $\mathcal{H}''$. We have, for example, for the empty state of $\mathcal{H}$

$$\varnothing \in (\varnothing \cap \{a,d,e\})^\diamond \cap (\varnothing \cap \{b,c,e\})^\diamond = \{\varnothing, \{b\}\} \cap \{\varnothing, \{a\}\} = \{\varnothing\}.$$

The states $\varnothing \in \mathcal{H}'$ and $\varnothing \in \mathcal{H}''$ are not independent, since we have

$$\begin{aligned} p'(\varnothing) \times p''(\varnothing) &= \big(p(\varnothing) + p\{b\}\big) \times \big(p(\varnothing) + p\{a\}\big) \\ &= (.04 + .06) \times (.04 + .10) \\ &= .014 \\ &\neq .04 = p(\varnothing). \end{aligned}$$

Problem 8 asks the reader to modify the state probabilities in this example in such a way that the the states $\varnothing \in \mathcal{H}'$ and $\varnothing \in \mathcal{H}''$ are independent, but the projections $\mathcal{H}'$ and $\mathcal{H}''$ themselves are not.

b) On the other hand, consider the knowledge structure $\{\varnothing, \{a\}, \{b\}, \{a, b\}\}$, with states probabilities

$$p(\varnothing) = .05, \quad p\{a\} = .15, \quad p\{b\} = .20, \quad p\{a, b\} = .60.$$

The two probabilistic projections induced on $\{a\}$ and $\{b\}$ are independent, and thus form a independent representation (Problem 9). The following result is a straightforward consequence of Definition 11.9.1.

11.9.3 Theorem. *If $(Q_i, \mathcal{K}_i, p_i)$, $i = 1, \ldots, k$ is an independent representation of a knowledge structure $(Q, \mathcal{K}, p)$, then*

$$\sum_{K \in \mathcal{K}} \prod_{i=1}^{k} p_i(K \cap Q_i) = 1.$$

PROOF. Writing $\mathbb{P}$ for the probability measure induced on $2^{\mathcal{K}}$, we obtain, using Equation (11.30),

$$1 = \sum_{K \in \mathcal{K}} p(K) = \sum_{K \in \mathcal{K}} \mathbb{P}(\{K\}) = \sum_{K \in \mathcal{K}} \prod_{i=1}^{k} \mathbb{P}([K/Q_i]) = \sum_{K \in \mathcal{K}} \prod_{i=1}^{k} p_i(K \cap Q_i).$$

□

This suggests the following construction. Consider a collection of projections $(Q_i, \mathcal{K}_i)$ forming a representation of a basic knowledge structure $(Q, \mathcal{K})$, and suppose that the state probabilities p_i of the projections are available. If one has reasons to believe that the projections are approximately independent, then the state probabilities $p(K)$ of the parent could be approximated by the formula

$$p(K) = \frac{\prod_{i=1}^{k} p_i(K \cap Q_i)}{\sum_{L \in \mathcal{K}} \prod_{i=1}^{k} p_i(L \cap Q_i)}.$$

The concept of independence introduced in this section is consistent with other, standard ones. As an example, we consider the correlation between items.

11.9.4 Definition. Take a probabilistic knowledge structure $(Q, \mathcal{K}, p)$ and suppose that the basic local independence model holds (cf. Definition 11.1.2). We thus have a collection of parameters $\beta_q, \eta_q \in [0,1[$, $q \in Q$, specifying the response function r of Equation (11.6). For any $q \in Q$, define a random variable

$$\mathbf{X}_q = \begin{cases} 1 & \text{if the subject's response is correct,} \\ 0 & \text{otherwise.} \end{cases}$$

The $\mathbf{X}_q$'s will be called *item indicator* random variables.

11.9.5 Theorem. *Assume the basic local independence model holds, and consider the following three conditions for two items q and q' with $q \neq q'$:*

(i) *the item indicator random variables $\mathbf{X}_q$, $\mathbf{X}_{q'}$ are independent;*
(ii) *their covariance vanishes: $Cov(\mathbf{X}_q, \mathbf{X}_{q'}) = 0$;*
(iii) *the probabilistic projections $\{\{\varnothing, \{q\}\}$ and $\{\varnothing, \{q'\}\}$ induced on $\{q\}$ and $\{q'\}$, respectively, are independent.*

Then (i) $\Leftrightarrow$ (ii) $\Rightarrow$ (iii).

PROOF. We write $\mathcal{K}$ for the knowledge structure, and $\mathbb{P}$ for the probability measure on $2^{\mathcal{K}}$. Let $\beta_q, \eta_q \in [0,1[$ be the response parameters. Since

$$\begin{aligned} \mathrm{Cov}(\mathbf{X}_q, \mathbf{X}_{q'}) &= \mathrm{E}(\mathbf{X}_q \mathbf{X}_{q'}) - \mathrm{E}(\mathbf{X}_q)\mathrm{E}(\mathbf{X}_{q'}) \\ &= \mathbb{P}(\mathbf{X}_q = 1, \mathbf{X}_{q'} = 1) - \mathbb{P}(\mathbf{X}_q = 1)\mathbb{P}(\mathbf{X}_{q'} = 1), \end{aligned} \tag{11.31}$$

the equivalence of (i) and (ii) is clear. Developing the r.h.s. of Equation (11.31), using Equations (11.5) and (11.6), we obtain

$$\begin{aligned} &\mathbb{P}(\mathcal{K}_q \cap \mathcal{K}_{q'})(1-\beta_q)(1-\beta_{q'}) + \mathbb{P}(\mathcal{K}_q \cap \mathcal{K}_{\overline{q'}})(1-\beta_q)\eta_{q'} \\ &\quad + \mathbb{P}(\mathcal{K}_{\overline{q}} \cap \mathcal{K}_{q'})\eta_q(1-\beta_{q'}) + \mathbb{P}(\mathcal{K}_{\overline{q}} \cap \mathcal{K}_{\overline{q'}})\eta_q\eta_{q'} \\ &\quad\quad - \big(\mathbb{P}(\mathcal{K}_q)(1-\beta_q) + \mathbb{P}(\mathcal{K}_{\overline{q}})\eta_q\big)\big(\mathbb{P}(\mathcal{K}_{q'})(1-\beta_{q'}) + \mathbb{P}(\mathcal{K}_{\overline{q'}})\eta_{q'}\big) \\ &= \big(\mathbb{P}(\mathcal{K}_q \cap \mathcal{K}_{q'}) - \mathbb{P}(\mathcal{K}_q)\mathbb{P}(\mathcal{K}_{q'})\big)(1-\beta_q)(1-\beta_{q'}) \\ &\quad + \big(\mathbb{P}(\mathcal{K}_q \cap \mathcal{K}_{\overline{q'}}) - \mathbb{P}(\mathcal{K}_q)\mathbb{P}(\mathcal{K}_{\overline{q'}})\big)(1-\beta_q)\eta_{q'} \\ &\quad\quad + \big(\mathbb{P}(\mathcal{K}_{\overline{q}} \cap \mathcal{K}_{q'}) - \mathbb{P}(\mathcal{K}_{\overline{q}})\mathbb{P}(\mathcal{K}_{q'})\big)\eta_q(1-\beta_{q'}) \\ &\quad\quad\quad + \big(\mathbb{P}(\mathcal{K}_{\overline{q}} \cap \mathcal{K}_{\overline{q'}}) - \mathbb{P}(\mathcal{K}_{\overline{q}})\mathbb{P}(\mathcal{K}_{\overline{q'}})\big)\eta_q\eta_{q'}. \end{aligned} \tag{11.32}$$

By definition, the projections $\{\varnothing, \{q\}\}$ and $\{\varnothing, \{q'\}\}$ are independent if and only if for every state K of the parent, $[K/\{q\}]$ and $[K/\{q'\}]$ are independent events, in other words if and only if

$$\begin{aligned} &\mathbb{P}(\mathcal{K}_q \cap \mathcal{K}_{q'}) - \mathbb{P}(\mathcal{K}_q)\mathbb{P}(\mathcal{K}_{q'}) = \mathbb{P}(\mathcal{K}_q \cap \mathcal{K}_{\overline{q'}}) - \mathbb{P}(\mathcal{K}_q)\mathbb{P}(\mathcal{K}_{\overline{q'}}) \\ &= \mathbb{P}(\mathcal{K}_{\overline{q}} \cap \mathcal{K}_{q'}) - \mathbb{P}(\mathcal{K}_{\overline{q}})\mathbb{P}(\mathcal{K}_{q'}) = \mathbb{P}(\mathcal{K}_{\overline{q}} \cap \mathcal{K}_{\overline{q'}}) - \mathbb{P}(\mathcal{K}_{\overline{q}})\mathbb{P}(\mathcal{K}_{\overline{q'}}) = 0. \end{aligned}$$

Substituting in the r.h.s. of Equation (11.32), and noticing that the values of the parameters can be chosen arbitrarily in the interval $[0,1[$, the implication (ii) $\Rightarrow$ (iii) follows. $\square$

11.10 Original Sources and Related Works

Probabilistic concepts were introduced in knowledge space theory by Falmagne and Doignon (1988a,b) (see also Villano, Falmagne, Johannesen, and Doignon, 1987; Falmagne, 1989a,b).

Several researchers have applied the basic local independence model to real knowledge space data, in particular: Falmagne et al. (1990); Villano (1991); Lakshminarayan (1995); Taagepera et al. (1997); Lakshminarayan and Gilson (1998). Villano (1991)'s work deserves a special mention because of the systematic way he constructed a large knowledge space by testing successive uniform extensions of smaller ones. The method of Cosyn and Thiéry (2000) reviewed in Chapter 15, which relies in part on Villano's techniques but also applies a sophisticated type of `QUERY` algorithm to question the experts, has been used successfully in the `ALEKS` system.

The Markov models described in Subsections 11.6.1 and 11.6.3 were proposed by Falmagne (1994). Some empirical tests of these models are described in Fries (1997).

All the Markov chain concepts used in this chapter are standard. For an introduction to Markov chain theory, we have already referred the reader to Feller (1968), Kemeny and Snell (1960), Parzen (1994), and Shyryayev (1960).

Problems

1. Modify the local independence assumption in such a manner that the probability of a response pattern R, given a state K varies with the subject. You should obtain an explicit formula for the probability $\rho(R)$ of obtaining a response pattern R in that case.

2. Compute the number of degrees of freedom of the chi-square for the basic probabilistic model, in the case of a 7 item test with open responses. What would this number be in a multiple choice situation, in which 5 alternatives are proposed for each question, and assuming that the options have been designed so as to make all guessing probabilities equal to $\frac{1}{5}$?

3. Generalize the simple learning model by assuming that the probability of mastering an item may depend upon past events, in particular, upon the last item mastered.

4. Suppose that the parent structure satisfies the simple learning model. Does any projection also satisfies that model? More generally, which of the models discussed in the chapter is preserved under projections?

5. Could a projection satisfy the simple learning model, while the parent structure does not satisfy that model?

6. Generalize Markov Chain Model 1, by assuming that subjects of the sample may have different learning step numbers. Specifically, assume that the learning step has a negative binomial distribution, and derive the predictions permitting a test of the model.

7. Develop Markov Chain Model 1 in order to predict the results of a sample of subjects tested at two different times (cf. Remark 11.6.2(c)).

8. Modify the state probabilities of the knowledge structure of the standard example, in such a manner that the states $\varnothing \in \mathcal{H}'$ and $\varnothing \in \mathcal{H}''$ are independent but the two projections themselves are not (cf. Example 11.9.2(a)).

9. Consider the knowledge structure $\{\varnothing, \{a\}, \{b\}, \{a,b\}\}$ used in Example 11.9.2(b), in which the states probabilities were $p(\varnothing) = .05$, $p\{a\} = .15$, $p\{b\} = .20$, $p\{a,b\} = .60$. Prove that the two probabilistic projections induced on $\{a\}$ and $\{b\}$ are independent, and thus form an independent representation.

10. In the Example of Problem 9, the nomenclature forming the representation was a classification (Definition 11.8.1). Is this condition necessary? Prove your response.

11. If $\Upsilon = (Q_i, \mathcal{K}_i, p_i)_{1 \le i \le k}$ is an independent representation of a probabilistic knowledge structure $(Q, \mathcal{K}, p)$, then the projections form a nomenclature and are independent in the sense of Definition 11.9.1. The latter condition means that the traces of any $K \in \mathcal{K}$ are independent. Does this imply that, for any distinct states K, K' of $\mathcal{K}$ and with $i \neq j$, $K \cap Q_i$ and $K' \cap Q_j$ are also independent?

12. Generalize the Simple Learning Model in the case of a knowledge structure which is not (necessarily) well-graded (cf. 11.4.1).

13. Show by an example that a knowledge structure may have a collection Υ of pairwise independent projections, without having Υ itself being independent in the sense of Definition 11.9.1.

14. Find an expression for $\mathrm{Cov}(\mathbf{X}_q, \mathbf{X}_{q'})$ when the knowledge structure is a chain, and the parameters $\beta_q, \beta_{q'}, \eta_q, \eta_{q'}$ vanish (cf. 11.9.1).

15. Develop the Markov Model 2 of 11.6.3 in the style used for the Markov Model 1. In particular, what does the transition matrix look like?

12

Stochastic Learning Paths*

The stochastic theory presented in this chapter is more ambitious than those examined in Chapter 11. The description of the learning process is more complete and takes place in real time, rather than in a sequence of discrete trials. This theory also contains a provision for individual differences. Nevertheless, its basic intuition is similar, in that any student progresses through some learning path. As time passes, the student successively masters the items encountered along the learning path. The exposition of the theory given here follows closely Falmagne (1993, 1996). As before, we shall illustrate the concepts of the theory in the framework of an example. We star the title of this chapter because its concepts and results are not used elsewhere in this book.

12.1 A Knowledge Structure in Euclidean Geometry

This empirical application is due to Lakshminarayan (1995), and was alluded to in 1.1.3. It involves five problems in high school geometry, which are displayed in Figure 12.1. These five items were part of a test given to 959 undergraduate students at the University of California, Irvine. There were two consecutive applications of the test, separated by a short lecture recalling some fundamental facts of Euclidean geometry. The problems in the two applications were pairwise equally informative but not identical. In the terminology introduced in 1.1.1, this means that the problems in the second application were different instances of the same five items. (For our use of the term 'equally informative' in this context, see 2.1.5.) The analysis of the data[1], for the five items a, b, c, d, and e, resulted in the knowledge structure below, which is represented by the graph of Figure 12.2.

$$\begin{aligned}\mathcal{L} = \big\{&\varnothing, \{a\},\{b\}, \{a,b\}, \{a,d\}, \{b,c\}, \{a,b,c\},\\ &\{a,b,d\}, \{b,c,d\}, \{a,b,c,d\}, \{a,b,c,d,e,\}\big\}.\end{aligned} \tag{12.1}$$

[1] The details of the Lakshminarayan (1995) study are reported later in this chapter.

J.-C. Falmagne, J.-P. Doignon, *Learning Spaces*,
DOI 10.1007/978-3-642-01039-2_12, © Springer-Verlag Berlin Heidelberg 2011

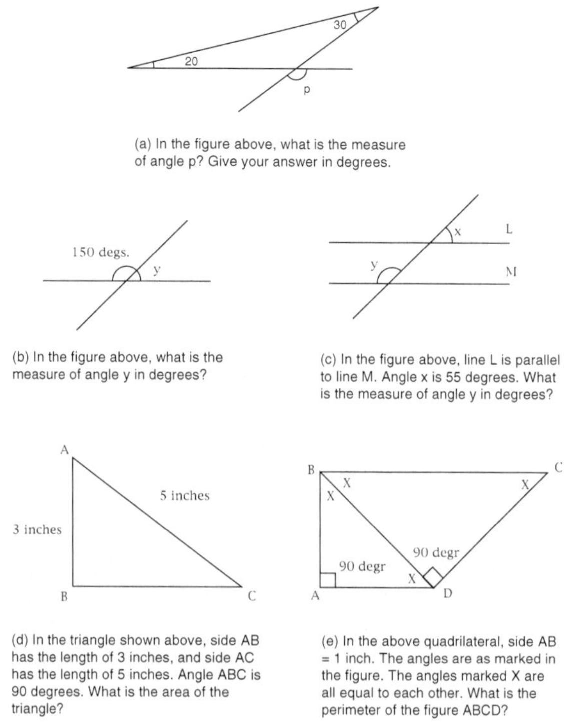

Figure 12.1. The five geometry items in Lakshminarayan's study.

12.2 Basic Concepts

The knowledge structure $\mathcal{L}$ in Equation (12.1) is a learning space with seven gradations. In general, the theory presented here is suitable for knowledge structures satisfying Axiom [L1] of a learning space[2], but not necessarily [L2]. So, all the learning paths are gradations but closure under union may not be satisfied. Students may differ from each other not only by the particular gradation followed, but also by their 'learning rate.' We assume that, for a given population of subjects, the learning rate is a random variable, which will be denoted by $\mathbf{L}$. A graphical illustration of a density function for this random variable $\mathbf{L}$ is given at the upper left of Figure 12.2. Consider a student equipped with a learning rate $\mathbf{L} = \lambda$, and suppose that this student, for some chance reasons, is engaged in the gradation

$$\varnothing \subset \{b\} \subset \{b, c\} \subset \{a, b, c\} \subset \{a, b, c, d\} \subset \{a, b, c, d, e\} \tag{12.2}$$

which is marked by the string of red arrows starting from the empty set in Figure 12.2.

[2] Cf. 2.2.1 on page 26. These knowledge structures are thus finite.

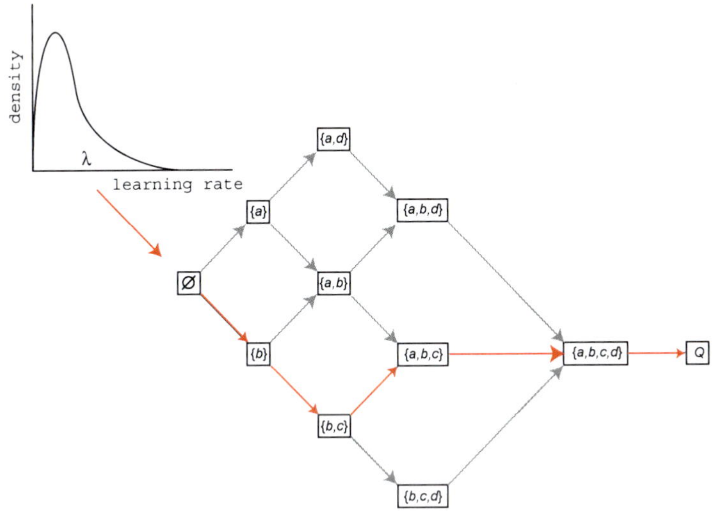

Figure 12.2. Indicated by the red arrows, the gradation followed in the learning space $\mathcal{L}$ by a certain student equipped with the learning rate λ.

We also assume that, at the outset, this student is naive. Thus, the initial state is the empty set $\varnothing$. Reaching the next state of that gradation requires mastering the concepts associated with item b. This may or not be laborious, depending on the learning rate of the student and the difficulty of the item. In the theory expounded in this chapter, the time required to reach state $\{b\}$ is a random variable, the distribution of which may depend upon these two factors: the difficulty of the item and the student's learning rate, represented here by the number λ. The item difficulty will be measured by a parameter of the theory.

The next state encountered along the gradation in Equation (12.2) is $\{b, c\}$. Again, the time required for a transition from state $\{b\}$ to state $\{b, c\}$ is a random variable, with a distribution depending upon λ and the difficulty of item c. And so on. If a test takes place, the student will respond according to her current state. However, we do not assume that the responses to items contained in the student's current state will necessarily be correct. As in Chapter 11, we suppose that careless errors and lucky guesses are possible. The rule specifying the probabilities of both kinds of events will be the same as before, namely, the rule defined by Equation (11.6). If a student is tested n times, at times $t_1, \ldots, t_n$, the n-tuple of observed patterns of responses will depend upon the n knowledge states occupied at the times of testing. As usual, we do not assume that the knowledge states directly observable. However, from the axioms of the theory, explicit formulas can be derived for the prediction of the joint probabilities of occurrence of all n-tuples of states.

A formal statement of the theory requires a few additional concepts.

12.2.1 Definition. Consider a fixed, finite, well-graded knowledge structure $(Q, \mathcal{K})$. For any real number $t \geq 0$, we denote by $\mathbf{K}_t$ the knowledge state at time t; thus, $\mathbf{K}_t$ is a random variable taking its values in $\mathcal{K}$. We write $\mathcal{G}_{\mathcal{K}}$ for the collection of all gradations in $\mathcal{K}$. We shall usually abbreviate $\mathcal{G}_{\mathcal{K}}$ as $\mathcal{G}$ when no ambiguity can arise. Let $K \subset K' \subset \cdots \subset K''$ be any chain of states. We write $\mathcal{G}(K, K', \ldots, K'')$ for the subcollection of all gradations containing all the states $K, K', \ldots, K''$. Obviously, $\mathcal{G}(\varnothing) = \mathcal{G}(Q) = \mathcal{G}$, and in the knowledge structure $(Q, \mathcal{L})$ specified by Equation (12.1) (cf. Figure 12.2), we have, with obvious abbreviations,

$$\mathcal{G}(\{a\}, \{a, b, c, d\}) = \{\nu_1, \nu_2, \nu_3\},$$
$$\nu_1 = adbce\,, \quad \nu_2 = abdce\,, \quad \nu_3 = abcde.$$

The gradation taken by a subject will also be regarded as a random variable. We write $\mathbf{C} = \nu$ to signify that ν is the gradation followed by the student. Thus, the random variable $\mathbf{C}$ takes its values in $\mathcal{G}$. The existence of a probability distribution on the set of all gradations is a fairly general hypothesis. It involves, as possible special cases, various mechanisms describing the gradation followed by the subject as resulting from a succession of choices along the way. We shall go back to this issue later in this chapter.

For convenience, the notation and all the concepts basic to this chapter are recalled below.

12.2.2 Notation.

Q	a finite, nonempty set, called the *domain*
$\mathcal{K}$	a well-graded knowledge structure on Q
$\mathbf{K}_t = K \in \mathcal{K}$	the subject is in state K at time $t \geq 0$
$\mathbf{R}_t = R \in 2^Q$	R is the set of correct responses given by the subject at time $t \geq 0$
$\mathbf{L} = \lambda$	the learning rate of the subject is equal to $\lambda \geq 0$
$\mathcal{G}$	the collection of all gradations
$\mathcal{G}(K, K', \ldots)$	the collection of all gradations containing $K, K', \ldots$
$\mathbf{C} = \nu \in \mathcal{G}$	the gradation of the subject is ν.

12.2.3 General Axioms. We begin by formulating four axioms specifying the general stochastic structure of the theory. Precise hypotheses concerning the distributions of the random variables $\mathbf{L}$ and $\mathbf{C}$ and the conditional probabilities of the response patterns, given the states, will be made in later sections of this chapter. Each axiom is followed by a paraphrase in plain language. These axioms define, up to some functional relations, the probability measure associated with the triple

$$((\mathbf{R}_t, \mathbf{K}_t)_{t \geq 0}, \mathbf{C}, \mathbf{L}).$$

The selection of a subject in the population corresponds to a choice of values $\mathbf{L} = \lambda$ and $\mathbf{C} = \nu$ for the learning rate and the gradation of the subject, respectively. In turn, these values specify a stochastic process

$$((\mathbf{R}_t, \mathbf{K}_t)_{t \geq 0}, \nu, \lambda)$$

describing the progress of the particular subject through the material, along gradation ν.

[B] Beginning State. For all $\nu \in \mathcal{G}$ and $\lambda \in \mathbb{R}_+$,

$$\mathbb{P}(\mathbf{K}_0 = \varnothing \,|\, \mathbf{L} = \lambda, \mathbf{C} = \nu) = 1.$$

With probability one, the initial state of the subject at time $t = 0$ is the empty state, independently of the gradation and of the learning rate[3].

[I] Independence of Gradation and Learning Rate. The random variables $\mathbf{L}$ and $\mathbf{C}$ are independent. That is, for all $\lambda \in \mathbb{R}_+$ and $\nu \in \mathcal{G}$,

$$\mathbb{P}(\mathbf{L} \leq \lambda, \mathbf{C} = \nu) = \mathbb{P}(\mathbf{L} \leq \lambda) \cdot \mathbb{P}(\mathbf{C} = \nu).$$

The gradation followed is independent of the learning rate.

[R] Response Rule. There is a function $r : 2^Q \times \mathcal{K} \to [0, 1]$ such that[4] for all $\lambda \in \mathbb{R}_+$, $\nu \in \mathcal{G}$, $n \in \mathbb{N}$, $R_n \in 2^Q$, $K_n \in \mathcal{K}$, $t_n > t_{n-1} > \cdots > t_1 \geq 0$, and any event $\mathcal{E}$ determined only by

$$\big((\mathbf{R}_{t_{n-1}}, \mathbf{K}_{t_{n-1}}), (\mathbf{R}_{t_{n-2}}, \mathbf{K}_{t_{n-2}}), \ldots, (\mathbf{R}_{t_1}, \mathbf{K}_{t_1})\big),$$

we have

$$\begin{aligned}\mathbb{P}(\mathbf{R}_{t_n} = R \,|\, \mathbf{K}_{t_n} = K, \mathcal{E}, \mathbf{L} = \lambda, \mathbf{C} = \nu) &= \mathbb{P}(\mathbf{R}_{t_n} = R \,|\, \mathbf{K}_{t_n} = K) \\ &= r(R, K).\end{aligned}$$

If the state at a given time is known, the probability of a response pattern at that time only depends upon that state through the function r. It is independent of the learning rate, the gradation, and any sequence of past states and responses.

[L] Learning Rule. There is a function $\ell e : \mathcal{K} \times \mathcal{K} \times \mathbb{R}^2 \times \mathcal{G} \to [0, 1]$, such that, for all $\lambda \in \mathbb{R}_+$, $\nu \in \mathcal{G}$, $n \in \mathbb{N}$, $R_n \in 2^Q$, $K_n, K_{n+1} \in \mathcal{K}$, $t_{n+1} > t_n > \cdots > t_1 \geq 0$, and any event $\mathcal{E}$ determined only by

$$\big((\mathbf{R}_{t_{n-1}}, \mathbf{K}_{t_{n-1}}), (\mathbf{R}_{t_{n-2}}, \mathbf{K}_{t_{n-2}}), \ldots, (\mathbf{R}_{t_1}, \mathbf{K}_{t_1})\big),$$

[3] As $\mathbf{L}$ is a random variable taking its values in $\mathbb{R}_+$, the event $\mathbf{L} = \lambda$ may have measure 0. As usual, the conditional probability in Axiom [B] and in other similar expressions in this chapter is defined by a limiting operation (cf. Parzen, 1994, p. 335, for example).

[4] We depart here from our usual convention of denoting by n the size $|\mathcal{K}|$ of $\mathcal{K}$.

we have

$$\begin{aligned}\mathbb{P}(\mathbf{K}_{t_{n+1}} = K_{n+1} \,|\, \mathbf{K}_{t_n} = K_n, \mathbf{R}_{t_n} = R, \mathcal{E}, \mathbf{L} = \lambda, \mathbf{C} = \nu) \\ = \mathbb{P}(\mathbf{K}_{t_{n+1}} = K_{n+1} \,|\, \mathbf{K}_{t_n} = K_n, \mathbf{L} = \lambda, \mathbf{C} = \nu) \\ = \ell e\,(K_n, K_{n+1}, t_{n+1} - t_n, \lambda, \nu).\end{aligned}$$

Moreover, the function ℓe is assumed to satisfy the following two conditions: For any $K, K' \in \mathcal{K}$, $\nu, \nu' \in \mathcal{G}$, $\delta > 0$ and any $\lambda \in \mathbb{R}_+$,

[LR1] $\quad \ell e\,(K, K', \delta, \lambda, \nu) = 0 \qquad$ if $\quad \nu \notin \mathcal{G}(K, K')$;

[LR2] $\quad \ell e\,(K, K', \delta, \lambda, \nu) = \ell e\,(K, K', \delta, \lambda, \nu') \qquad$ if $\quad \nu, \nu' \in \mathcal{G}(K, K')$.

The probability of a state at a given time only depends upon the last state recorded, the time elapsed, the learning rate and the gradation. It does not depend upon previous states and responses. This probability is the same for all gradations containing both the last recorded state and the new state, and vanishes otherwise.

12.2.4 Definition. A triple $\mathcal{S} = ((\mathbf{K}_t, \mathbf{R}_t)_{t \geq 0}, \mathbf{C}, \mathbf{L})$ satisfying the four Axioms [B], [I], [R] and [L] is a *system of stochastic learning paths*. The functions r and ℓe of Axioms [R] and [L] will be called the *response function* and the *learning function* of the system $\mathcal{S}$, respectively.

Of special interest is another condition on the learning function ℓe which essentially states that there is no forgetting. When ℓe also satisfies

[LR3] $\qquad \ell e\,(K, K', \delta, \lambda, \nu) = 0 \quad$ if $\quad K \not\subseteq K'$,

for all $K, K' \in \mathcal{K}$, $\nu \in \mathcal{G}$, $\delta > 0$ and $\lambda \in \mathbb{R}$, we shall say that $\mathcal{S}$ is *progressive*. Only progressive systems of stochastic learning paths will be considered in this chapter (see Problem 1, however).

12.2.5 Remarks. a) One may object to Axiom [B] that when a subject is tested, the time elapsed since the beginning of learning—the index t in $\mathbf{R}_t$ and $\mathbf{K}_t$—is not always known exactly. We may perhaps assume, for example, that the learning of mathematics begins roughly at age 2 or 3, but this may not be precise enough. In some situations, it may be possible to consider at least some of these indexed times as parameters to be estimated from the data.

b) Axiom [I], which concerns the independence of the learning rate and the learning path, was difficult to avoid—however unrealistic it may perhaps appear. Its strong appeal is that it renders the derivations relatively straightforward. For the time being, there does not seem to be any obvious alternative. Our hope is that the predictions of the model are robust with respect to that assumption.

c) Axiom [R] seems reasonable enough. This axiom will be specialized in Subsection 12.4.1 as the local independence assumption already encountered in Chapter 11 (see 11.1.2).

d) On the other hand, the assumptions embedded in [L] require a thorough examination. To begin with, one might argue that a two or k-dimensional version of the learning rate may be required in some cases. The number λ in the argument of the function ℓe would then be replaced by a real vector, the components of which would measure different aspect of the learning process. The theory could be elaborated along such lines if the need arises.

Also, one might be tempted to formulate, instead of Axiom [L], the much stronger Markovian condition formalized by the equation

$$\mathbb{P}(\mathbf{K}_{t_{n+1}} = K_{n+1} \,|\, \mathbf{K}_{t_n} = K_n, \mathbf{R}_{t_n} = R_n, \mathcal{E}) = \mathbb{P}(\mathbf{K}_{t_{n+1}} = K_{n+1} \,|\, \mathbf{K}_{t_n} = K_n),$$

with $\mathcal{E}$ as in Axiom [L]. Thus, the probability of a state at time t_{n+1} would only depend upon the last recorded state, and possibly, the time elapsed since that observation. This assumption is inappropriate because the more detailed history embedded in the event $\mathcal{E}$ in the l.h.s. of the above equation may provide information on the learning rate and the learning path, which in turn, may modify the probabilities of the states at time t_{n+1}. In Problem 2, we ask the reader to provide a numerical example falsifying this equation, based on the domain Q and the knowledge structure $\mathcal{L}$ of Equation (12.1) and Figure 12.2.

Much more can be said about Axiom [L]. It turns out that, in the framework of the other Axioms [B], [I] and [R], this axiom considerably restricts the distributional form of the latencies (see Remark 12.5.1). We postpone further discussion of Axiom [L] to derive a few results which only depend upon Axioms [B], [I], [R] and [L]. In other words, no assumptions are made regarding the functional form of r or ℓe.

12.3 General Results

12.3.1 Theorem. *For all integer $n > 0$, all real numbers $t_n > \cdots > t_1 \geq 0$, and any event $\mathcal{E}$ determined only by $(\mathbf{R}_{t_{n-1}}, \mathbf{R}_{t_{n-2}}, \ldots, \mathbf{R}_{t_1})$, we have*

$$\begin{aligned}\mathbb{P}(\mathbf{K}_{t_n} = K_n \,|\, \mathbf{K}_{t_{n-1}} = &K_{n-1}, \ldots, \mathbf{K}_{t_1} = K_1, \mathcal{E}) \\ &= \mathbb{P}(\mathbf{K}_{t_n} = K_n \,|\, \mathbf{K}_{t_{n-1}} = K_{n-1}, \ldots, \mathbf{K}_{t_1} = K_1).\end{aligned}$$

We postpone the proof for a moment (see 12.3.4).

12.3.2 Convention. To lighten the notation, we occasionally use the abbreviations:

$$\begin{aligned}\kappa_n &= \cap_{i=1}^n [\mathbf{K}_{t_i} = K_i], \\ \rho_n &= \cap_{i=1}^n [\mathbf{R}_{t_i} = R_i].\end{aligned}$$

Notice that the choice of the times $t_n > \cdots > t_1 \geq 0$ is implicit in this notation. We also write p_ν for $\mathbb{P}(\mathbf{C} = \nu)$. Our proof of Theorem 12.3.1 requires a preparatory lemma.

12.3.3 Lemma. *For any real number λ and any learning path ν, and with $\mathcal{E}$ only depending on ρ_n, we have*

$$\mathbb{P}(\mathcal{E} \mid \kappa_n, \mathbf{L} = \lambda, \mathbf{C} = \nu) = \mathbb{P}(\mathcal{E} \mid \kappa_n).$$

PROOF. By induction, using Axioms [R] and [L] in alternation (Problem 11). □

12.3.4 Proof of Theorem 13.3.1. By Lemma 12.3.3, we have for any positive integer n,

$$\begin{aligned}\frac{\mathbb{P}(\kappa_n, \mathcal{E} \mid \mathbf{L} = \lambda, \mathbf{C} = \nu)}{\mathbb{P}(\kappa_n, \mathcal{E})} &= \frac{\mathbb{P}(\mathcal{E} \mid \kappa_n, \mathbf{L} = \lambda, \mathbf{C} = \nu)\mathbb{P}(\kappa_n \mid \mathbf{L} = \lambda, \mathbf{C} = \nu)}{\mathbb{P}(\mathcal{E} \mid \kappa_n)\mathbb{P}(\kappa_n)} \\ &= \frac{\mathbb{P}(\mathcal{E} \mid \kappa_n)\mathbb{P}(\kappa_n \mid \mathbf{L} = \lambda, \mathbf{C} = \nu)}{\mathbb{P}(\mathcal{E} \mid \kappa_n)\mathbb{P}(\kappa_n)} \\ &= \frac{\mathbb{P}(\kappa_n \mid \mathbf{L} = \lambda, \mathbf{C} = \nu)}{\mathbb{P}(\kappa_n)} = g(\kappa_n, \lambda, \nu)\end{aligned}$$

the last equality defining the function g. With $p_\nu = \mathbb{P}(\mathbf{C} = \nu)$, we successively have

$$\begin{aligned}\mathbb{P}(\mathbf{K}_{t_n} = K_n \mid \kappa_{n-1}, \mathcal{E}) = \int_{-\infty}^{\infty} \sum_{\nu \in \mathcal{G}} &\mathbb{P}(\mathbf{K}_{t_n} = K_n \mid \kappa_{n-1}, \mathcal{E}, \mathbf{L} = \lambda, \mathbf{C} = \nu) \\ &\times \frac{\mathbb{P}(\kappa_{n-1}, \mathcal{E} \mid \mathbf{L} = \lambda, \mathbf{C} = \nu)}{\mathbb{P}(\kappa_{n-1}, \mathcal{E})} p_\nu \, d\mathbb{P}(\mathbf{L} \le \lambda)\end{aligned}$$

$$\begin{aligned}&= \int_{-\infty}^{\infty} \sum_{\nu \in \mathcal{G}} \mathbb{P}(\mathbf{K}_{t_n} = K_n \mid \kappa_{n-1}, \mathbf{L} = \lambda, \mathbf{C} = \nu) \, g(\kappa_{n-1}, \lambda, \nu) \, p_\nu \, d\mathbb{P}(\mathbf{L} \le \lambda) \\ &= \mathbb{P}(\mathbf{K}_{t_n} = K_n \mid \kappa_{n-1}).\end{aligned}$$

□

As noted earlier, the results concerning the observable patterns of responses are our prime concern. However, an examination of Axiom [R] suggests that these results could be derived from a study of the process $(\mathbf{K}_t, \mathbf{L}, \mathbf{C})_{t \ge 0}$. The next theorem makes this idea precise.

12.3.5 Theorem. *For any positive integer n, any response patterns $R_i \in 2^Q$, $1 \le i \le n$, and any real numbers $t_n > t_{n-1} > \cdots > t_1 \ge 0$, we have*

$$\mathbb{P}\left(\cap_{i=1}^n [\mathbf{R}_{t_i} = R_i]\right) = \sum_{K_1 \in \mathcal{K}} \cdots \sum_{K_n \in \mathcal{K}} \left(\prod_{i=1}^n r(R_i, K_i) \right) \mathbb{P}\left(\cap_{i=1}^n [\mathbf{K}_{t_i} = K_i]\right).$$

PROOF. We have

$$\mathbb{P}\big(\cap_{i=1}^{n}[\mathbf{R}_{t_i} = R_i]\big) = \mathbb{P}(\rho_n) = \sum_{(\kappa_n)} \mathbb{P}(\rho_n, \kappa_n). \tag{12.3}$$

Developing the general term in the summation, we obtain successively, using Axiom [R] and Theorem 12.3.1,

$$\begin{aligned}
&\mathbb{P}(\rho_n, \kappa_n) = \\
&\quad \mathbb{P}(\mathbf{R}_{t_n} = R_n \,|\, \rho_{n-1}, \kappa_n)\, \mathbb{P}(\mathbf{K}_{t_n} = K_n \,|\, \rho_{n-1}, \kappa_{n-1}) \\
&\quad \times \mathbb{P}(\mathbf{R}_{t_{n-1}} = R_{n-1} \,|\, \rho_{n-2}, \kappa_{n-1})\, \mathbb{P}(\mathbf{K}_{t_{n-1}} = K_{n-1} \,|\, \rho_{n-2}, \kappa_{n-2}) \times \cdots \\
&\quad\quad \times \mathbb{P}(\mathbf{R}_{t_2} = R_2 \,|\, \mathbf{K}_{t_1} = K_1, \mathbf{R}_{t_1} = R_1, \mathbf{K}_{t_2} = K_2) \\
&\quad\quad\quad \times \mathbb{P}(\mathbf{K}_{t_2} = K_2 \,|\, \mathbf{K}_{t_1} = K_1, \mathbf{R}_{t_1} = R_1) \\
&\quad\quad\quad\quad \times \mathbb{P}(\mathbf{R}_{t_1} = R_1 \,|\, \mathbf{K}_{t_1} = K_1)\mathbb{P}(\mathbf{K}_{t_1} = K_1) \\
&\quad = \big(r(R_n, K_n) r(R_{n-1}, K_{n-1}) \cdots r(R_1, K_1)\big)\, \mathbb{P}(\mathbf{K}_{t_n} = K_n \,|\, \kappa_{n-1}) \\
&\quad \times \mathbb{P}(\mathbf{K}_{t_{n-1}} = K_{n-1} \,|\, \kappa_{n-2}) \cdots \mathbb{P}(\mathbf{K}_{t_2} = K_2 \,|\, \mathbf{K}_{t_1} = K_1)\, \mathbb{P}(\mathbf{K}_{t_1} = K_1) \\
&\quad = \left(\prod_{i=1}^{n} r(R_i, K_i)\right) \mathbb{P}(\kappa_n).
\end{aligned} \tag{12.4}$$

The result follows from (12.3) and (12.4). □

Thus, the joint probabilities of the response patterns can be obtained from the joint probabilities of the states, and from the conditional probabilities captured by the function r. We now turn to a study of the process $(\mathbf{K}_t, \mathbf{L}, \mathbf{C})_{t \geq 0}$.

12.3.6 Theorem. *For all real numbers $\lambda, t > 0$, and all $\nu \in \mathcal{G}$,*

$$\mathbb{P}(\mathbf{K}_t = K \,|\, \mathbf{L} = \lambda, \mathbf{C} = \nu) = \ell e\,(\varnothing, K, t, \lambda, \nu).$$

PROOF. By Axiom [B], $\mathbb{P}(\mathbf{K}_0 = K' \,|\, \mathbf{L} = \lambda, \mathbf{C} = \nu) = 0$ for any $K' \neq \varnothing$. Accordingly, we obtain:

$$\begin{aligned}
&\mathbb{P}(\mathbf{K}_t = K \,|\, \mathbf{L} = \lambda, \mathbf{C} = \nu) \\
&\quad = \sum_{K' \in \mathcal{K}} \mathbb{P}(\mathbf{K}_t = K \,|\, \mathbf{K}_0 = K', \mathbf{L} = \lambda, \mathbf{C} = \nu)\, \mathbb{P}(\mathbf{K}_0 = K' \,|\, \mathbf{L} = \lambda, \mathbf{C} = \nu) \\
&\quad = \mathbb{P}(\mathbf{K}_t = K \,|\, \mathbf{K}_0 = \varnothing, \mathbf{L} = \lambda, \mathbf{C} = \nu) \\
&\quad = \ell e\,(\varnothing, K, t, \lambda, \nu).
\end{aligned}$$

□

12.3.7 Theorem. *For all integers $n > 0$ and times $t_n > \cdots > t_1 > t_0 = 0$, and with $K_0 = \varnothing$,*

$$\begin{aligned}
&\mathbb{P}(\mathbf{K}_{t_1} = K_1, \ldots, \mathbf{K}_{t_n} = K_n) \\
&\quad = \int_{-\infty}^{\infty} \sum_{\nu \in \mathcal{G}} \left(\prod_{i=0}^{n-1} \ell e\,(K_i, K_{i+1}, t_{i+1} - t_i, \lambda, \nu)\right) p_\nu \, d\,\mathbb{P}(\mathbf{L} \leq \lambda).
\end{aligned}$$

PROOF. With the notation κ_n as in Convention 12.3.2, we have

$$\mathbb{P}(\kappa_n) = \int_{-\infty}^{\infty} \sum_{\nu \in \mathcal{G}} \mathbb{P}(\kappa_n \,|\, \mathbf{L} = \lambda, \mathbf{C} = \nu)\, p_\nu \, d\,\mathbb{P}(\mathbf{L} \leq \lambda). \tag{12.5}$$

Using Axiom [L] and Theorem 12.3.6,

$$\begin{aligned}
&\mathbb{P}(\kappa_n \,|\, \mathbf{L} = \lambda, \mathbf{C} = \nu) \\
&\quad = \mathbb{P}(\mathbf{K}_{t_n} = K_{t_n} \,|\, \kappa_{n-1}, \mathbf{L} = \lambda, \mathbf{C} = \nu)\, \mathbb{P}(\mathbf{K}_{t_{n-1}} = K_{t_{n-1}} \,|\, \kappa_{n-2}, \mathbf{L} = \lambda, \mathbf{C} = \nu) \\
&\qquad \cdots \mathbb{P}(\mathbf{K}_{t_2} = K_2 \,|\, \mathbf{K}_{t_1} = K_1, \mathbf{L} = \lambda, \mathbf{C} = \nu)\, \mathbb{P}(\mathbf{K}_{t_1} = K_1 \,|\, \mathbf{L} = \lambda, \mathbf{C} = \nu) \\
&\quad = \prod_{i=0}^{n-1} \ell e\,(K_i, K_{i+1}, t_{i+1} - t_i, \lambda, \nu)\,.
\end{aligned}$$

The result obtains after substituting in (12.5). □

12.4 Assumptions on Distributions

This theory is of limited practical use without making specific hypotheses concerning the distributions of the random variable **L** measuring the learning rate, and the random variables implicit in the expression of the learning function ℓe of Axiom [L], which governs the time required to master the items. We also need to specify the response function r of Axiom [R], representing the conditional probabilities of the response patterns, given the states. We formulate here two axioms specifying the response function and the distribution of the learning rate. The distributions of the learning latencies are discussed in the next section. It is fair to say that our choice of axioms result from a compromise between realism and applicability. These axioms are by no means the only feasible ones. Different compromises, still in the framework of Axioms [B], [I], [R] and [L], could be adopted.

12.4.1 Axioms on r and L. The first axiom embodies the standard "local independence" condition of psychometric theory (see Lord and Novick, 1974). It has been used in Chapter 11 in the guise of Equation (11.6).

[N] LOCAL INDEPENDENCE. For each item q in Q, there is a parameter β_q, $0 \leq \beta_q < 1$, representing the probability of a careless error in responding to this item if it is present in the current knowledge state. There is also a collection of parameters η_q representing the probability of a lucky guess[5] for a response to an item $q \in Q$ not present in the current learning state.

[5] We recall that the terms "lucky guess" and "careless error" have no cognitive interpretation. They always refer to a current knowledge state. By convention, a correct response to an item not belonging to the knowledge state of reference is a lucky guess, and an incorrect response to an item belonging to that state is a careless error.

These parameters specify the function r of Axiom [R], that is, the probability of a response set R, conditional to a knowledge state K, according to the formula:

$$r(R,K) = \Big(\prod_{q\in K\setminus R} \beta_q\Big)\Big(\prod_{q\in K\cap R} (1-\beta_q)\Big)\Big(\prod_{q\in R\setminus K} \eta_q\Big)\Big(\prod_{q\in \overline{R\cup K}} (1-\eta_q)\Big),$$

in which the complement $\overline{R\cup K}$ in the last factor is taken with respect to the domain Q.

[A] Learning Ability. The random variable $\mathbf{L}$ measuring the learning rate is continuous, with a density function f, and a mass concentrated on the positive reals. Specifically, it is assumed that $\mathbf{L}$ is distributed gamma, with parameters $\alpha > 0$ and $\xi > 0$; that is:

$$f(\lambda) = \begin{cases} \dfrac{\xi^\alpha}{\Gamma(\alpha)}\lambda^{\alpha-1}e^{-\xi\lambda} & \text{for } \lambda > 0, \\ 0 & \text{for } \lambda \leq 0. \end{cases} \tag{12.6}$$

Thus, $E(\mathbf{L}) = \frac{\alpha}{\xi}$ and $Var(\mathbf{L}) = \frac{\alpha}{\xi^2}$ (where 'E' denotes, as usual, the expectation of the random variable $\mathbf{L}$).

12.5 The Learning Latencies

The four axioms [B], [R], [I] and [L] put severe constraints on the functional form of the learning latencies. For example, we may not simply assume that these latencies are distributed gamma[6]. We show below by a functional equation argument that the latency distributions must be exponential.

12.5.1 Remarks. Let us write $\mathbf{T}_{q,K,\lambda}$ for the random variable measuring the time required to master some new item q, for a subject with a learning rate λ, this subject being in some state K from which item q is learnable, that is, $K \cup \{q\}$ is a state in the structure. More specifically, for any gradation ν containing both K and $K \cup \{q\}$ and any $\tau > 0$, we infer from Axiom [L] that

$$\mathbb{P}(\mathbf{T}_{q,K,\lambda} \leq \tau) = \ell e\,(K, K\cup\{q\}, \tau, \lambda, \nu)\,,$$

or equivalently

$$\mathbb{P}(\mathbf{T}_{q,K,\lambda} > \tau) = \ell e\,(K, K, \tau, \lambda, \nu). \tag{12.7}$$

For any $t, \delta, \delta', \lambda > 0$, any state K in $\mathcal{K}$, and any gradation ν containing both K and $K \cup \{q\}$, we have

[6] This assumption was made by Falmagne (1993). As argued by Stern and Lakshminarayan (1995), it is incorrect. The history of this issue is summarized in the last section of this chapter.

$$\begin{aligned}
&\mathbb{P}(\mathbf{K}_{t+\delta+\delta'} = K \,|\, \mathbf{K}_t = K, \mathbf{L} = \lambda, \mathbf{C} = \nu) \\
&\quad = \mathbb{P}(\mathbf{K}_{t+\delta+\delta'} = K \,|\, \mathbf{K}_{t+\delta} = K, \mathbf{K}_t = K, \mathbf{L} = \lambda, \mathbf{C} = \nu) \\
&\qquad\qquad\qquad\qquad \times\ \mathbb{P}(\mathbf{K}_{t+\delta} = K \,|\, \mathbf{K}_t = K, \mathbf{L} = \lambda, \mathbf{C} = \nu) \\
&\quad +\ \mathbb{P}(\mathbf{K}_{t+\delta+\delta'} = K \,|\, \mathbf{K}_{t+\delta} \neq K, \mathbf{K}_t = K, \mathbf{L} = \lambda, \mathbf{C} = \nu) \\
&\qquad\qquad\qquad\qquad \times\ \mathbb{P}(\mathbf{K}_{t+\delta} \neq K \,|\, \mathbf{K}_t = K, \mathbf{L} = \lambda, \mathbf{C} = \nu).
\end{aligned}$$

Because we assume that the system of stochastic learning path is progressive (see Definition 12.2.4), the second term in the r.h.s. vanishes and the above equation simplifies into

$$\begin{aligned}
&\mathbb{P}(\mathbf{K}_{t+\delta+\delta'} = K \,|\, \mathbf{K}_t = K, \mathbf{L} = \lambda, \mathbf{C} = \nu) \\
&\quad = \mathbb{P}(\mathbf{K}_{t+\delta+\delta'} = K \,|\, \mathbf{K}_{t+\delta} = K, \mathbf{L} = \lambda, \mathbf{C} = \nu) \\
&\qquad\qquad\qquad\qquad \times\ \mathbb{P}(\mathbf{K}_{t+\delta} = K \,|\, \mathbf{K}_t = K, \mathbf{L} = \lambda, \mathbf{C} = \nu),
\end{aligned}$$

that is, in terms of the function ℓe,

$$\ell e\,(K, K, \delta + \delta', \lambda, \nu) = \ell e\,(K, K, \delta', \lambda, \nu)\, \ell e\,(K, K, \delta, \lambda, \nu).$$

In turn, this can be rewritten as

$$\mathbb{P}(\mathbf{T}_{q,K,\lambda} > \delta + \delta') = \mathbb{P}(\mathbf{T}_{q,K,\lambda} > \delta')\, \mathbb{P}(\mathbf{T}_{q,K,\lambda} > \delta). \tag{12.8}$$

Fixing q, K and λ and writing $v(\tau) = \mathbb{P}(\mathbf{T}_{q,K,\lambda} > \tau)$ for $\tau > 0$, we get

$$v(\delta + \delta') = v(\delta')v(\delta) \tag{12.9}$$

for any $\delta, \delta' > 0$. The function v is defined on $]0, \infty[$ and decreasing (because $1 - v$ is a distribution function). Standard functional equations results apply (cf. Aczél, 1966), yielding for any $\tau > 0$ and some $\vartheta > 0$

$$v(\tau) = e^{-\vartheta\tau}.$$

The constant ϑ may of course depend upon q, K and λ. We obtain

$$\mathbb{P}(\mathbf{T}_{q,K,\lambda} \leq \tau) = 1 - e^{-\vartheta(q,K,\lambda)\tau}. \tag{12.10}$$

Note that $E(\mathbf{T}_{q,K,\lambda}) = 1/\vartheta(q, K, \lambda)$.

The above argument includes the possibility that the distribution of the learning latencies may depend on the state through the parameters ϑ. For the rest of this section, however, we shall assume that $\vartheta(q, K, \lambda)$ does not depend on K for all states K such that q is learnable from K. In other words, K can be dropped in Equations (12.8) and (12.10). We discuss the validity of this assumption in a later section of this chapter; see 12.7.1.

12.5.2 Remark. It makes sense to require that a subject having twice the learning rate μ of some other subject would master, on the average, a given item in half the time. Thus, we want to have:

$$E(\mathbf{T}_{q,2\mu}) = \frac{1}{\vartheta(q,2\mu)} = \frac{E(\mathbf{T}_{q,\mu})}{2} = \frac{1}{2\vartheta(q,\mu)}.$$

Generalizing this observation leads to

$$\frac{\vartheta(q,\lambda\mu)}{\lambda} = \vartheta(q,\mu) \tag{12.11}$$

for all $\lambda > 0$. Setting $\mu = 1$ and $\gamma_q = 1/\vartheta(q,1)$ in Equation (12.11), we obtain: $\vartheta(q,\lambda) = \lambda/\gamma_q$. The distribution function of the learning latencies is thus

$$\mathbb{P}(\mathbf{T}_{q,\lambda} \leq \tau) = 1 - e^{-(\lambda/\gamma_q)\tau}, \tag{12.12}$$

with $E(\mathbf{T}_{q,\lambda}) = \gamma_q/\lambda$. The form of this expectation is appealing. It entails that the difficulties of the items encountered along a learning path are additive in the following sense. Suppose that a subject with learning rate λ successively solves items $q_1, \ldots, q_n$. The total time to master all of these items has expectation

$$E(\mathbf{T}_{q_1,\lambda} + \cdots + \mathbf{T}_{q_n,\lambda}) = \frac{\sum_{i=1}^{n} \gamma_{q_i}}{\lambda}. \tag{12.13}$$

In words: the average time required to solve successively a number of items is the ratio of the sum of their difficulties, to the learning rate of the subject.

At the beginning of this section, we stated that the four Axioms [B], [I], [R] and [L] implicitly specified the functional form of the learning latencies. In fact, for a variety of reasons, we cannot simply derive the form of the latency distributions as a theorem. For example, we could not deduce from the stated axioms that the random variables $\mathbf{T}_{q,\lambda}$ do not depend upon K and that Equation (12.12) holds for $\mathbf{T}_{q,K,\lambda} = \mathbf{T}_{q,\lambda}$, which is what we want and will make the theory manageable (in view of Equation (12.13)). Accordingly, we formulate below a new axiom specifying the form of these latency distributions. The following notation will be instrumental.

12.5.3 Definition. Let ν be a gradation containing a state $K \neq Q$. We write K^ν for the state immediately following state K in gradation ν. We thus have $K \subset K^\nu \in \nu$ with $|K^\nu \setminus K| = 1$, and for any $S \in \nu$ with $K \subset S$, we must have $K^\nu \subseteq S$.

[T] Learning Times. We assume that, for a subject with learning rate λ, the time required for the mastery of item q (this item being accessible from that subject's current state) is a random variable $\mathbf{T}_{q,\lambda}$, which has an exponential distribution with parameter λ/γ_q, where $\gamma_q > 0$ is an index measuring the difficulty of item q. Thus, $E(\mathbf{T}_{q,\lambda}) = \frac{\gamma_q}{\lambda}$ and $Var(\mathbf{T}_{q,\lambda}) = (\frac{\gamma_q}{\lambda})^2$. All these random variables (for all q and λ) are assumed to be independent. The items are mastered successively. This means that the total time to master successively items $q, q', \ldots$ encountered along some gradation ν is the sum $\mathbf{T}_{q,\lambda} + \mathbf{T}_{q',\lambda} + \cdots$ of exponentially distributed random variables, with parameters λ/γ_q, $\lambda/\gamma_{q'}, \ldots$ Formally, for any positive real numbers δ and λ, any learning path $\nu \in \mathcal{G}$, and any two states $K, K' \in \nu$ with $K \subseteq K'$, we have

$$\ell e\,(K, K', \delta, \lambda, \nu) = \begin{cases} \mathbb{P}(\mathbf{T}_{q,\lambda} > \delta) \;\text{ with } \{q\} = K^\nu \setminus K & \text{if } K = K' \neq Q, \\ \mathbb{P}(\sum_{q\in K'\setminus K} \mathbf{T}_{q,\lambda} \leq \delta) - \mathbb{P}(\sum_{q\in K'^\nu\setminus K} \mathbf{T}_{q,\lambda} \leq \delta) & \text{if } K \subset K' \neq Q, \\ \mathbb{P}(\sum_{q\in K'\setminus K} \mathbf{T}_{q,\lambda} \leq \delta) & \text{if } K \subset K' = Q, \\ 1 & \text{if } K = K' = Q, \\ 0 & \text{in all other cases.} \end{cases}$$

Thus, the total time required to solve all the items in a state K is the random variable $\sum_{q\in K} \mathbf{T}_{q,\lambda}$ which is distributed as a sum of $|K|$ independent exponential random variables.

12.6 Empirical Predictions

In this section and the next one, we suppose that we have a system of stochastic learning paths $\mathcal{S} = ((\mathbf{K}_t, \mathbf{R}_t)_{t\geq 0}, \mathbf{C}, \mathbf{L})$ with the distributions satisfying the three Axioms [N], [A] and [T]. The predictions given below were derived by Stern and Lakshminarayan (1995) and Lakshminarayan (1995). We begin with a well-known result (cf. Adke and Manshunath, 1984).

12.6.1 Theorem. *Let $\mathbf{T}_1, \mathbf{T}_2, \ldots, \mathbf{T}_n$ be jointly distributed independent exponential random variables with parameters $\lambda_1, \ldots, \lambda_n$ respectively, and suppose that $\lambda_i \neq \lambda_j$ for $i \neq j$. Then the density function $h_{\mathbf{T}}$ and the associated distribution function $H_{\mathbf{T}}$ of the sum $\mathbf{T} = \mathbf{T}_1 + \cdots + \mathbf{T}_n$ are given, for $t \geq 0$ and $\delta \geq 0$, by*

$$h_{\mathbf{T}}(t) = \left(\prod_{i=1}^{n} \lambda_i\right) \sum_{j=1}^{n} e^{-\lambda_j t} \left(\prod_{\substack{k=1\\k\neq j}}^{n} (\lambda_k - \lambda_j)\right)^{-1}, \tag{12.14}$$

$$H_{\mathbf{T}}(\delta) = \int_0^\delta h_{\mathbf{T}}(t)\,dt = \left(\prod_{i=1}^{n} \lambda_i\right) \sum_{j=1}^{n} \frac{1 - e^{-\lambda_j \delta}}{\lambda_j \prod_{\substack{k=1\\k\neq j}}^{n} (\lambda_k - \lambda_j)}\,dt. \tag{12.15}$$

Thus, the results in the r.h.s. of (12.14) and (12.15) do not depend upon the order assigned by the index i. For example, with $n = 3$, we get

$$h_{\mathbf{T}}(t) = \lambda_1\lambda_2\lambda_3\Bigg(\frac{e^{-\lambda_1 t}}{(\lambda_2-\lambda_1)(\lambda_3-\lambda_1)}+ \frac{e^{-\lambda_2 t}}{(\lambda_1-\lambda_2)(\lambda_3-\lambda_2)}+\frac{e^{-\lambda_3 t}}{(\lambda_1-\lambda_3)(\lambda_2-\lambda_3)}\Bigg). \tag{12.16}$$

Since none of the above references contains a proof, we give one below.

PROOF. The MGF (moment generating function) $M_{\mathbf{T}_i}(\vartheta)$ of an exponential random variable $\mathbf{T}_i$ with parameter λ_i is given by $M_{\mathbf{T}_i}(\vartheta) = \lambda_i(\lambda_i - \vartheta)^{-1}$. Accordingly, the MGF of the sum of n independent exponential random variables $\mathbf{T}_i$ with parameter λ_i, $1 \leq i \leq n$, is

$$M_{\mathbf{T}}(\vartheta) = \prod_{i=1}^{n} \frac{\lambda_i}{\lambda_i - \vartheta}. \tag{12.17}$$

The proof proceeds by showing that the MGF of the random variable with the density function specified by Equation (12.14) is that given by (12.17). The result follows from the fact that the MGF of a random variable uniquely determines its distribution. In passing, this would establish that the r.h.s.'s of (12.14) and (12.15) do not depend upon the order assigned by the index i. Multiplying by $e^{\vartheta t}$ in both sides of (12.14) and integrating from 0 to ∞ over t (assuming that $\vartheta < \lambda_j$ for all indices j) yields

$$E(e^{\vartheta \mathbf{T}}) = \left(\prod_{i=1}^{n}\lambda_i\right)\sum_{j=1}^{n}\Bigg((\lambda_j-\vartheta)\prod_{\substack{k=1\\k\neq j}}^{n}(\lambda_k-\lambda_j)\Bigg)^{-1} = \left(\prod_{i=1}^{n}\frac{\lambda_i}{\lambda_i-\vartheta}\right)\times D$$

with (see Problem 4)

$$D = \frac{\sum_{j=1}^{n}(-1)^{n-j}\left[\left(\prod_{i\neq j}(\lambda_i-\vartheta)\right)\left(\prod_{\substack{k=1\\k\neq j}}(\lambda_k-\lambda_l)\right)\right]}{\prod_{i<j}(\lambda_i-\lambda_j)}. \tag{12.18}$$

It suffices to show that $D = 1$ for all values of λ_i, $1 \leq i \leq n$ satisfying the conditions. The numerator in Equation (12.18) is a polynomial in ϑ of degree $n-1$. For $i = 1, 2, \ldots, n$, it is easy to check that the value of this polynomial evaluated at $\vartheta = \lambda_i$ is equal to the denominator of (12.18) (see Problem 5). Thus $D = 1$ for all values of ϑ, which completes the proof of Equation (12.14). Equation (12.15) follows easily. □

A similar result can be obtained in the case where some of the λ_i's have identical values. We shall not treat this case here.

12.6.2 Convention. For the remainder of this section, we suppose that all the difficulty parameters γ_q associated with the items have different values. Accordingly, the parameters λ/γ_q of the exponential learning latencies have also different values.

12.6.3 Definition. For any $S \subseteq Q$ and $\lambda > 0$, we denote by $g_{S,\lambda}$ the density function of the sum $\mathbf{T}_{S,\lambda} = \sum_{q \in S} \mathbf{T}_{q,\lambda}$, where each $\mathbf{T}_{q,\lambda}$ is an exponential random variable with parameter γ_q/λ and the random variables $\mathbf{T}_{q,\lambda}$ are pairwise independent. We also denote by $G_{S,\lambda}$ the corresponding distribution function. Thus, $g_{S,\lambda}$ and $G_{S,\lambda}$ are specified by Equations (12.14) and (12.15) up to the values of λ and the γ_q's. This notation is justified because any permutation of the index values in the r.h.s. of Equation (12.14) yield the same density function, and hence the same distribution function.

We will use Theorem 12.6.1 to obtain an explicit expression for the learning function ℓe in terms of the exponential distributions. The next theorem is a restatement of Axiom [T] following from Theorem 12.6.1 and Definition 12.6.3.

12.6.4 Theorem. *For any positive real numbers δ and λ, any learning path $\nu \in \mathcal{G}$, and any two states $K, K' \in \nu$, we have*

$$\ell e\,(K, K', \delta, \lambda, \nu) = \begin{cases} e^{-\frac{\lambda}{\gamma_q}\delta} \quad \text{with } \{q\} = K^{\nu} \setminus K & \text{if } K = K' \neq Q, \\ G_{K'\setminus K,\lambda}(\delta) - G_{K'^{\nu}\setminus K,\lambda}(\delta) & \text{if } K \subset K' \neq Q, \\ G_{K'\setminus K,\lambda}(\delta) & \text{if } K \subset K' = Q, \\ 1 & \text{if } K = K' = Q, \\ 0 & \text{in all other cases.} \end{cases}$$

We are now equipped to derive explicit predictions for the joint probabilities of the states occurring at successive times. The next theorem contains one example. As a convention, we set $G_{\varnothing,\lambda}(\delta) = 1$ and $G_{Q^{\nu}\setminus S,\lambda}(\delta) = 0$. Notice that $G_{S,\mathbf{L}}(\delta)$ is a random variable, with expectation

$$E\left[G_{S,\mathbf{L}}(\delta)\right] = \int_0^{\infty} G_{S,\lambda}(\delta) f(\lambda) d\lambda,$$

where f is the gamma density function defined by (12.6) in Axiom [A].

12.6.5 Theorem. *For all integers $n > 0$, all states $K_1 \subseteq \cdots \subseteq K_n$, and all real numbers $t_n > \cdots > t_1 \geq 0$, we have*

$$\begin{aligned} &\mathbb{P}(\mathbf{K}_{t_1} = K_1, \ldots, \mathbf{K}_{t_n} = K_n) \\ &= \sum_{\nu} p_{\nu} \left\{ \int_0^{\infty} \prod_{i=0}^{n-1} \left[G_{K_{i+1}\setminus K_i,\lambda}(t_{i+1} - t_i) - G_{K^{\nu}_{i+1}\setminus K_i,\lambda}(t_{i+1} - t_i) \right] f(\lambda) d\lambda \right\} \\ &= \sum_{\nu} p_{\nu}\, E \left\{ \prod_{i=0}^{n-1} \left[G_{K_{i+1}\setminus K_i,\mathbf{L}}(t_{i+1} - t_i) - G_{K^{\nu}_{i+1}\setminus K_i,\mathbf{L}}(t_{i+1} - t_i) \right] \right\}, \end{aligned}$$

with $t_0 = 0$, $K_0 = \varnothing$ and the sum extending over all $\nu \in \mathcal{G}(K_1, \ldots, K_n)$.

The predictions for the Case $n = 2$ are relevant to the application described in Section 12.10, which is due to Lakshminarayan (1995). We have clearly:

12.6.6 Theorem. *For any pair of response patterns* $(R_2, R_1) \in 2^Q \times 2^Q$, *and any real numbers* $t_2 > t_1 \geq 0$,

$$\begin{aligned}\mathbb{P}(\mathbf{R}_{t_1} &= R_1, \mathbf{R}_{t_2} = R_2)\\ &= \sum_{K_1 \in \mathcal{K}} \sum_{K_2 \in \mathcal{K}} r(R_1, K_1) r(R_2, K_2)\, \mathbb{P}(\mathbf{K}_{t_1} = K_1, \mathbf{K}_{t_2} = K_2) \qquad (12.19)\end{aligned}$$

with $r(R_1, K_1)$ *and* $r(R_2, K_2)$ *specified by Axiom* [N] *in terms of the parameters* β_q *and* η_q, *and* $\mathbb{P}(\mathbf{K}_{t_1} = K_1, \mathbf{K}_{t_2} = K_2)$ *as in Theorem* 12.6.5.

An explicit expression of the joint probabilities $\mathbb{P}(\mathbf{R}_{t_1} = R_1, \mathbf{R}_{t_2} = R_2)$ is now easy to obtain. To begin with, we can replace the values $r(R_i, K_i)$ of the response function by their expressions in terms of the probabilities of the careless errors and of the lucky guesses given by Axiom [N]. Next, the values of the joint probabilities of the state $\mathbb{P}(\mathbf{K}_{t_1} = K_1, \mathbf{K}_{t_2} = K_2)$ in terms of the distributions of the learning rate and the learning latencies can be obtained by routine integration via Axioms [A] and Theorems 12.6.6, 12.6.1, 12.6.4 and 12.6.5. We shall not spell out these results here (see Lakshminarayan, 1995, for details).

12.7 Limitations of this Theory

The assumption that the time required to master some item q does not depend upon the current state K—provided that item q can be learned from K—was made for the sake of simplicity, but is not immune from criticisms. One can easily imagine situations in which it might fail. Intuitively, if an item q can be learned from both K and K', a student in state K may conceivably be better prepared for the learning of q than a student in state K'. The argument in the discussion and the example below makes this idea concrete[7].

12.7.1 Remarks. We consider a knowledge space $\mathcal{K}$ which is a projection of some large, idealized knowledge structure $\mathring{\mathcal{K}}$ containing all the items in a given field of information, with domain $\mathring{Q}$. Let us denote by $\mathring{\mathcal{Q}}$ the surmise relation of $\mathring{\mathcal{K}}$ (cf. Definition 3.7.1); thus,

$$q' \mathring{\mathcal{Q}} q \quad \Longleftrightarrow \quad (\forall K \in \mathring{\mathcal{K}} : q \in K \Rightarrow q' \in K).$$

[7] This argument was suggested by Lakshminarayan (personal communication; see also Stern and Lakshminarayan, 1995). The details are as in Falmagne (1996).

We also define, for any q in $\mathring{Q}$ and any subset S of $\mathring{Q}$,

$$\mathring{\mathcal{Q}}^{-1}(q) = \{r \in \mathring{Q} \,|\, r\mathring{\mathcal{Q}}q\}$$
$$\mathring{\mathcal{Q}}^{-1}(S) = \{r \in \mathring{Q} \,|\, r\mathring{\mathcal{Q}}q, \text{ for some } q \in S\} = \cup_{q \in S} \mathring{\mathcal{Q}}^{-1}(q).$$

Consider a state K in $\mathcal{K}$ and suppose that items a and b can be learned from K in the sense that both a and b are in the outer fringe of K in $\mathcal{K}$. Thus, both $K \cup \{a\}$ and $K \cup \{b\}$ are states of $\mathcal{K}$. Since $\mathcal{K}$ is a knowledge space, $K \cup \{a, b\}$ is also a state of $\mathcal{K}$. This means that item a, for example, can be learned from either state K or state $K \cup \{b\}$. It makes sense to suppose that the difficulty of mastering a from state K must depend on the items from $\mathring{Q} \setminus K$ that must be mastered before mastering a; that is, it depends on the set

$$S(a, K) = \mathring{\mathcal{Q}}^{-1}(a) \setminus \mathring{\mathcal{Q}}^{-1}(K).$$

Similarly, the difficulty of mastering item a from the state $K \cup \{b\}$ depends on the set

$$S(a, K \cup \{b\}) = \mathring{\mathcal{Q}}^{-1}(a) \setminus \mathring{\mathcal{Q}}^{-1}(K \cup \{b\}).$$

The assumption that the difficulty of an item does not depend upon the state of the subject leads one to require that $S(a, K) = S(a, K \cup \{b\})$. In fact, we have by definition $S(a, K \cup \{b\}) \subseteq S(a, K)$ but the equality holds only in special circumstances. Indeed, some manipulation yields (cf. Problem 6):

$$S(a, K) \setminus S(a, K \cup \{b\}) \;=\; \mathring{\mathcal{Q}}^{-1}(a) \cap \overline{\mathring{\mathcal{Q}}^{-1}(K)} \cap \mathring{\mathcal{Q}}^{-1}(K \cup \{b\}).$$

Supposing that the intersection in the r.h.s. is empty means that there is no item q in $\mathring{Q}$ preceding both a and b, and not preceding at least one item in K. Such an assumption does not hold for general knowledge structures. We give a counterexample below.

12.7.2 Example. Consider the knowledge space

$$\mathring{\mathcal{K}} = \{\varnothing, \{c\}, \{d\}, \{c, d\}, \{a, c, d\}, \{b, c, d\}, \{a, b, c, d\}\}.$$

Thus $\mathring{Q} = \{a, b, c, d\}$. The projection of $\mathring{\mathcal{K}}$ on $Q = \{a, b, c\}$ is

$$\mathcal{K} = \{\varnothing, \{c\}, \{a, c\}, \{b, c\}, \{a, b, c\}\}.$$

If we take $K = \{c\} \in \mathcal{K}$, we get

$$\mathring{\mathcal{Q}}^{-1}(a) = \{a, c, d\}, \quad \mathring{\mathcal{Q}}^{-1}(K) = \{c\},$$
$$\overline{\mathring{\mathcal{Q}}^{-1}(K)} = \{a, b, d\}, \quad \mathring{\mathcal{Q}}^{-1}(K \cup \{b\}) = \{b, c, d\}$$

with

$$\mathring{\mathcal{Q}}^{-1}(a) \cap \overline{\mathring{\mathcal{Q}}^{-1}(K)} \cap \mathring{\mathcal{Q}}^{-1}(K \cup \{b\}) = \{d\}.$$

A cure is easy but costly. As argued earlier, we can replace all the difficulty parameters γ_q by parameters $\gamma_{q,K}$ explicitly depending on the current state K of the subject. Less prohibitive solutions would be preferable, of course. In practice, it may turn out that the dependence of the difficulty of mastering some item q on the state from which q is accessed, while theoretically justified, is in fact mild. The reason could either be that the estimates of the parameters $\gamma_{q,K}$ and $\gamma_{q,K'}$ do not differ much, or that this dependence only affects a small proportion of the states and items. This question was investigated empirically by Lakshminarayan (1995) who showed that a model with exponentially distributed learning latencies and with difficulty parameters γ_q that did not depend on the state was able to fit the data quite well.

In any event, our discussion raises the following problem: Under wich conditions on the knowledge structure $(\mathring{Q}, \mathring{\mathcal{K}})$ are all the set differences $S(a, K) \setminus S(a, K \cup \{b\})$ empty, for all substructures $(Q, \mathcal{K})$? It is clear that a sufficient condition is that $\mathring{\mathcal{K}}$ is a chain. This condition is not necessary, however. We leave this question as one of our open problems (see 18.4.2).

The above discussion concerning a possible lack of invariance of the difficulty parameters was organized around the concept of a projection $(Q, \mathcal{K})$ of a parent structure $(\mathring{Q}, \mathring{\mathcal{K}})$. In short, it was argued that the difficulty of acquiring a new item a accessible from two states K and $K \cup \{b\}$ in $\mathcal{K}$, could differ because the implicit paths in $\mathring{\mathcal{K}}$ leading from K to $K \cup \{a\}$ and from $K \cup \{b\}$ to $K \cup \{a, b\}$ could required the mastery of different items in $\mathring{Q} \setminus Q$. The same type of argument can be used to show that, from the standpoint of the learning latencies, it cannot be the case that the axioms of the model holds for both $\mathcal{K}$ and $\mathring{\mathcal{K}}$. Specifically, if the learning latencies are exponentially distributed in a model assumed to hold for the parent knowledge structure $(\mathring{Q}, \mathring{\mathcal{K}})$, then these latencies cannot, in general, be exponentially distributed for the substructure $(Q, \mathcal{K})$. Rather, they have to be sums of exponential random variables, or even mixtures of such sums. Consequently, because the exponential form is necessary, the model cannot in principle hold for $(Q, \mathcal{K})$. For a more detailed discussion on this point, see Stern and Lakshminarayan (1995).

12.8 Simplifying Assumptions

Another type of objection that may be raised against this theory is that the number of gradations may be very large. Because a probability is attached to each gradation, the number of parameters to be estimated from the data may be prohibitive. However, some fairly natural simplifying assumptions can be made which would result in a substantial decrease in the number of parameters attached to the gradations. We discuss an example.

12.8.1 Markovian learning paths. Suppose that the probability of a gradation

$$\varnothing \subset \{a\} \subset \{a,b\} \subset \{a,b,c\} \subset \{a,b,c,d\} \subset Q \qquad (12.20)$$

in some well-graded knowledge structure with domain $Q = \{a, b, c, d, e\}$ can be obtained by multiplying the successive transition probabilities from state to state along the gradation. Let us denote by $p_{K,K+\{q\}}$ the condition probability of a transition from a state K to state $K + \{q\}$, with K a non maximal state[8]. In the notation of 12.2.2 for the subsets of gradations, we have thus

$$p_{K,K+\{q\}} = \mathbb{P}(\mathbf{C} \in \mathcal{G}(K + \{q\})|\mathbf{C} \in \mathcal{G}(K)) = \frac{\mathbb{P}(\mathbf{C} \in \mathcal{G}(K, (K + \{q\})}{\mathbb{P}(\mathbf{C} \in \mathcal{G}(K))}.$$

Our simplifying assumption is that the probability of the gradation in (12.20) is given by the product

$$p_{\varnothing,\{a\}} \cdot p_{\{a\},\{a,b\}} \cdot p_{\{a,b\},\{a,b,c\}} \cdot p_{\{a,b,c\},\{a,b,c,d\}} \cdot$$

(Note that the conditional probability of a transition from state $\{a, b, c, d\}$ into state Q is equal to one.)

Let us generalize this example. We recall that, for any non maximal state K in gradation ν, K^ν stands for the state immediately following state K in ν. We now extend this notation. For any nonempty state K in gradation ν, we write ${}^\nu K$ for the state immediately preceding state K in ν.

[MLP] Markovian Assumption on the Learning Paths. For all $\nu \in \mathcal{G}$, we have:

$$\mathbb{P}(\mathbf{C} = \nu) = p_{\varnothing,\varnothing^\nu} \cdot p_{\varnothing^\nu,(\varnothing^\nu)^\nu} \cdot \ldots \cdot p_{{}^\nu Q,Q},$$

some of which may be equal to 1.

For instance, we must have $p_{{}^\nu Q,Q} = 1$, for all learning paths ν.

It is our experience that, for large typical well-graded knowledge structures, this assumption will dramatically reduce the number of parameters in the model.

Clearly, more extreme simplifying assumptions can be tested. We could assume, for instance, that p_{K,K^ν} only depends upon $K^\nu \setminus K$. In the case of a well-graded knowledge structure, this would reduce the number of transition probabilities to at most $|Q|$.

[8] For the gradation defined by (12.20), $p_{\varnothing,\{a\}}$ is thus the conditional probability that $\{a\}$ is first state visited after the initial empty state.

12.9 Remarks on Application and Use of the Theory

Suitable data for this theory consist in the frequencies of n-tuples of response patterns $R_1, \ldots, R_i, \ldots, R_n$, observed at times $t_1 < \ldots < t_i < \cdots < t_n$. Thus, each subject is tested n times with the full set Q of problems, and R_i denotes the subset of Q containing all the correct responses given by the subject at time t_i. Let us consider the case $n = 2$. Thus, a sample of subjects has been selected, and these subjects have been tested twice, at times t and $t + \delta$. We denote by $N(R, R')$, with $R, R' \subseteq Q$, the number of subjects having produced the two patterns of responses R and R' at times t and $t + \delta$.

12.9.1 A maximum likelihood procedure. The parameters may be estimated by maximizing the loglikelihood function

$$\sum_{R,R' \subseteq Q} N(R, R') \log \mathbb{P}(\mathbf{R}_t = R, \mathbf{R}_{t+\delta} = R') \tag{12.21}$$

in terms of the various parameters of the theory, namely, the response parameters β_q and η_q, the parameters α and ξ of the distribution of the learning rates, the item difficulty parameters γ_q, and the probabilities p_ν of the gradations. In some cases, it is also possible to consider the times t and $t + \delta$ as parameters. For example, the time t elapsed since the beginning of learning may be difficult to assess accurately. Applying a theory of such a complexity raises a number of issues, which we now address.

12.9.2 Remarks. a) To begin with, some readers may cringe at this plethora of parameters, and wonder whether an application of this theory is a realistic prospect. Actually, as indicated earlier in this chapter, the theory has been successfully applied to several sets of real data (see Arasasingham, Taagepera, Potter, and Lonjers, 2004; Arasasingham, Taagepera, Potter, Martorell, and Lonjers, 2005; Falmagne et al., 1990; Lakshminarayan, 1995; Taagepera et al., 1997; Taagepera and Noori, 2000; Taagepera, Arasasingham, Potter, Soroudi, and Lam, 2002; Taagepera, Arasasingham, King, Potter, Martorell, Ford, Wu, and Kearney, 2008), and also simulated data (cf. Falmagne and Lakshminarayan, 1994). Notice that, when a test is administered several times, the number of parameters does not increase, whereas the number of response categories—i.e. the number of degrees of freedom in the data—increases exponentially. If the number of gradations is not prohibitive, or if some Markovian assumptions in the spirit of the last section are satisfied, the complexity of the theory will remain well beneath that of the data to be explained. For example, under the simplest Markovian assumption mentioned at the end of the last section, the number of parameters of the theory is of the order of $|Q|$, while the number of response categories is of the order of $2^{n|Q|}$ (with n applications of the test to the same sample of subjects), which is a good return. The maximization of the loglikelihood function in Equation (12.21) may be achieved by a procedure such as the Conjugate Gradient Search algorithm by Powell (1964) which is available in form of the C subroutine `PRAXIS` (see

Gegenfurtner, 1992). In practice, the procedure is applied many times[9], with different starting values for the parameters, to ensure that the final estimates do not correspond to a local maximum.

b) Some of the applications mentioned above (for example Taagepera et al., 1997) were performed under the hypothesis that the learning latencies had general gamma (rather than the correct exponential) distributions. In 12.5.1, the gamma assumption was shown to be inconsistent with the other axioms. However, because this inconsistency does not propagate to the predictions (in the sense that Equation (12.19), for example, defines a genuine distribution on the set of all pairs (R_1, R_2) of patterns of responses), it does not preclude a good fit to the data. Moreover, these learning latencies affect the predictions only indirectly, after smearing by the learning rate random variable. This suggest that the predictions of the model may be robust to the particular assumptions made on the learning latencies.

c) For the theory to be applicable, a well-graded knowledge structure must be assumed. Methods for building a knowledge space have been considered in Chapter 7 (cf. Koppen, 1989; Koppen and Doignon, 1990; Falmagne et al., 1990; Kambouri, Koppen, Villano, and Falmagne, 1994; Müller, 1989; Dowling, 1991a,b, 1993a; Villano, 1991; Cosyn and Thiéry, 2000) and will be discussed again in more detail in Chapter 15. The special case of a learning space, which implies wellgradedness, is analyzed in Chapter 16. A reasonable procedure is to start the analysis with a tentative knowledge structure, presumed to contain all the right states, and possibly some subsets of questions which are not states. If the application of the model proves to be successful, this starting knowledge structure can be progressively refined by assuming that some learning paths have probability zero, thereby eliminating some states. This method is exemplified in Falmagne et al. (1990).

12.10 An Application of the Theory to the Case $n = 2$

The most ambitious application of the theory described in this chapter is due to Lakshminarayan (1995), and we summarize it here. Only the main lines of the analysis and of the results will be reported.

12.10.1 The items. The domain contains the five problems in high school geometry displayed in Figure 12.1, which deal with basic concepts such as angles, parallel lines, triangles and the Pythagorean Theorem. These problems are labeled a, b, c, d and e. Two versions (instances) of each problem were generated, forming two sets V1 and V2 of five problems which were applied at different times[10]. The differences between the two versions only concerned the particular numbers used as measures of angles or length.

[9] At least several hundred times.

[10] V1 and V2 were in fact embedded into two larger equivalent sets containing 14 problems each.

12.10.2 Procedure. The experiment had three phases:

1. PRETEST. The subjects were presented with one version, and asked to solve the problems. Approximately half of the subjects were given version V1, and the rest, version V2.

2. LESSON. After completion of Phase 1, the test sheets were removed and a 9-page booklet containing a lesson in high school geometry was distributed to the subjects, who were required to study it. The lesson was directly concerned with the problems to be solved.

3. POSTTEST. The subjects were given the other version of the test, and were asked to solve the problems.

There was no delay between the phases, and the subjects were allowed to spend as much time as they wanted on each problem. Typically, the students spent between 25 to 55 minutes to complete the three phases of the experiment.

The data consist in the observed frequencies of each of the $2^5 \times 2^5 = 1024$ possible pairs of response patterns. The subjects were undergraduate students of the University of California at Irvine; 959 subjects participated in the experiment.

12.10.3 Parameter estimation. The overall data set provided by the 959 subjects was split into two unequal part. Those of 159 subjects were set aside, and kept for testing the predictions of the model. Those of the remaining 800 subjects were used to uncover the knowledge structure and to estimate parameters. A tentative knowledge structure was initially postulated, based on the content of the items. This structure was then gradually refined on the basis of goodness-of-fit (likelihood ratio) calculation.

All the problems have open responses. Accordingly, all the guessing parameters η_q were set equal to zero. The remaining parameters were estimated by a maximum likelihood method.

12.10.4 Results. The final knowledge structure was a knowledge space included in that displayed in Figure 12.2. Two of the original gradations, namely *abcde* and *adbce* were dropped in the course of the analysis because their estimated probabilities were not significantly different from zero. As a consequence, the state $\{a, d\}$ was removed. The resulting learning space is

$$\{\varnothing, \{a\}, \{b\}, \{a, b\}, \{b, c\}, \{a, b, c\}, \\ \{a, b, d\}, \{b, c, d\}, \{a, b, c, d\}, \{a, b, c, d, e, \}\}.$$

This space has six gradations, five of which were assigned a non zero probability. All but one the remaining gradations had small probabilities ($< .10$). Gradation *bcade* occurred with an estimated probability of .728. The estimated probabilities of the five gradations are given in Table 12.1, together with the estimates of the other parameters of the model. Note that t and $t + \delta$

(the two times of testing) are regarded as parameters because the time elapsed since the beginning of learning geometry could not be assessed accurately. The unit of t and δ is the same as that of ξ and is arbitrary.

Parameters	Estimates
$\mathbb{P}(\mathbf{C} = abcde)$	.047
$\mathbb{P}(\mathbf{C} = bcdae)$	.059
$\mathbb{P}(\mathbf{C} = bcade)$	.728
$\mathbb{P}(\mathbf{C} = bacde)$	.087
$\mathbb{P}(\mathbf{C} = badce)$	.079
γ_a	11.787
γ_b	25.777
γ_c	11.542
γ_d	34.135
γ_e	90.529
α	113.972
t	11.088
δ	2.448
ξ	13.153
β_a	.085
β_b	.043
β_c	.082
β_d	.199
β_e	.169

Table 12.1. Estimates of the parameters. We recall that all the parameters η_q have been set equal to 0 a priori. The quantities t and $t + \delta$ representing the two times of testing are regarded as parameters in this application because the time t elapsed since the beginning of learning geometry could not be assessed accurately.

The fit of the model based of the value of the log-likelihood statistic was good. In general, the estimated values of the parameter seem reasonable. In particular, the values obtained for the β_q are small, which is consistent with the interpretation of these parameters as careless error probabilities. The value of δ is overly large compared to that of t. For example if for the sake of illustration we set the unit of t as equal to one year, then the estimated total time of learning until the first test amounts to eleven years and one month, while the time between the first and the second test is estimated to be approximately two years and five months, which is absurd.

The most likely explanation is that—relative to the two years and five months—the eleven years and one month is an overestimation reflecting the forgetting that took place for most students between the end of their learning geometry in high school and the time of the test. If the first test had occurred while the students were learning geometry in high school, the estimates of t and δ would presumably have been more consistent.

Another subset of 4 items was analyzed by the same methods by Lakshminarayan (1995), which yielded a much less satisfactory fit. At this point, deriving negative conclusions from these other results would be premature. As we mentioned earlier, fitting a model of that kind, with so many parameters, is a complex procedure which is not guaranteed to lead automatically to the best knowledge structure and the best set of estimates of the parameters for the data.

12.11 Original Sources and Related Works

In this chapter, we formalize the learning that takes place in the framework of a well-graded knowledge structure, as involving the choice of a gradation, paired with a stochastic process describing the progressive mastery of the items along that gradation. This concept is natural enough and was already exploited to some extent in the models discussed in Chapter 11. The model discussed here is to date the most elaborate implementation of this idea. As made clear by our presentation, this model attempts to give a realistic picture of learning by making an explicit provision for individual differences and by describing learning as a real time stochastic process. Thus, the target data for the model is made of n-tuples $(R_1, \ldots, R_n)$ of responses patterns observed at arbitrary times $t_1, \ldots, t_n$. This attempt was only partly successful. The potential drawbacks of the current model are summarized below together with the relevant literature.

A first pass at constructing such a model was made by Falmagne (1989a). Even though most of the ideas of this chapter were already used in that earlier attempt, the 1989 model was not fully satisfactory because it was not formulated explicitly as a stochastic process. The 1989 model was tested against a standard unidimensional psychometric model by Lakshminarayan and Gilson (1998), with encouraging results. In his 1993 paper published in the Journal of Mathematical Psychology[11], Falmagne developed what is essentially the model of this chapter, except that the learning latencies were assumed to be distributed as general gammas. In the notation of Equation 12.12, we had

$$\mathbb{P}(\mathbf{T}_{q,\lambda} \leq \delta) = \int_0^\delta \frac{\tau^{\gamma_q} \lambda^{\gamma_q - 1} e^{-\lambda\tau}}{\Gamma(\gamma_q)} d\tau, \tag{12.22}$$

instead of

$$\mathbb{P}(\mathbf{T}_{q,\lambda} \leq \delta) = 1 - e^{-(\lambda/\gamma_q)\delta} \tag{12.23}$$

as assumed in this chapter, the interpretation of the parameters being the same in both cases. A simulation study of the 1993 model was made by Falmagne and Lakshminarayan (1994). This model was successfully applied to real data by Taagepera et al. (1997) which gave a good account of the experimental results. However, around May 1994, Stern and Lakshminarayan (1995) discovered that the assumption that the learning latencies are distributed gamma was inconsistent with the axioms of the theory as stated here, and surmised that the appropriate distributions for these latencies had to be exponential (Stern and Lakshminarayan, 1995; Lakshminarayan, 1995; Falmagne, 1996). Our tentative conclusion is that the model is fairly robust with respect to specific hypotheses regarding the latency distributions.

[11] Falmagne (1993).

Problems

1. Formulate an axiom concerning the learning function ℓe resulting in a system of stochastic learning paths that is not progressive (cf. Definition 12.2.4). (Hint: Allow for the possibility of forgetting.)

2. Argue that the Markovian assumption

$$\mathbb{P}(\mathbf{K}_{t_{n+1}} = K_{n+1} \,|\, \mathbf{K}_{t_n} = K_n, \mathbf{R}_{t_n} = R_n, \mathcal{E}) = \mathbb{P}(\mathbf{K}_{t_{n+1}} = K_{n+1} \,|\, \mathbf{K}_{t_n} = K_n)$$

   (with $\mathcal{E}$ as in Axiom [L]) discussed in Remark 12.2.5 (d) is inappropriate. (Hint: Show by an example that the more detailed history embedded in the event $\mathcal{E}$ in the l.h.s. may provide information on the learning rate and/or the learning path, which in turn, may modify the probabilities of the states at time t_{n+1}.)

3. Investigate a possible reformulation of the learning function that would not imply, via Equations (12.8) and (12.9), that the learning latencies are distributed exponentially.

4. Verify Equation (12.18).

5. Establish all the assertions following Equation (12.18) in the proof of Theorem 12.6.1.

6. Verify the computation of $S(a, K) \setminus S(a, K \cup \{b\})$ in Remark 12.7.1.

7. Following up on the discussion of Remark 12.7.1, find a counterexample to the Condition $\mathring{\mathcal{Q}}^{-1}(a) \cap \overline{\mathring{\mathcal{Q}}^{-1}(K)} \cap \mathring{\mathcal{Q}}^{-1}(K \cup \{b\}) = \varnothing$, different from the one in Example 12.7.2.

8. In relation with Example 12.7.2 and Remark 12.7.1, find necessary and sufficient conditions on the knowledge structure $(\mathring{Q}, \mathring{\mathcal{K}})$ such that, for all substructures $(Q, \mathcal{K})$, all states $K \in \mathcal{K}$ and all pairs of items a and b, the set difference $S(a, K) \setminus S(a, K \cup \{b\})$ is empty. The result may evoke some ironical reflexions. Check this impulse. Instead, work on Problem 9.

9. Investigate realistic cures for the difficulty pointed out by Example 12.7.2.

10. For any $t > 0$ and any item q, let $N_{t,q}$ be the number of subjects having provided a correct response to item q in a sample of N subjects. Let $\overline{N}_{t,q}$ denote the number of subjects having provided an incorrect response to that item. Thus $N_{t,q} + \overline{N}_{t,q} = N$. Consider a situation in which a test has been applied twice to a sample of N subjects, at times t and $t + \delta$. Investigate the statistical properties of $\overline{N}_{t+\delta,q}/N_{t,q}$ as an estimator of the careless error probability β_q. Is this estimator unbiased, that is, do we have $E(\overline{N}_{t+\delta,q}/N_{t,q}) = \beta_q$?

11. Prove Lemma 12.3.3.

13

Uncovering the Latent State: A Continuous Markov Procedure

Suppose that, having applied the techniques described in the preceding chapters[1], we have obtained a particular knowledge structure. We now ask: how can we uncover, by appropriate questioning, the knowledge state of a particular individual? Two broad classes of stochastic assessment procedures are described in this chapter and the next one.

13.1 A Deterministic Algorithm

By way of introduction, we first consider a simple algorithm in the spirit of those discussed in Chapter 9 and which is suitable when there are no errors of any kind and no lucky guesses.

13.1.1 Example. The knowledge structure

$$\mathcal{K} = \{\varnothing, \{a\}, \{c\}, \{a,b\}, \{a,c\}, \{b,c\},\\ \{a,b,c\}, \{a,b,c,d\}, \{a,b,c,d,e\}\}, \tag{13.1}$$

with domain $Q = \{a,b,c,d,e\}$ displayed in Figure 13.1 will serve as an illustration for our discussion.

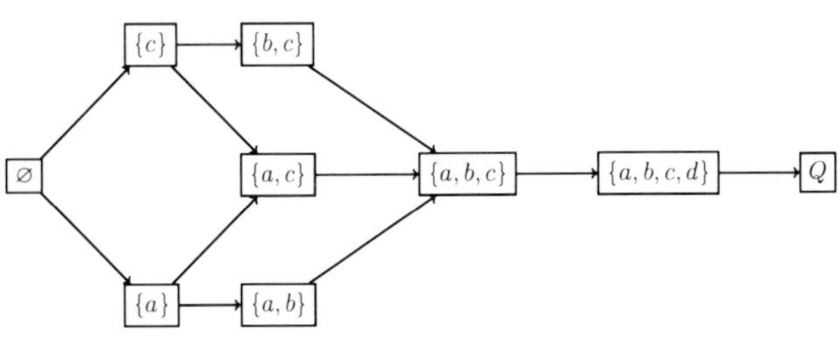

Figure 13.1. The learning space $\mathcal{K}$ of Equation (13.1).

[1] And some other techniques which we describe in Chapters 15 and 16.

J.-C. Falmagne, J.-P. Doignon, *Learning Spaces*,
DOI 10.1007/978-3-642-01039-2_13, © Springer-Verlag Berlin Heidelberg 2011

This knowledge structure is a learning space, with four gradations. These special features will play no role in the assessment procedure describe below[2]. We shall assume that a subject's response never results from a lucky guess or a careless error. (The impact of this assumption is discussed later in this chapter.) Consider an assessment in which Question b is the first question asked. If an incorrect response is obtained, only the states **not** containing b must be retained. We indicate this conclusion by marking with the symbol $\surd$ the remaining possible states in the second column of Table 13.1; they form the subfamily $\mathcal{K}_{\bar{b}}$ (cf. Definition 9.1.2 for this notation).

Problem Response	b 0	a 1	c 1
$\varnothing$	$\surd$		
$\{a\}$	$\surd$	$\surd$	
$\{c\}$	$\surd$		
$\{a,b\}$			
$\{a,c\}$	$\surd$	$\surd$	$\surd$
$\{b,c\}$			
$\{a,b,c\}$			
$\{a,b,c,d\}$			
$\{a,b,c,d,e\}$			

Table 13.1. Inferences from the successive responses of a subject for the knowledge structure $\mathcal{K}$ of Figure 13.1. The subject's response is marked as '0' for 'incorrect' and '1' for 'correct.' The states remaining after each response are indicated by the symbol '$\surd$'.

Next, Problem a is presented, and a correct response is recorded, eliminating the two states $\varnothing$ and $\{c\}$. Thus, the only feasible states after two questions are $\{a\}$ and $\{a,c\}$. The last problem asked is c, which elicits a correct response, eliminating the state $\{a\}$. In this deterministic framework, $\{a,c\}$ is the only state consistent with the data:

$$(b, \text{incorrect}), (a, \text{correct}), (c, \text{correct}).$$

Clearly, all the states can be uncovered by this procedure, which can be represented by a binary decision tree (see Figure 13.2). The procedure is certainly economical. For instance, if the states are equiprobable, a state can be uncovered by asking an average of $3\frac{2}{9}$ questions out of 5 questions.

This procedure was investigated from a formal viewpoint by Degreef et al. (1986) and was reviewed in Chapter 9. A major drawback of this type of algorithm is that it cannot effectively deal with a possible intrinsic randomness, or even instability, of the subject performance. Obvious examples of randomness are the careless errors and lucky guesses formalized in the models developed in Chapters 11 and 12. Another case of instability may arise if the subject's state

[2] However, remember that, in the case of a well-graded family, any knowledge state is defined by its fringes (cf. Remark 4.1.8(a)). So, the outcome of an assessment can be presented in the form of the fringes of the uncovered state. In the framework of the theory, the outer fringe of a state specifies what a subject in that state is ready to learn. This feature may play a critical role when the assessment is a placement test or a prelude to teaching.

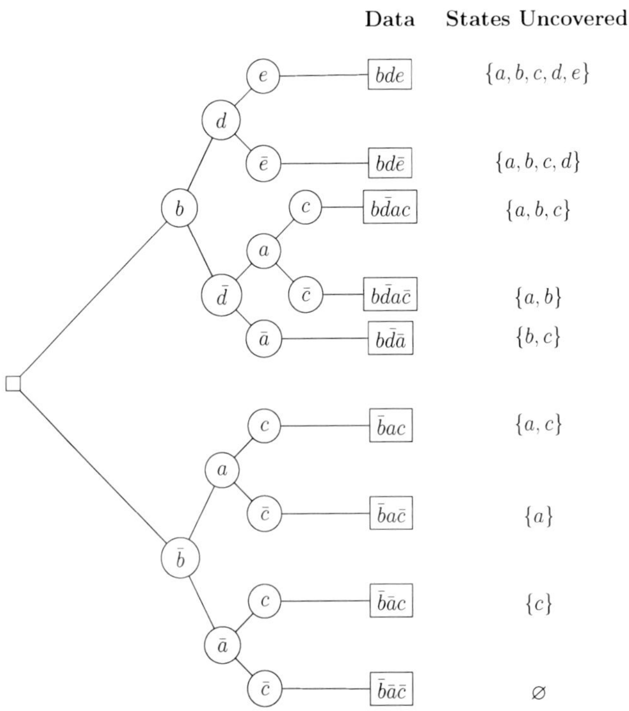

Figure 13.2. Binary decision tree for uncovering the states in Example 13.1.1.

varies in the course of the questioning. This might happen if the problems of the test cover concepts learned by the subject a long time earlier. The first few questions asked may jolt the subject's memory, and facilitate the retrieval of some material relevant to the last part of the test. In any event, more robust procedures are needed that are capable of uncovering a subject's state, or at least approaching it closely, despite noisy data.

13.2 Outline of a Markovian Stochastic Process

The Markov procedures described in this chapter and in Chapter 14 enter into a general framework illustrated by Figure 13.3.

At the beginning of step n of the procedure, all the information gathered from step 1 to step $n-1$ is summarized by a 'likelihood function' which assigns a positive real number—a 'likelihood value'—to each of the knowledge states in the structure. This likelihood function is used by the procedure to select the next question to ask. The mechanism of this selection lies in a 'questioning rule', an operator applied to the likelihood function, whose output is the

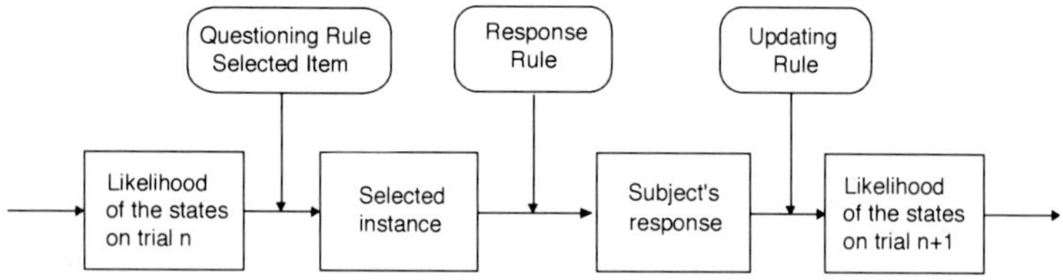

Figure 13.3. Transition diagram for the two classes of Markovian procedures.

question chosen. The subject's response is then observed, and it is assumed that it is generated by the subject's knowledge state, through a 'response rule.' In the simplest case, it is assumed that the response is correct if the question belongs to the subject's state, and incorrect otherwise. Careless error and lucky guess parameters may also be introduced at this stage. (These will play only a minor role in this chapter; however, see Remark 13.8.1.) Finally, the likelihood function is recomputed, through an 'updating rule', based on the question asked and the subject's observed response.

In this chapter, we consider a case in which the likelihood function is a probability distribution on the family of states. Our presentation follows closely Falmagne and Doignon (1988a). As in Chapters 11 and 12, we consider a probabilistic knowledge structure[3] $(Q, \mathcal{K}, L)$. However, the results of the present chapter also apply to any finite family $\mathcal{K}$ with $Q = \cup\mathcal{K}$ finite. For every state K, we denote by $L(K)$ the probability of state K in the population of reference. We assume $0 < L < 1$. For concreteness, we consider the knowledge structure $\mathcal{K}$ of Example 13.1.1 and Figure 13.1 and we suppose that the knowledge states have the probabilities represented by the histogram of Figure 13.4A. In each of the three graphs of Figure 13.4, the knowledge states are represented by squares, and the items by circles. A link between a square and a circle means that the state contains the corresponding item. For instance, the square at the extreme left of Figure 13.4A represents the knowledge state $\{a\}$. We have $L(\{a\}) = .10$, $L(\{a, b\}) = .05$, etc.

The probability distribution L of the probabilistic knowledge structure $(Q, \mathcal{K}, L)$ will be regarded as the *a priori* distribution representing the uncertainty of the assessment engine at the beginning of assessment. We set $L_1 = L$ to denote the likelihood function on step 1 of the procedure. As before, suppose that item b is presented to the subject, who fails to respond correctly. This information will induce a transformation of the likelihood L_1 by an operator, which will decrease the probabilities of all the states containing b, and increase the probabilities of all the states not containing that item. The resulting distribution L_2 is pictured by the histogram in Figure 13.4B. Next, items a and c are presented successively, eliciting two correct responses. The accu-

[3] Remember the assumptions "finite, partial" included in Definition 11.1.2 of this concept.

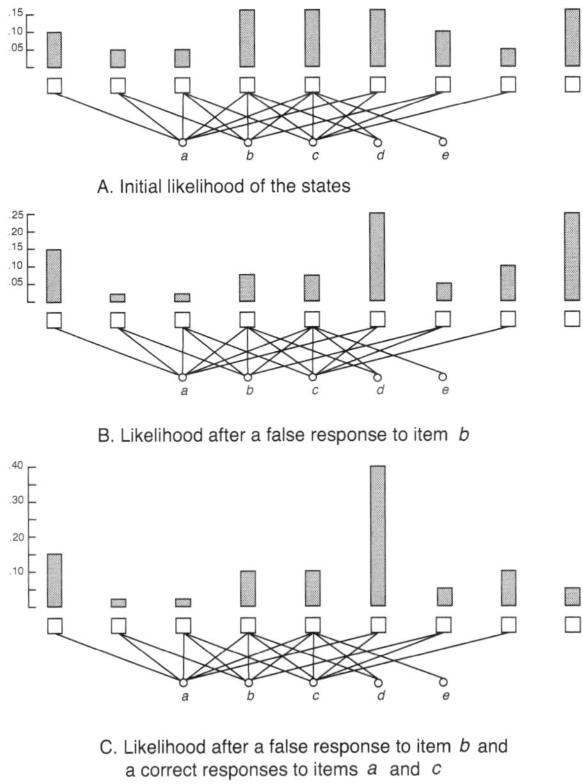

Figure 13.4. Successive transformations of the likelihood induced by the events: (item b, incorrect), (item a, correct), (item c, correct).

mulated effect of these events on the likelihood is depicted by the histogram in Figure 13.4C representing L_4, in which the probability of state $\{a, c\}$ is shown as much higher than that of any other state. This result is similar to what we had obtained, with the same sequence of events, from the deterministic algorithm represented in Figure 13.2, but much less brutal: a knowledge state is not suddenly eliminated; rather, its likelihood decreases. Needless to say, there are many ways of implementing this idea. Several implementations will be considered in this chapter.

A possible source of noise in the data obviously lies in the response mechanisms, which we have formalized by the Local Independence Axiom [N] of Chapter 12 (see Subsection 12.4.1 on page 224). For most of this chapter, we shall assume that such factors play a minor role and can be neglected. In other words, all the parameters β_q and η_q take value zero during the main phase of the assessment (thus there are no lucky guesses and no careless errors). Once the assessment algorithm has terminated, the result of the assessment can be refined by reviving these response mechanisms (cf. again our Remark 13.8.1).

In the meantime, the response rule will be simple. Suppose that the subject is in some knowledge state K_0 and that some question q is asked. The subject's response will be correct with probability one if $q \in K_0$, and incorrect with probability one in the other case.

13.3 Basic Concepts

13.3.1 Definitions. Let $(Q, \mathcal{K}, L)$ be some arbitrary probabilistic knowledge structure with a finite domain Q. The set of all positive probability distributions on $\mathcal{K}$ is denoted by Λ_+. (We thus have $L \in \Lambda_+$.) We suppose that the subject is, with probability one, in some unknown knowledge state K_0 which will be called *latent* and has to be uncovered.

Any application of an assessment procedure in the sense of this chapter is a realization of a stochastic process $(\mathbf{R}_n, \mathbf{Q}_n, \mathbf{L}_n)$, in which

n denotes the step number, or *trial* number, $n = 1, 2, \ldots$;

$\mathbf{L}_n$ is a random probability distribution on $\mathcal{K}$; we have $\mathbf{L}_n = L_n \in \Lambda_+$ (so $L_n > 0$) if L_n is the probability distribution on $\mathcal{K}$ at the beginning of trial n;

$\mathbf{L}_n(K)$ for $K \in \mathcal{K}$, is a r.v. measuring the probability of state K on trial n;

$\mathbf{Q}_n$ is a r.v. representing the question asked on trial n; we have $\mathbf{Q}_n = q \in Q$ if q is the question asked on trial n;

$\mathbf{R}_n$ is a r.v. coding the response on trial n:

$$\mathbf{R}_n = \begin{cases} 1 & \text{if the response is correct.} \\ 0 & \text{otherwise.} \end{cases}$$

The process begins, on trial 1, by setting $\mathbf{L}_1 = L \in \Lambda_+$. So, the initial probability distribution is the same for any realization. Any further trial $n > 1$ begins with a value $L_n \in \Lambda_+$ of the random distribution $\mathbf{L}_n$ updated as a function of the event on trial $n - 1$. We write for any $\mathcal{F} \subseteq \mathcal{K}$,

$$L_n(\mathcal{F}) = \sum_{K \in \mathcal{F}} L_n(K). \tag{13.2}$$

The second feature of a trial involves the question asked, that is, the value of the r.v. $\mathbf{Q}_n$. In general, the choice of a question is governed by a function $(q, L_n) \mapsto \Psi(q, L_n)$ mapping $Q \times \Lambda_+$ into the interval $[0, 1]$ and specifying the probability that $\mathbf{Q}_n = q$. The function Ψ will be called the *questioning rule*. Two special cases of this function are analyzed in this chapter (see Definitions 13.4.7 and 13.4.8 below).

The response on trial n is represented by a value of the r.v. $\mathbf{R}_n$. Only two cases are considered: (i) a correct response, which is coded as $\mathbf{R}_n = 1$; (ii) an error, which is coded as $\mathbf{R}_n = 0$. The probability of a correct response to question q is equal to one if $q \in K_0$, and to zero otherwise.

At the core of the procedure is a Markovian transition rule stating that, with probability one, the likelihood function L_{n+1} on trial $n+1$ only depends on the likelihood function L_n on trial n, the question asked on that trial, and the observed response. This transition rule is formalized by the equation[4]

$$\mathbf{L}_{n+1} \overset{\text{a.s.}}{=} u(\mathbf{R}_n, \mathbf{Q}_n, \mathbf{L}_n), \tag{13.3}$$

in which u is a function mapping $\{0,1\} \times Q \times \Lambda_+$ to Λ_+. The function u is referred to as the *updating rule*. Two cases of this function will be considered (see later Definitions 13.4.2 and 13.4.4). The diagram below summarizes these transitions:

$$(\mathbf{L}_n \to \mathbf{Q}_n \to \mathbf{R}_n) \to \mathbf{L}_{n+1}. \tag{13.4}$$

The Cartesian product $\Gamma = \{0,1\} \times Q \times \Lambda_+$ is thus the state space of the process, each trial n being characterized by a value of the triple $(\mathbf{R}_n, \mathbf{Q}_n, \mathbf{L}_n)$. We denote by Ω the sample space, that is, the set of all sequences of points in Γ. The complete history of the process from trial 1 to trial n is denoted by

$$\mathbf{W}_n = ((\mathbf{R}_n, \mathbf{Q}_n, \mathbf{L}_n), \ldots, (\mathbf{R}_1, \mathbf{Q}_1, \mathbf{L}_1)).$$

The notation $\mathbf{W}_0$ stands for the empty history.

In general, the assessment problem consists in uncovering the latent state K_0. This quest has a natural formalization in terms of the condition

$$\mathbf{L}_n(K_o) \overset{\text{a.s.}}{\longrightarrow} 1. \tag{13.5}$$

When this condition holds for some particular assessment process, we shall sometimes say that K_0 is *uncoverable* (by that procedure).

We recall that for any subset A of a fixed set S, the *indicator function* of A is a function $x \mapsto \iota_A(x)$ which is defined on S by

$$\iota_A(x) = \begin{cases} 1 & \text{if } x \in A \\ 0 & \text{if } x \in S \setminus A. \end{cases} \tag{13.6}$$

For convenience, all the main concepts are recalled in the list below.

[4] The specification 'a.s.' is an abbreviation of 'almost surely' and means that the equality holds with probability one. A similar remark holds for the '$\overset{\text{a.s.}}{\longrightarrow} 1$' convergence of Equation 13.5.

13.3.2 Notation.

$(Q, \mathcal{K}, L)$	a finite probabilistic knowledge structure;
Λ_+	the set of all positive probability distributions on $\mathcal{K}$;
Γ	the state space of the process $(\mathbf{R}_n, \mathbf{Q}_n, \mathbf{L}_n)_{n\in\mathbb{N}}$;
K_0	the latent knowledge state of the subject;
$\mathbf{L}_1 = L$	the initial probability distribution on $\mathcal{K}$, $0 < L < 1$;
$\mathbf{L}_n(K)$	a r.v. representing the probability of state K on trial n;
$\mathbf{Q}_n$	a r.v. representing the question asked on trial n;
$\mathbf{R}_n$	a r.v. representing the response given on trial n;
Ψ	$(q, \mathbf{L}_n) \mapsto \Psi(q, \mathbf{L}_n)$, the questioning rule;
u	$(\mathbf{R}_n, \mathbf{Q}_n, \mathbf{L}_n) \mapsto u(\mathbf{R}_n, \mathbf{Q}_n, \mathbf{L}_n)$, the updating rule;
$\mathbf{W}_n$	random history of the process from trial 1 to trial n;
ι_A	the indicator function of a set A.

13.3.3 General Axioms. The three axioms below concern a probabilistic knowledge structure $(Q, \mathcal{K}, L)$, the distinguished latent state K_0 of the subject, and the sequence of random triples $(\mathbf{R}_n, \mathbf{Q}_n, \mathbf{L}_n)$.

[U] **Updating Rule.** We have $\mathbb{P}(\mathbf{L}_1 = L) = 1$, and for any positive integer n and all measurable sets $B \subseteq \Lambda_+$,

$$\mathbb{P}(\mathbf{L}_{n+1} \in B \,|\, \mathbf{W}_n) = \iota_B\big(u(\mathbf{R}_n, \mathbf{Q}_n, \mathbf{L}_n)\big),$$

where u is a function mapping $\{0, 1\} \times Q \times \Lambda_+$ to Λ_+. Writing u_K for the coordinate of u associated with the knowledge state K, we thus have

$$\mathbf{L}_{n+1}(K) \overset{\text{a.s.}}{=} u_K(\mathbf{R}_n, \mathbf{Q}_n, \mathbf{L}_n).$$

Moreover, the function u satisfies the following condition:

$$u_K(\mathbf{R}_n, \mathbf{Q}_n, \mathbf{L}_n) \begin{cases} > \mathbf{L}_n(K) & \text{if } \iota_K(\mathbf{Q}_n) = \mathbf{R}_n, \\ < \mathbf{L}_n(K) & \text{if } \iota_K(\mathbf{Q}_n) \neq \mathbf{R}_n. \end{cases}$$

[Q] **Questioning Rule.** For all $q \in Q$ and all positive integers n,

$$\mathbb{P}(\mathbf{Q}_n = q \,|\, \mathbf{L}_n, \mathbf{W}_{n-1}) = \Psi(q, \mathbf{L}_n)$$

where Ψ is a function mapping $Q \times \Lambda_+$ to the interval $[0, 1]$.

[R] **Response Rule.** For all positive integers n,

$$\mathbb{P}(\mathbf{R}_n = \iota_{K_0}(q) \,|\, \mathbf{Q}_n = q, \mathbf{L}_n, \mathbf{W}_{n-1}) = 1$$

where K_0 is the latent state.

We recall that, as a knowledge structure, $\mathcal{K}$ contains at least two states, namely $\varnothing$ and Q.

13.3.4 Definition. We shall refer to a process $(\mathbf{R}_n, \mathbf{Q}_n, \mathbf{L}_n)$ satisfying the Axioms [U], [Q] and [R] in 13.3.3 as a *(continuous) stochastic assessment process for* $(Q, \mathcal{K}, L)$, *parametrized by* u, Ψ and K_0. The functions u, Ψ and ι_{K_0} are called the *updating rule*, the *questioning rule*, and the *response rule*, respectively.

13.3.5 Remarks. Axiom [R] is straightforward. Axioms [Q] and [U] govern, respectively, the choice of a question $\mathbf{Q}_n$ to be asked, and the reallocation of the mass of $\mathbf{L}_n$ on trial $n+1$ depending on the values of $\mathbf{Q}_n$ and $\mathbf{R}_n$. Axiom [U] states that if $\mathbf{L}_n = L_n$, $\mathbf{Q}_n = q$ and $\mathbf{R}_n = r$ then $\mathbf{L}_{n+1}$ is almost surely equal to $u(r, q, L_n)$. As a general scheme, this seems reasonable, since we want our procedure to specify the likelihood of each of the knowledge states on each trial. This axiom ensures that no knowledge state will ever have a likelihood of zero, and that the likelihood of any state K will increase whenever we observe either a correct response to a question $q \in K$, or an incorrect response to a question $q \notin K$, and decrease in the two remaining cases. Notice that the first two axioms pertain to the assessment process *per se*, while the third describes some hypothetical mechanism governing a student's response.

It is easily shown that each of $(\mathbf{R}_n, \mathbf{Q}_n, \mathbf{L}_n)$, $(\mathbf{Q}_n, \mathbf{L}_n)$, and $(\mathbf{L}_n)$ is a Markov process (see Theorem 13.5.2). An important question concerns general conditions on the functions u and Ψ under which $\mathbf{L}_n$ converges to some random probability distibution $\mathbf{L}$ on $\mathcal{K}$ independent of the initial distribution L. This aspect of the process will not be investigated here, however. Rather, we shall focus on the problem of defining useful procedures capable of uncovering the subject's knowledge state. Such procedures will be discussed in Section 13.6.

The class of processes defined by Axioms [U], [Q] and [R] is very large. Useful special cases can be obtain by specializing the questioning rule and the updating rule.

13.4 Special Cases

The initial likelihood L may be estimated, for example, by testing a representative sample of subjects from the population, using one of the models discussed in Chapters 11 and 12. In the absence of information on that initial likelihood, we may reasonably set

$$L(K) = \frac{1}{|\mathcal{K}|} \qquad (K \in \mathcal{K}).$$

13.4.1 Two examples of updating rule u. Suppose that some question q is presented on trial n, and that the subject's response is correct; thus, $\mathbf{Q}_n = q$ and $\mathbf{R}_n = 1$. Axiom [U] requires that the likelihood of any state containing q should almost surely increase, and the likelihood of any state not containing q should almost surely decrease. If the response is incorrect, the opposite result should obtain. Some questions may be judged more revealing than others. For instance, it may be argued that, since a correct response to a multiple choice question may be due to a lucky guess, it should not be given as much weight as, say, a correct numerical response resulting from a computation. Moreover, the response itself may be taken into account: a correct numerical response may signify the mastery of a question, but an error does not necessarily imply complete ignorance. These considerations will be implemented into two rather different exemplary updating rules, in which the reallocation of the mass of $\mathbf{L}_n = L_n$ on trial $n+1$ will be governed by a parameter which may depend upon the question asked and on the response given on trial n.

13.4.2 Definition. The updating rule u of Axiom [U] will be called *convex with parameters* $\theta_{q,r}$, where $0 < \theta_{q,r} < 1$ for $q \in Q$ and $r \in \{0,1\}$, if the function u of Axiom [U] satisfies the following condition:

For all $K \in \mathcal{K}$ and with $\mathbf{L}_n = L_n$, $\mathbf{R}_n = r$, and $\mathbf{Q}_n = q$,

$$u_K(r,q,L_n) = (1-\theta_{q,r})L_n(K) + \theta_{q,r}g_K(r,q,L_n) \tag{13.7}$$

where

$$g_K(r,q,L_n) = \begin{cases} r\frac{L_n(K)}{L_n(\mathcal{K}_q)}, & \text{if } K \in \mathcal{K}_q \\ (1-r)\frac{L_n(K)}{L_n(\mathcal{K}_{\bar{q}})}, & \text{if } K \in \mathcal{K}_{\bar{q}}. \end{cases}$$

Thus, the r.h.s. of Equation (13.7) specifies a convex combination between the current likelihood L_n and a conditional one, obtained from discarding all the knowledge states inconsistent with the observed response. The updating rule is *convex with constant parameter* θ if Equation (13.7) holds with $\theta_{q,r} = \theta$ for all $q \in Q$ and $r \in \{0,1\}$.

One objection to this particular form of the updating rule is that it is not 'commutative.' One could require that the likelihood on trial $n+1$ should not depend, as it does in Equation (13.7) (cf. Problem 1), on the order of the pairs of questions and responses up to that trial. Consider the two cases

1. $(\mathbf{Q}_{n-1} = q, \mathbf{R}_{n-1} = r)$, $(\mathbf{Q}_n = q', \mathbf{R}_n = r')$,
2. $(\mathbf{Q}_{n-1} = q', \mathbf{R}_{n-1} = r')$, $(\mathbf{Q}_n = q, \mathbf{R}_n = r)$.

It could be argued that, for a given value of the likelihood $\mathbf{L}_{n-1} = l$, the likelihood on trial $n+1$ should be the same in these two cases because they convey the same information. Slightly changing our notation by setting $\xi = (q,r)$, $\xi' = (q',r')$ and $F(l,\xi) = u(r,q,l)$, this translates into the condition

$$F\big(F(l,\xi),\xi'\big) = F\big(F(l,\xi'),\xi\big). \tag{13.8}$$

In the functional equation literature, an operator F satisfying (13.8) is called 'permutable' (Aczél, 1966, p. 270). In some special cases, permutability greatly reduces the possible form of an operator. However, the side conditions used by Aczél (1966) are too strong for our purpose (see Luce, 1964; Marley, 1967, in this connection). Nonetheless, this concept is of obvious relevance.

13.4.3 Definition. We shall call *permutable* an updating rule u with an operator F satisfying Equation (13.8).

An example of a permutable updating rule is given below.

13.4.4 Definition. The updating rule is called *multiplicative with parameters* $\zeta_{q,r}$, where $1 < \zeta_{q,r}$ for $q \in Q, r = 0, 1$, if the function u of Axiom [U] satisfies the condition: with $\mathbf{Q}_n = q$, $\mathbf{R}_n = r$, $\mathbf{L}_n = L_n$ and

$$\zeta_{q,r}^K = \begin{cases} 1 & \text{if } \iota_K(q) \neq r, \\ \zeta_{q,r} & \text{if } \iota_K(q) = r \end{cases} \tag{13.9}$$

we have

$$u_K(r, q, L_n) = \frac{\zeta_{q,r}^K L_n(K)}{\sum_{K' \in \mathcal{K}} \zeta_{q,r}^{K'} L_n(K')}. \tag{13.10}$$

It is easy to verify that this multiplicative rule is permutable (Problem 2).

Other updating rules applicable in different, but similar situations are reviewed by Landy and Hummel (1986). The two examples of updating rules introduced in 13.4.2 and 13.4.4 have been inspired by some operators used in mathematical learning theory (see the Sources Section 13.10 at the end of this chapter).

13.4.5 Remark. It was pointed out by Mathieu Koppen[5] that the multiplicative updating rule can be interpreted as a Bayesian updating. The latter occurs when the values of $\zeta_{q,r}^K$ are linked in a specific manner with the probabilities of respectively a careless error and a lucky guess in the answer for item q (for the introduction of these probabilities, see 11.1.2). Fixing question q, and slightly changing our notation, we write

$P_q(K)$	for the a priori probability of state K,
$P_q(K \mid r)$	for the a posteriori probability of state K after having observed response r,

with a similar interpretation for $P_q(r \mid K)$. From Bayes Theorem, we have

$$P_q(K \mid r) = \frac{P_q(r \mid K) P_q(K)}{\sum_{K' \in \mathcal{K}} P_q(r \mid K') P_q(K')}. \tag{13.11}$$

We see that Equations (13.10) and (13.11) have the same form except that $\zeta_{q,r}^K$ cannot be regarded as a conditional probability. In particular, we do not generally have

$$\zeta_{q,0}^K + \zeta_{q,1}^K = 1.$$

However, let us compare the multiplicative updating rule given in Equation (13.10) with the explicit form of the Bayesian rule. Equation (13.10)

[5] Personal communication.

requires that the values $L_{n+1}(K)$ be proportional to the values $\zeta^K_{q,r} L_n(K)$ (for a fixed question q and answer r), with

$$\zeta^K_{q,r} = \begin{cases} \zeta_{q,1} & \text{if } q \in K, r = 1, \\ 1 & \text{if } q \notin K, r = 1, \\ 1 & \text{if } q \in K, r = 0, \\ \zeta_{q,0} & \text{if } q \notin K, r = 0. \end{cases}$$

Similarly, Bayesian updating—Equation (13.11)—makes the values $L_{n+1}(K)$ proportional to the values $Z^K_{q,r} L_n(K)$, where the real numbers $Z^K_{q,r}$ are specified by

$$Z^K_{q,r} = \begin{cases} 1 - \beta_q & \text{if } q \in K, r = 1, \\ \gamma_q & \text{if } q \notin K, r = 1, \\ \beta_q & \text{if } q \in K, r = 0, \\ 1 - \gamma_q & \text{if } q \notin K, r = 0. \end{cases}$$

Thus, the multiplicative updating rule coincides with Bayesian updating if and only if for all items q

$$\frac{\zeta_{q,1}}{1 - \beta_q} = \frac{1}{\gamma_q} \quad \text{and} \quad \frac{1}{\beta_q} = \frac{\zeta_{q,0}}{1 - \gamma_q}.$$

These equations can be rewritten as

$$\zeta_{q,1} = \frac{1 - \beta_q}{\gamma_q} \quad \text{and} \quad \zeta_{q,0} = \frac{1 - \gamma_q}{\beta_q}$$

or as

$$\beta_q = \frac{\zeta_{q,1} - 1}{\zeta_{q,1}\zeta_{q,0} - 1} \quad \text{and} \quad \gamma_q = \frac{\zeta_{q,0} - 1}{\zeta_{q,1}\zeta_{q,0} - 1}.$$

13.4.6 Two examples of questioning rule. A simple idea for the questioning rule is to select, on any trial n, a question q that partitions the set $\mathcal{K}$ of all the states into two subsets $\mathcal{K}_q$ and $\mathcal{K}_{\bar{q}}$ with a mass as equal as possible; that is, such that $L_n(\mathcal{K}_q)$ is as close as possible to $L_n(\mathcal{K}_{\bar{q}}) = 1 - L_n(\mathcal{K}_q)$. Note in this connection that any likelihood L_n defines a set $S(L_n) \subseteq Q$ containing all those questions q minimizing

$$|2L_n(\mathcal{K}_q) - 1|.$$

Under this questioning rule, we must have $\mathbf{Q}_n \in S(L_n)$ with a probability equal to one. The questions in the set $S(L_n)$ are then chosen with equal probability.

13.4.7 Definition. The questioning rule [Q] will be called *half-split* when

$$\Psi(q, L_n) = \frac{\iota_{S(L_n)}(q)}{|S(L_n)|}. \tag{13.12}$$

Another method may be used, which is computationally more demanding and may seem at first blush more exact. The uncertainty of the assessment engine on trial n of the procedure may be evaluated by the entropy of the likelihood on that trial, that is, by the quantity

$$H(L_n) = -\sum_{K \in \mathcal{K}} L_n(K) \log_2 L_n(K).$$

It seems reasonable to choose a question so as to reduce that entropy as much as possible. For $\mathbf{Q}_n = q$ and $\mathbf{L}_n = L_n$ the expected value of the entropy on trial $n+1$ is given by the sum

$$\begin{aligned} \mathbb{P}(\mathbf{R}_n = 1 \,|\, \mathbf{Q}_n = q) H\big(u(1, q, L_n)\big) \\ + \mathbb{P}(\mathbf{R}_n = 0 \,|\, \mathbf{Q}_n = q) H\big(u(0, q, L_n)\big). \end{aligned} \tag{13.13}$$

But the conditional probability $\mathbb{P}(\mathbf{R}_n = 1 \,|\, \mathbf{Q}_n = q)$ of a correct response to question q is unknown, since it depends on the latent state K_0. Thus, (13.13) cannot be computed. However, it makes sense to replace, in the evaluation of (13.13), the conditional probability $\mathbb{P}(\mathbf{R}_n = 1 \,|\, \mathbf{Q}_n = q)$ by the likelihood $L_n(\mathcal{K}_q)$ of a correct response to question q. The idea is thus to minimize the quantity

$$\tilde{H}(q, L_n) = L_n(\mathcal{K}_q) H\big(u(1, q, L_n)\big) + L_n(\mathcal{K}_{\bar{q}}) H\big(u(0, q, L_n)\big), \tag{13.14}$$

over all possible $q \in Q$. Let $J(L_n) \subseteq Q$ be the set of questions q minimizing $\tilde{H}(q, L_n)$. The question asked on trial $n+1$ is then randomly selected in the set $J(L_n)$.

13.4.8 Definition. This particular form of the questioning rule, which is specified by the equation

$$\Psi(q, L_n) = \frac{\iota_{J(L_n)}(q)}{|J(L_n)|}, \tag{13.15}$$

will be referred to as *informative*. Note that, in Equation (13.15), the choice of a question varies with the updating rule. This is not the case for Equation (13.12). Surprisingly, for the convex updating rule with a constant parameter θ, the half-split and the informative questioning rule induce the same drawing of questions. We shall postpone the proof of this fact for the moment, however (see Theorem 13.6.6 whose proof is in 13.9.1).

13.5 General Results

13.5.1 Convention. In the rest of this chapter, we assume that $(\mathbf{R}_n, \mathbf{Q}_n, \mathbf{L}_n)$ is a stochastic assessment process for a probabilistic knowledge structure $(Q, \mathcal{K}, L)$ with $L > 0$, parametrized by u, Ψ and K_0. Special cases of this procedure will be specified whenever appropriate.

13.5.2 Theorem. *The stochastic process* $(\mathbf{L}_n)$ *is Markovian. That is, for any positive integer* n *and any measurable set* $B \subseteq \Lambda_+$

$$\mathbb{P}(\mathbf{L}_{n+1} \in B \,|\, \mathbf{L}_n, \ldots, \mathbf{L}_1) = \mathbb{P}(\mathbf{L}_{n+1} \in B \,|\, \mathbf{L}_n). \tag{13.16}$$

A similar property holds for the processes $(\mathbf{R}_n, \mathbf{Q}_n, \mathbf{L}_n)$ *and* $(\mathbf{Q}_n, \mathbf{L}_n)$.

PROOF. Using successively Axioms [U], [Q], and [R], we have

$$\begin{aligned}
&\mathbb{P}(\mathbf{L}_{n+1} \in B \,|\, \mathbf{L}_n, \ldots, \mathbf{L}_1)\\
&\quad = \sum_{(\mathbf{R}_n, \mathbf{Q}_n)} \mathbb{P}(\mathbf{L}_{n+1} \in B \,|\, \mathbf{R}_n, \mathbf{Q}_n, \mathbf{L}_n, \ldots, \mathbf{L}_1)\mathbb{P}(\mathbf{R}_n, \mathbf{Q}_n \,|\, \mathbf{L}_n, \ldots, \mathbf{L}_1)\\
&\quad = \sum_{(\mathbf{R}_n, \mathbf{Q}_n)} \iota_B\big(u(\mathbf{R}_n, \mathbf{Q}_n, \mathbf{L}_n)\big)\mathbb{P}(\mathbf{R}_n, \mathbf{Q}_n \,|\, \mathbf{L}_n, \ldots, \mathbf{L}_1)\\
&\quad = \sum_{(\mathbf{R}_n, \mathbf{Q}_n)} \iota_B\big(u(\mathbf{R}_n, \mathbf{Q}_n, \mathbf{L}_n)\big)\mathbb{P}(\mathbf{R}_n \,|\, \mathbf{Q}_n, \mathbf{L}_n \ldots \mathbf{L}_1)\mathbb{P}(\mathbf{Q}_n \,|\, \mathbf{L}_n, \ldots, \mathbf{L}_1)\\
&\quad = \sum_{(\mathbf{R}_n, \mathbf{Q}_n)} \iota_B\big(u(\mathbf{R}_n, \mathbf{Q}_n, \mathbf{L}_n)\big)\iota_{K_0}(\mathbf{Q}_n)\Psi(\mathbf{Q}_n, \mathbf{L}_n)
\end{aligned}$$

which only depends on the set B and on $\mathbf{L}_n$. We leave the two other cases to the reader (Problems 3 and 4). □

In general, a stochastic assessment process is not necessarily capable of uncovering a latent state K_0. The next theorem gathers some simple, but very general results in this connection. We recall that $\triangle$ denotes the symmetric difference between sets[6].

13.5.3 Theorem. *If the latent state is* K_0*, then for all positive integers* n*, all real numbers* ϵ *with* $0 < \epsilon < 1$ *and all states* $K \neq K_0$*, we have*

$$\mathbb{P}\big(\mathbf{L}_{n+1}(K_0) > \mathbf{L}_n(K_0)\big) = 1; \tag{13.17}$$

$$\mathbb{P}\big(\mathbf{L}_{n+1}(K_0) \geq 1 - \epsilon\big) \geq \mathbb{P}\big(\mathbf{L}_n(K_0) \geq 1 - \epsilon\big); \tag{13.18}$$

$$\mathbb{P}\big(\mathbf{L}_{n+1}(K) < \mathbf{L}_n(K)\big) = \mathbb{P}(\mathbf{Q}_n \in K \triangle K_0). \tag{13.19}$$

Moreover, we have

$$\lim_{n\to\infty} \mathbb{P}\big(\mathbf{L}_{n+1}(K_0) \geq 1 - \epsilon > \mathbf{L}_n(K_0)\big) = 0. \tag{13.20}$$

Equation (13.18) implies that the sequence $\mathbb{P}\big(\mathbf{L}_n(K_0) \geq 1 - \epsilon\big)$ converges.

PROOF. Equation (13.17) is an immediate consequence of Axioms [U] and [R]. It implies

$$\mathbb{P}\big(\mathbf{L}_{n+1}(K_0) \geq 1 - \epsilon \,|\, \mathbf{L}_n(K_0) \geq 1 - \epsilon\big) = 1. \tag{13.21}$$

6 Cf. 1.6.12.

Consequently,

$$\begin{aligned}
&\mathbb{P}\big(\mathbf{L}_{n+1}(K_0) \geq 1-\epsilon\big) \\
&\qquad = \mathbb{P}\big(\mathbf{L}_{n+1}(K_0) \geq 1-\epsilon \,\big|\, \mathbf{L}_n(K_0) \geq 1-\epsilon\big)\, \mathbb{P}\big(\mathbf{L}_n(K_0) \geq 1-\epsilon\big) \\
&\qquad\quad + \mathbb{P}\big(\mathbf{L}_{n+1}(K_0) \geq 1-\epsilon \,\big|\, \mathbf{L}_n(K_0) < 1-\epsilon\big)\, \mathbb{P}\big(\mathbf{L}_n(K_0) < 1-\epsilon\big) \\
&\qquad \geq \mathbb{P}\big(\mathbf{L}_n(K_0) \geq 1-\epsilon\big). \qquad (13.22)
\end{aligned}$$

This establishes Equation (13.18). Writing $\mu(\epsilon) = \lim_{n\to\infty} \mathbb{P}\big(\mathbf{L}_n(K_0) \geq 1-\epsilon\big)$, and taking limits on both sides of Equation (13.22), we obtain using (13.21) again

$$\mu(\epsilon) = \mu(\epsilon) + \lim_{n\to\infty} \mathbb{P}\big(\mathbf{L}_{n+1}(K_0) \geq 1-\epsilon > \mathbf{L}_n(K_0)\big),$$

yielding (13.20). The l.h.s. of (13.19) can be decomposed into

$$\begin{aligned}
&\mathbb{P}\big(\mathbf{L}_{n+1}(K) < \mathbf{L}_n(K) \,\big|\, \mathbf{Q}_n \in K \triangle K_0\big)\, \mathbb{P}(\mathbf{Q}_n \in K \triangle K_0) \\
&\qquad + \mathbb{P}\big(\mathbf{L}_{n+1}(K) < \mathbf{L}_n(K) \,\big|\, \mathbf{Q}_n \in \overline{K \triangle K_0}\big)\, \mathbb{P}(\mathbf{Q}_n \in \overline{K \triangle K_0}).
\end{aligned}$$

By Axioms [R] and [U], the factor $\mathbb{P}\big(\mathbf{L}_{n+1}(K) < \mathbf{L}_n(K) \,|\, \mathbf{Q}_n \in K \triangle K_0\big)$ in the first term is equal to one, and the last term vanishes. Thus, Equation (13.19) follows. □

13.6 Uncovering the Latent State

Under some fairly general conditions on the updating and the questioning rules, the latent state K_0 can be uncovered. These general conditions include the cases in which the updating rule is convex or multiplicative, and the questioning rule is half-split. We first consider an example using a convex updating rule with a constant parameter θ and the half-split questioning rule.

13.6.1 Example. Let $Q = \{a, b, c\}$ and

$$\mathcal{K} = \big\{\varnothing, \{a\}, \{b, c\}, \{a, c\}, \{a, b, c\}\big\},$$

with the latent state $K_0 = \{b, c\}$ and $L_1(K) = .2$ for all $K \in \mathcal{K}$. Since the questioning rule is half-split, and

$$|2L_1(\mathcal{K}_q) - 1| = .2$$

for all $q \in Q$, we have $S(L_1) = \{a, b, c\}$ (in the notation of 13.4.6). That is, on trial one, the questions are selected in $S(L_1)$ with equal probabilities. Notice that

$$L_1(\mathcal{K}_{\bar{a}}) = L_1(\mathcal{K}_b) = .4, \quad \text{while} \quad L_1(\mathcal{K}_c) = .6.$$

For the likelihood of the state $K_0 = \{b, c\}$ on trial two, we thus obtain, by the convex updating rule

$$L_2(K_0) = \begin{cases} (1-\theta)\,.2 + \theta\frac{.2}{.4} & \text{with probability } \frac{1}{3} \quad (a \text{ is chosen});\\ (1-\theta)\,.2 + \theta\frac{.2}{.4} & \text{with probability } \frac{1}{3} \quad (b \text{ is chosen});\\ (1-\theta)\,.2 + \theta\frac{.2}{.6} & \text{with probability } \frac{1}{3} \quad (c \text{ is chosen}). \end{cases}$$

In accordance with Equation (13.17) of Theorem 13.5.3, this implies

$$\mathbb{P}\big(\mathbf{L}_2(K_0) > \mathbf{L}_1(K_0)\big) = 1.$$

In fact, Theorem 13.6.7 shows that K_0 is uncoverable.

We now turn to a general result of convergence, based on some strengthening of the conditions defining a stochastic assessment process.

13.6.2 Definition. An updating rule u is called *regular* if there is a non-increasing function $v :]0,1[\to \mathbb{R}$ such that, for all $r \in \{0,1\}$, $q \in Q$, and $l \in \Lambda_+$,

(i) $v(t) > 1$ for all $t \in]0,1[$;
(ii) $u_K(r,q,l) \geq v\big(l(\mathcal{K}_q)\big)\, l(K)$, if $\iota_K(q) = r = 1$;
(iii) $u_K(r,q,l) \geq v\big(l(\mathcal{K}_{\bar{q}})\big)\, l(K)$, if $\iota_K(q) = r = 0$.

13.6.3 Theorem. *Both the convex and the multiplicative updating rules are regular.*

PROOF. For the convex updating rule, if $\iota_K(q) = r = 1$, Equation (13.7) can be rewritten as

$$u_K(r,q,L_n) = \left(1 + \theta_{q,r}\left(\frac{1}{L_n(\mathcal{K}_q)} - 1\right)\right) L_n(K).$$

We define $\theta = \min\{\theta_{q,r} \,|\, q \in Q, r \in \{0,1\}\}$. Note that $\theta > 0$. This gives

$$u_K(r,q,L_n) \geq \left(1 + \theta\left(\frac{1}{L_n(\mathcal{K}_q)} - 1\right)\right) L_n(K).$$

In the case $\iota_K(q) = r = 0$, we obtain similarly

$$\begin{aligned} u_K(r,q,L_n) &= \left(1 + \theta_{q,r}\left(\frac{1}{L_n(\mathcal{K}_{\bar{q}})} - 1\right)\right) L_n(K)\\ &\geq \left(1 + \theta\left(\frac{1}{L_n(\mathcal{K}_{\bar{q}})} - 1\right)\right) L_n(K). \end{aligned}$$

Thus, (i)–(iii) in Definition 13.6.2 are satisfied with

$$v(t) = 1 + \theta\left(\frac{1}{t} - 1\right), \quad \text{for } t \in]0,1[.$$

For the multiplicative updating rule, in the case $\iota_K(q) = r = 1$, we have

$$\begin{aligned} u_K(r, q, L_n) &= \frac{\zeta_{q,r}}{\zeta_{q,r} L_n(\mathcal{K}_q) + L_n(\mathcal{K}_{\bar{q}})} L_n(K) \\ &= \frac{\zeta_{q,r}}{1 + (\zeta_{q,r} - 1) l(\mathcal{K}_q)} L_n(K), \end{aligned}$$

and similarly, if $\iota_K(q) = r = 0$,

$$\begin{aligned} u_K(r, q, L_n) &= \frac{\zeta_{q,r}}{L_n(\mathcal{K}_q) + \zeta_{q,r} L_n(\mathcal{K}_{\bar{q}})} L_n(K) \\ &= \frac{\zeta_{q,r}}{1 + (\zeta_{q,r} - 1) l(\mathcal{K}_{\bar{q}})} L_n(K). \end{aligned}$$

For $q \in Q$ and $r \in \{0, 1\}$, each of the functions $v_{q,r} : t \mapsto v_{q,r}(t)$ defined for $\zeta_{q,r} > 1$ and $t \in]0, 1[$ by

$$v_{q,r}(t) = \frac{\zeta_{q,r}}{1 + (\zeta_{q,r} - 1)t}$$

is decreasing and takes values > 1. Thus, $v = \min\{v_{q,r} \,|\, q \in Q, r \in \{0, 1\}\}$ satisfies (i)–(iii) in Definition 13.6.2. □

As far as the questioning rule is concerned, it is intuitively clear that, from the standpoint of the observer, it would not be efficient to choose a question q with a likelihood $L_n(\mathcal{K}_q)$ of a correct response close to zero or one. Actually, it makes good sense to choose, as in the half-split questioning rule, a question q with $L_n(\mathcal{K}_q)$ as far as possible from zero or one. A much weaker form of this idea is captured in the next definition.

13.6.4 Definition. Let ν be a real valued function defined on the open interval $]0, 1[$, with two numbers $\gamma, \delta > 0$ satisfying $\gamma + \delta < 1$, such that

(i) ν is strictly decreasing on $]0, \gamma[$;

(ii) ν is strictly increasing on $]1 - \delta, 1[$;

(iii) $\nu(t) > \nu(t')$ whenever $\gamma \leq t' \leq 1 - \delta$ and either $0 < t < \gamma$ or $1 - \delta < t < 1$.

An example of such a function ν is shown in Figure 13.5.

Define, for $l \in \Lambda_+$,

$$S(\nu, l) = \{q \in Q \,|\, \nu(l(\mathcal{K}_q)) \leq \nu(l(\mathcal{K}_{q'})) \text{ for all } q' \in Q\}. \tag{13.23}$$

A questioning rule Ψ is said to be *inner* if such a function ν exists with

$$\Psi(q, l) = \frac{\iota_{S(\nu,l)}(q)}{|S(\nu, l)|}.$$

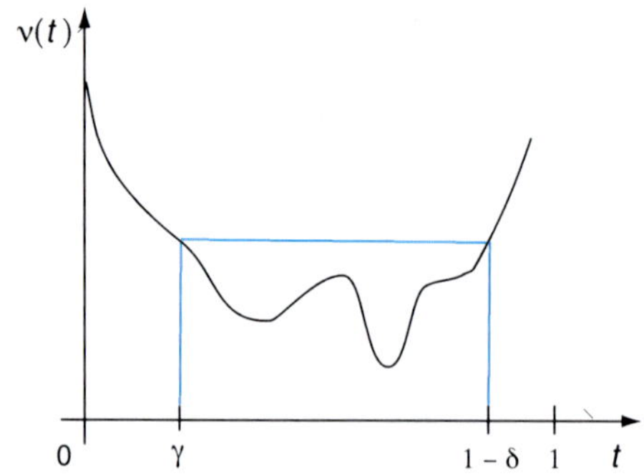

Figure 13.5. An example of the function ν defined in 13.6.4.

13.6.5 Theorem. *The half-split questioning rule is inner.*

This follows readily from the definitions, using $\nu(t) = |2t - 1|$.

13.6.6 Theorem. *The informative questioning rule is inner if the updating rule is convex with a constant parameter θ. Moreover, in this case, the informative and the half-split questioning rules induce the same drawing of questions.*

The proofs of this theorem and of the next one are gathered in a later section of this chapter, which contains starred material. The next theorem states the main result of this chapter.

13.6.7 Theorem. *Let $(\mathbf{R}_n, \mathbf{Q}_n, \mathbf{L}_n)$ be a stochastic assessment process parametrized by u, Ψ and K_0, with u regular and Ψ inner. Then, K_0 is uncoverable in the sense that:*

$$\mathbf{L}_n(K_0) \overset{\text{a.s.}}{\to} 1.$$

13.6.8 Corollary. *A latent knowledge state is uncoverable by a stochastic assessment process with an updating rule which is either convex or multiplicative, and a questioning rule which is half-split.*

Proof. This results from Theorems 13.6.3, 13.6.5, 13.6.7 and the definitions. □

13.6.9 Corollary. *A latent knowledge state is uncoverable by a stochastic assessment process with a convex updating rule having a constant parameter θ, and an informative questioning rule.*

Proof. Use Theorems 13.6.3, 13.6.6, 13.6.7 and the definitions. □

Notice that the results of this Chapter are not complete: we do not have a proof of the almost sure convergence $\mathbf{L}_n(K_0) \to 1$ for the fourth special case, namely, a multiplicative updating rule with an informative questioning rule.

13.7 A Two-Step Assessment Algorithm

The last section describes a stochastic assessment process for uncovering a latent state in a (finite, partial) probabilistic knowledge structure $(\mathcal{K}, Q)$. However, in a real-life applications such as those of the ALEKS system, the number of potential states—the size of $\mathcal{K}$—may sometimes exceed the capacity of the computer. To handle the problem, we sketched in Remark 2.4.13 a more elaborate algorithm consisting in the following two main steps (see 2.4.2 for the definitions and notation of the technical terms).

(1) Perform an assessment on a projection $\mathcal{K}_{|Q'}$ of $\mathcal{K}$ on a suitable subset Q' of Q. This step results in a state W of the projection $\mathcal{K}_{|Q'}$, where $W = K \cap Q'$ for some $K \in \mathcal{K}$.

(2) Perform an assessment on the Q'-child $\mathcal{K}_{[K]}$ of $\mathcal{K}$. This leads to a state M of $\mathcal{K}_{[K]}$, which is equal to $L \setminus \cap[K]$ for some state L of $\mathcal{K}$. The state L can then be taken as the final state obtained for the two-step assessment routine.

We devote this section to a more detailed description of this algorithm.

13.7.1 Definition. Let $(Q, \mathcal{K})$ be a finite, partial knowledge structure, and let Q' be a proper subset of Q. Take any element W of the projection of $\mathcal{K}$ on Q'. The family $\mathcal{K}(Q', W) = \{L \in \mathcal{K} \mid L \cap Q' = W\}$ is called the *ascendent* family of W. The *child of* W is the family $\mathcal{K}[Q'|W] = \{L \setminus (\cap\mathcal{K}(Q', W)) \mid L \in \mathcal{K}(Q', W)\}$. The family $\mathcal{K}(Q', W)$ is thus a subfamily of $\mathcal{K}$ and the child $\mathcal{K}[Q'|W]$ of W is exactly the Q'-child $\mathcal{K}_{[K]}$ of $\mathcal{K}$ (in the sense of 2.4.1), for any K in $\mathcal{K}(Q', W)$.

In all the algorithms of the present section, the word 'assessment' refers to an assessment subroutine producing a single state (for instance, the implementation of a stochastic assessment process culminating in the choice of a most probable state, in the sense of Definition 13.3.4 and Theorem 13.6.7). We first consider a routine and an algorithm that are applicable to general finite, partial knowledge structures. We then turn to knowledge spaces and learning spaces. In such special cases, more powerful dedicated routines become available to the algorithms.

13.7.2 Algorithm (`Two-step`, **Version 1**). A finite, partial knowledge structure $\mathcal{K}$ is given to the algorithm with $Q = \cup\mathcal{K}$.

STEP 1.
1.1. Choose a suitable subset Q' in Q—see Remark 13.7.3(b).
1.2. Build the projection $\mathcal{K}_{|Q'}$ of $\mathcal{K}$ on Q'.
1.3. Run an assessment on $\mathcal{K}_{|Q'}$ and get a state W of $\mathcal{K}_{|Q'}$.

STEP 2.
2.1. Build the child $\mathcal{K}[Q'|W]$ of W.
2.2. Run an assessment on $\mathcal{K}[Q'|W]$, and get an element X of $\mathcal{K}[Q'|W]$.
2.3. Return $W \cup (\cap\mathcal{K}(Q', W)) \cup X$.

Some comments on the algorithm are in order.

13.7.3 Remarks. a) We need to keep Q' small enough that the assessment can effectively be run on the projection $\mathcal{K}_{|Q'}$. For a similar reason, we also need to keep the child $\mathcal{K}(Q', W)$ small. Since these two requirements are opposites, the construction of Q' requires careful handling. In all the practical applications that we are aware of, the size of Q does not exceed a few hundred items and the choice of Q' is feasible. Note that the choice of Q' is made before the assessment and is not affected by it. So, Steps 1.1 and 1.2 can be performed by a preprocessor program days or months before any assessment is made. This means that the same subset Q' can be used for different students. Also, several interchangeable Q' subsets can be computed ahead of time, with the algorithm randomly choosing between them at the time of the assessment.

b) Since the final state selected is $W \cup (\cap\mathcal{K}(Q', W)) \cup X$, the second step of this algorithm never changes the status of any item lying in $Q' \cup (\cap\mathcal{K}(Q', W))$. It only adds to W the whole of $\cap\mathcal{K}(Q', W)$ and maybe further items from $Q \setminus (Q' \cup \cap\mathcal{K}(Q', W))$. Thus, this algorithm does not correct during Step 2 possible mistakes made by the algorithm on Step 1. In Algorithm 13.7.12, we propose a cure for this potentially serious drawback.

c) Another drawback is of a different nature. The computations are grounded on the whole family $\mathcal{K}$, which may be very large. For real applications, the base is much more accessible, and in any event considerably smaller, than $\mathcal{K}$ itself. In 13.7.8, we restate Algorithm 13.7.2 for the case in which the base of the knowledge space is given. A word of warning is in order about the information conveyed by the base. This information might simply consist in a listing of the elements of the base, each of which is an atom. We could also give, for each such atom, the list of items at which it is an atom. (See Definitions 3.4.5 for these concepts. As will be recalled before Algorithm 13.7.10, the second point of view is akin to the surmise relation.) We will adopt both points of view successively. They lead to different algorithms.

A few basic properties of the base with respect to projections are needed for our next algorithm. They are collected in Theorem 13.7.4, together with additional results, some of which are applicable to infinite spaces. We recall that a space $\mathcal{K}$ is finitary when the intersection of any chain of states in $\mathcal{K}$ is a state in $\mathcal{K}$ (cf. Definition 3.6.1; motivation for the concept is given at the end of Section 3.9).

13.7.4 Theorem. *Let $\mathcal{K}$ be a knowledge space on Q with base $\mathcal{B}$, and let Q' be a nonempty subset of Q. Denote by $\mathcal{K}' = \mathcal{K}_{|Q'}$ the knowledge space which is the projection of $\mathcal{K}$ on Q' (cf. Lemma 2.4.6(ii)), and by $\mathcal{G}$ the projection of the base $\mathcal{B}$ on Q'. Then the following three assertions are true:*

(i) *$\mathcal{G}$ spans $\mathcal{K}'$, but $\mathcal{K}'$ does not necessarily have a base;*
(ii) *if the base $\mathcal{B}'$ of $\mathcal{K}'$ exists, it satisfies $\mathcal{B}' \subseteq \mathcal{G}$. However, the converse inclusion does not necessarily hold;*
(iii) *if $\mathcal{K}$ is finitary, then so is $\mathcal{K}'$. In this case, $\mathcal{K}'$ has a base $\mathcal{B}'$ which satisfies $\mathcal{B}' \subseteq \mathcal{G}$ but not necessarily $\mathcal{G} \subseteq \mathcal{B}'$.*

PROOF. That $\mathcal{G}$ spans $\mathcal{K}'$ follows directly from the fact that $\mathcal{B}$ spans $\mathcal{K}$. Example 13.7.6 below completes the proof of Assertion (i).

Next, suppose that the base $\mathcal{B}'$ exists. Since $\mathcal{G}$ spans $\mathcal{K}'$, the inclusion announced in Assertion (ii) follows from Theorem 3.4.2. Example 13.7.5 establishes the second assertion in (ii).

To prove the first assertion in (iii), we consider a chain $(L_i)_{i\in I}$ of states in $\mathcal{K}'$. For each index value i in the index set I, there is a state K_i in $\mathcal{K}$ such that $L_i = K_i \cap Q'$. Since the family $(K_i)_{i\in I}$ is not necessarily a chain, we build a new family by setting $H_i = \cup\{\, K \in \mathcal{K} \mid K \cap Q' \subseteq L_i \,\}$, for each i in I. Then $(H_i)_{i\in I}$ is a chain of states of $\mathcal{K}$ which moreover satisfies $L_i = H_i \cap Q'$ (because $K_i \subseteq H_i$). By assumption, $\mathcal{K}$ is finitary. Hence $\cap\{\, H_i \mid i \in I \,\}$ is a state of $\mathcal{K}$, say H. From $\cap\{\, L_i \mid i \in I \,\} = H \cap Q'$, it follows that $\cap\{\, L_i \mid i \in I \,\}$ is a state of $\mathcal{K}'$. This establishes that $\mathcal{K}'$ is finitary. By Theorems 3.6.3 and 3.6.6, $\mathcal{K}'$ has then a base which we denote as before by $\mathcal{B}'$. From (ii), we know $\mathcal{B}' \subseteq \mathcal{G}$. We rely on Example 13.7.5 again to show that the reverse inclusion does not necessarily hold. □

13.7.5 Example. The family $\{\{a\}, \{b\}, \{a, b, c\}\}$ forms the base of a knowledge space on $Q = \{a, b, c\}$. Choose $Q' = \{a, b\}$. In the notation of Theorem 13.7.4, we get $\mathcal{B}' = \{\{a\}, \{b\}\} \subset \mathcal{G} = \{\{a\}, \{b\}, \{a, b\}\}$.

13.7.6 Example. Here is a case in which some knowledge space with a base has a projection without a base. This example supports the second assertion in Theorem 13.7.4(i). Let $\mathcal{O}$ be the collection of all open subsets of the real line. For each state O of $\mathcal{O}$, form the set $O \cup \{O\}$. The span of $\mathcal{G} = \{O \cup \{O\} \mid O \in \mathcal{O}\}$ is a knowledge space $\mathcal{K}$ on $\mathbb{R} \cup \mathcal{O}$. Since any two states in $\mathcal{G}$ are incomparable with respect to inclusion, the base of $\mathcal{K}$ must be $\mathcal{G}$. The projection of $\mathcal{K}$ on $\mathbb{R}$ coincides with $\mathcal{O}$ and consequently has no base.

13.7.7 Remark. Apart from the finitary assumption on $\mathcal{K}$ we do not know of any interesting, sufficient conditions on the knowledge space $\mathcal{K}$ in Theorem 13.7.4 implying that the trace $\mathcal{K}'$ has always a base (for any nonempty subset Q' of Q). A necessary and sufficient condition would of course be even preferable; see Open Problem 18.1.3.

The next algorithm is similar to Algorithm 13.7.2 but accepts as input the base $\mathcal{B}$ rather than the knowledge space $\mathcal{K}$ itself[7]. As in Algorithm 13.7.2, the word "assessment" refers to any assessment routine producing a single state, but this time we may use a (more powerful) routine that only works on spaces specified by their base.

As a child of a knowledge space is always a partial knowledge space but not necessarily a knowledge space (see Lemma 2.4.6 and Example 2.4.3), a minor extension of the concept of 'base' is required. The *base*† of a partial knowledge space $\mathcal{K}$ is a family $\mathcal{B}$ of sets which spans† $\mathcal{K}$ and is moreover

[7] This algorithm still does not correct on Step 2 possible mistakes made on Step 1.

minimal with respect to inclusion for having this property. Notice that $\varnothing \in \mathcal{B}$ if and only if $\varnothing \in \mathcal{K}$. Also, if $\varnothing \in \mathcal{K}$ (that is, if $\mathcal{K}$ is a space), then $\mathcal{B}$ is a base† of $\mathcal{K}$ if and only if $\mathcal{B} \setminus \{\varnothing\}$ is a base of $\mathcal{K}$. And if $\varnothing \notin \mathcal{K}$, then $\mathcal{B}$ is a base† of $\mathcal{K}$ if and only if $\mathcal{B}$ is a base of $\mathcal{K} \cup \{\varnothing\}$. To simplify notation, we omit the $\dagger$ superscrit in the sequel and rely on the context to discriminate between the two closely related concepts of a base.

We refer to Algorithm 3.5.5 and Remark 3.5.8 for constructing the knowledge space spanned by a family of sets.

When the input data consist of the base, we need a subroutine to generate economically the ascendent family $\mathcal{K}(Q', W) = \{L \in \mathcal{K} \,|\, L \cap Q' = W\}$ (where W is the state of $\mathcal{K}_{|Q'}$ uncovered in Step 1). While it is clear that $\mathcal{K}(Q', W)$ is a partial knowledge space, and so has a unique base $\mathcal{F}$, it is not obvious how $\mathcal{F}$ can be efficiently built from the base $\mathcal{B}$ of $\mathcal{K}$. There is however a simple way to produce $\mathcal{F}$, albeit not an efficient one, that goes as follows. Starting from the base $\mathcal{B}$ of $\mathcal{K}$ and the element W of $\mathcal{K}_{|Q'}$, we first define

$$\mathcal{B}(W) = \{B \in \mathcal{B} \,|\, B \cap Q' \subseteq W\}. \tag{13.24}$$

The span of $\mathcal{B}(W)$ contains not only all the sets we want, but in fact many more, for it contains all the states K in $\mathcal{K}$ such that $K \cap Q' \subseteq W$ (while we aim at $K \cap Q' = W$). To obtain the desired states, we generate the span of $\mathcal{B}(W)$ (as in Algorithm 3.5.5) and then screen the span to eliminate from it any state K that satisfies $K \cap Q' \subset W$. This is the crux of Step 2.2 in the next algorithm. Additional comments on the algorithm are provided in Remarks 13.7.9 below.

13.7.8 Algorithm (`Two-step`, Version 2). The input consists of the base $\mathcal{B}$ of a finite knowledge space $\mathcal{K}$.

STEP 1.
1.1. Set $Q = \cup\mathcal{B}$, and then choose a suitable subset Q' in Q.
1.2. Build the projection $\mathcal{G}$ of the base $\mathcal{B}$ on Q'.
1.3. Run Algorithm 3.5.5 to get the knowledge space $\mathcal{K}_{|Q'}$ spanned by $\mathcal{G}$.
1.4. Run the assessment on $\mathcal{K}_{|Q'}$ and get a state W of $\mathcal{K}_{|Q'}$.

STEP 2.
2.1. Form the collection $\mathcal{B}(W) = \{B \in \mathcal{B} \,|\, B \cap Q' \subseteq W\}$.
2.2. Compute the ascendent family $\mathcal{L} = \{K \in \mathcal{K} \,|\, K \cap Q' = W\}$ (see text before algorithm).
2.3. Compute $\mathcal{M} = \{L \setminus \cap\mathcal{L} \,|\, l \in \mathcal{L}\}$.
2.4. Compute the base of the partial knowledge space $\mathcal{M}$.
2.5. Perform an assessment on the partial knowledge space $\mathcal{M}$ (summarized by its base), and get an element X of $\mathcal{M}$.
2.6. Return $W \cup (\cap\mathcal{L}) \cup X$.

This algorithm prompts several remarks.

13.7.9 Remarks. a) Note that $\mathcal{B}(W)$ also contains all elements B of $\mathcal{B}$ such that $B \cap Q' = \varnothing$.

b) In Step 2.2, $\mathcal{L}$ is exactly the ascendent family $\mathcal{K}(Q', W)$. In Step 2.3, $\mathcal{M}$ is the child $\mathcal{K}[Q'|W]$ of W. In Step 2.4, Algorithm 3.5.1 can be used.

c) As regards the word "suitable" in Step 1.1, we pointed out in Remark 13.7.3(a) that the family $\mathcal{K}_{|Q'}$ had to be small enough for Algorithm 13.7.2 to be applicable. The same remark also applies to Algorithm 13.7.8, not only with respect to $\mathcal{K}_{|Q'}$ but also to the family $\mathcal{L}$.

d) The assessment routine used in Algorithm 13.7.8, at Steps 1.4 and 2.4, should be designed to accept any knowledge space as input. Indeed, both collections $\mathcal{K}_{|Q'}$ and $\mathcal{M}$ form knowledge spaces.

The information in the base of a knowledge space can be obtained from another source. We suppose for the rest of this section that the knowledge space $\mathcal{K}$ is provided to the algorithm in the form of its surmise relation σ. Remember from Definition 5.1.2 and Theorem 5.2.5 that, for any p in Q, $\sigma(p)$ is the collection of all the atoms of $\mathcal{K}$ at p. So, we have $\cup_{p\in Q}\sigma(p) = \mathcal{B}$. However, σ contains more information than the base. This added information calls to a slight modification of the algorithm. The new Algorithm 13.7.10 relies on the following facts and notation. If $Q' \subseteq Q$, then $\cup_{p\in Q'}\{B \cap Q' \mid B \in \sigma(p)\}$ spans the projection of $\mathcal{K}$ on Q'. In the definitions below, the set Q' is supposed to be clearly indicated by the context and is omitted in the notation. Given $W \in \mathcal{K}_{|Q'}$ and $p \in W$, we write:

(i) $\sigma(p, W)$ for $\{B \in \sigma(p) \mid B \cap Q' \subseteq W\}$,
(ii) $\mathcal{B}(W)$ for $\{B \in \mathcal{B} \mid B \cap Q' \subseteq W \text{ and } \forall q \in W : B \notin \sigma(q, W)\}$,
(iii) $\mathcal{K}(W)$ for the span of $\mathcal{B}(W)$.

We may build the ascendent family $\mathcal{L} = \{K \in \mathcal{K} \mid K \cap Q' = W\}$ in another way, via the following trick: if $W = \{q_1, q_2, \ldots, q_k\}$, then collect all the unions of an element in $\mathcal{K}(W)$ with one element of $\sigma(q_i, W)$ for each $i = 1, 2, \ldots, k$. There are a priori

$$|\mathcal{K}(W)| \times |\sigma(q_1, W)| \times |\sigma(q_2, W)| \times \cdots \times |\sigma(q_k, W)|$$

such unions. However there is no need to keep duplicated states in the collection. Problem 11 asks the reader to find a method for handling such duplications.

13.7.10 Algorithm (`Two-step`, Version 3). The input consists of the surmise function σ of a finite knowledge space $\mathcal{K}$.

STEP 1.
1.1. Set $Q = \cup_{p\in Q}\sigma(p)$, and then choose a suitable subset Q' in Q.
1.2. Set $\mathcal{H} = \cup_{p\in Q'}\{B \cap Q' \mid B \in \sigma(p)\}$, and compute the span $\mathcal{K}_{|Q'}$ of $\mathcal{H}$ (by running Algorithm 3.5.5).
1.3. Run the assessment on $\mathcal{K}_{|Q'}$ and get a state W of $\mathcal{K}_{|Q'}$.

STEP 2. 2.1. Compute all collections $\sigma(p, W)$ for $p \in W$.
2.2. Compute the subcollection $\mathcal{L} = \{ K \in \mathcal{K} \mid K \cap Q' = W \}$ of $\mathcal{K}$ by applying the trick described before the algorithm.
2.3. Compute $\mathcal{M} = \{L \setminus \cap\mathcal{L} \mid L \in \mathcal{L}\}$.
2.4. Perform an assessment on the knowledge space $\mathcal{M}$, and get an element X of $\mathcal{M}$.
2.5. Return $W \cup (\cap\mathcal{L}) \cup X$.

13.7.11 Remarks. a) Regarding the word 'suitable' in Step 1.1, the comments in Remark 13.7.9(c) are also applicable here.

b) The assessment routine used in Algorithm 13.7.10, at Steps 1.3 and 2.4, should accept as input any knowledge space. Indeed, both collections $\mathcal{K}_{|Q'}$ and $\mathcal{M} = \{L \setminus \cap\mathcal{L} \mid L \in \mathcal{L}\}$ form knowledge spaces.

In Remark 13.7.3(c), we mentioned a serious defect of Algorithm 13.7.2: it does not contain in Step 2 any mechanism for correcting possible errors made in selecting the set W produced in Step 1. Algorithms 13.7.8 and 13.7.10 share the same defect. We devote the rest of this section to a sketch of an improved procedure.

The last two-step algorithm of this section amends the defect just mentioned. It does so by a rule of thumb using balls or neighborhoods in the sense of Definitions 3.4.5 and 4.1.6. The idea is to replace, before starting Step 2, the set W selected in Step 1 with some kind of (modified) neighborhood around W in the projection $\mathcal{K}_{|Q'}$ and then to infer a more general family than the antecedent one $\mathcal{K}(Q', W)$, or better to directly replace the antecedent family $\mathcal{K}(Q', W)$ with an appropriate subfamily of $\mathcal{K}$. For convenience, we denote the latter, new family as $\mathcal{L}(W)$. Many possible ways exist for building $\mathcal{L}(W)$, and consequently (at least) as many ways of counter-acting errors made in producing W. We will come back to this in Remarks 13.7.13(a) and (b) below.

13.7.12 Algorithm (`Two-step`, Version 4). The algorithm receives the surmise function σ of some knowledge space $\mathcal{K}$.

STEP 1. 1.1. Set $Q = \cup_{p \in Q} \sigma(p)$, and then choose a subset Q' in Q.
1.2. Set $\mathcal{H} = \cup_{p \in Q'} \{B \cap Q' \mid B \in \sigma(p)\}$, and compute the span $\mathcal{K}_{|Q'}$ of $\mathcal{H}$ (by running Algorithm 3.5.5).
1.3. Run the assessment on $\mathcal{K}_{|Q'}$ and get a state W of $\mathcal{K}_{|Q'}$.

STEP 2. 2.1. In the space $\mathcal{K}$ build the approximating set $\mathcal{L}(W)$ (see text before algorithm and also the Remarks 13.7.13).
2.2. Perform an assessment on the family $\mathcal{L}(W)$ and get an element X of $\mathcal{L}$.
2.3. Return X.

13.7.13 Remarks. a) For Algorithm 13.7.12 to be applicable, both $\mathcal{K}_{|Q'}$ and $\mathcal{L}(W)$ must be acceptable by our assessment routine. Moreover, it must be possible to build $\mathcal{L}(W)$ in a reasonable amount of time. Here is a possibility for $\mathcal{L}(W)$ which seems to fit these requirements. Take a ball centered at W in $\mathcal{K}_{|Q'}$, say $\mathcal{N}(W,k)$ (for some adequate, small value of k; see 4.1.6 for the terminology used here[8]). Then set $\mathcal{L}(W) = \{K \in \mathcal{K} \mid K \cap Q' \in \mathcal{N}(W,k)\}$. To build the latter family $\mathcal{L}(W)$ in Step 2.2, we could make a repeated use of the trick described just before Algorithm 13.7.10 (and used there). We apply the trick to each element W' of $\mathcal{N}(W,k)$ individually, and then form the union $\mathcal{L}(W)$ of all the resulting families $\mathcal{K}(W')$.

b) It would be interesting to further investigate the possibilities for $\mathcal{L}(W)$, in particular in the case in which the applicability of the assessment routine is restricted to some types of structures. In this regard, notice that balls do not share all the properties we would like them to have. In particular, they are generally not $\cup$-closed nor well-graded—see our next example. Notice that the definition of $\mathcal{L}(W)$ does not require the detour to a neighborhood in $\mathcal{K}_{|Q'}$.

13.7.14 Example. In the knowledge space $\mathcal{K}$ with

$$\mathcal{K} = \{\varnothing, \{a\}, \{b\}, \{a,b\}, \{a,c\}, \{b,c\}, Q\}, \tag{13.25}$$

where $Q = \{a,b,c\}$, the ball $\mathcal{N}(\varnothing, 2)$ equals $\mathcal{K} \setminus \{Q\}$ and is neither $\cup$-closed nor well-graded (in view of its two elements $\{a,c\}$ and $\{b,c\}$). It is as easy to build similar examples with a ball centered at a nonempty state and/or with a larger radius. On the other hand, a ball of radius 1 is automatically well-graded, but not necessarily $\cup$-closed.

13.8 Refining the Assessment

13.8.1 Remark. In practice, the output of an assessment algorithm in the style of this chapter takes the form of a probability distribution L_n (for some final trial number n) on the collection of states. In principle, such a probability distribution will eventually have a mass concentrated on one or a few closely related states. For example, the procedure may have selected three largely overlapping knowledge states K_1, K_2 and K_3. As an illustration, suppose that

$$K_1 = \{a,b,d,e\}, \quad K_2 = \{b,c,d\}, \quad K_3 = \{a,b,c,e\},$$

with

$$L_n(K_1) = L_n(K_2) = .25, \quad L_n(K_3) = .40,$$

and the rest of the mass of L_n being scattered among the remaining states. On the basis of this information, the best bet for the state of the subject

[8] Other concepts of “neighborhood” of W in $\mathcal{K}_{|Q'}$ could also be used.

is K_3. However, one may wish to refine this assessment by reconsidering at that time the full sequence of questions asked and responses observed during the application of the procedure. The selection of one among these three states could be based on a Bayesian heuristic. Recalling the local independence assumption (Axiom [N] in 12.4.1), we can recompute the conditional likelihood of observing each of the selected states K_1, K_2 and K_3, given the actual sequence of responses to the questions asked. The states yielding the greatest likelihood can then be taken as the final result of the assessment. Obviously, such a computation makes sense only when good estimates are available for the probabilities β_q and η_q of careless errors and lucky guesses, respectively.

Suppose that the subject has been asked questions c, d, and e, and has provided an incorrect response to c, and a correct response to both d and e. We denote these data by the letter 'D'. By the local independence Axiom [N], the conditional probabilities of these data, given the three states, are

$$P(D \mid K_1) = (1 - \eta_c)(1 - \beta_d)(1 - \beta_e), \tag{13.26}$$
$$P(D \mid K_2) = \beta_c(1 - \beta_d)\eta_e, \tag{13.27}$$
$$P(D \mid K_3) = \beta_c\eta_d(1 - \beta_e). \tag{13.28}$$

Using a Bayesian rule to recompute the probabilities of the states, we get for $i = 1, 2, 3$:

$$P(K_i \mid D) = \frac{P(D \mid K_i)\mathbf{L}_n(K_i)}{\sum_{j=1}^{3} P(D \mid K_j)\mathbf{L}_n(K_j)}. \tag{13.29}$$

Let us assume that, from previous analyses, the following estimates have been obtained for the careless error and lucky guess parameters:

$$\widehat{\beta}_c = \widehat{\beta}_d = .05, \qquad \widehat{\beta}_e = .10,$$
$$\widehat{\eta}_c = \widehat{\eta}_d = .10, \qquad \widehat{\eta}_e = .05.$$

Replacing the $P(D \mid K_j)$ in (13.29) by their estimates in terms of the $\widehat{\beta}_q$'s and $\widehat{\eta}_q$'s, via Equations (13.26), (13.27) and (13.28), we obtain approximately

$$\widehat{P}(K_1 \mid D) = .988,$$
$$\widehat{P}(K_2 \mid D) = .003,$$
$$\widehat{P}(K_3 \mid D) = .009.$$

The picture resulting from such a computation[9] is quite different from that based on L_n alone: the overwhelmingly most plausible state is now K_1.

[9] Note that, even though this Bayesian computation is heuristically defensible, it is not strictly founded in theory; see Problem 6 in this connection.

13.9 Proofs*

13.9.1 Proof of Theorem 13.6.6. (All logarithms are in base 2.) By definition of the informative questioning rule in 13.4.8 and Equation (13.14), we have to minimize, over all q in Q, the quantity (writing $l = L_n$ for simplicity)

$$\tilde{H}(q,l) = l(\mathcal{K}_q)H\big(u(1,q,l)\big) + l(\mathcal{K}_{\bar{q}})H\big(u(0,q,l)\big), \tag{13.30}$$

where

$$l(\mathcal{K}_q) = \sum_{K\in\mathcal{K}_q} l(K), \quad l(\mathcal{K}_{\bar{q}}) = \sum_{K\in\mathcal{K}_{\bar{q}}} l(K).$$

This quantity depends upon the updating rule u. As the updating rule is assumed to be convex with a constant parameter θ, we obtain from Equation (13.7)

$$\begin{aligned} H\big(u(1,q,l)\big) &= -\sum_{K\in\mathcal{K}} u_K(1,q,l)\log u_K(1,q,l) \\ &= -\sum_{K\in\mathcal{K}_q} l(K)\left(1-\theta+\frac{\theta}{l(\mathcal{K}_q)}\right)\left(\log l(K) + \log\left(1-\theta+\frac{\theta}{l(\mathcal{K}_q)}\right)\right) \\ &\qquad - \sum_{J\in\mathcal{K}_{\bar{q}}} l(J)(1-\theta)\big(\log l(J) + \log(1-\theta)\big) \end{aligned} \tag{13.31}$$

and

$$\begin{aligned} H\big(u(0,q,l)\big) &= -\sum_{J\in\mathcal{K}_{\bar{q}}} l(J)\left(1-\theta+\frac{\theta}{l(\mathcal{K}_{\bar{q}})}\right)\left(\log l(J) + \log\left(1-\theta+\frac{\theta}{l(\mathcal{K}_{\bar{q}})}\right)\right) \\ &\qquad - \sum_{K\in\mathcal{K}_q} l(K)(1-\theta)\big(\log l(K) + \log(1-\theta)\big). \end{aligned} \tag{13.32}$$

Using Equations (13.30), (13.31) and (13.32) and grouping appropriately leads to

$$\begin{aligned} \tilde{H}(q,l) = H(l) &- 2l(\mathcal{K}_q)l(\mathcal{K}_{\bar{q}})(1-\theta)\log(1-\theta) \\ &- l(\mathcal{K}_q)\big((1-\theta)l(\mathcal{K}_q)+\theta\big)\log\left(1-\theta+\frac{\theta}{l(\mathcal{K}_q)}\right) \\ &- l(\mathcal{K}_{\bar{q}})\big((1-\theta)l(\mathcal{K}_{\bar{q}})+\theta\big)\log\left(1-\theta+\frac{\theta}{l(\mathcal{K}_{\bar{q}})}\right). \end{aligned}$$

That is, with $l(\mathcal{K}_q) = t$ and for $t \in]0,1[$

$$g(t) = -2t(1-t)(1-\theta)\log(1-\theta) - t\big((1-\theta)t+\theta\big)\log\left(1-\theta+\frac{\theta}{t}\right) \\ -(1-t)\big((1-\theta)(1-t)+\theta\big)\log\left(1-\theta+\frac{\theta}{1-t}\right), \tag{13.33}$$

we have

$$\tilde{H}(q,l) = H(l) + g(t). \tag{13.34}$$

Notice that g is symmetric around $\frac{1}{2}$, that is, $g(t) = g(1-t)$ for $0 < t < 1$. To establish the two assertions of the Theorem, it suffices now to prove that the function g is convex on $]0,1[$ and has a strict extremum at $\frac{1}{2}$. (Thus, g will serve as the function ν in Definition 13.6.4.) Since g is symmetric around $\frac{1}{2}$, we only have to show that the second derivative is strictly positive on $]0,\frac{1}{2}[$. We shall derive this from the fact that $g'''(t) < 0$ for $0 < t < \frac{1}{2}$, together with $g''(\frac{1}{2}) > 0$. To compute the derivatives, we simplify the expression of g. With the notation

$$a(t) = (1-\theta)t, \qquad b(t) = (1-\theta)t+\theta,$$
$$f(t) = -a(t)b(t)\log\frac{b(t)}{a(t)},$$

Equation (13.33) simplifies into

$$g(t) = -\log(1-\theta) + \frac{1}{1-\theta}\big(f(t)+f(1-t)\big). \tag{13.35}$$

Using $a'(t) = b'(t) = 1-\theta$, we obtain for the derivatives of f

$$f''(t) = (1-\theta)^2\left(\theta\frac{a(t)+b(t)}{a(t)\,b(t)} - 2\log\frac{b(t)}{a(t)}\right),$$
$$f'''(t) = -\frac{(1-\theta)^3\,\theta^3}{a(t)^2\,b(t)^2} < 0.$$

Equation (13.35), implies that $g'''(t) < 0$ for $0 < t < \frac{1}{2}$.

On the other hand, with

$$h(\theta) = \frac{2\theta}{1-\theta^2} - \log\frac{1+\theta}{1-\theta},$$

we have

$$g''\left(\frac{1}{2}\right) = 4(1-\theta)h(\theta).$$

Since $\lim_{\theta\to 0^+} h(\theta) = 0$ and $h'(\theta) > 0$ for $0 < \theta < 1$, we have $g''(\frac{1}{2}) > 0$. □

13.9.2 Proof of Theorem 13.6.7 Let $\widetilde{\Omega}$ be the set of all realizations ω for which, for every trial n,

(i) $\mathbf{Q}_n \in S(\nu, \mathbf{L}_n)$, with S as in Definition 13.6.4;
(ii) $\mathbf{R}_n = \iota_{K_0}(\mathbf{Q}_n)$.

Notice that $\widetilde{\Omega}$ is a measurable set of the sample space Ω, and that $\mathbb{P}(\widetilde{\Omega}) = 1$. Writing $\mathbf{L}_n^\omega(K_0)$ for the value of the random variable $\mathbf{L}_n(K_0)$ at the point $\omega \in \Omega$, we only have to establish that for any point $\omega \in \widetilde{\Omega}$, we have

$$\lim_{n\to\infty} \mathbf{L}_n^\omega(K_0) = 1.$$

Take $\omega \in \widetilde{\Omega}$ arbitrarily. It follows readily from the assumptions that $\mathbf{L}_n^\omega(K_0)$ is nondecreasing, and thus converges. Therefore, it suffices to show that $\mathbf{L}_{n_i}^\omega(K_0) \to 1$ for at least one subsequence $s = (n_i)$ of the positive integers. Since $\mathcal{K}$ is finite and $\mathbf{L}_n(K) \in]0,1[$, we can take $s = (n_i)$ such that $\mathbf{L}_{n_i}^\omega(K)$ converges for all $K \in \mathcal{K}$. In the rest of this proof, we consider a fixed subsequence s satisfying those conditions.

We define a function $f_{\omega,s} : Q \to [0,1]$ by

$$f_{\omega,s}(q) = \begin{cases} \lim_{i\to\infty} \mathbf{L}_{n_i}^\omega(\mathcal{K}_q) & \text{if } q \in K_0; \\ \lim_{i\to\infty} \mathbf{L}_{n_i}^\omega(\mathcal{K}_{\bar{q}}) & \text{if } q \notin K_0. \end{cases}$$

We also define

$$\widetilde{Q}_{w,s} = \{q \in Q \,|\, f_{\omega,s}(q) < 1\}.$$

If $\widetilde{Q}_{w,s}$ is empty, the fact that $\mathbf{L}_{n_i}^\omega(K_0) \to 1$ follows readily from Lemma 3 (see below). The core of the proof of Theorem 13.6.7 consists in establishing that $\widetilde{Q}_{w,s} = \varnothing$, which is achieved by Lemma 2.

LEMMA 1. *If $f_{\omega,s}(q) < 1$, then $\{i \in \mathbb{N} \,|\, \mathbf{Q}_{n_i}(\omega) = q\}$ is a finite set.*

Proof. Assume that $q \in K_0$. Since $f_{\omega,s}(q) < 1$, there is $\epsilon > 0$ such that $f_{\omega,s}(q) + \epsilon < 1$. If $\mathbf{Q}_{n_i}(\omega) = q$ and i is large enough to ensure that $\mathbf{L}_{n_i}^\omega(\mathcal{K}_q) \leq f_{\omega,s}(q) + \epsilon$, we derive

$$\begin{aligned} \mathbf{L}_{n_{i+1}}^\omega(K_0) &\geq \mathbf{L}_{n_i+1}^\omega(K_0) \\ &\geq v\big(\mathbf{L}_{n_i}^\omega(\mathcal{K}_q)\big)\mathbf{L}_{n_i}^\omega(K_0) && \text{(by Definition 13.6.2)} \\ &\geq v\big(f_{\omega,s}(q) + \epsilon\big)\mathbf{L}_{n_i}^\omega(K_0). \end{aligned}$$

Since $v\big(f_{\omega,s}(q) + \epsilon\big) > 1$ does not depend on i and $\mathbf{L}_{n_i}^\omega(K_0) \leq 1$, we may have $\mathbf{Q}_{n_i}(\omega) = q$ for at most a finite number of values of i. The proof is similar in the case $q \notin K_0$. ◊

LEMMA 2. $\widetilde{Q}_{w,s} = \{q \in Q \,|\, f_{\omega,s}(q) < 1\} = \varnothing$.

Proof. We proceed by contradiction. If $f_{\omega,s}(q) < 1$ for some $q \in Q$, we can assert the existence of some positive integer j and some $\epsilon > 0$ such that whenever $i > j$, we have

$$\begin{array}{lll} \text{either} & 0 < l_1(K_0) \leq \mathbf{L}^{\omega}_{n_i}(\mathcal{K}_q) < f_{\omega,s}(q) + \epsilon < 1 & \text{if } q \in K_0, \\ \text{or} & 0 < l_1(K_0) \leq \mathbf{L}^{\omega}_{n_i}(\mathcal{K}_{\bar{q}}) < f_{\omega,s}(q) + \epsilon < 1 & \text{if } q \notin K_0. \end{array}$$

This means that for $i > j$, both $\mathbf{L}^{\omega}_{n_i}(\mathcal{K}_q)$ and $\mathbf{L}^{\omega}_{n_i}(\mathcal{K}_{\bar{q}})$ remain in some interval $]\gamma'_q, 1 - \delta'_q[$ with $0 < \gamma'_q, \delta'_q$. The above argument applies to all $q \in \widetilde{Q}_{w,s}$. In view of the finiteness of $\widetilde{Q}_{w,s}$, the index q may be dropped in γ'_q, δ'_q. Moreover, referring to Definition 13.6.4, we can assert the existence of $\bar{\gamma}$ and $\bar{\delta}$ such that $0 < \bar{\gamma} < \gamma$, $0 < \bar{\delta} < \delta$ and $\mathbf{L}^{\omega}_{n_i}(\mathcal{K}_q) \in]\bar{\gamma}, 1 - \bar{\delta}[$ for $i > j$ and $q \in \widetilde{Q}_{w,s}$.

Since $\widetilde{Q}_{w,s}$ is finite, Lemma 1 may be invoked to infer the existence of a positive integer k such that $\mathbf{Q}_{n_i}(\omega) \notin \widetilde{Q}_{w,s}$ whenever $i > k$. Note that, by definition of $\widetilde{Q}_{w,s}$, we have $f_{\omega,s}(q') = 1$ for all $q' \notin \widetilde{Q}_{w,s}$. Since $\mathbf{L}^{\omega}_{n_i}(\mathcal{K}_{q'})$ converges, there is $i^\star > j, k$ such that neither $\mathbf{L}^{\omega}_{n_i}(\mathcal{K}_{q'})$ nor $\mathbf{L}^{\omega}_{n_i}(\mathcal{K}_{\overline{q'}})$ are points of $]\bar{\gamma}, 1 - \bar{\delta}[$ for all $i > i^\star$ and $q' \notin \widetilde{Q}_{w,s}$. By definition of $\widetilde{\Omega}$, we must have $\mathbf{Q}_{n_i}(\omega) \in \widetilde{Q}_{w,s}$, contradicting $i > k$. ◊

Define, for every $K \in \mathcal{K}$,

$$A_{\omega,s}(K) = \{q \in K \,|\, f_{\omega,s}(q) = 1\}.$$

LEMMA 3. *Suppose that, for some $K \in \mathcal{K}$, $A_{\omega,s}(K) \neq A_{\omega,s}(K_0)$. Then*

$$\lim_{i\to\infty} \mathbf{L}^{\omega}_{n_i}(K) = 0.$$

Proof. Assume that there is some $q \in K_0 \setminus K$ such that $f_{\omega,s}(q) = 1$; that is $\lim_{i\to\infty} \mathbf{L}^{\omega}_{n_i}(\mathcal{K}_q) = 1$. This implies $\lim_{i\to\infty} \mathbf{L}^{\omega}_{n_i}(\mathcal{K}_{\bar{q}}) = 0$ and the thesis since $K \in \mathcal{K}_{\bar{q}}$. The other case, $q \in K \setminus K_0$, is similar. ◊

By Lemma 2, $K \neq K_0$ implies $A_{\omega,s}(K) \neq A_{\omega,s}(K_0)$. It follows from Lemma 3 that, for all $K \neq K_0$, $\lim_{i\to\infty} \mathbf{L}^{\omega}_{n_i}(K) = 0$, yielding $\mathbf{L}^{\omega}_{n_i}(K_0) \to 1$.

This concludes the proof of Theorem 13.6.7. □

13.9.3 Remark. A careful study of the above proof shows that the assumption that the questioning rule is inner can be replaced by the following one. For any $\gamma, \delta \in]0, 1[$, denote by $E_{n,q}(\gamma, \delta)$ the event that $\gamma < \mathbf{L}_n(\mathcal{K}_q) < 1 - \delta$, and let $E_n(\gamma, \delta) = \cup_{q \in Q} E_{n,q}(\gamma, \delta)$. The condition states that there exists $\sigma > 0$ such that, for all $\gamma, \delta \in]0, \sigma[$,

$$\mathbb{P}(\mathbf{Q}_n = q' \,|\, \overline{E_{n,q'}(\gamma, \delta)} \cap E_n(\gamma, \delta)) = 0.$$

In other words, and somewhat loosely: no question q will be chosen with $\mathbf{L}_n(\mathcal{K}_q)$ in a neighborhood of one or zero when this can be avoided.

13.10 Original Sources and Related Works

Except for Section 13.7, the algorithms of which are new, this chapter follows closely Falmagne and Doignon (1988a). The first applications of the other algorithms described in this chapter were made by M. Villano, who has tested them extensively in his dissertation (Villano, 1991); see also Villano, Falmagne, Johannesen, and Doignon (1987) and Kambouri (1991). These algorithms form a key component of the knowledge assessment engine of the ALEKS system briefly described in Chapter 1. Various results concerning the predictive power (or validity) of such assessments have been obtained. Some of these results are summarized in Chapter 17.

As mentioned earlier, the updating operators involved in the algorithms have been inspired by some operators of mathematical learning theory (see Bush and Mosteller, 1955; Norman, 1972). Specifically, the convex updating rule is related to a Bush and Mosteller learning operator (Bush and Mosteller, 1955), and the multiplicative updating rule is close to the learning operator of the so-called beta learning model of Luce (1959) (also relevant are Luce, 1964; Marley, 1967). Bayesian updating rules in intelligent tutoring systems have been discussed by Kimbal (1982).

Problems

1. Show that the convex updating rule defined by Equation (13.7) is not permutable in the sense of Equation (13.8).
2. Check that the multiplicative updating rule of Equation (13.9) is permutable in the sense of Equation (13.8).
3. Complete the proof of Theorem 13.5.2 and show that the stochastic process $(\mathbf{R}_n, \mathbf{Q}_n, \mathbf{L}_n)$ is Markovian.
4. (Continuation.) Prove that the process $(\mathbf{Q}_n, \mathbf{L}_n)$ is also Markovian.
5. In the knowledge structure of Example 13.6.1, suppose that the half-split questioning rule and the multiplicative updating rule with a constant parameter $\zeta_{q,r}$ have been used. Verify Equations (13.17), (13.18) and (13.19) for $n = 1, 2$. You should assume that $L_1(K) = .2$ for any state K.
6. Discuss the Bayesian computation proposed in Remark 13.8.1 for refining the assessment from a theoretical viewpoint.
7. Give a detailed proof of Corollary 13.6.8.
8. Give a detailed proof of Corollary 13.6.9.

9. Let $Q = \{a, b, c\}$ and $\mathcal{K} = \{\varnothing, \{a, b\}, \{b, c\}, \{a, c\}, Q\}$. Assume that the subject's state varies randomly between trials according to the probability distribution ϕ defined by $\phi(\{a, b\}) = \phi(\{a, c\}) = \phi(Q) = \frac{1}{3}$. Suppose that the half-split questioning rule and the convex updating rule with a constant parameter θ are used. Prove that $\lim_{n\to\infty} E(\mathbf{L}_n(\{b, c\}) > 0$ (or better, compute this limit). Argue then that even the domain of ϕ cannot be uncovered by the assessment process.

10. Suppose that the subject's knowledge state changes once during the assessment. Discuss in detail the impact of such a change on the efficiency and the accuracy of the multiplicative assessment process.

11. Examine the problem of the duplicates in Steps 2 and 3 of Algorithm 13.7.10. Either find an algorithm for the economic removal of such duplicate, or come up with another solution.

14

A Markov Chain Procedure

This chapter discusses an assessment procedure that is similar in spirit to those described in Chapter 13, but different in a key aspect: it is based on a finite Markov chain rather than on a Markov process with an uncountable set of Markov states. As a consequence, the procedure requires less storage and computation. It can thus be implemented on a small machine. A multi-step application of the procedure described in this chapter can be set up exactly as we did for the procedure of the previous chapter.

14.1 Outline

A fixed, finite knowledge structure $(Q, \mathcal{K})$ is used by the assessment engine. Later in this chapitre, we will assume that $(Q, \mathcal{K})$ is well-graded. We suppose that, on any trial in the course of the assessment, some of the knowledge states in $\mathcal{K}$ are considered as plausible from the standpoint of the assessment engine. These 'marked states' are collected in a family which is regarded as a value of a random variable $\mathbf{M}_n$, where the index $n = 1, 2, \ldots$ indicates the trial number. During the first phase of the procedure, this family decreases in size until a single marked state remains. In the second phase, the single 'marked state' evolves in the structure. This last feature allows the assessment engine, through a statistical analysis of the observed sequence of problems and answers, to estimate the 'true' state (or states, if the subject knowledge state varies somewhat from trial to trial; we give a formal definition of 'true' states in 14.2.2.) Note that, in some cases, a useful estimate can be obtained even if the 'true' state estimate is not part of the structure. Before getting into technicalities, we will illustrate the basic ideas by tracing an exemplary realization of the Markov chain to be described.

14.1.1 Example. We take the same knowledge structure as in Example 13.1.1 (cf. Figure 13.1), that is:

$$\mathcal{K} = \big\{\varnothing, \{a\}, \{c\}, \{a,c\}, \{b,c\}, \{a,b\}, \{a,b,c\}, \{a,b,c,d\}, \{a,b,c,d,e\}\big\}. \tag{14.1}$$

J.-C. Falmagne, J.-P. Doignon, *Learning Spaces*,
DOI 10.1007/978-3-642-01039-2_14, © Springer-Verlag Berlin Heidelberg 2011

We suppose that the assessment engine is initially unbiased in the sense that all the knowledge states are regarded as plausible. Thus, all nine states of $\mathcal{K}$ are marked and we set by convention $\mathbf{M}_1 = \mathcal{K}$. (In some situations, a smaller subset of states could be marked that would reflect some a priori information on the student population.) During the first phase of the procedure, each question asked is selected in such a manner that, no matter which response is given (correct or incorrect), the number of marked states is decreased as much as possible. This goes on until only a single marked state remains. For example, suppose that $\mathbf{M}_n = \mathcal{M}$ is the set of marked states on trial n. Assume that some question q is chosen on that trial and that a correct response is given. The set of states marked on trial $n+1$ would be $\mathcal{M}_q$, the subset of states of $\mathcal{M}$ containing q. If the response is incorrect, then the set of marked states on trial $n+1$ would be $\mathcal{M}_{\bar{q}}$, that is, the subset of states of $\mathcal{M}$ not containing q. It makes sense to select q in order to minimize the maximum of the two possible numbers of states kept. This clearly amounts to selecting a question q that divides as equally as possible the currently marked states into those containing q and those not containing q. So, q should render

$$\big||\mathcal{M}_q| - |\mathcal{M}_{\bar{q}}|\big|$$

as small as possible. In our example, we have for $q = a$ and $\mathcal{M} = \mathcal{K}$

$$\big||\mathcal{K}_a| - |\mathcal{K}_{\bar{a}}|\big| = |6 - 3| = 3.$$

Similar calculations for the other questions give the counts:

$$\begin{aligned} &\text{for } b: && |5-4| = 1,\\ &\text{for } c: && |6-3| = 3,\\ &\text{for } d: && |2-7| = 5,\\ &\text{for } e: && |1-8| = 7. \end{aligned}$$

Thus, b should be the first question asked. Denoting the question asked on trial n by $\mathbf{Q}_n$, a random variable, we set $\mathbf{Q}_1 = b$ with probability 1. Suppose that we observe an incorrect answer. We denote this fact by writing $\mathbf{R}_1 = 0$. In general, we define $\mathbf{R}_n$ as a random variable taking the value 1 if the question asked on trial n is correctly answered, and 0 otherwise. In our example, the family of marked states becomes on trial 2

$$\mathbf{M}_2 = \big\{\varnothing, \{a\}, \{c\}, \{a, c\}\big\}.$$

The new counts for selecting the next question are:

$$\begin{aligned} &\text{for } a: && |2-2| = 0,\\ &\text{for } b: && |0-4| = 4,\\ &\text{for } c: && |2-2| = 0,\\ &\text{for } d: && |0-4| = 4,\\ &\text{for } e: && |0-4| = 4. \end{aligned}$$

Thus, either a or c should be asked. We choose randomly between them, with equal probabilities: $\mathbb{P}(\mathbf{Q}_2 = a) = \mathbb{P}(\mathbf{Q}_2 = b) = .5$. Suppose that we ask item a and get a correct answer, that is $\mathbf{Q}_2 = a$ and $\mathbf{R}_2 = 1$. The family of marked states on trial 3 is

$$\mathbf{M}_3 = \{\{a\}, \{a, c\}\}.$$

With probability 1, we get $\mathbf{Q}_3 = c$. If $\mathbf{R}_3 = 1$, we are left with the single marked state $\{a, c\}$. That is, we have

$$\mathbf{M}_4 = \{\{a, c\}\}.$$

In this miniature example, the second phase of the procedure starts on trial 4. From here on, the set of marked states will always contain a single state which may vary from trial to trial according to the question asked and the response given. The choice of a question on any trial $n \geq 4$ is based on the set of all the states in the ball or neighborhood of the current single marked state. Specifically, we use the ball formed by all states in $\mathcal{K}$ situated at a distance at most 1 from that marked state (cf. 4.1.6). In the case of the single marked state $\{a, c\}$, this ball is

$$\mathcal{N}(\{a, c\}, 1) = \{\{a, c\}, \{a\}, \{c\}, \{a, b, c\}\}.$$

As in Phase 1, the next question q is selected in order to split as equally as possible this set of states into those which contain q and those which do not contain q. Here, a, b or c will be randomly selected (with equal probability). If the answer collected confirms the current single marked state, we keep it as the only marked state. Otherwise, we change this state into another one, according to the new information. For concreteness, we consider the four generic cases given in Table 14.1. Suppose first that $\mathbf{Q}_4 = a$ and $\mathbf{R}_4 = 0$ (row 1 of Table 14.1). As the response to a is incorrect, we remove a from the single marked state $\{a, c\}$, which yields the single marked state $\{c\}$, with $\mathbf{M}_5 = \{\{c\}\}$.

Table 14.1. Four generic cases producing $\mathbf{M}_5$.

$\mathbf{M}_4$	$\mathbf{Q}_4$	$\mathbf{R}_4$	$\mathbf{M}_5$
$\{\{a, c\}\}$	a	0	$\{\{c\}\}$
$\{\{a, c\}\}$	a	1	$\{\{a, c\}\}$
$\{\{a, c\}\}$	b	0	$\{\{a, c\}\}$
$\{\{a, c\}\}$	b	1	$\{\{a, b, c\}\}$

In row 2 of Table 14.1, $\mathbf{Q}_4 = a$ and a is correctly solved. Thus $\{a, c\}$ is confirmed and we keep it as the single marked state, that is $\mathbf{M}_5 = \{\{a, c\}\}$.

In row 3, we also end up with the same single marked state $\{a, c\}$ as a result of a confirmation, but this time b is asked and the answer is incorrect. Finally, in row 4, b is also asked but yields a correct answer. As a result, we add b to the current single marked state. A third possible question on trial 4 is c; we leave this case to the reader. Table 14.2 summarizes an exemplary realization of the process in the early trials. By convention, we set $\mathbf{M}_1 = \mathcal{K}$. Rows 4, 5 in Table 14.2 correspond to row 1 in Table 14.1.

Table 14.2. An exemplary realization of the process in the early trials.

trial n	$\mathbf{M}_n$	$\mathbf{Q}_n$	$\mathbf{R}_n$
1	$\mathcal{K}$	b	0
2	$\{\varnothing, \{a\}, \{c\}, \{a, c\}\}$	a	1
3	$\{\{a\}, \{a, c\}\}$	c	1
4	$\{\{a, c\}\}$	a	0
5	$\{\{c\}\}$	...	...

14.1.2 Remarks. a) In this example, it is easy to check that the marking rule sketched here will always yield a single marked state after trial 4, whatever the question asked and the answer observed. It is clear that some assumptions about the knowledge structure are needed in order to establish these results in general. We will assume the structure to be well-graded and that the question is selected in the fringe of the single marked state (cf. Definitions 2.2.2, 4.1.6, and Theorem 4.1.7). Under these conditions, only one state will remain marked (see Theorem 14.3.3).

b) Our rationale for adopting these transition rules is dictated by caution. Some misleading answers by the subject may occur in the course of the procedure (due to lucky guesses or to careless errors, for example), resulting in a failure to uncover the 'true' state among those in the structure. It is also possible that the knowledge structure itself is mistaken to some extent: some states might be missing. Ideally, the final outcome of our procedure should compensate for both kinds of errors. This can be achieved by analyzing the data collected during the second phase of the procedure, after the single state has been reached. For instance, suppose that some particular state S has been omitted from the structure $\mathcal{K}$ used by the assessment engine. This missing state may nevertheless closely resemble some of the states in $\mathcal{K}$, say K_1, K_2 and K_3. In such a case, the sequence of single marked states, in the second phase of the procedure, may very well consist in transitions between these three states. The intersection and the union of these three states may provide lower and upper bounds for the state S to be uncovered.

On the other hand, if we believe that the state to be uncovered is one of the states visited during the second phase, then the choice between them may be dictated by standard statistical methods. In the style of Section 13.8, we may simply chose that state maximizing the likelihood of the sequence of responses observed, using the conditional response probabilities β_q and η_q introduced in Chapter 11 in the context of the local independence assumption (Definition 11.1.2). In general, the single marked states visited during the second phase of the procedure may be used to estimate the true state whether or not this state is contained in the structure used by the assessment engine.

Using the balls around states to guide both the choice of the question and the determination of the marked states is a sound idea not only for the second phase, but in fact for the whole process. The axioms given in the next section formalize this concept. The procedure described here shares many features with those presented in Chapter 13. The diagram displayed in Figure 14.1 highlights the differences and the similarities.

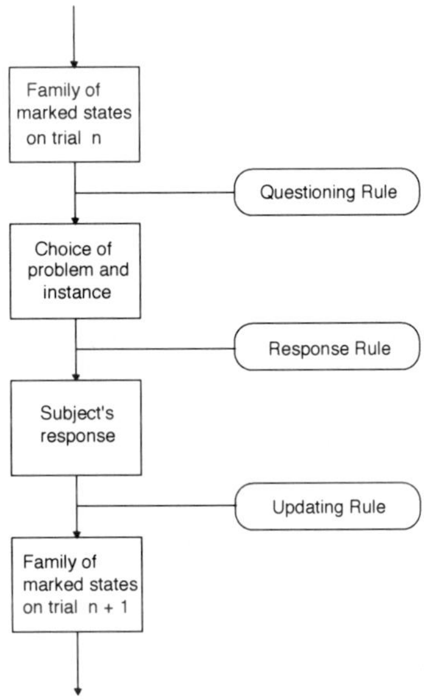

Figure 14.1. Diagram of the transitions for the Markov chain procedures.

14.2 The Stochastic Assessment Process

The stochastic assessment procedure is described by four sequences of jointly distributed random variables $(\mathbf{R}_n, \mathbf{Q}_n, \mathbf{K}_n, \mathbf{M}_n)$, where $n = 1, 2, \ldots$ stands for the trial number.

The subject's unobservable state on trial n is denoted by $\mathbf{K}_n$. We assume that $\mathbf{K}_n$ takes its values in a fixed finite knowledge structure $\mathcal{K}$ with domain Q. (Examples will be given later in which this assumption is weakened; see 14.6.3.) The question asked and the response observed on trial n are denoted by $\mathbf{Q}_n$ and $\mathbf{R}_n$; these two random variables take their values respectively in Q and $\{0, 1\}$, with 0 standing for incorrect and 1 for correct. Finally, the random variable $\mathbf{M}_n$ stands for the family of *marked states* on trial n. As indicated in our introductory section, these marked states are those which are regarded as plausible candidates for the subject's unknown state. Thus, the values of $\mathbf{M}_n$ lie in $2^{\mathcal{K}}$. The stochastic process is defined by the sequence of quadruples $(\mathbf{R}_n, \mathbf{Q}_n, \mathbf{K}_n, \mathbf{M}_n)$, $n = 1, 2, \ldots$ The complete history of the process from trial 1 to trial n is abbreviated as

$$\mathbf{W}_n = \big((\mathbf{R}_n, \mathbf{Q}_n, \mathbf{K}_n, \mathbf{M}_n), \ldots, (\mathbf{R}_1, \mathbf{Q}_1, \mathbf{K}_1, \mathbf{M}_1)\big),$$

with $\mathbf{W}_0$ denoting the empty history.

The four axioms given below specify the probability measure recursively. We first give general requirements, involving several unspecified functions and parameters. In the next section, severe restrictions will be put on these functions and parameters. Note that, in the spirit of Chapter 11, we use a local independence assumption (cf. Definition 11.1.2) to specify the conditional response probabilities in Axiom [RM].

14.2.1 Axioms for the Markov chain procedure. The Markov chain is specified by four axioms.

[K] KNOWLEDGE STATE RULE. There is a fixed probability distribution π on $\mathcal{K}$ such that for all natural numbers n,

$$\mathbb{P}(\mathbf{K}_n = K \,|\, \mathbf{M}_n,\, \mathbf{W}_{n-1}) \;=\; \pi(K).$$

In words: *The knowledge state of the subject varies from trial to trial according to a probability distribution π on $\mathcal{K}$, independent of the trial number.*

[QM] QUESTIONING RULE. There is a *questioning function* $\tau : Q \times 2^{\mathcal{K}} \to [0,1]$ such that, for all natural numbers n,

$$\mathbb{P}(\mathbf{Q}_n = q \,|\, \mathbf{K}_n,\, \mathbf{M}_n,\, \mathbf{W}_{n-1}) \;=\; \tau(q, \mathbf{M}_n).$$

That is: *The question asked on trial n depends only on the marked states.*

[RM] RESPONSE RULE. Two parameters $0 \leq \beta_q < 1$ and $0 \leq \eta_q < 1$ are attached to each item q, such that for all nonnegative integers n,

$$\mathbb{P}(\mathbf{R}_n = 1 \,|\, \mathbf{Q}_n = q,\, \mathbf{K}_n = K,\, \mathbf{M}_n, \mathbf{W}_{n-1}) = \begin{cases} 1 - \beta_q & \text{if } q \in K; \\ \eta_q & \text{if } q \notin K. \end{cases}$$

Accordingly: *The response on trial n only depends upon the knowledge state and the question asked on that trial via the parameters β_q and η_q, respectively.*

[M] MARKING RULE. There is a *marking function*

$$\mu : 2^{\mathcal{K}} \times \{0,1\} \times Q \times 2^{\mathcal{K}} \to [0,1]$$

such that

$$\mathbb{P}(\mathbf{M}_{n+1} = \Psi \,|\, \mathbf{W}_n) \;=\; \mu(\Psi, \mathbf{R}_n, \mathbf{Q}_n, \mathbf{M}_n).$$

Thus: *The marked states on trial $n+1$ only depend upon the following events on trial n: the marked states, the question asked and the response collected.*

14.2.2 Definition. A process $(\mathbf{R}_n, \mathbf{Q}_n, \mathbf{K}_n, \mathbf{M}_n)$ satisfying Axioms [K], [QM], [RM], and [M] is a *(discrete) stochastic assessment process parametrized by* π, τ, β, η, and μ. The case in which $\eta_q = 0$ for each question q is called *fair*. This case deserves special attention in the theory since, in many practical applications, the questions may be designed so as to make lucky guesses impossible or negligibly rare. If in addition $\beta_q = 0$ for each q, then this case is *straight*.

The knowledge states K for which $\pi(K) > 0$ will be called the *true* states; they form the *support* of π, denoted by $supp(\pi)$. Our tentative view is that in practice the support will only contain a small number of states, which moreover are 'close to each other' in a sense made precise in the next section. If the support contains only one state, this state is the *unit support* of π.

As indicated in Axioms [QM] and [M], the functions τ and μ are referred to as the questioning function and the marking function, respectively. Special cases of the process defined by the four general axioms will arise from particularizing the questioning function and the marking function. In general, as indicated by a cursory examination of these axioms, the process $(\mathbf{R}_n, \mathbf{Q}_n, \mathbf{K}_n, \mathbf{M}_n)$ is a Markov chain. The same remark holds for various other subprocesses, such as $(\mathbf{M}_n)$ and $(\mathbf{Q}_n, \mathbf{M}_n)$. Note in passing the implicit assumption that the subject's state distribution is not affected by the questioning procedure. The Markov chain $(\mathbf{M}_n)$ will be referred to as the *marking process* and is of central interest. Its behavior is affected by several sources of errors (or randomness), in particular the error probabilities β_q, the guessing probabilities η_q, and the subject's distribution π on the family of states.

14.3 Combinatorial Assumptions on the Structure

To implement the concepts introduced in Example 14.1.1, some combinatorial machinery is required which extends the tools introduced in Definition 4.1.6. We consider a finite, discriminative knowledge structure $(Q, \mathcal{K})$. As before, we measure the distance between two states K and L in $\mathcal{K}$ by their symmetric difference distance $d(K, L) = |K \bigtriangleup L|$, which counts the number of items by which K and L differ. We now generalize the notion of neighborhood of a state (Definition 4.1.6) to that of neighborhood of a collection of states.

14.3.1 Definition. The *ε-neighborhood* of any subcollection Ψ of $\mathcal{K}$ is defined by the equation

$$\mathcal{N}(\Psi, \varepsilon) = \{K' \in \mathcal{K} \,|\, d(K, K') \le \varepsilon, \text{ for some } K \text{ in } \Psi\}.$$

The states in $\mathcal{N}(\Psi, \epsilon)$ are called *ε-neighbors* of Ψ. Those ε-neighbors containing item q are *(q, ε)-neighbors* of Ψ and similarly those that do not contain q are *$(\bar{q}, \varepsilon)$-neighbors*. We define

$$\mathcal{N}_q(\Psi, \varepsilon) = \mathcal{N}(\Psi, \varepsilon) \cap \mathcal{K}_q \qquad \text{and} \qquad \mathcal{N}_{\bar{q}}(\Psi, \varepsilon) = \mathcal{N}(\Psi, \varepsilon) \cap \mathcal{K}_{\bar{q}},$$

calling these sets respectively the (q,ε)-*neighborhood* and $(\bar{q},\varepsilon)$-*neighborhood* of Ψ. When $\Psi = \{K\}$, we abbreviate $\mathcal{N}(\{K\},\varepsilon)$ into $\mathcal{N}(K,\varepsilon)$. In the same vein, we also write in shorthand $\mathcal{N}_q(K,\varepsilon)$ and $\mathcal{N}_{\bar{q}}(K,\varepsilon)$.

To exercise these concepts, here are a few straightforward facts whose proofs we leave to the reader as Problem 4. For $y = q$ or $y = \bar{q}$, we always have $\mathcal{N}_y(\Psi,0) = \Psi_y \subseteq \Psi$. The last inclusion is strict except in two cases: (i) $y = q \in \cap\Psi$; or (ii) $y = \bar{q}$ and $q \notin \cup\Psi$. Also, $\mathcal{N}_y(\Psi,0)$ is empty if $y = q \notin \cup\Psi$, or $y = \bar{q}$ and $q \in \cap\Psi$.

These neighborhood concepts will be used to specify the questioning function and the marking function along the lines introduced in Example 14.1.1. Let $(\mathbf{R}_n, \mathbf{Q}_n, \mathbf{K}_n, \mathbf{M}_n)$ be a stochastic assessment process parametrized by π, τ, β, η, and μ. We impose that the process selects the question q asked on trial n on the basis of the set $\mathbf{M}_n$ of marked states. To this aim, the process builds the ε-neighborhood $\mathcal{N}(\mathbf{M}_n,\varepsilon)$, where ε may depend on the size of $\mathbf{M}_n$. Loosely speaking, q is chosen so as to split this ε-neighborhood into two subsets $\mathcal{N}_q(\mathbf{M}_n,\varepsilon)$ and $\mathcal{N}_{\bar{q}}(\mathbf{M}_n,\varepsilon)$ as equal in size as feasible. We base the choice of the question on $\mathcal{N}(\mathbf{M}_n,\varepsilon)$ rather than on the potentially smaller set $\mathbf{M}_n$ on account of the possibility of errors committed by the assessment procedure on earlier steps. These ideas will be implemented in a particular form of the questioning function τ of the process (cf. Axiom [QM]).

14.3.2 Definition. We now make $\varepsilon : \mathbb{N} \cup \{0\} \to \mathbb{R}^+ \cup \{0\}$ a function with values varying with the size of the set of marked states. Denote by $\mathbf{T}_n$ the set of all items q in Q which, on trial n, minimize the quantity

$$\nu_q(\mathbf{M}_n, k) = \Big| |\mathcal{N}_q(\mathbf{M}_n, \varepsilon(|\mathbf{M}_n|))| - |\mathcal{N}_{\bar{q}}(\mathbf{M}_n, \varepsilon(|\mathbf{M}_n|))| \Big|.$$

(As before, we simplify the writing and take $\nu_q(K,j)$ to mean $\nu_q(\{K\},j)$.) We say that the questioning function τ in Axiom [QM] is ε-*half-split* iff

$$\tau(q, \mathbf{M}_n) = \frac{\iota_{\mathbf{T}_n}(q)}{|\mathbf{T}_n|},$$

where ι_A is the indicator function of the set A. Note in passing that if $|\mathbf{M}_n| = 1$ and $\varepsilon(1) = 0$, or if $\mathbf{M}_n = \varnothing$, then $\mathbf{T}_n = Q$; in these cases, all the questions in Q have the same probability of being chosen.

The assumption $\varepsilon(1) = 1$ will be central in the sequel, together with the closely related concept of fringe $K^{\mathcal{F}}$ of a state K. This concept was defined in 4.1.6 as the union $K^{\mathcal{F}} = K^{\mathcal{I}} \cup K^{\mathcal{O}}$, where $K^{\mathcal{I}}$ and $K^{\mathcal{O}}$ are respectively the inner and outer fringes of K, that is:

$$K^{\mathcal{I}} = \{q \in K \mid K \setminus \{q\} \in \mathcal{K}\}, \tag{14.2}$$
$$K^{\mathcal{O}} = \{q \in Q \setminus K \mid K \cup \{q\} \in \mathcal{K}\}. \tag{14.3}$$

14.3.3 Theorem. *For any state K, any $q \in K^{\mathcal{F}}$ and any $r \in Q \setminus K^{\mathcal{F}}$, we have:*

$$\nu_q(K,1) \;=\; |\mathcal{N}(K,1)| - 2 \;<\; \nu_r(K,1) \;=\; |\mathcal{N}(K,1)|. \tag{14.4}$$

Moreover, if $K^{\mathcal{F}} \neq \varnothing$ and the questioning function τ is ε-half-split with $\varepsilon(1) = 1$, then for any positive integer n we have

$$\mathbb{P}\big(\mathbf{Q}_n = q \,|\, \mathbf{K}_n, \mathbf{M}_n = \{K\}, \mathbf{W}_{n-1}\big) \;=\; \begin{cases} 1/|K^{\mathcal{F}}| & \text{if } q \in K^{\mathcal{F}}; \\ 0 & \text{otherwise,} \end{cases}$$

and so

$$\mathbb{P}\big(\mathbf{Q}_n \in K^{\mathcal{F}} \,|\, \mathbf{K}_n, \mathbf{M}_n = \{K\}, \mathbf{W}_{n-1}\big) \;=\; 1.$$

Proof. Since the second assertion follows readily from the first one, we only establish Equation (14.4). We first deal with $\nu_q(K,1)$. We consider two cases. Suppose $q \in K^{\mathcal{F}}$. If $q \in K$, then there is a state L such that $K = L \cup \{q\}$ and $\mathcal{N}_{\bar{q}}(K,1) = \{L\}$. This yields

$$\nu_q(K,1) = |\mathcal{N}(K,1)| - 2. \tag{14.5}$$

If $q \notin K$, then the same equality obtains with this time $L = K \cup \{q\}$ and $\mathcal{N}_q(K,1) = \{L\}$.

Turning to $\nu_r(K,1)$, suppose that $r \in Q \setminus K^{\mathcal{F}}$. We have also two cases: $r \notin \cup\mathcal{N}(K,1)$ or $r \in \cap\mathcal{N}(K,1)$. In the first case, we get

$$\mathcal{N}_{\bar{r}}(K,1) = \mathcal{N}(K,1), \qquad \mathcal{N}_r(K,1) = \varnothing$$

and

$$\nu_r(K,1) = |\mathcal{N}(K,1)|. \tag{14.6}$$

The second case yields

$$\mathcal{N}_r(K,1) = \mathcal{N}(K,1), \qquad \mathcal{N}_{\bar{r}}(K,1) = \varnothing, \tag{14.7}$$

yielding again Equation (14.6). The first assertion of the statement follows from Equations (14.5) and (14.6). □

We now turn to the marking function. The idea here is that suggested by Example 14.1.1, namely, we retain as marked states on trial $n+1$ only those states in a δ-neighborhood of $\mathbf{M}_n$ which are consistent with the question asked and the response observed. Thus, with probability one,

$$\mathbf{M}_{n+1} = \mathcal{N}_y(\mathbf{M}_n, \delta), \tag{14.8}$$

where

$$y = \begin{cases} q & \text{if the answer to } q \text{ is correct} \\ \bar{q} & \text{otherwise.} \end{cases} \tag{14.9}$$

The value of the parameter δ in Equation (14.8) may vary with the size of $\mathbf{M}_n$. It may also depend upon whether the question $\mathbf{Q}_n$ belongs to at least one of the marked states in $\mathbf{M}_n$ and whether the response was correct. This generates the four cases in the next definition. The motivation for enlarging the family of marked states is the same as before: the true state could have escaped the actual collection of marked states, and we would like the procedure to correct this omission.

14.3.4 Definition. Let δ_1, δ_2, $\bar{\delta}_1$, and $\bar{\delta}_2$ be four functions defined on the nonnegative integers, with nonnegative real values. The marking function μ of Axiom [M] is *selective with parameter* $\delta = (\delta_1, \delta_2, \bar{\delta}_1, \bar{\delta}_2)$ if it satisfies:

$$\mu(\mathbf{M}_{n+1}, \mathbf{R}_n, \mathbf{Q}_n, \mathbf{M}_n) = \begin{cases} 1 & \text{in the four cases (i)–(iv) below;} \\ 0 & \text{in all other cases.} \end{cases}$$

$$\begin{aligned}
&\text{(i) } \mathbf{R}_n = 1,\ \mathbf{Q}_n = q \in \bigcup \mathbf{M}_n, \ \text{ and } \ \mathbf{M}_{n+1} = N_q(\mathbf{M}_n, \delta_1(|\mathbf{M}_n|));\\
&\text{(ii) } \mathbf{R}_n = 1,\ \mathbf{Q}_n = q \notin \bigcup \mathbf{M}_n, \ \text{ and } \ \mathbf{M}_{n+1} = N_q(\mathbf{M}_n, \delta_2(|\mathbf{M}_n|));\\
&\text{(iii) } \mathbf{R}_n = 0,\ \mathbf{Q}_n = q \in \bigcup \mathbf{M}_n, \ \text{ and } \ \mathbf{M}_{n+1} = N_{\bar{q}}(\mathbf{M}_n, \bar{\delta}_1(|\mathbf{M}_n|));\\
&\text{(iv) } \mathbf{R}_n = 0,\ \mathbf{Q}_n = q \notin \bigcup \mathbf{M}_n, \ \text{ and } \ \mathbf{M}_{n+1} = N_{\bar{q}}(\mathbf{M}_n, \bar{\delta}_2(|\mathbf{M}_n|)).
\end{aligned}$$

Note that this requirement could be generalized by letting the functions δ_1, δ_2, $\bar{\delta}_1$, and $\bar{\delta}_2$ depend on the question q in Q.

We now apply Definition 14.3.4 in the case in which some state K_0 is the unit support and $\mathbf{M}_n = \{K\}$, for some K, $K_0 \in \mathcal{K}$.

14.3.5 Theorem. *Suppose that the marking rule μ is selective with parameter $\delta = (\delta_1, \delta_2, \bar{\delta}_1, \bar{\delta}_2)$, and moreover*

$$\delta_1(1) = \bar{\delta}_2(1) = 0, \qquad \delta_2(1) = \bar{\delta}_1(1) = 1. \tag{14.10}$$

Writing $A_n(K, K_0)$ for the joint event $(\mathbf{M}_n = \{K\}, \mathbf{K}_n = K_0)$, we have then

$$\begin{aligned}
\mathbb{P}(\mathbf{M}_{n+1} = \{K\} \,|\, \mathbf{Q}_n \in K^{\mathcal{I}} \cap K_0, A_n(K, K_0)) &= 1 - \beta_{\mathbf{Q}_n};\\
\mathbb{P}(\mathbf{M}_{n+1} = \{K \setminus \mathbf{Q}_n\} \,|\, \mathbf{Q}_n \in K^{\mathcal{I}} \cap K_0,\ A_n(K, K_0)) &= \beta_{\mathbf{Q}_n};\\
\mathbb{P}(\mathbf{M}_{n+1} = \{K\} \,|\, \mathbf{Q}_n \in K^{\mathcal{I}} \setminus K_0,\ A_n(K, K_0)) &= \eta_{\mathbf{Q}_n};\\
\mathbb{P}(\mathbf{M}_{n+1} = \{K \setminus \mathbf{Q}_n\} \,|\, \mathbf{Q}_n \in K^{\mathcal{I}} \setminus K_0,\ A_n(K, K_0)) &= 1 - \eta_{\mathbf{Q}_n};\\
\mathbb{P}(\mathbf{M}_{n+1} = \{K \cup \mathbf{Q}_n\} \,|\, \mathbf{Q}_n \in K^{\mathcal{O}} \cap K_0,\ A_n(K, K_0)) &= 1 - \beta_{\mathbf{Q}_n};\\
\mathbb{P}(\mathbf{M}_{n+1} = \{K\} \,|\, \mathbf{Q}_n \in K^{\mathcal{O}} \cap K_0,\ A_n(K, K_0)) &= \beta_{\mathbf{Q}_n};\\
\mathbb{P}(\mathbf{M}_{n+1} = \{K \cup \mathbf{Q}_n\} \,|\, \mathbf{Q}_n \in K^{\mathcal{O}} \setminus K_0,\ A_n(K, K_0)) &= \eta_{\mathbf{Q}_n};\\
\mathbb{P}(\mathbf{M}_{n+1} = \{K\} \,|\, \mathbf{Q}_n \in K^{\mathcal{O}} \setminus K_0,\ A_n(K, K_0)) &= 1 - \eta_{\mathbf{Q}_n}.
\end{aligned}$$

The proof is left as Problem 5. Note also that if $K^{\mathcal{F}} \neq \varnothing$ and the questioning function is ε-halfsplit with $\varepsilon(1) = 1$, Theorems 14.3.3 and 14.3.5 allow us to calculate all the possible transition probabilities from $\mathbf{M}_n = \{K\}$ to $\mathbf{M}_{n+1} = \{K'\}$.

14.4 Markov Chains Terminology

Our results concern Markov chains. To avoid ambiguities, we will use the term *m-state* to refer to the Markov states of these chains, reserving the expression '(knowledge) states' for the elements of $\mathcal{K}$. For Markov chain concepts, we point the reader to Kemeny and Snell (1960), Chung (1967), Feller (1968), Parzen (1994) or Barucha-Reid (1997). Except when otherwise indicated, we follow the somewhat idiosyncratic terminology of Kemeny and Snell (for example, we say 'ergodic' rather than 'recurrent' or 'persistent').

Here is a brief glossary of the terminology, recalling some concepts encountered in Chapter 11.

14.4.1 Definition. Let $(\mathbf{X}_n)_{n\in\mathbb{N}}$ be a Markov chain on a finite set E of m-states, with *transition probability matrix* $M = (M_{ij})_{i,j\in E}$ and *initial probability distribution* $v = (v_i)_{i\in E}$; thus

$$\begin{aligned} v_i &= \mathbb{P}(\mathbf{X}_1 = i), && i \in E; \\ M_{ij} &= \mathbb{P}\big(\mathbf{X}_{n+1} = j \,|\, \mathbf{X}_n = i\big), && n = 1, 2, \ldots \end{aligned}$$

An m-state j is *reachable* from an m-state i when there is a natural number n such that $(M^n)_{ij} > 0$ (note that j is not necessarily reachable from itself). A subset C of E is a *closed set* of m-states if any m-state outside C cannot be reached from any m-state in C. An m-state is *absorbing* when it is the single element of a closed set. A *class (in a Markov chain)* is a subset C of E which is maximal for the property that, for any two m-states i, j in C, j is reachable from i; in particular, j is reachable from itself.

Because of the finiteness of E, we may define an *ergodic m-state* as an m-state that belongs to some closed class. A closed class is sometimes called an *ergodic set*. An m-state which is not ergodic is *transient*.

The chain $(\mathbf{X}_n)$ is *regular* when all its m-states form a single class and $(\mathbf{X}_n)$ is not *periodic*, the latter requirement meaning that $(M^n)_{ij} > 0$ for some $n \in \mathbb{N}$ and all $i, j \in E$. Finally, a probability distribution p on E is *stationary* (or *invariant*) when $\sum_{i\in E} p_i M_{ij} = p_j$.

It is well-known that if the chain $(\mathbf{X}_n)$ is regular, then it has a unique stationary distribution p which is called the *limit* or *asymptotic distribution*, with

$$p_j = \lim_{n\to\infty} \mathbb{P}(\mathbf{X}_n = j) = \lim_{n\to\infty} \sum_{i\in E} v_i (M^n)_{ij}.$$

The limit distribution p does not depend on the initial probability distribution v on E.

14.5 Results for the Fair Case

As before, let $\mathcal{K}$ be a knowledge structure with finite domain Q. We suppose that $(\mathbf{R}_n, \mathbf{Q}_n, \mathbf{K}_n, \mathbf{M}_n)$ is a stochastic assessment procedure parametrized by π, τ, β, η and μ. We begin the investigation of its behavior with a simple general result for the straight case (that is, both the error probability β_q and the guessing probability η_q are equal to zero for all questions q in Q).

14.5.1 Theorem. *In the straight case, assume that there is only one true state K_0 and that the marking rule is selective. Then, for all $n \in \mathbb{N}$, we have:*

$$\mathbb{P}(K_0 \in \mathbf{M}_{n+1} \,|\, K_0 \in \mathbf{M}_n) \;=\; 1. \tag{14.11}$$

In other words, the set of all m-states containing K_0 is a closed set of the Markov chain $(\mathbf{M}_n)$.

To establish the result, it suffices to prove Equation 14.11, which is left as Problem 6. We consider next a situation in which the parameters ε and δ are chosen so as to narrow down quickly (during Phase 1, as we considered in Example 14.1.1) the set of marked states.

14.5.2 Theorem. *Suppose that the questioning rule is ε-half-split, that the marking rule is selective with parameter δ, and that $\varepsilon(j) = \delta_k(j) = \bar{\delta}_k(j) = 0$ for $k = 1$, 2, and all integers $j > 1$. We have then*

$$\mathbb{P}\big(|\mathbf{M}_{n+1}| < |\mathbf{M}_n| \,\big|\, |\mathbf{M}_n| > 1\big) \;=\; 1. \tag{14.12}$$

If moreover $\delta_1(1) = 0$, $\delta_2(1) \leq 1$, $\bar{\delta}_1(1) \leq 1$, and $\bar{\delta}_2(1) = 0$, then for some natural number r we have for all $n \geq r$

$$\mathbb{P}(|\mathbf{M}_n| \leq 1) \;=\; 1. \tag{14.13}$$

In particular, in the straight case, if $\delta_2(1) = \bar{\delta}_1(1) = 0$ and K_0 is the unit support, then there is a positive integer r such that whenever $n \geq r$,

$$\mathbb{P}(\mathbf{M}_n = \{K_0\} \,|\, K_0 \in \mathbf{M}_1) \;=\; 1, \tag{14.14}$$

and

$$\lim_{n\to\infty} \mathbb{P}(\mathbf{M}_n = \varnothing) \;=\; 1 \quad\iff\quad K_0 \notin \mathbf{M}_1. \tag{14.15}$$

In fact, the Markov chain $(\mathbf{M}_n)$ has exactly two absorbing m-states which are $\{K_0\}$ and $\varnothing$.

Proof. Equation (14.12) results immediately from the axioms and the hypotheses. Since $\mathbf{M}_n$ is finite for every positive integer n, Equation (14.13) follows. Applying Theorem 14.5.1 gives (14.14), from which (14.15) follows easily. □

The assumption $\delta_k(1) = \bar{\delta}_k(1) = 0$ practically locks the set of marked states as soon as not more than one such state remains. Thus, the unit support is either found quickly if it belongs to $\mathbf{M}_1$, or missed otherwise. We now study a more flexible approach that allows in Phase 2 a single marked state K to evolve in the structure, thus making possible a gradual construction (in the straight case) or approximation (in the fair case) of the unit support by K. For convenience, we give below a label to this set of conditions. Notice the requirement $\varepsilon(1) = 1$ which was motivated in Example 14.1.1 and allows only small changes of the single marked state K.

14.5.3 Definition. Let $(\mathbf{R}_n, \mathbf{Q}_n, \mathbf{K}_n, \mathbf{M}_n)$ be a stochastic assessment procedure parametrized by π, τ, β, η and μ, with an ε-half-split questioning function and a marking function which is selective with parameter δ. Suppose that the following conditions are satisfied:

(i) the knowledge structure $(Q, \mathcal{K})$ is a well-graded space;
(ii) $\varepsilon(1) = 1$ and $\varepsilon(n) = 0$ for $n > 1$;
(iii) $\delta = 0$ except in two cases: $\delta_2(1) = 1$, and $\bar{\delta}_1(1) = 1$.

Then $(\mathbf{R}_n, \mathbf{Q}_n, \mathbf{K}_n, \mathbf{M}_n)$ is called *unitary*.

14.5.4 Convention. In the rest of this section, we consider a fair, unitary stochastic assessment procedure $(\mathbf{R}_n, \mathbf{Q}_n, \mathbf{K}_n, \mathbf{M}_n)$ in the sense (and with the notation) of Definition 14.5.3. We assume that there is a unique support K_0; accordingly, the certain event $\mathbf{K}_n = K_0$ will not be mentioned in the statement of results.

Terms such as 'm-state' and 'ergodic set' refer to the Markov chain $\mathbf{M}_n$, which is our principal object of investigation. This chain thus satisfies Equations (14.12) and (14.13). As soon as some state K remains the single marked state (which is bound to happen in view of Theorem 14.5.2), question $\mathbf{Q}_n$ will be drawn in the fringe of K. This fringe $K^{\mathcal{F}}$ is nonempty because $\mathcal{K}$ is well-graded (cf. Theorem 4.1.7(iv)). We first give the possible transitions from $\mathbf{M}_n = \{K\}$, leaving the proof to the reader (see Problem 8).

14.5.5 Theorem. *There is a natural number n_0 such that for all $n \geq n_0$*

$$\mathbb{P}(|\mathbf{M}_n| = 1) = 1. \tag{14.16}$$

Moreover, for any natural number n, we have

$$\mathbb{P}(|\mathbf{M}_{n+1} = 1| \,\big|\, |\mathbf{M}_n| = 1) = 1.$$

More precisely, with

$$A = |K^{\mathcal{O}} \setminus K_0| + \sum_{q \in K^{\mathcal{O}} \cap K_0} \beta_q + \sum_{q \in K^{\mathcal{I}} \cap K_0} (1 - \beta_q),$$

we get

$$\mathbb{P}(\mathbf{M}_{n+1} = \{K'\} \,|\, \mathbf{M}_n = \{K\}) =$$
$$\begin{cases} (1/|K^{\mathcal{F}}|)(1-\beta_q) & \text{if } K' = K \cup \{q\}, \text{ with } q \in K^{\mathcal{O}} \cap K_0, \\ (1/|K^{\mathcal{F}}|) & \text{if } K' = K \setminus \{q\}, \text{ with } q \in K^{\mathcal{I}} \setminus K_0, \\ (1/|K^{\mathcal{F}}|)\beta_q & \text{if } K' = K \setminus \{q\}, \text{ with } q \in K^{\mathcal{I}} \cap K_0, \\ (1/|K^{\mathcal{F}}|)A & \text{if } K = K', \\ 0 & \text{in all other cases.} \end{cases}$$

In particular, we thus have for any $K \not\subseteq K_0$,

$$\mathbb{P}(\mathbf{M}_{n+1} = \{K\} \,|\, \mathbf{M}_n = \{K_0\}) = 0.$$

In the straight case, we have $\beta_{\mathbf{Q}_n} = 0$ which implies that with probability one we have $d(\mathbf{M}_{n+1}, K_0) \leq d(\mathbf{M}_n, K_0)$ when $|\mathbf{M}_n| = 1$. In general, if Ψ is a nonempty family of subsets of Q and K a subset of Q, we define as customary

$$d(\Psi, K) = \min\{d(K', K) \,|\, K' \in \Psi\}.$$

14.5.6 Theorem. *In the straight case, for any choice of a nonempty $\mathbf{M}_1$, we have*

$$\lim_{n\to\infty} \mathbb{P}(\mathbf{M}_n = \{K_0\}) = 1. \tag{14.17}$$

Moreover, for any state K with $d(K, K_0) = j > 0$ and any $\ell, n \in \mathbb{N}$, we have

$$\mathbb{P}(\mathbf{M}_{\ell+n} = \{K_0\} \,|\, \mathbf{M}_n = \{K\}) \geq \sum_{k=0}^{\ell-j} \binom{j+k-1}{k} \lambda^j (1-\lambda)^k \tag{14.18}$$

where λ is defined by

$$\lambda = \min \frac{|(K \triangle K') \cap K^{\mathcal{F}}|}{|K^{\mathcal{F}}|},$$

for $K, K' \in \mathcal{K}$ with $K \neq K'$.

Proof. We first prove Equation (14.18). To this end, we consider a sequence of Bernoulli trials with 'success' meaning a step of size one towards K_0. Thus, a success on trial n means that $d(\mathbf{M}_{n+1}, K_0) = d(\mathbf{M}_n, K_0) - 1$. If the event $\mathbf{M}_n = \{K\}$ is realized, and $\mathbf{K}_n = K_0$ with $d(K_0, K) = j$, then at least j successes are necessary in the following ℓ trials to achieve $\mathbf{M}_{\ell+n} = \{K_0\}$, with the probability of each success being at least λ. The wellgradedness implies that $\lambda > 0$. Now, in a sequence of Bernoulli trials each having a probability of success equal to λ, the number $j+k$ of trials required to achieve exactly j successes has a probability specified by

$$\binom{j+k-1}{k} \lambda^j (1-\lambda)^k.$$

Thus, k is a value of a random variable, which is distributed as a negative binomial with parameters j and λ. Consequently, Equation (14.18) follows from Equation (14.16). Since the r.h.s. of (14.18) tends to 1 as ℓ tends to ∞, Theorem 14.5.5 implies Equation (14.17). □

We also have the following result:

14.5.7 Theorem. *There exists an integer $n_0 > 0$ such that $\mathbb{P}(|\mathbf{M}_n| = 1) = 1$ for any $n \geq n_0$. Moreover, the Markov chain $(\mathbf{M}_n)$ has a unique ergodic set E_0 which contains $\{K_0\}$ and possibly some m-states $\{K\}$ such that $K \subseteq K_0$, but no other m-states. If, in addition $\beta_q > 0$ for all $q \in K_0$, then E_0 is in fact the family of all those m-states $\{K\}$ such that $K \subseteq K_0$.*

PROOF. By Equation (14.16) of Theorem 14.5.5, an ergodic m-state contains exactly one knowledge state. Using the transition probabilities described in Theorem 14.5.5 and the wellgradedness of $\mathcal{K}$, we see that it is possible to reach the m-state $\{K_0\}$ from any m-state $\{K\}$. This clearly implies the uniqueness of the ergodic set E_0, with moreover $\{K_0\} \in E_0$. The remaining assertions also follow readily from Theorem 14.5.5. □

14.6 Uncovering a Stochastic State: Examples

Most of the results presented in the previous section suppose a unit support. The case in which the probability distribution π on the family $\mathcal{K}$ is not concentrated on a single state is also worth considering.

In an ideal situation, the Markov chain $\mathbf{M}_n$ admits a unique ergodic set ξ that contains $supp(\pi)$, the support of the probability distribution π. A sensible strategy is then to analyze the statistics of occupation times of the m-states in ξ in order to assess the probability $\pi(K)$ of each true state K. (In practice, because we cannot ask many questions, we aim only at ballpark estimates of these probabilities.) This strategy will be illustrated in two examples. The first one is based on a discriminative chain $\mathcal{K}$ of states. Its items are thus linearly ordered[1].

In all the examples of this section, we consider a fair, unitary stochastic assessment procedure (cf. Definitions 14.2.2 and 14.5.3).

14.6.1 Example. Suppose that $\mathcal{K}$ is a chain of states

$$L_0 = \varnothing,\ L_1 = \{q_1\},\ \ldots,\ L_m = \{q_1, \ldots, q_m\}.$$

By Equation (14.13) of Theorem 14.5.2, all the m-states of the Markov chain $(\mathbf{M}_n)$ containing more than one knowledge state are transient. Any m-state containing a single true state is ergodic. (All assertions left unproved in this

[1] Needless to say, this special case could have been treated by other methods (such as "tailored testing"; see Lord, 1974; Weiss, 1983).

example are dealt with by Problem 9.) The same holds for any m-state $\{K\}$ such that $K' \subseteq K \subseteq K''$ for two true states K' and K''. If $\beta_q > 0$ for each q in $\cap\, supp(\pi)$, then $\{K\}$ is an ergodic m-state for any state K included in a true state. On the other hand, if $\beta_q = 0$ for all $q \in \cap\, supp(\pi)$, then the only ergodic m-states are of the form $\{K\}$ with K being a state between two true states. Thus, in the straight case, if the true states form a subchain $L_j, L_{j+1}, \ldots, L_k$ of $\mathcal{K}$ for some j, k with $0 \leq j \leq k \leq m$, the ergodic m-states are essentially the true states. As indicated by the example below, the statistics of the occupation times of the states in the recurrent class can be used to estimate the probabilities $\pi(K)$. Suppose that we have exactly three true states L_{i-1}, L_i, and L_{i+1}, with $1 < i < m$. Thus, setting

$$\pi_{i-1} = \pi(L_{i-1}) \qquad \text{and} \qquad \pi_{i+1} = \pi(L_{i+1}),$$

we assume $\pi(L_i) = 1 - \pi_{i-1} - \pi_{i+1} > 0$, and $\pi_{i-1} > 0$, $\pi_{i+1} > 0$. Suppose that we also have $\beta_q = 0$ for any $q \in Q$. (We are thus in the straight case.) From these assumptions, it follows that there are three ergodic m-states, namely $\{L_{i-1}\}$, $\{L_i\}$, and $\{L_{i+1}\}$. The probabilities for the observable variables on trial n are given by the tree-diagram of Figure 14.2 for the transient m-states $\{L_j\}$, for $0 < j < i - 1$, and $\{L_k\}$, for $i + 1 < k < m$ (we leave the cases of $\{\varnothing\}$ and $\{Q\}$ to the reader). A similar tree-diagram for the ergodic m-states is provided in Figure 14.3. The possible transitions between all m-states of the form $\{K\}$, with their probabilities, are shown in Figure 14.4.

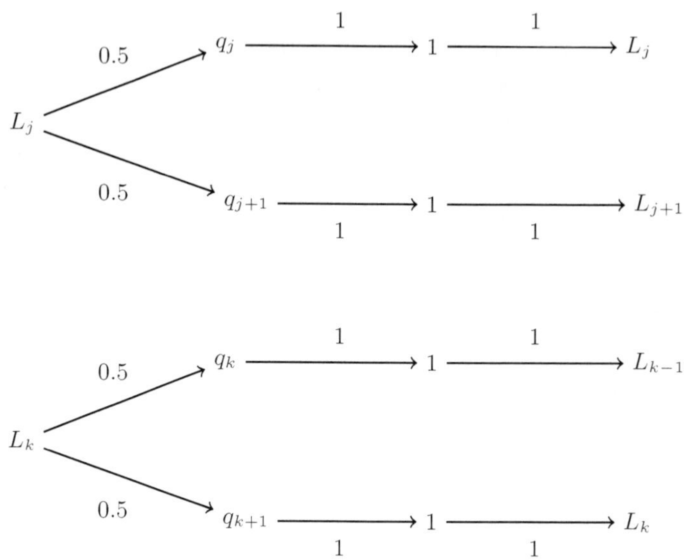

Figure 14.2. Probabilities of the observable variables on trial n in Example 14.6.1, when leaving a transient state $\{L_j\}$ or $\{L_k\}$, for $0 < j < i - 1$ or $i + 1 < k < m$.

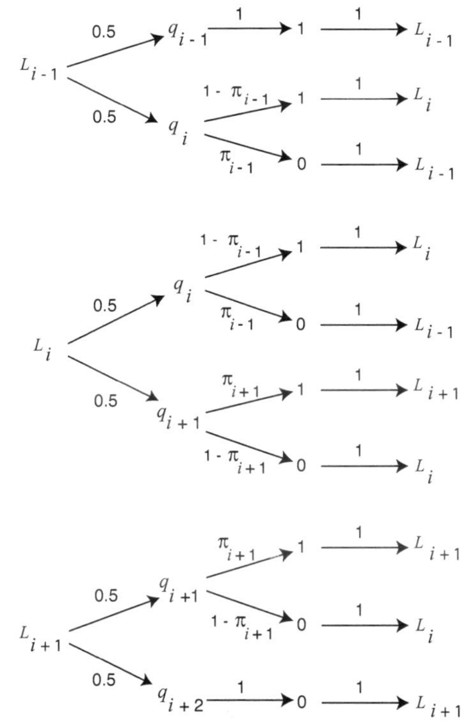

Figure 14.3. Probabilities for the observable variables at trial n in Example 14.6.1, when leaving an ergodic state $\{L_{i-1}\}$, $\{L_i\}$, or $\{L_{i+1}\}$.

Asymptotically, we have a Markov chain on the three m-states L_{i-1}, L_i, and L_{i+1}, which is regular. Setting

$$p_j = \lim_{n\to\infty} \mathbb{P}(\mathbf{M}_n = \{L_j\}),$$

we thus have $p_j \neq 0$ iff $i-1 \leq j \leq i+1$. The stationary distribution of this chain on three states is the unique solution to the following system of linear equations:

$$(p_{i-1}, p_i, p_{i+1}) \cdot \begin{pmatrix} \frac{1+\pi_{i-1}}{2} & \frac{1-\pi_{i-1}}{2} & 0 & 1 \\ \frac{\pi_{i-1}}{2} & \frac{2-\pi_{i-1}-\pi_{i+1}}{2} & \frac{\pi_{i+1}}{2} & 1 \\ 0 & \frac{1-\pi_{i+1}}{2} & \frac{1+\pi_{i+1}}{2} & 1 \end{pmatrix} = (p_{i-1}, p_i, p_{i+1}, 1).$$

As the first equation gives

$$\frac{1+\pi_{i-1}}{2} p_{i-1} + \frac{\pi_{i-1}}{2} p_i = p_{i-1},$$

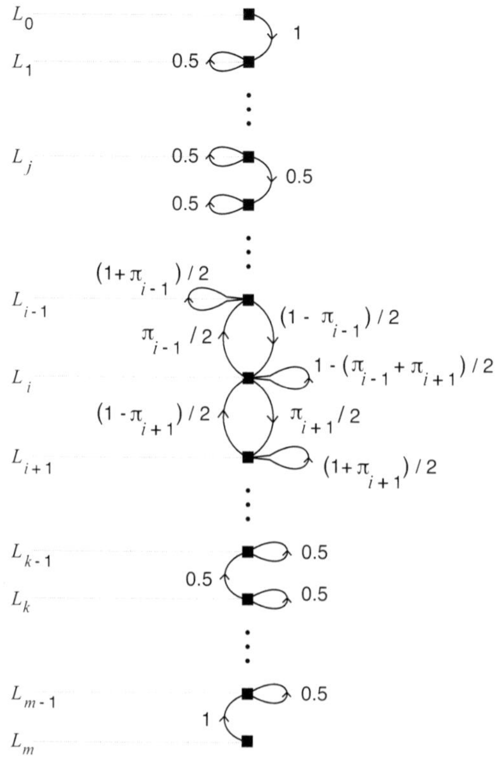

Figure 14.4. Transitions with their probabilities between the m-states of the form $\{L_j\}$, see Example 14.6.1.

we infer

$$\pi_{i-1} = \frac{p_{i-1}}{p_{i-1} + p_i}.$$

A similar computation with the third equation yields

$$\pi_{i+1} = \frac{p_{i+1}}{p_i + p_{i+1}}.$$

To be complete, we also give

$$\pi(L_i) = \frac{p_i^2 - p_{i-1}p_{i+1}}{p_i + p_{i-1}p_{i+1}}.$$

Since in this case the asymptotic probabilities of the knowledge states can be estimated from the proportions of visits to the m-states in any realization, we have the possibility of obtaining at least a rough estimate of the unknown probabilities $\pi(L_j)$ from the data of any particular subject.

Our next example shows that these techniques are not restricted to the case of a chain of states.

14.6.2 Example. Let $Q = \{a, b, c, d, e\}$ be a set of five items, with a knowledge space $\mathcal{K}$ derived from the partial order in Figure 14.5 (a). We thus have

$$\mathcal{K} = \{\varnothing, \{c\}, \{e\}, \{b, c\}, \{c, e\}, \{a, b, c\}, \{b, c, e\}, \{c, d, e\}, \{a, b, c, e\}, \{b, c, d, e\}, Q\}.$$

Suppose that the straight case holds, and that $\{c\}$ and $\{e\}$ are the only true states. Setting $\alpha = \pi(e)$, we thus have $\pi(c) = 1 - \alpha$. As we assume that the assessment procedure is straight, we conclude that there is only one ergodic set, namely $\{\{\varnothing\}, \{\{c\}\}, \{\{e\}\}, \{\{c, e\}\}\}$.

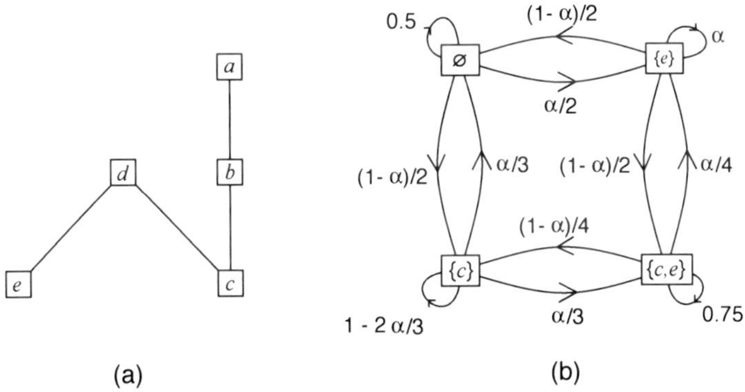

Figure 14.5. (a) The Hasse diagram of the partial order in Example 14.6.2. (b) Transition probabilities among the four ergodic m-states in that Example. (One layer of brackets is omitted, i.e. $\{a\}$ stands for $\{\{a\}\}$.)

Figure 14.5 (b) provides the transition probabilities among the four ergodic m-states. Setting, for $\{K\} \in \{\{\varnothing\}, \{\{c\}\}, \{\{e\}\}, \{\{c, e\}\}\}$,

$$p(K) = \lim_{n\to\infty} \mathbb{P}(\mathbf{M}_n = \{K\}),$$

we obtain for example

$$p(\varnothing) = \frac{1}{2}p(\varnothing) + \frac{1}{3}\alpha p(\{c\}) + \frac{1}{2}(1-\alpha)p(\{e\}).$$

Solving for α, we get

$$\alpha = \frac{p(\varnothing) - p(\{e\})}{(2/3)p(\{c\}) - p(\{e\})}.$$

Here again, the unknown quantity α can be estimated from the asymptotic probabilities of the ergodic m-states of $(\mathbf{M}_n)$, via the proportions of visits of the m-states in any realization of the procedure. In the next section, we consider some circumstances in which such an estimation is theoretically feasible.

We now turn to an example in which the knowledge structure is not described accurately at the start. For instance, the student's state could have been omitted.

14.6.3 Example. With the same structure as in Example 14.6.2, we suppose that the subject has only mastered items c and d. Thus, the 'knowledge state' of that student is $\{c, d\}$ which is not a state of $\mathcal{K}$. When running a fair, unitary procedure to assess the knowledge of this subject, we assume that the response is correct if questions c or d were asked, and incorrect otherwise. From an analysis of this case, it turns out that $(\mathbf{M}_n)$ is a Markov chain; Figure 14.6 displays its reachability relation, together with the transition probabilities. There are two absorbing m-states, namely $\{\{c\}\}$ and $\{\{c, d, e\}\}$, which are marked by the thick red frames in the figure. Thus, depending on the starting point, the chain will end remaining in one of these two m-states with probability one. As a conclusion, we see that observing one realization of the process may lead the observer to diagnose (unavoidably) an incorrect state, but one that is nevertheless not too far from the student's actual state: we have

$$d(\{c\}, \{c, d\}) = d(\{c, d, e\}, \{c, d\}) = 1.$$

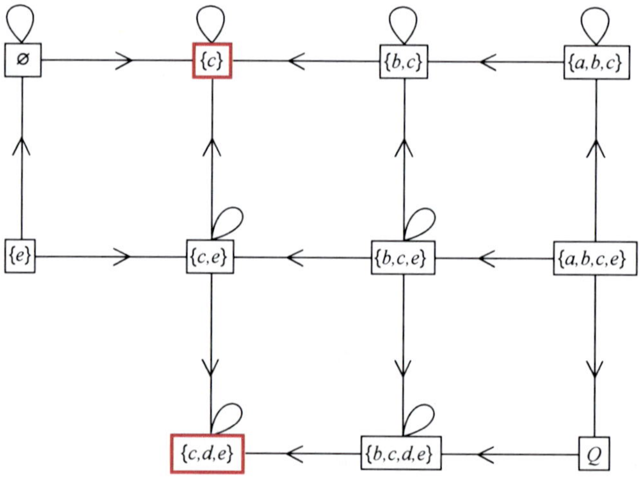

Figure 14.6. Possible transitions among the m-states from Example 14.6.3. (One layer of brackets is omitted.)

14.7 Intractable Cases

The results of this section were included for completeness. They are revealing without being very useful from the standpoint of the applications because they put too stringent conditions on the knowledge structure. We deal with the case in which the subject's knowledge state may be randomly varying in the course of the procedure according to a fixed probability distribution π on the knowledge structure $\mathcal{K}$ on a domain $Q = \{q_1, \ldots, q_k\}$. Thus, the distribution π is specific to that subject. An exemplary result is that, for the probability distribution π to be recoverable by a stochastic assessment process in the sense of this chapter, the number of nonempty states of $\mathcal{K}$ cannot exceed the number of items. This could happen if the knowledge structure $\mathcal{K}$ is a discriminative maximal chain

$$\varnothing \subset \{q_1\} \subset \{q_1, q_2\} \subset \ldots \subset \{q_1, \ldots, q_k\}.$$

Such a chain is the only possibility if we require the knowledge structure to be a learning space. (Problem 14 asks the reader to prove this fact.)

We suppose that the error probabilities β_q and guessing probabilities η_q are zero for all $q \in Q$ (straight case). The observable probabilities $\rho(q)$ of correct answers to questions q are then completely determined by the subject's distribution π through the formula:

$$\sum_{K \in \mathcal{K}_q} \pi(K) = \rho(q), \qquad \text{for } q \in Q \tag{14.19}$$

(where as before $\mathcal{K}_q = \{K \in \mathcal{K} \,|\, q \in K\}$). This is a system of linear equations in the unknown quantities $\pi(K)$, and with the constant terms $\rho(q)$ in the right hand sides. Notice that the coefficients of the unknowns take only the values 0 or 1.

14.7.1 Definition. The *incidence matrix* of a collection $\mathcal{K}$ of subsets of the finite domain Q of questions is the matrix $\mathbf{M} = (M_{q,K})$ whose rows are indexed by items q in Q, columns are indexed by states K in $\mathcal{K}$, and

$$M_{q,K} = \begin{cases} 1 & \text{if } q \in K, \\ 0 & \text{otherwise.} \end{cases}$$

Let $\rho = (\rho(q))_{q \in Q}$ be a vector of correct response probabilities. We ask: under which condition does there exists exactly one vector solution $\pi = (\pi(K))$ (with $K \in \mathcal{K}$) to Equation (14.19) that moreover satisfies $\pi(K) \geq 0$ and $\sum_{K \in \mathcal{K}} \pi(K) = 1$? In the next theorem, we suppose that some vector solution $\hat{\pi}$ to Equation (14.19) exists, with $0 < \hat{\pi}(K) < 1$ for all $K \in \mathcal{K}$. The theorem specifies the condition for the uniqueness of such a solution.

14.7.2 Theorem. *In the straight case, a strictly positive probability distribution $\hat{\pi}$ on the knowledge structure $\mathcal{K}$ can be recovered from the vector of correct response probabilities through Equation* (14.19) *if and only if the rank of the incidence matrix of the collection $\mathcal{K}^\bullet = \mathcal{K} \setminus \{\varnothing\}$ is equal to $|\mathcal{K}^\bullet|$.*

PROOF. Denote by Λ the simplex formed by all the probability distributions π on $\mathcal{K}$; thus Λ lies in the vector space $\mathbb{R}^{\mathcal{K}}$, where a vector has one real coordinate for each element in $\mathcal{K}$. The strictly positive probability distribution $\hat{\pi}$ mentioned in the statement of the theorem is a relative interior point of Λ.

Notice that the vector subspace S_0 consisting of all solutions π (without a positive condition) to the homogeneous system

$$\sum_{K \in \mathcal{K}_q} \pi(K) = 0, \qquad \text{for } q \in Q, \tag{14.20}$$

always contains the vertex of Λ corresponding to the distribution having all its mass concentrated on the empty set. Hence this subspace S_0, which contains also the origin of $\mathbb{R}^{\mathcal{K}}$, is never parallel to the hyperplane defined by $\sum_{K \in \mathcal{K}} \pi(K) = 1$.

The affine subspace S of all solutions to Equation (14.19) is the translate of S_0 passing through the point $\hat{\pi}$. As $\hat{\pi}$ is a relative interior point of Λ, we see that $\hat{\pi}$ is uniquely determined iff the dimension of S_0 is 1. This happens exactly if the incidence matrix of $\mathcal{K}$ has rank $|\mathcal{K}| - 1$, or equivalently the incidence matrix of $\mathcal{K}^\bullet$ has rank $|\mathcal{K}^\bullet|$. □

14.7.3 Remark. Theorem 14.7.2 covers the case in which the support $supp(\pi)$ equals $\mathcal{K}$. Dropping this condition but assuming a fixed support $supp(\pi)$ containing $\varnothing$, the same argument shows that we cannot recover any latent distribution π from its induced response probabilities ρ when the rank of the incidence matrix of $supp(\pi)$ is not equal to $|supp(\pi)| - 1$.

These considerations lead to the following problem: characterize those collections $\mathcal{K}$ of nonempty subsets of a finite domain Q for which the incidence matrix has rank over the reals equal to $|\mathcal{K}|$. As mentioned in the introductory paragraph of this section, an obvious necessary condition is $|Q| \geq |\mathcal{K}|$, that is: for any probability distribution on $\mathcal{K}$ to be recoverable by a stochastic assessment procedure, the number of nonempty states cannot exceed the number of questions. While many types of families satisfy this condition—see below—only one satisfies the axioms of a learning space, namely: a maximal discriminative chain (cf. Problem 14).

14.7.4 Theorem. *If $\mathcal{P}$ is a partial order on the finite domain Q, then the family*

$$\mathcal{I} = \{\{q \in Q \mid q\mathcal{P}r\} \mid r \in Q\}$$

of its principal ideals has an incidence matrix of rank $|Q|$.

PROOF. By a well-known result of Szpilrajn (1930), there exists a linear extension T of $\mathcal{P}$ (see also Trotter, 1992). List the columns of the incidence matrix in the order of T. Since the principal ideal generated by an item r in Q is the first element of $\mathcal{I}$ that contains r, the column indexed by r is linearly independent from the preceding columns. □

14.8 Original Sources and Related Works

This chapter closely follows the paper by Falmagne and Doignon (1988b). The combinatorial part of this paper was presented in Chapter 4.

Problems

1. Work out several other realizations of the assessment procedure in Example 14.1.1, by making various choices of question when possible.

2. By carefully selecting an exemplary knowledge structure different from those encountered in the chapter, explain why the wellgradedness assumption is crucial. Give examples in which the Markov chain $(\mathbf{M}_n)$ can reach an m-state of the form $\{K\}$, with K not a true state, and remain there with probability 1. Use your examples to find out which results in this chapter remain true for knowledge structures that are not well-graded.

3. Show that the various processes mentioned at the end of Definition 14.2.2, such as $(\mathbf{M}_n)$ and $(\mathbf{Q}_n, \mathbf{M}_n)$, are indeed Markov chains.

4. Prove the following assertions made on page 279 (just after Definition 14.3.1. For $y = q$ or $y = \bar{q}$, we always have $\mathcal{N}_y(\Psi, 0) = \Psi_y \subseteq \Psi$, the last inclusion being strict except in two cases: (i) $y = q \in \bigcap \Psi$; or (ii) $y = \bar{q}$ and $q \notin \bigcup \Psi$. Also, $\mathcal{N}_y(\Psi, 0)$ is empty if $y = q \notin \bigcup \Psi$, or $y = \bar{q}$ and $q \in \bigcap \Psi$.

5. Provide proofs for the various cases of Theorem 14.3.5.

6. Prove Equation (14.11).

7. Assume a unitary assessment procedure is ran on a well-graded knowledge structure. Is the following statement true or false: *If there are two true states incomparable for inclusion, then the Markov chain* $(\mathbf{M}_n)$ *has at least three ergodic m-states*? Prove your answer.

8. Give a proof of Theorem 14.5.5.

9. Establish all the assertions left unproved in Example 14.6.1.

10. Analyze the case of Example 14.6.2 in which the support is a different family of states, for example containing three states.

11. Analyze the case of Example 14.6.3 in which the student's knowledge is another subset of Q, also not belonging to $\mathcal{K}$.

12. Evaluate the computer storage needed to implement the stochastic assessment procedure from this chapter (assuming student's answers are collected at the keyboard). Compare the memory required with that needed by the procedure from Chapter 13.

13. Determine all the collections $\mathcal{K}$ of nonempty subsets on a two-element set whose incidence matrix has rank over the reals equal to 2. Try to find (up to isomorphism) all the similar collections on a three-element domain with rank 3.

14. In Section 14.7 (Theorem 14.7.2) we proved that if the subject's knowledge state varies randomly across trials according to a fixed distribution π on the knowledge structure $\mathcal{K}$, then π is recoverable by an stochastic assessment structure in the sense of this chapter only if the number of nonempty states does not exceed the number of items. Prove that the only type of learning spaces satisfying this condition is a chain.

15

Building a Knowledge Space

In the two preceding chapters, we have described assessment procedures for uncovering the knowledge state of a student in a scholarly topic. Such a knowledge state is one among possibly many states forming a knowledge structure for the topic. We now turn to the problem of building a knowledge structure in practice. In this chapter, we deal with the case of knowledge spaces, that is, union-closed knowledge structures. Methods for building learning spaces are considered in Chapter 16. They are based in part on the techniques of this chapter, which is sensible because learning spaces are particular knowledge spaces (cf. Theorem 2.2.4).

Our basic tool for building a knowledge space, the '`QUERY`' routine, is due to Koppen (1993) and Müller (1989) (cf. also Dowling, 1994). The `QUERY` routine proceeds by building first an entailment, in the sense of Definition 7.1.4. The input to the routine can be a collection of responses of experts to particular 'queries.' This input can also take the form of student assessment statistics providing essentially the same nuggets of information regarding the structure (see Remark 15.4.7 for details). The output is the knowledge space derived from the entailment (cf. Theorem 7.1.5 and Definition 7.1.6). This knowledge space is thus not necessarily a learning space.

This chapter has three main parts. We begin by describing the `QUERY` routine itself. We then discuss an application of `QUERY` reported by Kambouri, Koppen, Villano, and Falmagne (1994) (see also Kambouri, 1991). The experts questioned by the routine were four experienced teachers and the experimenter herself, Maria Kambouri. The items were taken from the standard high school curriculum in mathematics. Our presentation of this study follows closely Kambouri et al. (1994). We will see that this application of the `QUERY` routine was only partly successful. One explanation may be found in the nature of the task, which requires painstaking work over dozens of hours spread over several days. It may be difficult for an expert to be consistently reliable throughout. The work of Cosyn and Thiéry (2000) attempts to palliate this defect by postponing the implementation of the responses to some query until they are confirmed by later responses. Otherwise, those responses are discarded. The result of their work is the `PS-QUERY` routine (for 'Pending-Status-`QUERY`), which we describe in the last part of the chapter.

J.-C. Falmagne, J.-P. Doignon, *Learning Spaces*,
DOI 10.1007/978-3-642-01039-2_15, © Springer-Verlag Berlin Heidelberg 2011

15.1 Background to the QUERY routine

When the input to the QUERY routine comes from experts in the domain, their opinions take the form of responses to queries such as:

[Q1] *Suppose that a student under examination has just provided wrong responses to all the questions in some set A. Is it practically certain that this student will also fail item q? Assume that the conditions are ideal in the sense that errors and lucky guesses are excluded.*

The experts are assumed to be capable of faithfully answering such queries[1]. The set A appearing in [Q1] will often be called an *antecedent* set. This type of query may be denoted by (A, q), and abbreviated as: *Does failing all the questions in some antecedent set A entail failing also item q?*

We have encountered queries such as [Q1] before. In Chapter 7, they were the motivation for the concepts of entailments and entail relations. There, we established the equivalence of two seemingly quite different concepts: on the one hand the knowledge spaces, and on the other hand the entailments for Q. The latter are the relations $\mathcal{P} \subseteq (2^Q \setminus \{\varnothing\}) \times Q$ that satisfy the following two conditions: for all $q \in Q$ and $A, B \in 2^Q \setminus \{\varnothing\}$,

(i) if $q \in A$, then $A\mathcal{P}q$;
(ii) if $A\mathcal{P}b$ holds whenever $b \in B$, and moreover $B\mathcal{P}q$, then $A\mathcal{P}q$

(see Definition 7.1.4). The unique entailment $\mathcal{P}$ derived from some particular space $\mathcal{K}$ on Q is defined by the formula

$$A\mathcal{P}q \iff (\forall K \in \mathcal{K} : A \cap K = \varnothing \Rightarrow q \notin K), \tag{15.1}$$

where $A \in 2^Q \setminus \{\varnothing\}$ and $q \in Q$ (cf. Theorem 7.1.5). The unique knowledge space $\mathcal{K}$ on Q deriving from a given entailment $\mathcal{P}$ on Q is defined by:

$$K \in \mathcal{K} \iff (\forall (A, q) \in \mathcal{P} :\ A \cap K = \varnothing \Rightarrow q \notin K). \tag{15.2}$$

It is clear from (15.1) that, for a given knowledge structure $\mathcal{K}$, the responses to all the queries of the form [Q1] are determined. By hypothesis, the student mentioned in query [Q1] must be in some state not intersecting the antecedent set A. If none of those states having an empty intersection with A contains q, the expert responds 'yes'; otherwise, he responds 'no.' Conversely, if the responses to all such queries are given, then by Theorem 7.1.5 the unique knowledge space $\mathcal{K}$ specified by Equation (15.2) obtains. In the terminology of Definition 7.1.6, we say then that this particular knowledge space is derived from the given entailment. These considerations suggest a practical technique for constructing a knowledge space. We suppose that, when asked a query of the form [Q1], an expert relies (explicitly or implicitly) on a personal knowledge space to manufacture a response. It thus suffices to ask the expert enough

[1] A test of this assumption is reported in Section 15.3.

queries of the form [Q1] to uncover his latent knowledge space. However, an obvious difficulty is that the number of possible queries of the form [Q1] is considerable: if Q contains m items, then there are $(2^m - 1) \cdot m$ queries of the form [Q1]. In practice, however, only a minute fraction of these queries need to be asked because many responses to new queries are either trivial or can be inferred from previous responses. The inference mechanism lies at the core of the `QUERY` procedure and is one of the features that makes `QUERY` applicable in real-life situations.

To help the reader's intuition, we first outline a naïve approach to the construction of a knowledge space from responses to queries. As usual we denote by Q the domain (the set of items), which we assume in this chapter to be finite.

15.1.1 Algorithm (A naive querying algorithm).

STEP 1. Draw up the list of all the subsets of Q.

STEP 2. Successively submit all the queries (A, q) of the form [Q1]. Whenever a positive response $A\mathcal{P}q$ is observed, remove from the list of remaining subsets all the sets containing q and disjoint from A.

Step 2 implements the requirement captured by Equation (15.2). The unique knowledge space consistent with the query responses is ultimately generated by this procedure.

15.1.2 Example. An example of elimination of potential states by an application of the naïve routine 15.1.1 is given in Table 15.1. The domain is the set $Q = \{a, b, c, d, e\}$, and we suppose that only six queries led to positive responses, namely:

$$\{a\}\mathcal{P}b, \quad \{e\}\mathcal{P}a, \quad \{e\}\mathcal{P}b, \tag{15.3}$$

$$\{a, d\}\mathcal{P}c, \quad \{b, d\}\mathcal{P}c, \quad \text{and} \quad \{d, e\}\mathcal{P}c. \tag{15.4}$$

We have gathered the six positive responses $A\mathcal{P}q$ according to the size of A. As we will later see, `QUERY` relies heavily on such gathering by asking first all the queries (A, q) with $|A| = 1$, then all those queries with $|A| = 2$, etc. Accordingly, we say that `QUERY` proceeds by 'blocks': first Block 1, then Block 2, etc. In general, we refer to responses $A\mathcal{P}q$ with $|A| = k$ as coming from *Block k* of the `QUERY` routine. In this example, only Block 1 and Block 2 were needed to construct the knowledge space. The elimination of potential states is marked in Table 15.1 by the '×' symbol. We see that three more states were eliminated during the second block. The final states remaining at the end are indicated by the symbol '√'.

The inclusion diagram of the power set of the set $\{a, b, c, d, e\}$ is represented in Figure 15.1, with the sets eliminated by the two blocks represented in the ovals, shaded respectively in red and blue.

Table 15.1. Elimination of potential states from the six responses listed in (15.3) and (15.4). The symbol '×' marks the states eliminated by one of these responses. Thus, the response $\{a\}\mathcal{P}b$ eliminates the set $\{b\}$, which is also eliminated by the response $\{e\}\mathcal{P}b$. Many other sets are also eliminated by the two responses. The left part of the table (in red) refers to Block 1 of the **QUERY** procedure, the right one (in blue) to Block 2. The sets remaining after Block 1 or Blocks 1 and 2 are indicated by the black symbol '$\surd$' in the fifth and ninth columns of the table. Three more states are eliminated in Block 2, namely $\{c\}$, $\{c,e\}$, and $\{a,c,e\}$.

	$\{a\}\mathcal{P}b$	$\{e\}\mathcal{P}a$	$\{e\}\mathcal{P}b$		$\{a,d\}\mathcal{P}c$	$\{b,d\}\mathcal{P}c$	$\{d,e\}\mathcal{P}c$	
$\varnothing$				$\surd$				$\surd$
$\{a\}$		×						
$\{b\}$	×		×					
$\{c\}$				$\surd$	×	×	×	
$\{d\}$				$\surd$				$\surd$
$\{e\}$				$\surd$				$\surd$
$\{a,b\}$		×	×					
$\{a,c\}$		×				×	×	
$\{a,d\}$		×						
$\{a,e\}$				$\surd$				$\surd$
$\{b,c\}$	×		×		×		×	
$\{b,d\}$	×		×					
$\{b,e\}$	×							
$\{c,d\}$				$\surd$				$\surd$
$\{c,e\}$				$\surd$	×	×		
$\{d,e\}$				$\surd$				$\surd$
$\{a,b,c\}$		×	×				×	
$\{a,b,d\}$		×	×					
$\{a,b,e\}$				$\surd$				$\surd$
$\{a,c,d\}$		×						
$\{a,c,e\}$				$\surd$		×		
$\{a,d,e\}$				$\surd$				$\surd$
$\{b,c,d\}$	×		×					
$\{b,c,e\}$	×				×			
$\{b,d,e\}$	×							
$\{c,d,e\}$				$\surd$				$\surd$
$\{a,b,c,d\}$		×	×					
$\{a,b,c,e\}$				$\surd$				$\surd$
$\{a,b,d,e\}$				$\surd$				$\surd$
$\{a,c,d,e\}$				$\surd$				$\surd$
$\{b,c,d,e\}$	×							
$\{a,b,c,d,e\}$				$\surd$				$\surd$

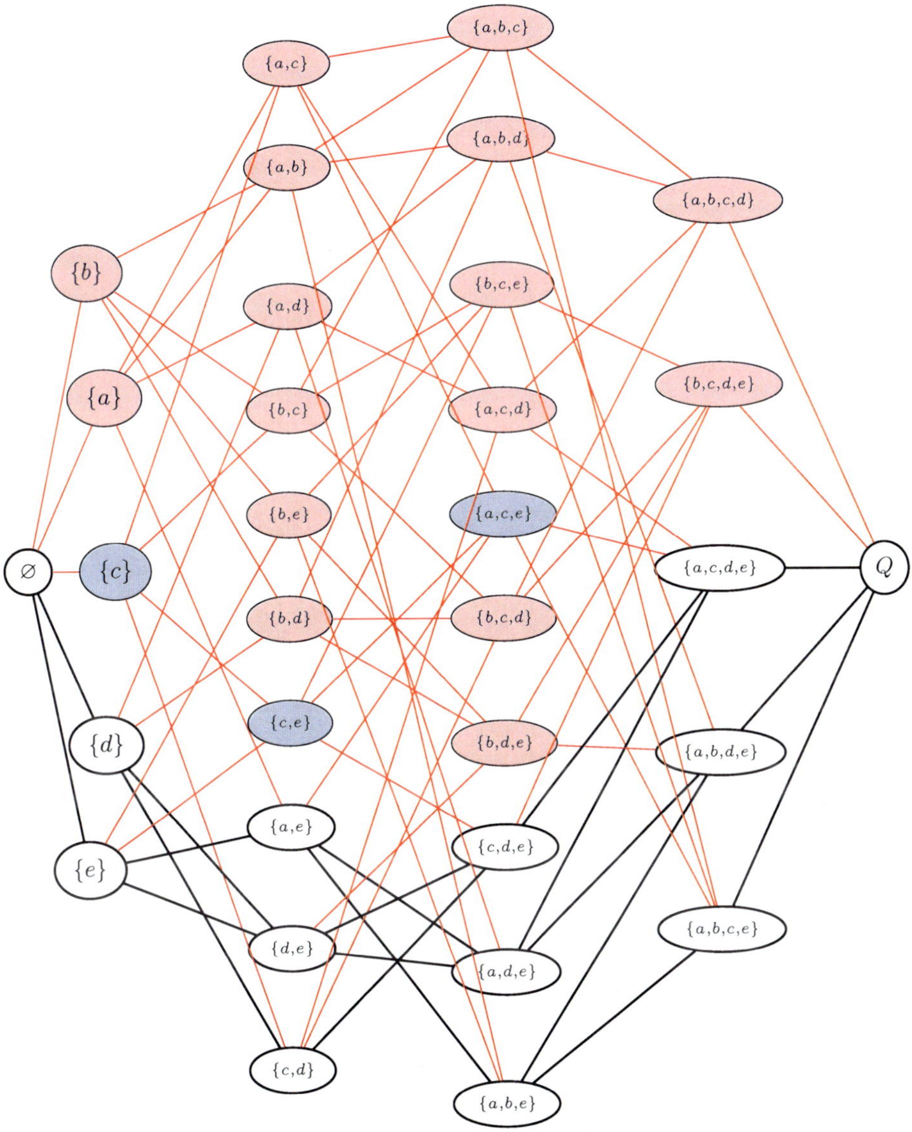

Figure 15.1. In heavy black and white, the inclusion graph of the knowledge space constructed in Table 15.1. The letter Q stands for the domain $\{a, b, c, d, e\}$. The sets eliminated in Block 1 of the naïve **QUERY** routine (cf. 15.1.1) are marked in red shaded ellipses, those eliminated in Block 2 in blue shaded ellipses.

15.1.3 Comments on the naive algorithm. This approach suffers from two algorithmic drawbacks. The first one is that listing all the subsets of Q and reviewing all the potential queries become infeasible as soon as the size m of Q grows large. Indeed, the number of subsets is exponential in m and the number of queries super-exponential. The `QUERY` procedure described in the next section avoids listing all subsets of Q by focusing solely on a clever management of queries. This procedure also eliminates a second drawback of the naive algorithm, which does not contain any mechanism for skipping redundant responses. In fact, in most applications (especially those we observe in real situations), a large fraction of the potential queries can be omitted. There are various reasons for this.

For example, for q in A, failing all the items in the set A trivially implies failing q. Hence the query (A, q) with $q \in A$ always elicits the response 'yes.' Consequently, we should never ask such a query. (In any event, the 'yes' responses they would receive would not eliminate any subsets.) This cuts the number of possible queries in half, down to $2^{m-1} \cdot m$ (see Problem 1).

Note also that, from a positive answer to the query (A, q), we may immediately infer that all queries (B, q) with $A \subseteq B$ will also generate a positive answer: this is easily derived from Equation (15.2) (Problem 2). Moreover, all the sets that could be eliminated from a positive response $B\mathcal{P}q$ have already vanished on the basis of the positive response $A\mathcal{P}q$. Indeed, this elimination concerns all the sets S containing q and having a nonempty intersection with B. Since $S \cap B = \varnothing$ and $A \subseteq B$ implies $S \cap A = \varnothing$, a positive response to the query (B, q) would not eliminate any new set. There is thus no reason to let `QUERY` test for $B\mathcal{P}q$. The query (B, q) is thus said to be *superfluous* (in view of the query (A, q)), and so are all queries (A, q) with $q \in A$.

We shall see in Section 15.2 how `QUERY` is able to further decrease the number of queries requiring submissions to the expert or confrontation with data, thus making realistic applications possible.

15.2 Koppen's Algorithm

Our discussion of the `QUERY` Algorithm will cover the main ideas. Full details can be found in Koppen (1993).

We write (A_1, q_1), (A_2, q_2), $\ldots$, (A_i, q_i), $\ldots$ for the sequence of queries submitted to the expert (or tested on existing data). The query (A_i, q_i) asks the expert whether a student failing all the items in the antecedent set A_i will also fail q_i (cf. Section 15.1). We denote by $\mathcal{P}^{\mathsf{yes}}_{i-1}$ and $\mathcal{P}^{\mathsf{no}}_{i-1}$ the subsets of queries among (A_1, q_1), (A_2, q_2), $\ldots$, (A_{i-1}, q_{i-1}) that elicited a 'yes' or 'no' response, respectively (thus $\mathcal{P}^{\mathsf{yes}}_{i-1} \cap \mathcal{P}^{\mathsf{no}}_{i-1} = \varnothing$); by convention, we set initially $\mathcal{P}^{\mathsf{yes}}_0 = \mathcal{P}^{\mathsf{no}}_0 = \varnothing$. We sometime omit the index when the relevant information is provided by the context (as in $\{e\}\mathcal{P}^{\mathsf{yes}}a$, for example). In the ideal situation taken for granted here, the expert generates the responses from his latent entailment $\mathcal{P}$ (which is derived from his latent knowledge space) so that, for any $i = 1, 2, \ldots,$

$$\mathcal{P}^{\text{yes}}_{i-1} \subseteq \mathcal{P}, \qquad \text{and} \qquad \mathcal{P}^{\text{no}}_{i-1} \subseteq \overline{\mathcal{P}}, \tag{15.5}$$

where $\overline{\mathcal{P}} = \big((2^Q \setminus \{\varnothing\}) \times Q\big) \setminus \mathcal{P}$, the negation of $\mathcal{P}$. (In general, however $\mathcal{P}^{\text{no}}_{i-1}$ is a proper subset of the negation of $\mathcal{P}^{\text{yes}}_{i-1}$.) Hence

$$A\overline{\mathcal{P}}q \quad \Longleftrightarrow \quad \begin{cases} \text{there is a (latent) state containing } q \\ \text{and none of the items in } A. \end{cases} \tag{15.6}$$

We now show how the `QUERY` routine is able to infer responses to some queries from those given to other queries. Examining Columns 2–4 of Table 15.1, we may check that all potential states that are eliminated by the response $\{e\}\mathcal{P}^{\text{yes}}b$ are in fact also eliminated by the responses $\{a\}\mathcal{P}^{\text{yes}}b$ or $\{e\}\mathcal{P}^{\text{yes}}a$. A general explanation of the rationale for such eliminations is given below. By convention, we abbreviate $\{p\}\mathcal{P}q$ into $p\mathcal{P}q$.

15.2.1 Examples of inferences. a) If the positive responses $p\mathcal{P}^{\text{yes}}q$ and $q\mathcal{P}^{\text{yes}}r$ have been observed, the query $(\{p\}, r)$ should not be asked because the restriction of the (latent) relation $\mathcal{P}$ to item pairs is transitive (see Problem 3). In any event, the positive response $p\mathcal{P}^{\text{yes}}r$ would not lead to any new elimination of sets (Problem 3).

b) The transitivity of the relation $\mathcal{P}$ restricted to item pairs also permits inferences to be drawn from negative responses. As in Example (a), we start with the observation $p\mathcal{P}^{\text{yes}}q$, but then observe $p\mathcal{P}^{\text{no}}r$. The routine should not ask whether failing q entails failing r since, by the above argument, a positive response would lead to infer $p\mathcal{P}r$, contradicting $p\mathcal{P}^{\text{no}}r$.

c) Obviously, inferences can also be drawn from other inferences. For example, from $p\mathcal{P}^{\text{yes}}q$, $q\mathcal{P}^{\text{yes}}r$ we infer $p\mathcal{P}r$; if we then observe $r\mathcal{P}^{\text{yes}}s$, we may also infer $p\mathcal{P}s$. In general, we repeat deriving inferences until no more pair can be inferred.

d) Examples (a), (b) and (c) describe cases concerning pairs of items, but can be generalized. Suppose that we have observed the positive responses $A\mathcal{P}^{\text{yes}}p_1$, $A\mathcal{P}^{\text{yes}}p_2$, …, $A\mathcal{P}^{\text{yes}}p_k$. Thus, failing all the items in the set A entails failing also $p_1, p_2, \ldots, p_k$. Suppose moreover that the expert has provided the positive response $\{p_1, p_2, \ldots, p_k\}\mathcal{P}^{\text{yes}}q$. As $\mathcal{P}$ is an entailment, we may then infer $A\mathcal{P}q$ and omit the corresponding query.

The types of inferences we need are more easily specified in terms of entail relations, as we proceed to do.

15.2.2 Entail relations. We recall from Chapter 7 (cf. Theorem 7.2.1 and Definition 7.2.2) that for any entailment $\mathcal{P} \subseteq (2^Q \setminus \{\varnothing\}) \times Q$ there is a unique entail relation $\hat{\mathcal{P}}$ on $2^Q \setminus \{\varnothing\}$ which is defined by the equivalence:

$$A\hat{\mathcal{P}}B \quad \Longleftrightarrow \quad \forall b \in B : A\mathcal{P}b \qquad (A, B \in 2^Q \setminus \{\varnothing\}). \tag{15.7}$$

15.2.3 Convention. In the sequel, we shall drop the 'hat' specifying the entail relation $\hat{\mathcal{P}}$ associated with an entailment $\mathcal{P}$, and use the same notation for both relations. As the entail relation extends the entailment in an obvious way, this abuse of notation will never be a source of ambiguity. Note that in practice we shall always write $A\mathcal{P}q$ instead of $A\mathcal{P}\{q\}$.

We pursue our analysis in terms of the entail relation $\mathcal{P}$ and its negation $\overline{\mathcal{P}}$. Note that the inference

$$\text{whenever } A\mathcal{P}B \text{ and } B\mathcal{P}q, \text{ then } A\mathcal{P}q \tag{15.8}$$

follows directly from the transitivity of the entail relation $\mathcal{P}$. The special case

$$p\mathcal{P}q \quad \text{and} \quad q\mathcal{P}r \quad \text{imply} \quad p\mathcal{P}r$$

was introduced in Example 15.2.1 (a).

We give one last example.

15.2.4 Example. Condition (15.8) is logically equivalent to

$$\text{whenever } A\mathcal{P}B \text{ and } A\overline{\mathcal{P}}q, \text{ then } B\overline{\mathcal{P}}q. \tag{15.9}$$

Accordingly, when the expert has provided the positive responses coded as $A\mathcal{P}B$ and the negative response $A\overline{\mathcal{P}}q$, the negative response $B\overline{\mathcal{P}}q$ may be inferred, and the corresponding query should not be asked.

15.2.5 Inferences. The five examples in 15.2.1 and 15.2.4 illustrate the types of inferences used in the `QUERY` routine. We base the inferences on the four rules given in Table 15.2, which can be derived from the transitivity of $\mathcal{P}$ and the implication

$$A\mathcal{P}b \quad \text{and} \quad A \cup \{b\}\mathcal{P}C \quad \text{imply} \quad A\mathcal{P}C.$$

We leave the justification of these four rules to the reader (Problems 4 to 7).

Table 15.2. The four rules of inference [IR1]–[IR4] permitting the deletion of redundant queries to the expert (see 15.2.5).

	From	we can infer	when it has been established that
[IR1]	$A\mathcal{P}p$	$B\mathcal{P}q$	$(A \cup \{p\})\mathcal{P}q$ and $B\mathcal{P}A$
[IR2]	$A\mathcal{P}p$	$B\overline{\mathcal{P}}q$	$A\overline{\mathcal{P}}q$ and $(A \cup \{p\})\mathcal{P}B$
[IR3]	$A\mathcal{P}p$	$B\overline{\mathcal{P}}q$	$B\overline{\mathcal{P}}p$ and $(B \cup \{q\})\mathcal{P}A$
[IR4]	$A\overline{\mathcal{P}}p$	$B\overline{\mathcal{P}}q$	$(B \cup \{q\})\mathcal{P}p$ and $A\mathcal{P}B$

As we indicated earlier, we repeatedely apply the inference rules of this table until no more new query pair can be produced. Specifically, we denote by $\mathcal{P}_{i-1}$ and $\mathcal{P}^-_{i-1}$ the collections of all pairs found, on any step up to $i-1$, to be in $\mathcal{P}$ and $\overline{\mathcal{P}}$, respectively. These collections are built by enlarging $\mathcal{P}^{\texttt{yes}}_{i-1}$ and $\mathcal{P}^{\texttt{no}}_{i-1}$ via repeated applications of the inferences. We have thus $\mathcal{P}^{\texttt{yes}}_{i-1} \subseteq \mathcal{P}_{i-1}$ and $\mathcal{P}^{\texttt{no}}_{i-1} \subseteq \mathcal{P}^-_{i-1}$. We first enlarge $\mathcal{P}^{\texttt{yes}}_{i-1}$ by applications of a property discussed in Section 15.1.3, namely:

$$\forall A \in 2^Q \setminus \{\varnothing\},\ \forall p \in Q: \quad p \in A \Rightarrow A\mathcal{P}p. \tag{15.10}$$

Thus, we begin by adding all pairs (A, p), with $p \in A$, to $\mathcal{P}^{\texttt{yes}}_{i-1}$. We then manufacture $\mathcal{P}_{i-1}$ and $\mathcal{P}^-_{i-1}$ by repeated applications of the inferences of Table 15.2.

An efficient way for the algorithm to compute, on step $i-1$, the relations $\mathcal{P}_{i-1}$ and $\mathcal{P}^-_{i-1}$ proceeds as follows. First, initialize $\mathcal{P}_{i-1}$ to $\mathcal{P}_{i-2} \cup \mathcal{P}^{\texttt{yes}}_{i-1}$ and $\mathcal{P}^-_{i-1}$ to $\mathcal{P}^-_{i-2} \cup \mathcal{P}^{\texttt{no}}_{i-1}$, and then repeatedly apply the inferences to the resulting relations until the latter are stabilized.

Table 15.3 summarizes the notation to be used in the description of the `QUERY` procedure. In a computer implementation of `QUERY`, a single program variable $\mathcal{P}^{\texttt{yes}}$ would be used with $\mathcal{P}^{\texttt{yes}}_{i-1}$ denoting its successive values during execution time. (A similar remark holds for each of $\mathcal{P}^{\texttt{no}}_{i-1}$, $\mathcal{P}_{i-1}$ and $\mathcal{P}^-_{i-1}$).

Table 15.3. Summary of terms and notation for `QUERY`

	positive	negative
collected responses up to the $(i-1)$-th query	$\mathcal{P}^{\texttt{yes}}_{i-1}$	$\mathcal{P}^{\texttt{no}}_{i-1}$
all inferences from $\mathcal{P}^{\texttt{yes}}_{i-1}$ and $\mathcal{P}^{\texttt{no}}_{i-1}$	$\mathcal{P}_{i-1}$	$\mathcal{P}^-_{i-1}$
i-th query	(A_i, q_i)	

It is clear that implementing these inference mechanisms should result in a marked improvement on the naïve approach of Algorithm 15.1.1. In fact, the new procedure to be defined shortly will skip all queries whose responses can be anticipated. However, several other improvements can still be made. We sketch three of them in the rest of this section.

The following lemma will be instrumental:

15.2.6 Lemma. *For each $i = 1, 2, \ldots,$ the relation $\mathcal{P}_{i-1}$ is an entailment.*

PROOF. Definition 7.1.4 for an entailment involves two conditions; we check them in turn. First, $A\mathcal{P}_{i-1}p$ holds for any subset A of Q and any item p in A because we added such a pair (A, p) to $\mathcal{P}^{\texttt{yes}}_{i-1}$. Second, if A, $B \in 2^Q \setminus \{\varnothing\}$ and $p \in Q$, then

$$(B\mathcal{P}a \text{ for all } a \in A) \text{ and } A\mathcal{P}p \text{ imply } B\mathcal{P}p.$$

This follows from the rule in first row of Table 15.2, by taking $p = q$ (and remembering that $B\mathcal{P}A$ means $B\mathcal{P}a$ for all a in A). □

15.2.7 Generating the space from the entailment table. The procedure keeps track of the queries asked and the responses given in the form of a subset-by-item table[2]. Just before the i-th query is submitted, the table records for each pair (A, p) whether $A\mathcal{P}_{i-1}q$, or $A\mathcal{P}_{i-1}^{-}q$, or whether the response to the query (A, p) is still unknown. Such a table conveys all the information required for constructing the knowledge space relevant to step $i-1$. The reason for this lies in Lemma 15.2.6 and Theorem 7.1.5: the relation $\mathcal{P}_{i-1}$ obtained at the completion of step $i-1$ by the algorithm is an entailment, from which a knowledge space is derivable.

This table contains $(2^m - 1) \times m$ entries and may be inconveniently large. As shown in the next subsection, however, the full table is redundant and all its information can be recovered from a much smaller subtable, using the inferences discussed above. We first explain here how the knowledge space can be generated from the table, assuming that the full, final table recording the relation $\mathcal{P}$ is available.

Each row of the table is indexed by a nonempty subset A of Q. The columns correspond to the items. For each row A, the entry in the cell (A, q) contains either $A\mathcal{P}q$ or $A\overline{\mathcal{P}}q$. We set

$$A^+ = \{q \in Q \mid A\mathcal{P}q\} \tag{15.11}$$

$$A^- = \{q \in Q \mid A\overline{\mathcal{P}}q\}. \tag{15.12}$$

Thus, A^+ contains all the items that the subject would fail who is known to have failed all the items in A. We have necessarily

$$A \subseteq A^+, \qquad A^+ \cap A^- = \varnothing, \quad \text{and} \quad A^+ \cup A^- = Q.$$

It is easy to see that not only is the set A^- a feasible knowledge state, but also that **any** state K must be equal to some set A^- defined as in Equation (15.12) (cf. Equation (7.6) and Problem 2 in Chapter 7; see also Problem 9 at the end of this chapter). Generating the knowledge space from the available table is thus straightforward. Note that there is no need to invoke Equation (15.2).

If we replace $\mathcal{P}$ with $\mathcal{P}_{i-1}$, the same explanation applies to the table existing before the production of the i-th query (all of $\mathcal{P}_{i-1}^{-}$ and the missing information must then be replaced with $\overline{\mathcal{P}_{i-1}}$).

15.2.8 Constructing a manageable subtable. As we have mentioned in our discussion of Example 15.1.2, the subtable of inferences is organized into blocks which are generated successively. Block 1 contains all the responses of the expert to questions of the form "Does failing p entails failing q?" The information in Block 2 concerns the generic question "Does failing both p_1 and p_2 entails failing q?" In general, Block k is defined by the number k of items in the antecedent set A involved in queries of type [Q1]: "Does failing all the items

[2] Programmers may prefer more elaborate data structures, such as linked lists.

in the set A entails failing also item q?" This numbering of the blocks reflects the order in which the queries are asked to the expert by the `QUERY` routine. In other words, the queries in Block 1 are asked first, then come the queries in Block 2, and so on. This ordering of the queries is sensible. The routine starts with the queries which are the easiest ones for the expert to resolve: Block 1 only involves two items and a possible relationship between the two. Block 2 concerns more difficult queries, with three items and the possible relationship between two of these items, forming the antecedent set, and the third item. As the block number increases, the expert's judgements become gradually more difficult. However, the data collected in the early blocks yield inferences affecting later blocks, removing thus from the list of open queries some that are among the most difficult for the expert to answer. The impact of these inferences may be dramatic. For example, in the application described later in this chapter, which involves 50 items, the final knowledge space of each of the five experts was obtained in less than 6 blocks. Moreover, most of the construction was already accomplished after 3 blocks (see Table 15.4)

The required queries in Block k are generated only after all the responses to queries from previous blocks have been collected or inferred. The information from these previous blocks is then used to construct the queries in the new block whose response cannot be inferred from the responses (manifested or inferred) from previous blocks. In practice, only a fraction of the total number of queries with antecedent set of size k will remain in Block k. Note that this does not apply to Block 1. Since this block is constructed from scratch, all the queries with antecedent set of size one must be *a priori* considered. The routine asks the queries of this block and draws inferences along the way.

The process terminates when the newly constructed block does not contain any open query. This indicates that the subtable constructed so far contains all the needed information to build the knowledge space corresponding to the entailment (see Kambouri et al., 1994). This knowledge space is defined as the collection of all sets $A^- = \{p \in Q \mid A\overline{\mathcal{P}}p\}$, where A runs through the rows of this subtable (cf. 15.2.7). We have not discussed here how the information in the previous blocks determines which queries will appear in the new block, or how the postponed inferences for the new block are found. All these specifics are provided in Koppen (1993).

Figure 15.2 displays a flowchart representing the overall design of the algorithm sketched above, which is by no means a complete account of the subtleties presented in Koppen (1993)'s paper. By Block 0, we mean the initial empty block. In general, on step k, the variable `Block` is first used to record all queries still to be asked which have an antecedent of size k. Executing the instruction "`Construct`" means "construct all the queries that were not inferred at the previous instruction in the flowchart", which in particular means that all the positive and negative inferences with the new antecedent size have to be computed.

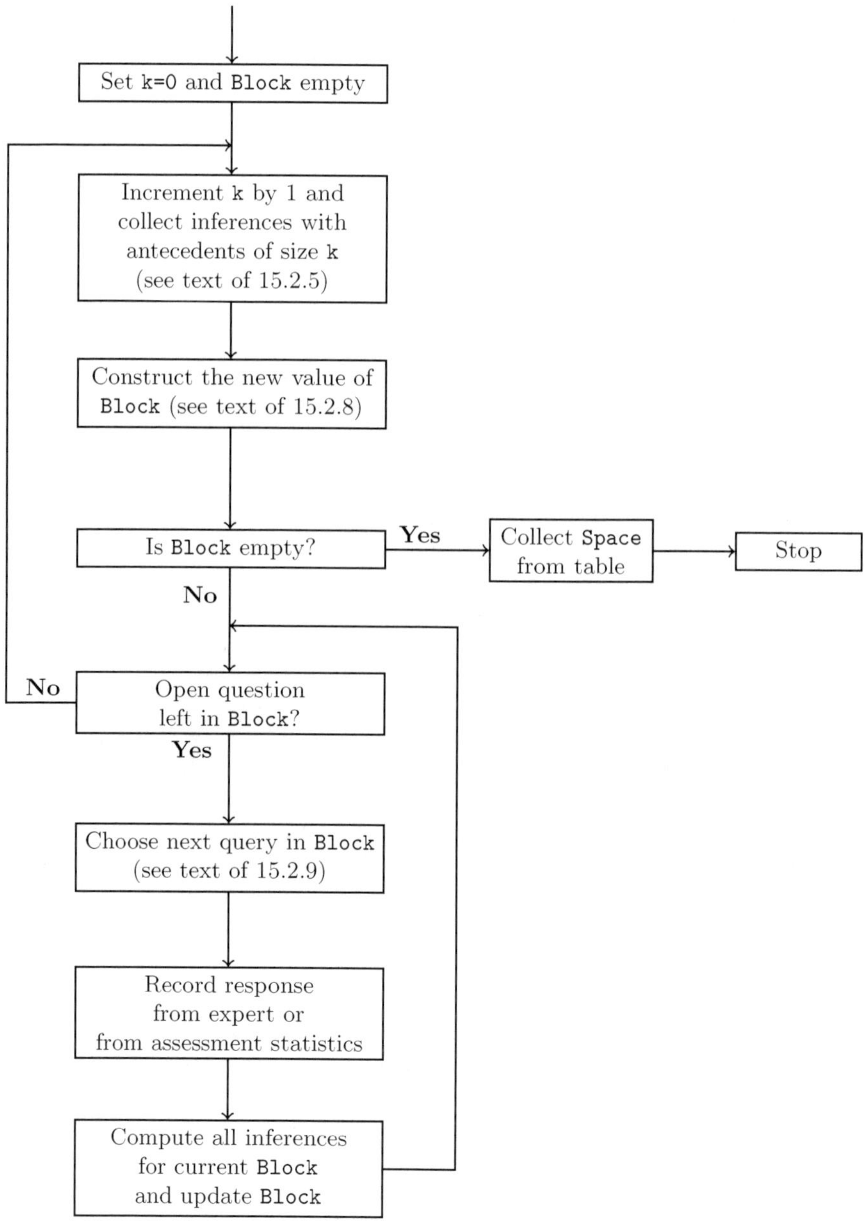

Figure 15.2. Overall design of the QUERY algorithm.

Koppen (1993) describes further ways of reducing the amount of storage needed by QUERY. Examining how the knowledge space is produced according to Subsection 15.2.7, we see that the same state $K = A^-$ may be obtained from different subsets A in Equation (15.11). There is a way of reducing the number of queries (A, p) needing to be stored until the generation of the states occurs, and this can even be harmonized with the block approach. Again, we refer the reader to Koppen (1993)'s paper for details.

15.2.9 Choosing the next query. As we have seen in the preceding subsection, the order in which the queries are asked depends to a considerable degree on the block structure of the procedure. Within a block, however, the choice of the next query is arbitrary. Any of the remaining open queries can be selected. We can take advantage of this latitude. In particular, it makes sense to select the next query so as to minimize the number of remaining open queries. Conceivably, this may be achieved by maximizing (in some sense) the total number of inferences that can be made from the response to the queries. Let us examine some possibilities along this line.

Since it is difficult to look more than one step ahead, the choice of a query by the routine is only guided by the number of inferences that would be derived if the query was asked at that moment. Because the inference rules are only applied in the current block (see 15.2.7), "total number of inferences" in the previous sentence is to be interpreted with this qualification. However, even in this restricted sense, we do not know how many inferences a query will yield, because we do not know what response the expert will give. The inferences following a positive response will differ from those after a negative response. Accordingly, in the sequel, we take into consideration, for any potential new query, the two classes of inferences associated with the positive or negative responses to the query, and more specifically, of the numbers of inferences in each of the two classes. There are various ways in which these two numbers may be used to determine which query is the best one to ask. For example, one may try to maximize the expected gain. In the absence of any information, we may assume that the expert is equally likely to give a positive or a negative response. This implies that the expected number of inferences associated with a query is given by the mean of the numbers of inferences in the two classes. This means that the sum of the two numbers becomes the criterion: we choose a query for which this sum is maximal over all remaining open queries.

In the application presented in the next section, another selection rule was adopted, based on a 'maximin' criterion. The purpose is not so much to maximize the direct gain as to minimize the possible cost of the 'bad' (in terms of efficiency) response to the query. For each query, the number of inferences for a positive response and that for a negative response are computed. First, we only consider the minimum of these two numbers: query (A, q) is chosen over query (B, p) if this minimum number for (A, q) is higher than that for (B, p). Only when the minimum for the two queries is the same do we look at the other number and we choose the query for which this other number is higher. For instance, suppose that query (A, q) yields 3 inferences in the case of a positive response and 2 inferences for a negative, while for queries (B, p) and (C, r) these numbers are 1 and 7, and 2 and 4, respectively. Then query (A, q) is chosen over query (B, p) since $2 = \min\{3, 2\}$ is greater than $1 = \min\{1, 7\}$, but query (C, r) is chosen over (A, q) since they have the same minimum 2, but $4 = \max\{2, 4\}$ is greater than $3 = \max\{3, 2\}$. In short, we select a query with the best worse case and, from among these, one with the best better case.

It is not always feasible to apply this selection process to all open queries. If there are too many open queries, we just pick a pseudorandom sample and choose the best query from this sample. We suspect that we do not lose much this way, since the range of the numbers of inferences from one query is rather restricted. Therefore, if there are many open queries, there will be many that are "best" or approximately so and the important thing is to avoid a particularly poor choice.

15.3 Kambouri's Experiment

The `QUERY` routine was used to construct the knowledge spaces of five subjects: four expert-teachers of high school mathematics, and the experimenter, Maria Kambouri (whose name is abbreviated M.K. in the sequel). In this section we describe the specific domain chosen to apply this routine, the subjects employed for this task, the procedure used, and the results obtained from this application of the routine. (For a more detailed presentation, see Kambouri, 1991, which is her Ph.D. dissertation).

15.3.1 Domain. The domain is within high school mathematics. The items chosen are standard problems in arithmetic, algebra and geometry from the 9th, 10th and 11th grade curriculum of New York State high schools. Specifically, the list of items was built around the Regents Competency Test in Mathematics (RCT). Passing this examination is a minimum requirement for graduation from New York State high schools. It is usually given at the end of the 9th grade. Statistics from the New York City Board of Education show that almost 70% of the students pass the test the first time around; those who fail are allowed to retake it a few times (sometimes up to the 12th grade).

At the time of the study, the full test contained a total of sixty items and was divided into two parts. The first one consisted of twenty completion (open-ended) items, while the second was made of forty multiple choice items for which the student had to select one answer from four alternatives. To pass, students were required to provide correct solutions to at least 39 problems. There was no time limit for this test. Most students returned their copy after less than three hours. For the purpose of the study, only the first part of the test, containing the open-ended items, was of interest. The twenty open-ended problems of the June 1987 Math RCT provided the core of the material on which experts were to be questioned by the `QUERY` procedure. In order to span a broad range of difficulty, this set of problems was extended to 50 by adding 10 simple problems of arithmetic and 20 more complex problems of first-year algebra. The 30 additional items were reviewed by one of the experts, who had extensive experience tutoring students for this type of material. A sample of the 50 items is contained in the next subsection. The topics covered by the final set of 50 problems included: addition, subtraction, multiplication and division of integers, decimals and ratios; percentage word problems; evaluating expressions; operations with signed numbers; elementary geometry and simple graphs; radicals; absolute values; monomials; systems of linear equations; operations with exponents and quadratic equations (by factorization).

15.3.2 Some sample items. Two examples among the ten easy items:

1. *Add:* $34 + 21 = \ ?$
2. *What is the product when 3 is multiplied by 5?*

Three of the 20 open items from the June 1987 RCT:

3. *Write the numeral for twelve thousand thirty seven.*
4. *In the triangle ABC, the measure of angle A is 30 degrees and the measure of angle B is 50 degrees. What is the number of degrees in the measure of angle C?*
5. *Add:* $546 + 1{,}248 + 26 = \ ?$

Three of the 20 more difficult items:

6. *Write an equation of the line which passes through* $(1,0);(3,6)$.
7. *Solve for x:*
$$\frac{2\,x+1}{5} + \frac{3\,x-7}{2} = 7.$$
8. *What is the product of* $6\,x^3$ *and* $12\,x^4$*?*

15.3.3 The experts. Three highly qualified and experienced teachers and one graduate student who had had extensive practice in tutoring mathematics were selected as experts. The teachers received payments at an hourly rate. The experimenter was included as a fifth expert. All five experts were accustomed to interacting with students in one-to-one settings. Moreover, each expert had taught students coming from widely different populations, ranging from gifted children to students with learning disabilities. At the time of this study, the three teachers were working in the New York City school system. A more detailed description of each of the five subjects follows. The initials of all subjects but the experimenter have been changed to protect their identity.

(**A.A.**) This subject had an M.Sc. in Biology and an M.A. in Education (Teachers College), both from Columbia University. He had taught in a public high school in Harlem for a total of 9 years: 2 in mathematics and the rest in science. He had taught a remedial RCT class (students of grades 9-12) and a Fundamental Mathematics class (9th grade), as a preparation for the June 1987 RCT exam. A.A. also taught summer schools and had been on the RCT grading committee for the three summers preceding this study.

(**B.B**) This subject had a B.A. in Psychology and Education from Brooklyn College and an M.A. in Special Education for the emotionally handicapped from Teachers College of Columbia University. She also held a Post-Graduate Certificate on Education, Administration and Supervision. She had over ten years of experience in teaching both mathematics and reading to students with different kinds of learning problems including the learning disabled and those with autistic tendencies. At the time of this study, she was also working as a teacher-trainer in special education and a consultant (Staff Development) for the N.Y.C. Board of Education. Her work involved tutoring students (up to 18 years old) in mathematics. B.B. helped teachers set up diagnostic tests and plan the curriculum at the beginning of the school year.

(**C.C.**) This expert held a Bachelor's degree in Political Science (Russian) from Barnard School at Columbia University and an M.A. in School Psychology from N.Y.U. She had 15 years of experience tutoring students (mainly high school) in mathematics: algebra, geometry, trigonometry, pre-calculus, as well as other topics. In addition, C.C. had been a consultant in various pediatric psychological centers, N.Y.C schools and maintained a private practice, working with learning disabled individuals of all ages on general organizational skills and study techniques.

The other two subjects who served as experts, a graduate student and the experimenter, originated from different educational systems. However, they had studied (and had tutored) all the topics covered by the 50 items chosen for this project.

(**D.D.**) This subject held a B.Sc. in Pure Mathematics from Odessa University (in the former USSR). At the time of the experiment, she was a student in the master's program at the Courant Institute of Mathematical Sciences, New York University. D.D. has taught a class of 25 gifted children aged 13 to 15 for three years, in such topics as calculus, number theory and logic. She also had experience tutoring high-school students who needed additional help in their Geometry and Algebra classes.

(**M.K.**) The experimenter held a B.Sc. and a M.Sc. in Statistics from the University of London. She had some experience in tutoring high school Mathematics and had been a teaching assistant for undergraduate and graduate statistics and probability courses.

15.3.4 Method. The experts were asked to respond to a series of questions generated by a computerized version of the `QUERY` routine. The `QUERY` routine was programmed in `Pascal` by Mathieu Koppen, who is one of the authors of the source paper Kambouri et al. (1994). The user-interface was written in `C` using the `curses` screen optimization library (Arnold, 1986) by another author, Michael Villano.

Each step began with the presentation, on a computer screen, of a query of the form [Q1]. The displayed question was accompanied by a request for the expert's response on the same screen. After the response had been entered at the keyboard, the program computed the query for the next step. An example of a typical screen is shown in Figure 15.3.

a. $\begin{array}{r} 318 \\ \times 605 \\ \hline ? \end{array}$ **b.** $58.7 \times 0.97 = ?$

c. $\frac{1}{2} \times \frac{5}{6} = ?$ **d.** What is 30% of 34 ?

Suppose that a student under examination has just provided a wrong response to problems **a, b, c** and **d.**

e. Gwendolyn is $\frac{3}{4}$ as old as Rebecca
Rebecca is $\frac{2}{3}$ as old as Edwin
Edwin is 20 years old.
How old is Gwendolyn?

Is it practically certain that this student will also fail Problem **e**? **Rating:**

Figure 15.3. A typical screen for the `QUERY` routine.

The top part of the screen displays the problem or problems that a hypothetical student has failed to solve (in this case, the four problems a to d). Four problems can be displayed simultaneously in this section of the display. If the number of these problems exceeds four[3], the expert can toggle the display of the remaining items by pressing a key. In the middle of the screen, the

[3] We shall see that no expert required more than 5 blocks to complete the task.

text of the query regarding the antecedent items is displayed. This is followed by the new problem (e). At the bottom of the screen, the remaining text of the query appears, along with the prompt for the expert's response, labeled "Rating:". The question for the expert in Figure 15.3 is whether the information about failing the four items a to d is compelling the conclusion that the student would also fail the new item e. Based on the expert's answer, the routine computes the next query to ask and, after a short delay (typically on the order of a few seconds), presents the expert with the next screen. Especially after the third block, the delay between blocks was much longer due to the large amount of computation necessary to prepare the open queries for the next block.

Although the `QUERY` routine consists in asking the experts the queries of the type [Q1], which involve a dichotomous (yes-no) decision, a rating scale was used to ensure that the 'yes' responses to be used by the procedure corresponded to a practical certainty on the part of the experts. Therefore, an unequally spaced 3-point scale was adopted with the numbers 1, 4 and 5. A '5' represented a firm 'yes' answer to an instance of [Q1], a '4' indicated less certainty, and a '1' represented a resounding 'no.' Only the '5' response was interpreted by the program as a 'yes' (positive) answer to a query of type [Q1]. Both '4' and '1' were converted to a 'no' (negative). The points 2 and 3 were not accepted by the program as valid responses.

Prior to the start of the experiment, the experts were given the text of the 50 items to review at home. Before they began the task, they received a set of written instructions which included a short explanation of the purpose of this experiment. Next, the list of the 50 items was displayed to acquaint the experts with the appearance of the items on the screen. The procedure was then explained and exemplified. A short, on-line training session followed, consisting of 10 sample steps intended to familiarize the subjects with the task, and in particular with the use of the 3 point scale. After each example, the expert received feedback concerning the interpretation of the rating scale. This introduction to the task lasted approximately thirty minutes. Before starting the main phase of the experiment, the experts were given time to ask queries about the procedure and discuss any difficulties they might have in answering the queries with a rating.

The experts determined the number of steps for a given session. They were advised to interrupt the experiment whenever they felt that their concentration was decreasing. At the beginning of the next session, the program would return them to the same query from which they had exited. At any step, the experts were given the option to go back one query and reconsider their response. They also had the choice to skip a query that they found particularly hard to answer. A skipped query could either come back at a later step, or be eliminated by inference from some other query. Depending on the expert, the task took from 11 to 22 hours (not including breaks).

15.4 Results

15.4.1 Number of queries asked. Table 15.4 displays the number of queries effectively asked by the procedure. The first two columns show the block numbers and the theoretical maximum numbers of queries in each corresponding block. This maximum number of queries would be asked if all the 2^{50} subsets are states. The remaining five columns contain, for each expert, the actual number of queries they were asked. Compared to the theoretical maximum, the reduction is spectacular.

Expert B.B. did not complete the procedure: as shown in Table 15.4, the number of queries asked in the third block was still increasing (the other experts' results show a gradual decrease after the second block). Furthermore, we observed that the fourth block of queries to be answered by expert B.B. was even larger than the third. It was therefore decided to interrupt the routine at that point. Of the remaining experts, one (D.D.) finished after 4 blocks, and the other three (A.A., C.C. and M.K.) after 5 blocks.

Table 15.4. The second columns provides the maximum numbers of queries to be asked in each of Blocks 1-5 through the **QUERY** procedure for 50 items. (These maxima obtain when all the subsets are states. The five last columns contain the actual numbers of queries asked the five experts for each of the five blocks. None of the experts required more than five blocks to terminate the questioning. The symbol '-' in a cell indicates that the procedure was interrupted (see the text above).

Block number	Maximum number of queries	Number of queries asked				
		A.A.	**B.B.**	**C.C.**	**D.D.**	**M.K.**
1	$\binom{50}{1} \times 49 = 2{,}450$	932	675	726	664	655
2	$\binom{50}{2} \times 48 = 58{,}800$	992	1,189	826	405	386
3	$\binom{50}{3} \times 47 = 921{,}200$	260	1,315	666	162	236
4	$\binom{50}{4} \times 46 \approx 10^7$	24	-	165	19	38
5	$\binom{50}{5} \times 45 \approx 10^8$	5	-	29	0	2

We noted in 15.2.7 that, at each intermediate step, the current family of remaining states constitutes a space. The algorithm described in the previous section can be adapted to construct such intermediate spaces. Conceptually, this is achieved by replacing, at the chosen point, the real expert with a fictitious one who only provides negative responses. That is, from this point on we let the procedure run with automatic negative responses to all open queries. This allows the construction of the knowledge spaces after each block (including the one for the interrupted expert B.B. after Block 3, representing the final data in her case).

15.4.2 Number of states. Table 15.5 presents the gradual reduction of the number of states in each block. At the outset, all 2^{50} ($\approx 10^{15}$) subsets of the 50 items set are considered as potential states. For experts A.A. and B.B. the number of states after the first block was over 100,000. At the end, the number of remaining states ranges from 881 to 3,093 (and 7,932 for the unfinished space of B.B.). This amounts to less than one billionth of one percent of the 2^{50} potential states considered initially. It is noteworthy that the reduction after Block 3 is minimal for the four experts having completed the task.

Table 15.5. Number of knowledge states remaining at the end of each block. The initial number is 2^{50} for each expert. (As in Table 15.4, the symbol '-' in a cell indicate that the procedure was interrupted.)

Block number	Number of states per expert after each block				
	A.A.	**B.B.**	**C.C.**	**D.D.**	**M.K.**
1	> 100,000	> 100,000	93,275	7,828	2,445
2	3,298	15,316	9,645	1,434	1,103
3	1,800	7,932	3,392	1,067	905
4	1,788	-	3,132	1,058	881
5	1,788	-	3,093	1,058	881

15.4.3 Comparisons of the data across experts. Despite a generally good overall agreement between the experts, there are substantial discrepancies concerning the details of their performances. In particular, we shall see that the five final knowledge spaces differ markedly.

We first examine the correlation between the ratings. The queries asked by the `QUERY` procedure depend on the responses previously given by the expert. In Block 1, however, the number of common queries asked for any two experts was sufficient to provide reliable estimates of the correlation between the ratings. For any two particular experts, such a correlation is computed from a 3×3 contingency table containing the observed responses to the queries that the experts had in common. The correlation was estimated, for each pair of experts, using the polychoric coefficient (see Tallis, 1962; Drasgow, 1986). This correlation coefficient is a measure of bivariate association which is appropriate when ordinal scales are used for both variables. The values obtained for this coefficient ranged between .53 and .63, which are disappointingly low values hinting at a possible lack of reliability or validity of the experts.

These results only concerned the common queries that any two experts were asked. Obviously, the same kind of correlation can also be computed on the basis of the inferred responses, provided that we only consider the two categories 'Yes-No' used internally by the `QUERY` routine. In fact, the tetrachoric coefficient was computed by Kambouri for all the responses—manifest or inferred—to all $2,450 = 50 \cdot 49$ theoretically possible queries in Block 1. In this case, the data for each of the 10 pairs of experts form a 2×2 contingency table in which the total of all 4 entries is equal to 2,450. The entries correspond to the 4 cases: both experts responded positively (YY); one expert responded positively, and the other negatively (2 cases, YN and NY); both experts responded negatively (NN). As indicated in Table 15.6, the values of the tetrachoric coefficient are then higher, revealing a better agreement between the experts[4]. We note in passing that the number of NN pairs far exceeds the number of YY pairs. A typical example is offered by the contingency matrix of Experts A.A. and B.B. We find 523 YY pairs, 70 YN pairs, 364 NY pairs, and 1,493 NN pairs. (There were thus 1,493 queries—out of 2,450—to which both Experts A.A. and B.B. responded negatively.)

Table 15.6. Values of the tetrachoric coefficient between the experts' ratings for all the responses (manifest and inferred) from the full Block 1 data. Each correlation is based on a 2×2 table, with a total cell count of 2,450.

	A	B	C	D	K
A	-	.62	.61	.67	.67
B	-	-	.67	.74	.73
C	-	-	-	.72	.73
D	-	-	-	-	.79

15.4.4 Descriptive analysis by item. Experienced teachers should be expected to be good judges of the relative difficulty of the items. The individual knowledge spaces provide an implicit evaluation of item difficulty. Consider some item q and the knowledge space of a particular expert. This item is contained in a number of states. Suppose that the smallest of these states contains k items. This means that at least $k - 1$ items must be mastered before mastering q. The number $k - 1$ constitutes a reasonable index of the difficulty of the item q, as reflected by the knowledge space of the expert. In general, we shall call the *height* of an item q the number $h(q) = k - 1$, where k is the number of items in a minimum state containing the item q. Thus, a height of zero for an item means that there is a state containing just that item.

[4] We recall, however, that the tetrachoric coefficient is regarded to be a generous index of correlation as compared, for example, to the phi-coefficient (cf. Chedzoy, 1983; Harris, 1983, in Volumes 6 and 9, respectively, of the Encyclopedia of Statistical Sciences).

Kambouri checked whether the experts generally agreed in their assessment of the difficulty of the items, evaluated by their heights. The height of each of the 50 items was obtained from the knowledge spaces of the five experts. The correlation between these heights was then computed, for each pair of experts. The results are contained in Table 15.7.

Table 15.7. Correlations (Pearson) between the heights of the 50 items computed for each of the 5 experts.

	A	B	C	D	K
A	-	.81	.79	.86	.81
B	-	-	.80	.85	.87
C	-	-	-	.83	.77
D	-	-	-	-	.86

The correlations between the indices of items in Table 15.7 are remarkably high. It may then come as a surprise that a comparison between the knowledge spaces exposes sharp disparities.

15.4.5 Comparison of the knowledge spaces. As indicated in Table 15.5, the number of states differs substantially across the 5 constructed knowledge spaces. In itself, however, this is not necessarily a sign of strong disagreement. For instance, it might happen that all the states recognized by one expert are also states in another expert's space. Unfortunately, the picture is not that simple.

The source paper contains the distribution of the size of the states in the different knowledge spaces. For each space, the number of knowledge states containing k items was computed, for $k = 0, \ldots, 50$. Two exemplary histograms of these distributions, concerning the two experts C.C. and A.A. are displayed in Figure 15.4. Notice that the histogram of A.A. is bimodal. This is the case for 4 of the 5 experts. No explanation was given for this fact which, in any event, points to a noticeable difference between one expert and the others.

The source paper also contains a comparison of the knowledge spaces of the five experts based on a computation of a 'discrepancy index' computed from the symmetric difference distance $d(A, B) = |A \triangle B|$ between sets A and B (cf. 1.6.12). Consider two arbitrary knowledge spaces $\mathcal{K}$ and $\mathcal{K}'$. If these two knowledge spaces resemble each other, then, for any knowledge state K in $\mathcal{K}$, there should be some state K' in $\mathcal{K}'$ which is either identical to K or does not differ much from it; that is, a state K' such that $d(K, K')$ is small. This suggests, for any state K in $\mathcal{K}$, to compute for all states K' in $\mathcal{K}'$, the distance $d(K, K')$, and then to take the minimum of all such distances. In set theory,

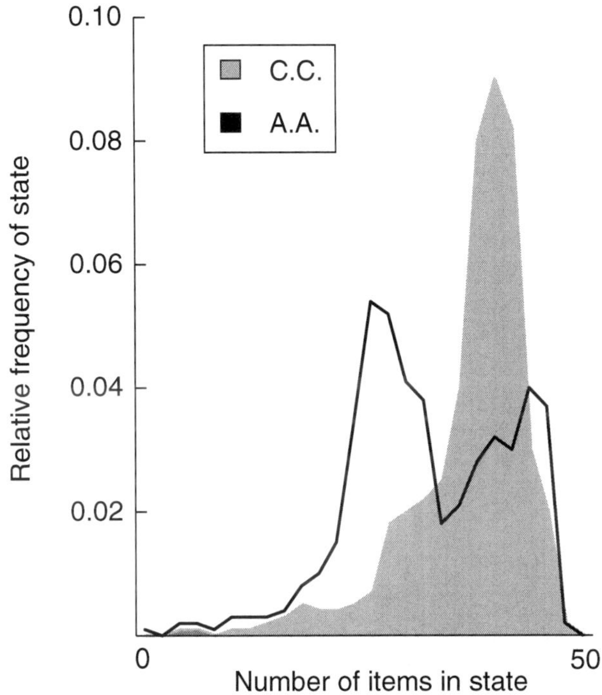

Figure 15.4. Histograms of the relative frequencies of states containing a given number of items, for the two experts C.C. and A.A.

the minimum distance between a set K and a family $\mathcal{K}$ is sometimes referred to as the *distance between* K *and* $\mathcal{K}'$ and is then denoted by $d(K, \mathcal{K}')$ (thus extending the notation d used for the distance between sets).

As an illustration, take the two knowledge spaces

$$
\begin{aligned}
\mathcal{K} &= \{\varnothing, \{c\}, \{d\}, \{c,d\}, \{b,c,d\}, \{a,b,c,d\}\}, \\
\mathcal{K}' &= \{\varnothing, \{a\}, \{a,b\}, \{a,b,c\}, \{a,b,c,d\}\ \}.
\end{aligned}
\tag{15.13}
$$

on the same domain $\{a,b,c,d\}$. For the state $\{c\}$ of $\mathcal{K}$, we have

$$
\begin{aligned}
d(\{c\}, \mathcal{K}') &= \min\{d(\{c\}, \varnothing),\ d(\{c\}, \{a\}),\ d(\{c\}, \{a,b\}), \\
&\qquad\qquad d(\{c\}, \{a,b,c\}),\ d(\{c\}, \{a,b,c,d\})\} \\
&= \min\{1, 2, 3\} = 1.
\end{aligned}
$$

Performing this computation for all the states of $\mathcal{K}$, we obtain a frequency distribution of these minimum distances. (We have one such distance for each state of $\mathcal{K}$.) Thus, this frequency distribution concerns the number $f_{\mathcal{K},\mathcal{K}'}(n)$

of states of $\mathcal{K}$ lying at a minimum distance n, with $n = 0, 1, \ldots$, to any state of $\mathcal{K}'$. For two identical knowledge spaces, this frequency distribution is concentrated at the point 0; that is,

$$f_{\mathcal{K},\mathcal{K}}(n) = \begin{cases} |\mathcal{K}| & \text{if } n = 0, \\ 0 & \text{if } n > 0. \end{cases}$$

In general, for a knowledge structure $\mathcal{K}$ on Q, the distance between any state of $\mathcal{K}$ and a knowledge structure $\mathcal{K}'$ on the same domain Q is at most one half the number of items. So, $f_{\mathcal{K},\mathcal{K}'}(n)$ is defined only for $0 \leq n \leq h(Q) = \lfloor |Q|/2 \rfloor$ (where $\lfloor r \rfloor$ is the largest integer smaller or equal to the number r). Two examples of such frequency distributions are given in Figure 15.5 for the two knowledge spaces of Equation (15.13). We see that three of the states of $\mathcal{K}$ lie at a minimum distance of 1 to any state of $\mathcal{K}'$, namely $\{d\}$, $\{c\}$ and $\{d, c, b\}$.

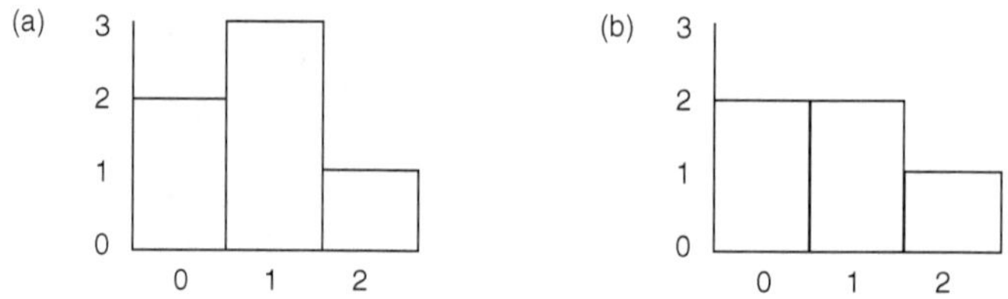

Figure 15.5. Two frequency distributions of the distances: (a) from $\mathcal{K}$ to $\mathcal{K}'$; and (b) from $\mathcal{K}'$ to $\mathcal{K}$; cf. Equation (15.13).

Note that the frequency distribution $f_{\mathcal{K},\mathcal{K}'}$ of the minimum distances from $\mathcal{K}$ to $\mathcal{K}'$ in Figure 15.5 (a) is distinct from the frequency distribution pictured in Figure 15.5 (b), which is that of the minimum distances from states of $\mathcal{K}'$ to states of $\mathcal{K}$ and is denoted by $f_{\mathcal{K}',\mathcal{K}}$.

Such frequency distributions were computed for all 20 pairs of spaces produced by the five subjects in the Kambouri (1991) study. For comparison purposes, it is sensible to carry out a normalization, and to convert all the frequencies into relative frequencies. In the case of $f_{\mathcal{K},\mathcal{K}'}$ this involves dividing all the frequencies by the number of states of $\mathcal{K}$. This type of distribution of relative frequencies will be referred to as the *discrepancy distribution from* the knowledge space $\mathcal{K}$ *to* the knowledge space $\mathcal{K}'$.

A *discrepancy index from* $\mathcal{K}$ to $\mathcal{K}'$ is obtained by computing the mean

$$di(\mathcal{K}, \mathcal{K}') = \frac{1}{|\mathcal{K}|} \sum_{k=0}^{h(Q)} k f_{\mathcal{K},\mathcal{K}'}(k) \qquad \text{(with } h(Q) = \lfloor |Q|/2 \rfloor) \tag{15.14}$$

of the discrepancy distribution from $\mathcal{K}$ to $\mathcal{K}'$, where Q is the common domain of $\mathcal{K}$ and $\mathcal{K}'$. The standard deviations of such discrepancy distributions are also informative.

Table 15.8. Means (and standard deviations) of the discrepancy distributions for all pairs of knowledge spaces. The entry 4.3(1.6) in the second row of the first column of the table refers to the mean (and the standard deviation) of the discrepancy distribution from the space of expert B.B. to the space of expert A.A. The last column contains the means and standard deviations of the discrepancy distributions from the knowledge spaces of each expert to the 'random' knowledge space [K].

	A	B	C	D	K	[K]
A	-	3.0(1.3)	4.8(2.0)	3.4(1.4)	4.3(1.6)	13.4(5.5)
B	4.3(1.6)	-	4.6(1.4)	4.3(1.6)	4.3(1.6)	15.9(3.8)
C	4.1(1.8)	4.0(1.4)	-	5.3(1.5)	5.6(1.7)	10.8(4.4)
D	3.2(1.3)	2.6(1.2)	4.7(1.6)	-	4.0(1.7)	13.2(5.6)
K	3.5(1.5)	2.4(1.1)	4.7(1.8)	3.6(1.7)	-	13.2(6.2)

The first five columns of Table 15.8 contain the computed means and standard deviations (in parentheses) of all 20 discrepancy distributions[5].

These are high numbers. As a baseline for evaluating these results, Kambouri (1991) also computed the discrepancy distributions from the knowledge space of each expert to a 'random' knowledge space, the construction of which is explained below. It was sensible to take, for such comparison purposes, a knowledge space with the same structure as that of one of the experts.
The knowledge space of M.K. was selected, but all 50 items were arbitrarily relabeled. To minimize the risk of choosing some atypical relabeling, 100 permutations on the domain were randomly selected. The discrepancy distribution from the knowledge spaces of each of expert to each of these 100 random knowledge spaces was computed. Then, the average relative frequencies were computed. In other words, a mixture of the resulting 100 discrepancy distributions was formed. The numbers in the last column of Table 15.8 are the means and the standard deviations of these mixture distributions. These means are considerably higher that those appearing in the first five columns. (Note that, since a knowledge structure contains at least the empty set and the domain, the distance between any state of one knowledge space, to some other knowledge space is at most 25, half the number of items.)

15.4.6 Discussion. On the basis of the data collected from the five experts, it must be concluded that the `QUERY` routine has proved to be applicable in a realistic setting. This was far from obvious a priori, in view of the enormous number of queries that had to be responded to. However, a close examination of the data reveals mixed results. On the positive side, there is a good agreement between experts concerning gross aspects of the results.

[5] We recall that we have two discrepancy distributions for each (unordered) pair of experts.

In particular:

1. the sizes of knowledge spaces have the same order of magnitude: a few thousand states—from around 900 to around 8,000—for the 50 items considered (see Table 15.5);
2. there is a good consistency across experts concerning the rating responses given for the same queries asked by the routine (Table 15.6);
3. high correlations between experts are also obtained for the difficulty of the items, as evaluated by their heights in the various spaces (Table 15.7).

Nevertheless, the discrepancy distributions reveal considerable differences between the knowledge spaces constructed by the `QUERY` routine for the 5 experts. For example, we see from Table 15.8 that most of the means of the discrepancy distributions exceed 4. (In other words, a state in one knowledge space differs, on average, by at least 4 items from the closest state in some other knowledge space.) This seems large considering that the domain has only 50 items.

A sensible interpretation of these results is that there are considerable individual differences between experts concerning either their personal knowledge spaces or at least their ability to perform the task proposed to them by the `QUERY` routine, or both of these factors. It must be realized that this task is intellectually quite demanding. In the context of the `QUERY` routine, the notion of 'expertise' has, in fact, two components. First, the expert has to be very familiar with the domain and the chosen population of students, so as to be (at least implicitly) aware of which knowledge states may appear in practice. Second, the expert must also be able to transmit this knowledge structure faithfully through his answers to the queries of the form [Q1]. Experts may differ on either of these components.

Assuming that some carefully selected expert is subjected to the `QUERY` routine, the questioning would not result in the correct knowledge space if the responses to the questions asked by `QUERY` do not, for some reason or other, faithfully stem from that expert's awareness of the feasible states. For example, it is conceivable that, when a query is perceived as too cognitively demanding, a tired expert may resort to some kind of shortcut strategy. An example given by Kambouri et al. (1994) involves the query displayed in Figure 15.3. An expert confronted with that query must examine the items a, b, c, d and e, and decide whether failing a to d would imply a failure on e. Rather than relying on the exact content of each of a to d the expert could just scan these items and arrive, somehow, at an estimate of some overall "difficulty level" for the set of items failed. Similarly, instead of looking at the precise content of the other item e, the expert might collapse the information into some "difficulty level" and then respond on the basis of a comparison of the two difficulty levels.

A variant of this strategy mentioned by Kambouri et al. (1994) may also play a role in which an expert would simply rely on the number of items failed. Referring again to the situation of Figure 15.3, an expert may be led to decide that a fifth item would also be failed, irrespective of its content. This tendency may be reinforced by the fact that in the `QUERY` routine this question is only asked when the expert has, before, given negative responses (directly or through inferences) to all queries involving e with a strict subset of the items a to d. So, the expert might read this query as a repetition, with an implied request to "finally say yes." (Notice, in this respect, that, with the one exception, all experts finished within 5 blocks, so the situation in Figure 15.3 is about as bad as it gets.)

These examples illustrate how experts seemingly well informed about the domain and the chosen population of students can nevertheless produce different knowledge structures by giving invalid responses to some of the questions posed by the `QUERY` routine. This phenomenon may explain some of the differences between experts that were observed in the experiment described in this chapter. This raises the question whether, in further use of this routine, anything can be done to reduce these effects. An answer to this question can go into one of two directions: (i) try to detect and correct such invalid responses, and (ii) try to avoid them as much as possible. Kambouri et al. (1994) discuss both of these possibilities in detail. We only outline the main ideas here.

A practical way of detecting at least some invalid responses is to postpone the implementation of the inferences associated with a response until either a confirmation or a contradiction of that response arises from a new response. Only the confirmed responses would be regarded as valid, and their inferences implemented. Although this may not necessarily detect all invalid responses (a confirmation of an erroneous response may still occur), it should certainly decrease the frequency of those erroneous responses due to the unreliability of the subject. As for avoiding invalid responses, we noticed earlier that the questions asked were not all of equal difficulty. The questions from Block 1 are certainly easier to answer than those from Block 5, say. We may be able to avoid some invalid responses by limiting the application of the `QUERY` routine to Block 1. We would end up with a knowledge space that might be much larger than the real one, but that could contain more valid states, possibly most of the states of the target structure. This large knowledge structure could then be reduced from student data, using a technique from Villano (1991) based on the elimination of states occurring infrequently in practice.

The last part of this chapter is devoted to the work of Cosyn and Thiéry (2000) who investigate some of these ideas, and show that they lead to a feasible overall procedure.

15.4.7 Remark. As already mentioned in 3.2.3, the `QUERY` routine can be used effectively with a different kind of data. We give more details on the method here. Suppose that a preliminary knowledge space has been constructed on the basis of the responses of expert teachers limited to Block 1. The resulting knowledge space, which is thus an ordinal space, may be adequate enough to be used with students. Suppose also that a large number of assessments has been performed, based on this ordinal space. In some applications of learning space theory, such as the ALEKS system, a randomly selected 'extra question' $\boldsymbol{p}$ is asked in any assessment. This question is not part of the assessment. However, it is instrumental in the computation of an index of its validity. The related technique is used extensively in Chapter 17. These extra questions can be used to estimate the conditional probabilities:

$$\mathbb{P}(\text{failing the extra question } \boldsymbol{p} \,|\, \text{failing all the items in a set } A) > \delta, \quad (15.15)$$

where the numerical value of the parameter δ depends upon various factors, such as the estimated probability of a careless error to the item $\boldsymbol{p}$. Let us show how such conditional probabilities can lead to a definition of the relation $\mathcal{P}$, thus replacing an expert's judgement by assessment statistics.

We introduce a notation. We write

$$\mathbf{R}_q = \begin{cases} 1 & \text{if the response to question } q \text{ is correct,} \\ 0 & \text{otherwise.} \end{cases}$$

Using this notation, we can define

$$A\mathcal{P}q \quad \Longleftrightarrow \quad \mathbb{P}\left(\mathbf{R}_q = 0 \,|\, \forall p \in A,\ \mathbf{R}_p = 0\right) > \delta \qquad (15.16)$$

where the r.h.s. of the equivalence is a more precise rewriting of Equation (15.15). Problem 10 asks the reader to examine the feasibility of the construction of a space by this method[6].

15.5 Cosyn and Thiéry's Work

Cosyn and Thiéry began where Kambouri et al. (1994) left off. They developed a procedure based on an improvement of QUERY mentioned in our discussion of the Kambouri et al. (1994) paper in Subsection 15.4.6.

15.5.1 The PS-QUERY, or Pending-Status-QUERY routine. This routine is a modification of QUERY in which a 'pending status' is conferred to any response provided by the expert to a new query. The key mechanisms involve two buffers and two tables in which such a response is temporarily stored, together with the positive and negative inferences arising from that response[7]. The buffers hold inferences immediately drawn from the combination of a new response with confirmed information, so that contradictions can be spotted. The tables hold inferences which did not produce contradiction but are still pending confirmation by a later response. We only give an outline of the algorithm (for details, see Cosyn and Thiéry, 2000).

[6] The knowledge spaces used in the ALEKS system are built by this method, which is based on the combination of human expertise with assessment statistics, grounded on Formula (15.16). We describe an application in Chapter 17.

[7] The inferences have the same meaning as in Subsection 15.2.1 and Table 15.2.

As before, we write Q for the domain and $(A_1, q_1), \ldots, (A_i, q_i), \ldots$ for the sequence of tested queries. To manage the operations at step i, we rely on several relations from $2^Q \setminus \{\varnothing\}$ to Q. The relations $\mathcal{C}_{i-1}$ and $\mathcal{C}^-_{i-1}$ store the positive and negative inferences, respectively, that were **confirmed** in a sense made clear below[8]. We denote by W^T_{i-1} and W^{-T}_{i-1} the two tables containing the pending positive and negative inferences before the i-th query is asked. These inferences are those from all the responses up to step $i-1$ that did not result in a contradiction but are still awaiting confirmation. Finally, we denote by F^B_i and F^{-B}_i the buffers holding fresh positive and negative inferences, respectively drawn from $\mathcal{C}_{i-1}$ and $\mathcal{C}^-_{i-1}$ and the expert's response to query (A_i, q_i). (Note that F^B_i contains the positive response to query (A_i, q_i), if any, and that F^{-B}_i contains the negative response to (A_i, q_i), if any.) Table 15.9 summarizes the terminology and notation.

Table 15.9. Summary of terms and notation for PS-QUERY

	positive	negative
Confirmed inferences (end of step $i-1$)	$\mathcal{C}_{i-1}$	$\mathcal{C}^-_{i-1}$
Tables of pending inferences (end of step $i-1$)	W^T_{i-1}	W^{-T}_{i-1}
i-th query	(A_i, q_i)	
buffers of fresh inferences (step i)	F^B_i	F^{-B}_i

At the beginning of step i, PS-QUERY collects the response to the query (A_i, q_i) and computes all the inferences (positive and negative) from $\mathcal{C}_{i-1}$, $\mathcal{C}^-_{i-1}$ and the response. The routine places these inferences in the buffers F^B_i and F^{-B}_i. These values are then compared with those in $\mathcal{C}_{i-1}$, $\mathcal{C}^-_{i-1}$, W^T_{i-1} and W^{-T}_{i-1}. A contradiction is detected by PS-QUERY in two cases: (i) when a fresh positive inference, i.e., one stored in F^B_i, is found to be already either in the table W^{-T}_{i-1} of pending negative inferences or in the relation $\mathcal{C}^-_{i-1}$; (ii) when a fresh negative inference appearing in F^{-B}_i is found to be either in the table W^T_{i-1} of pending positive inferences or in $\mathcal{C}_{i-1}$. If a contradiction of Type (i) occurs, then all the pairs lying in $F^B_i \cap W^{-T}_{i-1}$ are withdrawn from W^{-T}_{i-1} in order to yield W^{-T}_i. (This means in particular that the last response is discarded.) Similarly, if a contradiction of Type (ii) occurs, then all the pairs lying in $F^{-B}_i \cap W^T_{i-1}$ are withdrawn from W^T_{i-1}, which yields W^T_i. In these two cases, neither $\mathcal{C}_{i-1}$ nor $\mathcal{C}^-_{i-1}$ is modified to obtain $\mathcal{C}_i$ and $\mathcal{C}^-_i$.

If no contradiction is detected, PS-QUERY looks for possible confirmations. Each pair belonging to $F^B_i \cap W^T_{i-1}$ (and so being confirmed) is moved from W^T_{i-1} to $\mathcal{C}_{i-1}$. Thus $\mathcal{C}_i$ is initially set equal to $\mathcal{C}_{i-1} \cup \left(F^B_i \cap W^T_{i-1}\right)$ and W^T_i to $W^T_{i-1} \setminus F^B_i$. Similarly, each pair belonging to $F^{-B}_i \cap W^{-T}_{i-1}$ is moved from

[8] The relations $\mathcal{C}_{i-1}$ and $\mathcal{C}^-_{i-1}$ are similar to the relations $\mathcal{P}_{i-1}$ and $\mathcal{P}^-_{i-1}$ for the QUERY routine defined in Subsection 15.2.5. However, $\mathcal{C}_{i-1}$ and $\mathcal{C}^-_{i-1}$ record only the confirmed inferences, while $\mathcal{P}_{i-1}$ and $\mathcal{P}^-_{i-1}$ record all the inferences.

W_{i-1}^{-T} to $\mathcal{C}_{i-1}^{-}$ to get an initial $\mathcal{C}_i^{-}$. Then the procedure repeatedly computes all inferences from $\mathcal{C}_i$ and $\mathcal{C}_i^{-}$, adding them to the appropriate relation, until the two relations are stabilized. In all cases, the new buffers F_{i+1}^{B} and F_{i+1}^{-B} are reset to the empty value before the next step $i+1$ begins.

Cosyn and Thiéry (2000) defines an exception to the above rules. A query response which induces no new inference other than the response itself is always immediately added to $\mathcal{C}_i$ or $\mathcal{C}_i^{-}$ (the two relations being then closed for inferences).

15.5.2 Remark. As for QUERY, we may stop PS-QUERY at the completion of any step i and build a knowledge space $\mathcal{K}(i)$. The reason is that $\mathcal{C}_i$ is an entailment, from which a knowledge space $\mathcal{K}(i)$ can be derived. If the expert were responding according to some latent entailment $\mathcal{P}$ corresponding to a knowledge space $\mathcal{K}$, then we would have $\mathcal{C}_i \subseteq \mathcal{P}$, and therefore $\mathcal{K}(i) \supseteq \mathcal{K}$. (This reverse inclusion can be easily established; see also 8.6.2). Hence, we would end up with too many states, but no state would be missing.

15.5.3 Simulation of PS-QUERY. In their paper, Cosyn and Thiéry compare the performance of QUERY and PS-QUERY by computer simulations. The target knowledge space, that is the one from which derives the latent entailment of the virtual expert, has a domain of 50 items covering the arithmetic curriculum from grade 4 to grade 8 (the actual knowledge space was constructed, using QUERY, by a real human expert). We denote by $\mathcal{K}^r$ this *reference structure*, and by $\mathcal{K}^{r,1}$ the superset of $\mathcal{K}^r$ obtained from the data of Block 1 of the real expert. Thus, $\mathcal{K}^{r,1}$ is a quasi ordinal knowledge space (cf. Subsection 3.8.1), which we call the *reference order*. Note that $\mathcal{K}^r$ and $\mathcal{K}^{r,1}$ contain 3,043 and 14,346 states, respectively. The simulated expert was assumed to use $\mathcal{K}^r$ to provide the response to the questions asked by PS-QUERY.

Two kinds or error might occur in responding to the questions asked by QUERY or PS-QUERY : (1) the *false positive* responses, consisting in responding 'Yes' when the correct response according to the reference structure is 'No'; (2) the *misses*, which are erroneous 'No' responses. In Cosyn and Thiéry's simulation, the probabilities of two kinds of errors were set equal to 0.05.

The first block of both QUERY and PS-QUERY was simulated for nineteen fictitious experts. On the average, the number of queries required for terminating the first block of PS-QUERY was roughly twice that needed for terminating the first block of QUERY: 1,480 against 662. These numbers are displayed in the second column of Table 15.10, with the corresponding standard deviations in parentheses. The third and fourth column of the table contain the statistics for the discrepancy indices. The meaning of these statistics is as follows. Each of the nineteen simulations delivered a quasi ordinal space[9] $\mathcal{K}_i^{\mathrm{E},1}$ $(1 \leq i \leq 19)$. Using Formula (15.14), the two discrepancy indices $di(\mathcal{K}_i^{\mathrm{E},1}, \mathcal{K}^r)$ and $di(\mathcal{K}^r, \mathcal{K}_i^{\mathrm{E},1})$ were computed for each of the nineteen cases, yielding thus

[9] Because only Block 1 of the QUERY routine was simulated. The superscript 1 in $\mathcal{K}_i^{\mathrm{E},1}$ refers to Block 1.

two frequency distributions. The first number in each cell of Columns 3 and 4 refers to the mean of these two distributions. The standard deviations are indicated in parentheses. The summary notation $\mathcal{K}^{E,1}$ used in the heading of Table 15.10 refers to the variable 'uncovered quasi ordinal space', with values $\mathcal{K}_1^{E,1}, \ldots, \mathcal{K}_{19}^{E,1}$.

The critical column of the table is the third one. The number 1.51 in the second row indicates that the states in $\mathcal{K}^r$ differ on the average by 1.51 items from those contained in the quasi ordinal knowledge structures generated from the first block of QUERY. By contrast, the number 0.16 in the last line shows that, when the quasi ordinal spaces are generated by PS-QUERY, the states in $\mathcal{K}^r$ only differ on the average by 0.16 from those of $\mathcal{K}^{E,1}$. In other words, $\mathcal{K}^r$ is almost included in $\mathcal{K}^{E,1}$. Thus, in principle, almost all the states of the target structure $\mathcal{K}^r$ can be recovered by suitably selecting from the states of $\mathcal{K}^{E,1}$. How such a selection might proceed is discussed in the next section.

Table 15.10. The second column contains the means and standard deviations (in parentheses) of the number of queries required to terminate the first block of QUERY and PS-QUERY with the two error probabilities set equal to 0.05. The third and fourth columns display the means and the standard deviations of the distribution of the discrepancy indices. All the statistics are based on nineteen simulations.

	No. of steps	$di(\mathcal{K}^r, \mathcal{K}^{E,1})$	$di(\mathcal{K}^{E,1}, \mathcal{K}^r)$
QUERY	662 (39)	1.51 (0.65)	3.00 (1.21)
PS-QUERY	1,480 (65)	0.16 (0.11)	1.43 (0.17)

A more precise picture of the situation is provided by Figure 15.6 which is adapted from Cosyn and Thiéry's paper. This figure displays two histograms. To the left of the zero point on the abscissa, we have the histogram of the average discrepancy from $\mathcal{K}^r$ to the nineteen quasi ordinal spaces in $\mathcal{K}^{E,1}$. We see from the graph that, on the average (computed over the 19 simulations), of the 3,043 states of $\mathcal{K}^r$ about 2,600 are also in the expert's space, and about 600 are at a distance of 1 to that space. These number are consistent with the mean 0.16 in the last line of the third column of Table 15.10. To the right of the zero point in the abscissa of Figure 15.6, we have a similar histogram for the average discrepancy from the spaces in $\mathcal{K}^{E,1}$ to $\mathcal{K}^r$.

In the next section, we examine how Cosyn and Thiéry go about refining the ordinal knowledge space obtained from Block 1 of one expert so as to obtain a knowledge structure closely approximating[10] the target structure $\mathcal{K}^r$.

[10] In view of the false positive and misses responses to the queries, the probabilities of which were set to 0.05 in both cases, there is no guarantee that the target target structure $\mathcal{K}^r$ can be obtained exactly.

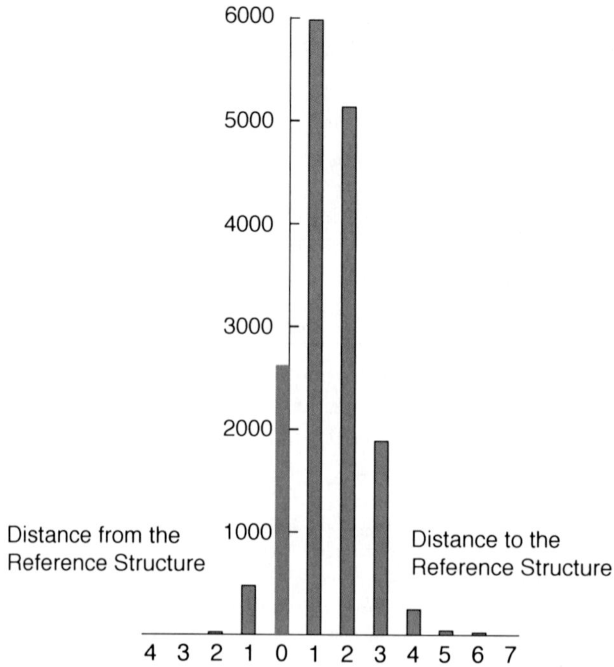

Figure 15.6. Discrepancy distributions $f_{\mathcal{K}^r,\mathcal{K}^{\mathrm{E},1}}$ (left) and $f_{\mathcal{K}^{\mathrm{E},1},\mathcal{K}^r}$ (right) obtained from the nineteen simulations of PS-QUERY, with the two error probabilities set equal to 0.05 (adapted from Cosyn and Thiéry, 2000). The means and standard deviations of these two distributions are displayed in the last line of Table 15.10.

15.6 Refining a Knowledge Structure

Cosyn and Thiéry start from the assumption that, using PS-QUERY or some other technique, a knowledge structure $\mathcal{K}^{e,0}$ has been obtained (from a single expert), that is a superset[11] of the target structure $\mathcal{K}^r$. They apply then a procedure due to Villano (1991). The idea is to use $\mathcal{K}^{e,0}$ for assessing students in a large enough sample from the population, and to use their data to prune $\mathcal{K}^{e,0}$ by removing states with low probabilities of occurrence. Cosyn and Thiéry simulation of this method show, surprisingly, that this can be achieved with a number of assessed students considerably smaller than the number of states in the structure to be pruned. (Plausible reasons for this fact are discussed later in this chapter.) The assessment procedure used is that described in Chapter 13, with the multiplicative updating rule (cf. 13.4.4).

[11] This would happen with an expert who would make no error in responding to the questions asked by PS-QUERY, an unrealistic assumption. Cosyn and Thiéry nevertheless make that assumption so as to clearly separate the investigation of the refinement procedure from the analysis of PS-QUERY.

The refinement is achieved in two steps. In the first step, a 'smoothing rule' is applied which transforms some initial probability distribution[12] on the set of states into another one which takes into account the results of the assessment of the students in the sample. In the second step, a 'pruning rule' is used which removes all the states—except the empty states and the domain—having a probability lower than some critical cut-off value. The probabilities of the remaining states are then normalized so as to obtained a probability distribution on the subset of remaining states. (Essentially, this amounts to computing the probabilities of the remaining states conditional to the event that one of them occurs.) The next two sections contain the details.

15.6.1 The Smoothing Rule. Suppose that a large number of students $s_1, \ldots, s_h$ from a representative sample have been assessed by the procedure of Chapter 13, and that the assessment has provided corresponding probability distributions $\ell_1, \ldots, \ell_h$ on the collection of states. Each of these probability distributions has most of its mass concentrated on one or a few states, and summarizes the assessment for one student. These probability distributions are used to transform an initial probability distribution φ_0. For concreteness, we may take φ_0 to be the uniform distribution U on some initial knowledge structure $\mathcal{K}^{e,0}$. For example, if $\mathcal{K}^{e,0}$ is the knowledge structure deduced from the responses of an expert to the PS-QUERY procedure and containing n states, then $\varphi_0(K) = \frac{1}{n}$ for any K in $\mathcal{K}^{e,0}$. We recall that $\mathcal{K}^{e,0}$ is assumed to be a superset of the target structure $\mathcal{K}^r$.

Keeping track of all the probability distributions ℓ_j for $j = 1, \ldots, h$ is cumbersome if h is large. Accordingly, the effects of $\ell_1, \ldots, \ell_h$ on φ_0 are computed successively. Cosyn and Thiéry (2000) use the transformations:

$$\varphi_0 = U, \tag{15.17}$$

$$\varphi_{j+1} = \frac{j\varphi_j + \ell_j}{j+1}, \qquad \text{for } 1 \leq j \leq h-1. \tag{15.18}$$

Thus, each of the probability distributions φ_{j+1} is a mixture of φ_j and ℓ_j with coefficients $\frac{j}{j+1}$ and $\frac{1}{j+1}$, respectively. Note that for an initial knowledge structure $\mathcal{K}^{e,0}$ and any $j = 1, \ldots, h$: if $\sum_{K\in\mathcal{K}^{e,1}} \varphi_j(K) = 1$, then

$$\sum_{K\in\mathcal{K}^{e,0}} \varphi_{j+1}(K) = \frac{j\sum_{K\in\mathcal{K}^{e,1}} \varphi_j(K) + \sum_{K\in\mathcal{K}^{e,1}} \ell_j(K)}{j+1} = 1.$$

Presumably, if a sufficiently large number h of students are assessed, then φ_{h+1} will differ from φ_1 in that most states in $\mathcal{K}^{e,0}\setminus\mathcal{K}^r$ will have a lower probability than most states in $\mathcal{K}^r$. Thus, by removing all those states K of $\mathcal{K}^{e,0}$ with $\varphi_h(K)$ smaller than some appropriately chosen threshold τ (excepting the empty state and the domain), one can hope to uncover the structure $\mathcal{K}^r$ to a satisfactory approximation.

[12] This may be the uniform distribution on $\mathcal{K}^r$.

15.6.2 The Pruning Rule. The pruning rule

$$\upsilon : (\mathcal{K}, \varphi, \tau) \mapsto \upsilon_\tau(\mathcal{K}, \varphi)$$

acts on any triple formed by a knowledge structure $\mathcal{K}$ on some domain Q, a probability distribution φ on $\mathcal{K}$ and a real number $\tau \in [0, 1]$. It assigns to this triple the subfamily of $\mathcal{K}$ defined by the equation

$$\upsilon_\tau(\mathcal{K}, \varphi) = \{K \in \mathcal{K} \,|\, \varphi(K) \geq \tau\} \cup \{\varnothing, Q\}. \tag{15.19}$$

Thus, $\upsilon_\tau(\mathcal{K}, \varphi)$ is a knowledge structure. A probability distribution φ' on $\upsilon_\tau(\mathcal{K}, \varphi)$ can be defined by the normalization

$$\varphi'(K) = \frac{\varphi(K)}{\sum_{L \in \upsilon_\tau(\mathcal{K}, \varphi)} \varphi(L)}.$$

Note that the order in which the successive transformations of φ_j are carried out is immaterial because the operator combining the successive distributions ℓ_j—which is implicitly defined by Equation (15.18)—is commutative (cf. Problem 11). The last student assessed thus has the same effect on building the resulting knowledge structure as the first one.

It remains to show that this scheme can work in practice. Again, Cosyn and Thiéry answer this question by a computer simulation. In passing they determine a suitable value for the threshold τ. Some of their analyses are based on computing the *quadratic discrepancy index* between a structure $\mathcal{K}$ and a structure $\mathcal{K}'$, that is, the quadratic mean

$$di_2(\mathcal{K}, \mathcal{K}') = \sqrt{di^2(\mathcal{K}, \mathcal{K}') + di^2(\mathcal{K}', \mathcal{K}')} \tag{15.20}$$

between the two discrepancy indices for these structures.

15.7 Simulations of Various Refinements

Cosyn and Thiéry (2000) consider a knowledge structure $\mathcal{K}^r$ representing the full set of knowledge states in some fictitious population of reference. They suppose that an expert has been questioned by `PS-QUERY`, and that the first block of responses provided a knowledge structure $\mathcal{K}^{e,1}$ (which is thus quasi ordinal). Specifically, they take $\mathcal{K}^r$ to be the reference knowledge structure used in the simulation of `PS-QUERY` and described in Subsection 15.5.3, and they suppose that $\mathcal{K}^{e,1} = \mathcal{K}^{r,1}$. Thus, $\mathcal{K}^r$ and $\mathcal{K}^{e,1}$ have $3,043$ and $14,346$ states, respectively.

Various samples of fictitious subjects were generated by random sampling in the set of $3,043$ states of $\mathcal{K}^r$. The probability distribution used for this sampling was the uniform distribution $\varphi_1 = U$ on $\mathcal{K}^r$. The sample sizes ranged from 1,000 to 10,000. Each of these fictitious subjects was tested by the assessment procedure of Chapter 13, using the multiplicative rule (cf. 13.4.4). It was assumed that these subjects were never able to guess the response of an item that was not in their state. On the other hand, the careless errors probability for any item q was assumed to be either 0 or 0.10. (In the notation of 12.4.1, we thus have $\eta_q = 0$ and either $\beta_q = 0$ or $\beta_q = 0.10$.) For each of the samples, the transformed distribution φ_h (where h is the size of the sample) was computed by the Smoothing Rule defined by Equations (15.17)–(15.18).

15.7.1 The value of the threshold. The first task was to determine an adequate value for the threshold τ of the Pruning Rule, according to which a state K of $\mathcal{K}^{e,1}$ is retained if and only if either $\varphi_k(K) \geq \tau$, or K is the domain, or K is empty. Various values for the threshold τ were compared using a sample of 1,000 fictitious subject. The knowledge structure generated by pruning $\mathcal{K}^{e,1}$ with a threshold τ is thus $\mathcal{V}_\tau(1,001) = v_\tau(\mathcal{K}^{e,1}, \varphi_{1,001})$. The criterion used for the comparison was the quadratic discrepancy index $di_2(\mathcal{K}^r, \mathcal{V}_\tau(1,001))$ between $\mathcal{K}^r$ and $\mathcal{V}_\tau(1,001)$, that is, the quadratic mean between the two discrepancy indices for these structures defined by Equation (15.20). This index was computed for τ varying over a wide range. It was found that $di_2(\mathcal{K}^r, \mathcal{V}_\tau(1,001))$ was very nearly minimum for $\tau = 1/|\mathcal{K}^{e,1}|$, the value of the uniform distribution on $\mathcal{K}^{e,1}$. This estimate of τ was adopted for all further simulations reported in Cosyn and Thiéry (2000). Accordingly, we drop the notation for the threshold and write $\mathcal{V}(h) = \mathcal{V}_\tau(h)$ in the sequel.

15.7.2 The number of subjects. The effect of the number of subjects assessed on the accuracy of the recovery of the target structure was also investigated by simulation. The number h of subjects was varied from 0 to 10,000, for two careless error probabilities: $\beta = 0$ and $\beta = 0.10$. The two discrepancy indices $di(\mathcal{K}^r, \mathcal{V}(h))$ and $di(\mathcal{V}(h), \mathcal{K}^r)$ were computed in all cases and form the basis of the evaluation.

15.7.3 Conclusions. The two discrepancy indices decrease as the number h of subjects increase from $h = 100$ upward. With the careless error probability $\beta = 0$, the discrepancy index $di(\mathcal{K}^r, \mathcal{V}(h))$ decreases from 0.62 initially, to 0.01 after about 3,000 simulated subjects. With $\beta = 0.10$, as many as 8,000 subjects are required for this index to reach the same value of 0.01.

The behavior of the other discrepancy index $di(\mathcal{V}(h), \mathcal{K}^r)$ is different. Its value decreases mostly for h between 0 and 1,000 and appears to reach an asymptote at about $h = 1,500$. The asymptotic values of the index for $\beta = 0$ and $\beta = 0.10$ are around 0.2 and 0.4, respectively. The results indicate that, in both cases, most of the superfluous states were discarded at about $h = 1{,}000$.

Overall, the refinement procedure studied by Cosyn and Thiéry (2000) manage to recover 92% of the states of the target structure. The role of the careless error rate is worth noticing. On the one hand, the asymptotic value of $\overline{di}(\mathcal{V}(h), \mathcal{K}^r)$ increases with the error probability β. On the other hand,

$di_2(\mathcal{K}^r, \mathcal{V}(h))$ appears to tend to zero regardless of the error rate. It appears thus that, while a large error rate of the subjects may produce a large refined structure, most of the referent states will nevertheless be recovered if enough subjects are tested.

15.8 Original Sources and Related Works

The bulk of this chapter comes from the three papers by Koppen (1993), Kambouri et al. (1994) and Cosyn and Thiéry (2000). Concepts similar to those used by Mathieu Koppen were developed by Cornelia Dowling in Müller (1989) (see also Dowling, 1994). In another paper (Dowling, 1993a), she combines these ideas into an algorithm that exploits the base to store a space economically. In Heller (2004), the reader can find a formal approach in terms of a generalization of closure spaces that brings much insight in the design of querying algorithms of the type discussed here.

Problems

1. Certain queries need not be asked by the `QUERY` routine because the responses are known a priori. For instance, any positive response $A\mathcal{P}q$ with $q \in A$ must be taken for granted. Why is that reducing the number of possible queries by one-half? (Cf. the comments in 15.1.3(a).)
2. Prove using Equation (15.1) that if $A \subseteq B \subseteq Q$, $q \in Q$ and $A\mathcal{P}q$, then $B\mathcal{P}q$ (see Example 15.2.1(a)).
3. Show that the relation $\mathcal{P}$ defined by (15.1) restricted to pairs of items is transitive.

   In the four following problems, we ask the reader to provide a formal justification for each of the four rules of inference contained in Table 15.2.
4. Prove [IR1].
5. Prove [IR2]
6. Prove [IR3].
7. Prove [IR4].
8. Is it true that any learning space on a domain containing exactly three items is an ordinal space? Prove this fact or give a counterexample. In the latter case, is the counterexample essentially unique? If so, what can you say about the corresponding entailment?

9. Let $\mathcal{P}$ be the unique entail relation corresponding to a knowledge space $(Q, \mathcal{K})$, and let A^+, A^- and $\overline{\mathcal{P}}$ be defined as in Equations (15.11), (15.12) and (15.5), respectively. Prove the following facts:
   (i) $A \subseteq A^+$;
   (ii) $A^+ \cap A^- = \varnothing$;
   (iii) $A^+ \cup A^- = Q$;
   (iv) A^- is a knowledge state in $\mathcal{K}$;
   (v) any knowledge state in $\mathcal{K}$ must be equal to some set A^-.
   (You may find Theorem 7.1.5 useful for answering the last two questions.)

10. Is the relation $\mathcal{P}$ defined by the equivalence (15.16) an entailment, that is, does $\mathcal{P}$ necessarily satisfy Conditions (i) and (ii) of Theorem 7.1.3? How critical are these condition for the construction of a space?

11. Show that the operator implicitly defined by Equation (15.18) and combining any two distribution ℓ_j and ℓ_{j+1} is commutative.

16

Building a Learning space

An application of either QUERY or its extension PS-QUERY results in a knowledge space, that is, a structure closed under union which is not necessarily a learning space. In many practical situations, however, the essential properties of learning spaces are regarded as crucial. In particular, the fringes enables a compact, precise delineation of any knowledge state (cf. Theorems 4.1.7 and 2.2.4(iii)). This property plays a key role in providing a meaningful summary of an assessment. Moreover, in the guise of the outer fringe, it opens the path to further learning. This raises the following problem: assuming that, except for errors, the responses to the queries are dictated by a latent learning space $\mathcal{L}$, can a learning space approximating $\mathcal{L}$ be derived by the querying method through some elaboration of QUERY? In this chapter, we outline two quite different procedures to achieve this goal.

The first one is a modification of QUERY inspired by the fact, established by Theorem 2.2.4, that a knowledge space $(Q, \mathcal{K})$ is a learning space if and only if it satisfies Axiom [MA] for an antimatroid:

[MA] If K is a nonempty subset of Q belonging to the family $\mathcal{K}$, then there is some q in K such that $K \setminus \{q\}$ is a state of $\mathcal{K}$.

This axiom suggests the following revision of QUERY. We start with a collection of (potential) states which forms some initial learning space $\mathcal{L}_0$. For example, $\mathcal{L}_0$ could be the power set of Q, or an ordinal space obtained from implementing Block 1 of QUERY. (We know by Theorem 4.1.10 that ordinal spaces are learning spaces.) We assume that the responses to the queries are in principle based on a latent learning space $\mathcal{L} \subseteq \mathcal{L}_0$. Whenever a positive response to a query (A, q) is observed, we delete from the current learning space, starting with $\mathcal{L}_0$, all the states contradicting this response only when the resulting structure satisfies Axiom [MA]. A simple test to this effect will be given in Theorem 16.1.6.

The defect of the above test is that it involves the whole collection of states which can be unmanageably large. Hence, the test and the whole procedure may not be applicable in many practical cases[1].

[1] The original QUERY procedure avoids this drawback by storing an entailment rather than the whole collection of (potential) states.

J.-C. Falmagne, J.-P. Doignon, *Learning Spaces*,
DOI 10.1007/978-3-642-01039-2_16, © Springer-Verlag Berlin Heidelberg 2011

Nevertheless, the idea is a sound one and it can be adapted: rather than removing states from the learning space itself, we can operate in a similar fashion on the base of the learning space or on the surmise function, both of which are typically much smaller structures. The relevant result is Condition (iii) in Theorem 5.4.1, which is akin to Axiom [MA]. This condition is recalled below in the form of an axiom for learning spaces.

[L3] For any clause C for an item $\{r\}$ in a knowledge space $\mathcal{K}$, the set $C \setminus \{r\}$ is a state of $\mathcal{K}$.

Rephrasing a previous result in terms of [L3], we have then:

> **Theorem 5.4.1(iii).** *A knowledge space is a learning space if and only if it satisfies Axiom* [L3].

This observation leads to an algorithm managing the surmise function appropriately. At the last stage of the algorithm, a learning space is constructed as the collection spanned by the clauses in the final surmise function. We describe this algorithm in Section 16.2.

The second procedure for building a learning space by the querying procedure, which is due to David Eppstein (see Eppstein, Falmagne, and Uzun, 2009; Eppstein, 2010), is quite different from the first one, which essentially (if not literally) proceed by a gradual elimination of potential states. In a first step, a knowledge space is built by a standard application of `QUERY` or `PS-QUERY`. A learning space is then constructed by judiciously adding states until wellgradedness is satisfied, while preserving $\cup$-closure. We sketch Eppstein method in Section 16.3.

16.1 Preparatory Concepts and an Example

Axiom [MA] suggests the concept of a 'critical' state, whose removal would result in a violation of the axiom.This concept is defined below, together with two related ones.

16.1.1 Definition. A nonempty state L in a knowledge structure $\mathcal{K}$ is *hanging* if its inner fringe $L^{\mathcal{I}}$ is empty[2]. The state L is *almost hanging* if it contains more than one item, but its inner fringe consists of a single item. Denoting the latter item by p, we then say that the state $L \setminus \{p\}$ is *critical* in $\mathcal{K}$ for L.

So, an almost hanging state defines exactly one critical state. However, a state may be critical for several almost hanging states.

16.1.2 Example. Consider the knowledge space with domain $Q = \{a, b, c, d\}$ and collection of states

$$\mathcal{L} = \{\varnothing, \{a\}, \{b\}, \{a, b\}, \{a, c\}, \{a, d\}, \{a, b, c\}, \{a, b, d\}, \{a, c, d\}, Q\}.$$

[2] We recall that the inner fringe $L^{\mathcal{I}}$ of a state L is the set of all items q in L such that $L \setminus \{q\}$ is a state; cf. Definition 4.1.6.

There are in $\mathcal{L}$ two almost hanging states, namely $\{a, c\}$ and $\{a, d\}$, while only $\{a\}$ is critical. Moreover, there is no hanging state. Accordingly—see the observation below—$\mathcal{L}$ is a learning space.

16.1.3 Lemma. *In a learning space, any almost hanging state belongs to the base of the space.*

We leave the proof as Problem 2. Notice a simple rephrasing of the equivalence (i) $\Leftrightarrow$ (ii) in Theorem 2.2.4:

16.1.4 Observation. A finite knowledge space $\mathcal{K}$ is a learning space if and only if it has no hanging state.

A query (A, q) with $q \in A$ always produces a positive answer and does not eliminate any state. We therefore assume $q \notin A$ for all queries (A, q) in this chapter. To analyze the effect of a positive answer to a query, we need some further notation.

16.1.5 Definition. Let $(Q, \mathcal{K})$ be a knowledge space and let (A, q) be any query with $\varnothing \neq A \subset Q$ and $q \in Q \setminus A$. For any subfamily $\mathcal{F}$ of $\mathcal{K}$, we define

$$\mathcal{D}_{\mathcal{F}}(A, q) = \{K \in \mathcal{F} \,|\, A \cap K = \varnothing \text{ and } q \in K\}. \tag{16.1}$$

Thus, $\mathcal{D}_{\mathcal{K}}(A, q)$ is the subfamily of all those states of $\mathcal{K}$ that would be removed by a positive response $A\mathcal{P}q$ to the query (A, q) in the framework of the `QUERY` routine.

16.1.6 Theorem. *For any knowledge space $\mathcal{K}$ and any query (A, q), the family of sets $\mathcal{K} \setminus \mathcal{D}_{\mathcal{K}}(A, q)$ is a knowledge space. If $\mathcal{K}$ is a learning space, then $\mathcal{K} \setminus \mathcal{D}_{\mathcal{K}}(A, q)$ is a learning space if and only if there is no almost hanging state L in $\mathcal{K}$ such that $A \cap L = L^{\mathfrak{I}}$ and $q \in L$.*

The proof of this theorem given in 16.1.8 shows that the state L of the theorem becomes hanging in the space $\mathcal{K} \setminus \mathcal{D}_{\mathcal{K}}(A, q)$; notice that $L \setminus L^{\mathfrak{I}}$ is critical for L in $\mathcal{K}$ and is removed from $\mathcal{L}$, while L itself is not removed.

16.1.7 Example. We consider the learning space

$$\mathcal{L} = \{\varnothing, \{a\}, \{b\}, \{a, b\}, \{a, c\}, \{a, d\}, \{a, b, c\}, \{a, b, d\}, \{a, c, d\}, Q\},$$

from Example 16.1.2, and examine three possible queries.

If we observe a positive response to the query $(\{a\}, b)$, only the state $\{b\}$ is to be removed; that is, $\mathcal{D}_{\mathcal{L}}(\{a\}, b) = \{\{b\}\}$. As $\{b\}$ is not critical in $\mathcal{L}$, its removal does not create any hanging state, so $\mathcal{L} \setminus \mathcal{D}_{\mathcal{L}}(\{a\}, b) = \mathcal{L} \setminus \{\{b\}\}$ is a learning space.

The positive response to the query $(\{c\}, a)$, prompts the removal of four states: $\mathcal{D}_{\mathcal{L}}(\{c\}, a) = \{\{a\}, \{a, b\}, \{a, d\}, \{a, b, d\}\}$. The remaining states form a knowledge space which is not a learning space. Theorem 16.1.6 applies to the almost hanging state $L = \{a, c\}$ (which becomes hanging in the new structure).

Finally, if the query $(\{b\}, a)$ results in a positive response, then the remaining states $\{\varnothing, \{b\}, \{a, b\}, \{a, b, c\}, \{a, b, d\}, Q\}$ form a learning space. Notice that the critical state $\{a\}$ is removed from $\mathcal{L}$, however the two states for which it was critical, namely $\{a, c\}$ and $\{a, d\}$, are removed at the same time.

16.1.8 Proof of Theorem 16.1.6. To prove that $\mathcal{K} \setminus \mathcal{D}_{\mathcal{K}}(A, q)$ is a knowledge space, notice first that neither $\varnothing$ not Q belongs to $\mathcal{D}_{\mathcal{K}}(A, q)$. Then, let $\mathcal{E}$ be any subcollection of $\mathcal{K} \setminus \mathcal{D}_{\mathcal{K}}(A, q)$, and set $L = \cup\mathcal{E}$. Then $L \in \mathcal{K}$ because $\mathcal{K}$ is a space. Moreover, $L \notin \mathcal{D}_{\mathcal{K}}(A, q)$. Indeed, if $A \cap L = \varnothing$, then $A \cap E = \varnothing$ for all E in $\mathcal{E}$, and as $E \notin \mathcal{D}_{\mathcal{K}}(A, q)$, we must have $q \notin E$. This in turn implies $q \notin L$, and so $L \in \mathcal{K} \setminus \mathcal{D}_{\mathcal{K}}(A, q)$.

Assume now that $\mathcal{K}$ is a learning space. If $\mathcal{K} \setminus \mathcal{D}_{\mathcal{K}}(A, q)$ is not a learning space, then the latter structure does not satisfy Axiom [MA]. So, it contains a hanging state L (Observation 16.1.4). As that state L was not hanging in $\mathcal{K}$, there is an item p in the inner fringe of L in $\mathcal{K}$ with $L \setminus \{p\} = K$, a state in $\mathcal{K}$. The state K must have been deleted by the positive response to the query (A, q); we have thus $A \cap K = \varnothing$ and $q \in K$. Since L lies in $\mathcal{K} \setminus \mathcal{D}_{\mathcal{K}}(A, q)$ and contains q, we get $L \cap A \neq \varnothing$. So, we have $A \cap L = \{p\}$. Because p was some item in $L^{\mathcal{I}}$, we derive $L^{\mathcal{I}} = \{p\}$. Thus L is almost hanging and it satisfies moreover the other conditions in the statement. Conversely, it is easily verified that if such a state L exists in $\mathcal{K}$, then $\mathcal{K} \setminus \mathcal{D}_{\mathcal{K}}(A, q)$ does not satisfy Axiom [MA] because L is hanging in it. □

The key concept of Theorem 16.1.6 deserves a name.

16.1.9 Definition. Let $(Q, \mathcal{L})$ be a learning space. A query (A, q) is *hanging-safe* if there is no clause C for some item r such that $A \cap C = \{r\}$ and $q \in C$. A query (A, q) is called *operative* (for $\mathcal{L}$) if $\mathcal{L} \setminus \mathcal{D}_{\mathcal{L}}(A, q) \subset \mathcal{L}$.

Thus the collection $\mathcal{L} \setminus \mathcal{D}_{\mathcal{L}}(A, q)$ is a learning space if and only if the query (A, q) is hanging-safe. The concepts of hanging-safety and operativeness are independent: a query (A, q) can be hanging-safe without being operative, and vice versa. A positive response to a hanging-safe, operative query (A, q) can be implemented, leading to a reduction of the learning space.

Theorem 16.1.6 suggests the next algorithm, which is in the spirit of the Naïve Algorithm 15.1.1.

16.1.10 Algorithm (A naive querying algorithm).

STEP 1. Draw up the collection $\mathcal{L}$ of all the subsets of Q.

STEP 2. Successively submit all the queries $(A_1, q_1), \ldots, (A_i, q_i), \ldots$ of the form [Q1] (with $q_i \notin A_i$). Whenever $A_i \mathcal{P} q_i$ is observed, check whether the current collection $\mathcal{L}$ contains an element L satisfying the conditions of Theorem 16.1.6, that is:

1. L is almost hanging in $\mathcal{L}$;
2. $A \cap L = L^{\mathfrak{I}}$ and $q \in L$;

if such an L exists, discard response $A_i \mathcal{P} q_i$, otherwise replace the current collection $\mathcal{L}$ with $\mathcal{L} \setminus \mathcal{D}_{\mathcal{L}}(A_i, q_i)$.

The routine stops when all the queries have been considered. Because the input to the routine is the power set of Q, this input $\mathcal{L}$ is a learning space. The test made on each response $A_i \mathcal{P} q_i$ ensures that $\mathcal{L}$ will remain a learning space throughout all of the algorithm execution.

16.1.11 Remark. This algorithm also works when, in Step 1, the collection $\mathcal{L}$ is initialized to any learning space—for example, to an ordinal space with domain Q. This is the case in Example 16.1.13 below.

Algorithm 16.1.10—which is rather trivial in that it does not use any of the inference mechanisms discussed in 15.2.1—has two other serious defects.

16.1.12 Drawbacks of the naive algorithm. a) The routine does not necessarily remove all the states that could be removed. Some query (A_i, q_i) may be discarded because there is a state K containing q_i with $K \cap A_i = \varnothing$ that is critical for some almost hanging state $L = K \cup \{p\}$ with $p \in A$. However, it is possible that the almost hanging state L is later removed by some query (A_{i+k}, q_{i+k}), rendering state K not critical and so removable. Our next example will provide a couple of such cases.

b) The second drawback is the same as that directed at the Naïve Algorithm 15.1.1 by Comment 15.1.3 (a): keeping track of all the remaining subsets of Q is infeasible for a large domain Q.

We deal with the first of these criticisms in our treatment of the simple example below[3]. Rather than discarding the query (A_i, q_i) of 16.1.12(a), we assign a 'pending status' to such a query, and to any other of that kind. At the end of the first querying round—we call it 'the first stage'—, we start a new round, the 'second stage', with all the queries in pending status. In the case of our example, only one pass in the second stage suffices to build the learning space. We will comment on this point after the example.

16.1.13 Example. The domain is the set $Q = \{a, b, c, d, e, f\}$ and we suppose that Block 1 of `QUERY` has uncovered the ordinal space $\mathcal{L}_0$ displayed in Figure 16.1. The figure shows the Hasse diagram of the partial order defined by the inclusion relation on the set $\mathcal{L}_0$ of states. The ordinal space $\mathcal{L}_0$ is also specified by the Hasse diagram of the corresponding partial order on Q displayed on the upper left of the figure (cf. Birkhoff's Theorem 3.8.3). The elimination of some of the states of $\mathcal{L}_0$ by the routine 16.1.10 is described by Table 16.1, and Figures 16.1 and 16.2. Here, as in Chapter 15, the queries (A, q) are gathered into blocks according to the size of the antecedent set A

[3] The second drawback is dealt with in Section 16.2.

(cf. Example 15.1.2 and Subsection 15.2.8). As before, the blocks are dealt with successively, with increasing value of the antecedent size. Table 16.1 is in the style of Table 15.1, with the following differences. We suppose that Block 2 consists in the positive responses to the four queries

$$(\{a,f\},b), \quad (\{a,e\},f), \quad (\{b,d\},c), \quad \text{and} \quad (\{a,d\},e) \tag{16.2}$$

and only those states affected by those responses are listed in the lines of the table. These states are represented in the red ovals in Figure 16.1. There are no other blocks, and the results of Block 1 are not recorded in the table.

Figure 16.2 is almost identical to Figure 16.1, the only difference being that the ovals of some of the states have been shaded to indicate the two stages of the elimination process. Our discussion of this case is based on both Table 16.1 and Figure 16.2.

The response $\{a,f\}\mathcal{P}b$ eliminates the four states $\{b\}$, $\{b,c\}$, $\{b,e\}$ and $\{b,c,e\}$ (Theorem 16.1.6 applies here). These states are among the ones marked by the red shading of their ellipses in Figure 16.2. The straightforward elimination of these states is marked by the four '$\times$' symbols in the second column of the table. The second response $\{a,e\}\mathcal{P}f$ is more problematic, since it would eliminate the states $\{c,f\}$ and $\{b,c,f\}$. However, $\{b,c,f\}$ is now—because of the earlier removal of $\{b,c,e\}$—critical for $\{b,c,e,f\}$ and so cannot be removed without leaving $\{b,c,e,f\}$ hanging, thus creating a violation of Axiom [MA]. We can't remove $\{c,f\}$ either since it is critical for $\{b,c,f\}$. We thus keep the two states $\{c,f\}$ and $\{b,c,f\}$. (To be sure to maintain stability under union when dealing with possible implementation of a positive response to the query (A,q) on the current learning space $\mathcal{L}$, we either remove from $\mathcal{L}$ all the states in $\mathcal{D}_{\mathcal{L}}(A,q)$, or none of them.) We mark these two states by the grey shaded ovals in Figure 16.2 and the two '$\times$' symbols in the third column of the table.

These two states are tagged for a possible subsequent elimination (from Figure 16.1, we see that if the state $\{b,c,e,f\}$ were removed by a later response, the elimination would indeed become possible). The next response $\{b,d\}\mathcal{P}c$ would eliminate the four states $\{c\}$, $\{a,c\}$, $\{c,f\}$ and $\{a,c,f\}$. But the removal of state $\{c,f\}$ would render the state $\{b,c,f\}$ hanging. Thus, we assign a pending status to $\{b,d\}\mathcal{P}c$ and write four $\times$'s in the appropriate cells of the fourth column of Table 16.1. Finally, the last response $\{a,d\}\mathcal{P}e$ eliminates the state $\{b,c,e,f\}$. This terminates the first stage. We summarize the results by marking, in the appropriate lines of column **R**/**P**, either $\times$ or $\times$ respectively for the removed states and for the states pending elimination.

The second stage of the algorithm considers, successively, the two responses currently in a pending status, namely $\{a,e\}\mathcal{P}f$ and $\{b,d\}\mathcal{P}c$. The first response now eliminates the two states $\{c,f\}$ and $\{b,c,f\}$ (this is made possible by the earlier removal of state $\{b,c,e,f\}$). Then similarly the second response, $\{b,d\}\mathcal{P}c$, eliminates the three states $\{c\}$, $\{a,c\}$ and $\{a,c,f\}$. We indicate the ultimate elimination of the overall five states by writing $\times$ in the cells of the

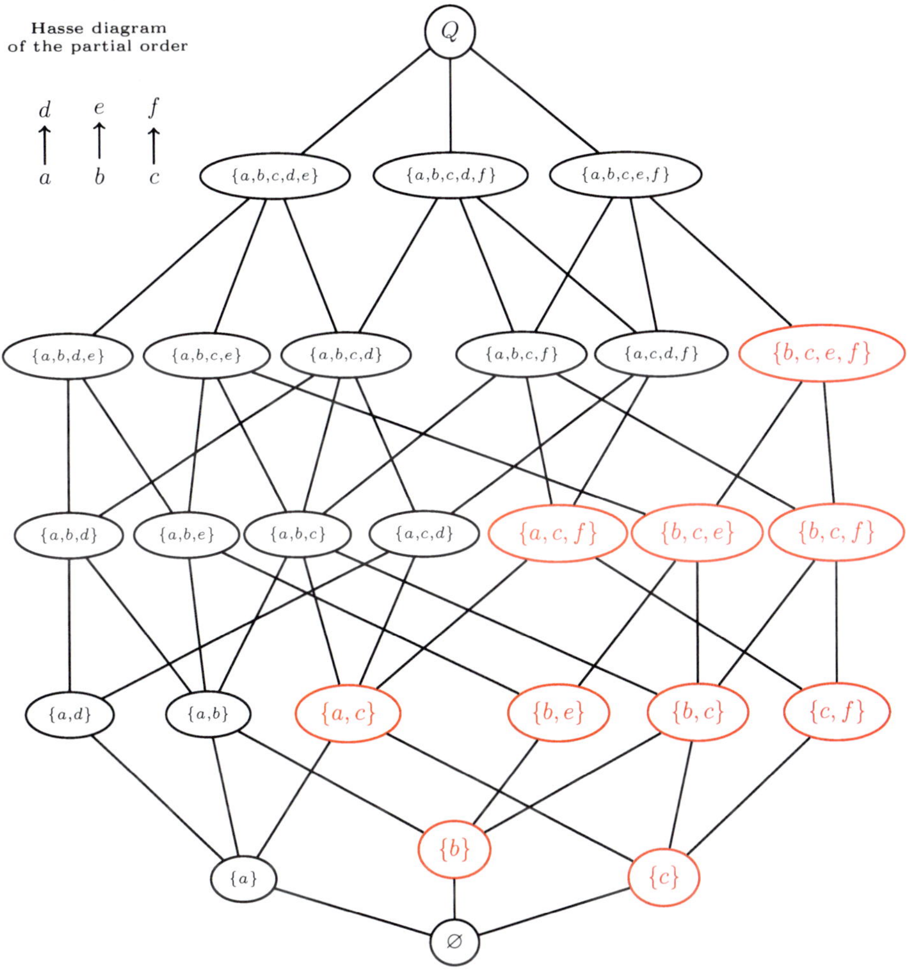

Figure 16.1. Inclusion graph of the ordinal space $\mathcal{L}_0$ on $Q = \{a, b, c, d, e, f\}$ hypothetically constructed by Block 1 of **QUERY**. The corresponding Hasse diagram is pictured on the upper left corner of the figure. The states relevant to the elimination in Block 2, via the four positive responses to the queries $(\{a, f\}, d)$, $(\{a, e\}, f)$, $(\{b, d\}, c)$, and $(\{a, d\}, e)$ are inscribed in the red ovals.

last column. Note that the final learning space obtained is not ordinal since neither $\{a, b, c\} \cap \{a, c, d\}$ nor $\{a, b, c, f\} \cap \{a, c, d, f\}$ are states.

Example 16.1.13 illustrates the importance of temporarily storing under a pending status a positive response to a query that was initially not applicable. It suggests a general algorithm in which pending queries are systematically revisited until none of them is applicable—in other words, until the collection of remaining states is stabilized. (In Example 16.1.13, only one pass in the

Table 16.1. Elimination table: Block 2 of the routine from Example 16.1.13 is applied to the ordinal space $\mathcal{L}_0$ that derives from the Hasse diagram on the right. Only four responses are considered for that block, namely $\{a, f\}\mathcal{P}b$, $\{a, e\}\mathcal{P}f$, $\{b, d\}\mathcal{P}c$, and $\{a, d\}\mathcal{P}e$. The symbols × and × mark a removal action due to the response heading the column. The removal is only potential in the case of the black symbol ×. The column headed by **R**/**P** summarizes the results of the first stage of the algorithm, with × marking the pending status of the response, and × the actual removal. The **R** column contains the final elimination results.

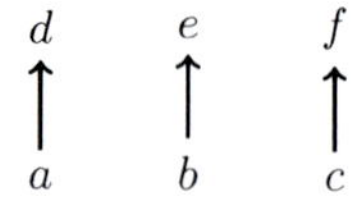

	First Pass					**Second Pass**		
	$\{a, f\}\mathcal{P}b$	$\{a, e\}\mathcal{P}f$	$\{b, d\}\mathcal{P}c$	$\{a, d\}\mathcal{P}e$	**R**/**P**	$\{a, e\}\mathcal{P}f$	$\{b, d\}\mathcal{P}c$	**R**
$\varnothing$								
...	...	...	...	...	...	...	...	
$\{b\}$	×				×			×
$\{c\}$			×		×		×	×
$\{a, c\}$			×		×		×	×
$\{b, c\}$	×				×			×
$\{b, e\}$	×				×			×
$\{c, f\}$		×	×		×	×		×
$\{a, c, f\}$			×		×		×	×
$\{b, c, e\}$	×				×			×
$\{b, c, f\}$		×			×	×		×
$\{b, c, e, f\}$				×	×			×
...	...	...	...	...	...	...	...	
Q								

second stage sufficed.) Theorem 16.1.16 asserts that such an algorithm cannot get jammed in an 'untrue' learning space—at least if the responses observed truthfully reflect a latent learning space. In this case, the algorithm will always output that learning space.

Note in passing that Theorem 16.1.16 asserts even more: if the responses observed reflect a latent knowledge space $\mathcal{K}$, then the algorithm outputs a learning space which is minimal among the learning spaces that contain $\mathcal{K}$.

To establish Theorem 16.1.16, we use a result from Edelman and Jamison (1985) (see also Caspard and Monjardet, 2004) that we rephrase as Theorem 16.1.15 below. The proof (which is the same as that of Edelman and Jamison, 1985, but formulated here for union-closed structures) is based on the following lemma.

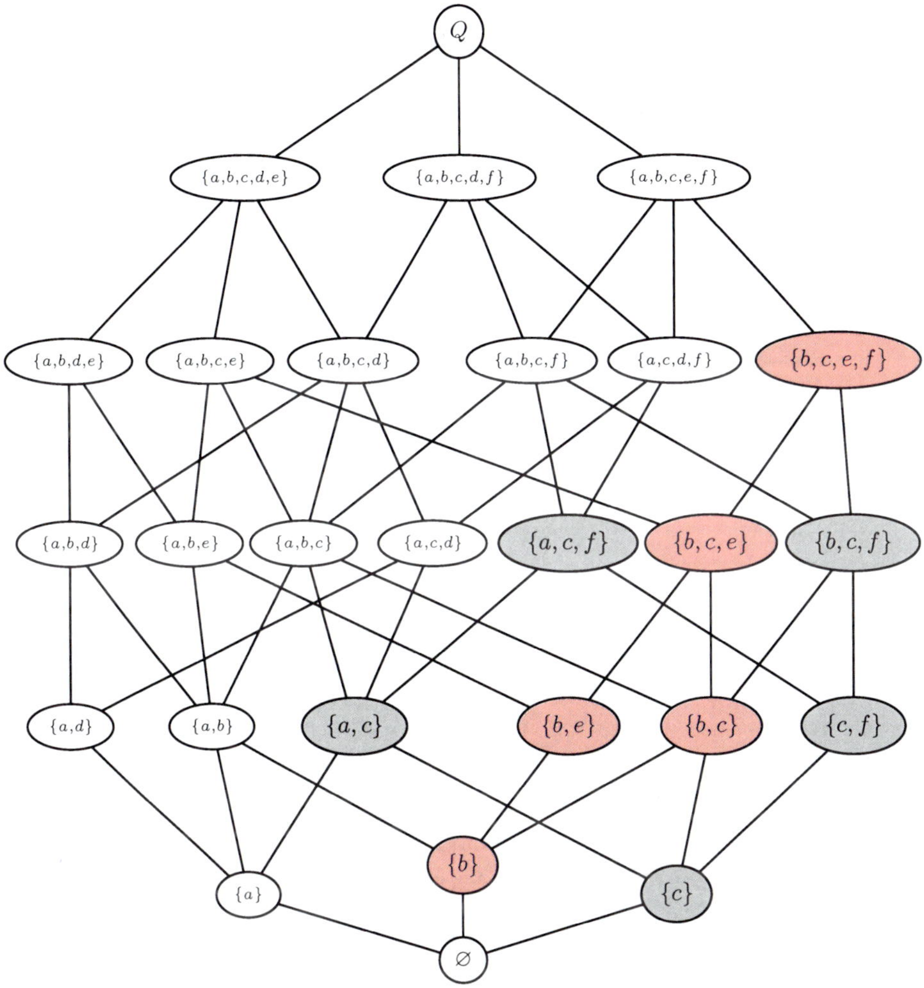

Figure 16.2. First stage of the algorithm eliminating some of the states of the ordinal space $\mathcal{L}_0$ on $Q = \{a, b, c, d, e, f\}$ hypothetically constructed by Block 1 of the procedure. The responses $\{a, f\}\mathcal{P}b$ and $\{a, d\}\mathcal{P}e$ eliminate the five states $\{b\}$, $\{b, c\}$, $\{b, e\}$, $\{b, c, e\}$ and $\{b, c, e, f\}$ which are represented in the red shaded ovals. The gray shadings of other ovals indicate the states that would be eliminated by the answers $\{a, e\}\mathcal{P}f$ and $\{a, e\}\mathcal{P}c$—these answers are set in pending status. See Table 16.1 and the text for details.

16.1.14 Lemma. (i) *For any state K in a finite knowledge space $\mathcal{K}$, the family $\mathcal{K} \setminus \{K\}$ is a knowledge space if and only if K is in the base of $\mathcal{K}$.*

(ii) *For any state L in a learning space $\mathcal{L}$, the family $\mathcal{L} \setminus \{L\}$ is a learning space if and only if L is a state of the base of $\mathcal{L}$ which is not critical in $\mathcal{L}$.*

PROOF. (i) Let $\mathcal{B}$ be the base of $\mathcal{K}$. If $\mathcal{K} \setminus \{K\}$ is a knowledge space, then K cannot be a union of other states in $\mathcal{K}$, and so $K \in \mathcal{B}$. Conversely, if $K \in \mathcal{B}$, then the union of any subfamily of $\mathcal{K} \setminus \{K\}$ belongs to $\mathcal{K}$ and cannot be equal to K, so $\mathcal{K} \setminus \{K\}$ is union-closed.

(ii) If $\mathcal{L} \setminus \{L\}$ is a learning space, then by the first statement, L must be in the base $\mathcal{B}$ of $\mathcal{L}$. Moreover, L cannot be critical for any almost hanging state M, since then M would be hanging in $\mathcal{L} \setminus \{L\}$. Conversely, let L be a state in $\mathcal{B}$ which is not critical for any state in $\mathcal{L}$. Then $\mathcal{L} \setminus \{L\}$ is union-closed and does not contain any hanging state. So, $\mathcal{L} \setminus \{L\}$ is a learning space. □

16.1.15 Theorem. *Let $\mathcal{J}$ and $\mathcal{L}$ be two learning spaces on the same domain, with $\mathcal{B}$ denoting the base of $\mathcal{L}$. Suppose that $\mathcal{J}$ is covered by $\mathcal{L}$ (that is, $\mathcal{J} \subset \mathcal{L}$ and there exists no learning space $\mathcal{M}$ satisfying $\mathcal{J} \subset \mathcal{M} \subset \mathcal{L}$). Then $\mathcal{L} \setminus \mathcal{J} = \{B\}$ for some B in $\mathcal{B}$.*

PROOF. The hypotheses imply that there exists some state L in $\mathcal{L} \setminus \mathcal{J}$. As L is the union of states from the base $\mathcal{B}$ of $\mathcal{L}$, there must exist some state B in $\mathcal{B} \setminus \mathcal{J}$ (otherwise, $\mathcal{J}$ would not be union-closed). We may suppose that B is maximal for inclusion in $\mathcal{B} \setminus \mathcal{J}$. We show that B is not critical in $\mathcal{L}$. Suppose, by contraposition, that B is critical for some almost hanging state K in $\mathcal{L}$. By Lemma 16.1.3 any almost hanging state in $\mathcal{L}$ necessarily belongs to $\mathcal{B}$. Hence, by the maximality of B in $\mathcal{B} \setminus \mathcal{J}$, we must have $K \in \mathcal{J}$, with K hanging in $\mathcal{J}$, contradicting our hypothesis that $\mathcal{J}$ is a learning space. So, B is not critical in $\mathcal{L}$. By Theorem 16.1.14(ii), $\mathcal{L} \setminus \{B\}$ is a learning space with $\mathcal{J} \subseteq \mathcal{L} \setminus \{B\} \subset \mathcal{L}$. Our assumption that $\mathcal{L}$ covers $\mathcal{J}$ implies $\mathcal{J} = \mathcal{L} \setminus \{B\}$. □

16.1.16 Theorem. *Let $\mathcal{K}$ be a knowledge space and $\mathcal{L}$ be a learning space on the same domain, with $\mathcal{K} \subseteq \mathcal{L}$. Suppose moreover that $\mathcal{L}$ is not minimal among the learning spaces including $\mathcal{K}$. Then there exists some query (A, q) such that $A\mathcal{P}q$ for the entailment relation $\mathcal{P}$ derived from $\mathcal{K}$, and moreover the collection $\mathcal{M} = \mathcal{L} \setminus \mathcal{D}_{\mathcal{L}}(A, q)$ is a learning space satisfying $\mathcal{K} \subseteq \mathcal{M} \subset \mathcal{L}$.*

Accordingly, if $\mathcal{K}$ is a learning space and the query responses are truthful with respect to $\mathcal{K}$, then the `QUERY` procedure will ultimately uncover $\mathcal{K}$.

PROOF. Let Q be the common domain of $\mathcal{K}$ and $\mathcal{L}$. Our hypotheses imply that there exists some learning space $\mathcal{J}$ such that $\mathcal{K} \subseteq \mathcal{J} \subset \mathcal{L}$ and $\mathcal{J}$ is covered by $\mathcal{L}$. By Theorem 16.1.15, $\mathcal{L} \setminus \mathcal{J} = \{B\}$ for some state B in the base of $\mathcal{L}$. Since $B \notin \mathcal{J}$, the largest state M of $\mathcal{J}$ that is included in B is distinct from B. (Notice that M may be empty.) Set $A = Q \setminus B$ and pick an item q in $B \setminus M$. Because $A \cap B = \varnothing$ and $q \in B$, we must have $B \in \mathcal{D}_{\mathcal{L}}(A, q)$. In fact, we have $\mathcal{D}_{\mathcal{L}}(A, q) = \{B\}$. Indeed, any state J of $\mathcal{L}$ disjoint from A is included in B; if J is not equal to B, then it belongs to $\mathcal{J}$ and is thus included in M. Consequently, $q \notin J$ and thus $J \notin \mathcal{D}_{\mathcal{L}}(A, q)$, so $\mathcal{D}_{\mathcal{L}}(A, q) = \{B\}$. Hence, we obtain

$$\mathcal{L} \setminus \mathcal{D}_{\mathcal{L}}(A, q) = \mathcal{L} \setminus \{B\} = \mathcal{J}.$$

We may thus set $\mathcal{M} = \mathcal{J}$. □

16.2 Managing the Surmise Function

We now tackle the second drawback of the naïve algorithm mentioned in 16.1.12, which remains a weakness of our method for dealing with Example 16.1.13 and also of the general approach suggested before Theorem 16.1.16. Namely, we cannot realistically keep track of all the states of the learning space under construction: even if the input to the algorithm is an ordinal space on say 50 items, the number of states may be on the order of several millions. In this section, we describe an algorithm that acts on the surmise function rather than on the learning space itself.

We know from Definition 5.2.1 that any finite knowledge space $(Q, \mathcal{K})$ has a surmise function $\sigma : Q \to 2^{2^Q}$. For any item q in Q, $\sigma(q)$ denotes the collection of all the atoms at q, that is, the states in $\mathcal{K}$ that are minimal for the property of containing q. The states in $\sigma(q)$ are also called the clauses for q in $\mathcal{K}$. The space $\mathcal{K}$ is a learning space if and only Axiom [L3] is satisfied[4]:

[L3] For any clause C for an item $\{r\}$ in a knowledge space $\mathcal{K}$, the set $C \setminus \{r\}$ is a state of $\mathcal{K}$.

This axiom is the cornerstone of the next algorithm because, by Theorem 5.4.1, a knowledge space is a learning space if and only if it satisfies Axiom [L3]. Moreover, an efficient test can be derived from [L3]. The general idea of the algorithm is that, before implementing a positive response to a query, the test is performed on the current surmise function to verify whether removing the relevant states would result in a knowledge space that still satisfies [L3], and so is a learning space. If the result of the test is positive, the algorithm updates the surmise function by replacing suitable clauses by new ones. When no more positive response can be implemented, the algorithm outputs the learning space spanned by the final collection of clauses.

We now relate the two concepts of almost hanging state and clause. The first assertion in the next theorem implies that, in a knowledge space, any clause is either a one-element set, a hanging state or an almost hanging state. We can then characterize learning spaces by stating that the clauses containing more than one item are exactly the almost hanging states; cf. (ii) $\Leftrightarrow$ (iii) below.

16.2.1 Theorem. *Let $(Q, \mathcal{K})$ be a knowledge space with surmise function σ. Then, for any state K in $\mathcal{K}$, we have*

$$K \in \sigma(p) \quad \Longrightarrow \quad K^{\mathcal{I}} \subseteq \{p\}. \tag{16.3}$$

If $(Q, \mathcal{K})$ is finite, then the following three conditions are equivalent:

(i) *for any state K in $\mathcal{K}$, if K is in $\sigma(p)$, then $K^{\mathcal{I}} = \{p\}$;*
(ii) *for any state K in $\mathcal{K}$ and any item p in Q,*
$K \in \sigma(p)$ *and* $|K| \geq 2 \iff K$ *is almost hanging in* $\mathcal{K}$ *with* $K^{\mathcal{I}} = \{p\}$;
(iii) *the space $(Q, \mathcal{K})$ is a learning space.*

---

[4] Cf. the restatement of Theorem 5.4.3(iii) on page 336.

The essence of Theorem 16.2.1 is that the base of a finite learning space consists of all the one-element states and the almost hanging states.

PROOF. Assume that K is in $\sigma(p)$. If $K^{\mathcal{I}}$ contained an item q distinct from p, then $K \setminus \{q\}$ would be a state such that $p \in K \setminus \{q\} \subset K$, contradicting the minimality of the clause K.

(i)⇒(ii). The implication from left to right in the equivalence in (ii) follows from (i) and the definition of an almost hanging state. Conversely, assume that K is almost hanging with $K^{\mathcal{I}} = \{p\}$. Then by definition $|K| \geq 2$. Suppose that K is not in $\sigma(p)$. Then, there would be a maximal state M such that $p \in M \subset K$. As $|K \setminus M| = \{r\}$ for some $r \neq p$ would contradict our assumption that $K^{\mathcal{I}} = \{p\}$, there are (at least) two items in $K \setminus M$, say r and q. Some clause C for r is included in K. This clause must contain q, otherwise we would get $M \subset M \cup C \subset K$ contradicting the maximality of M. By (i), the set $C \setminus \{r\}$ is a state. The union $M \cup (C \setminus \{r\})$ contains p and is another state also contradicting the maximality of M. So K must be a clause for p.

(ii)⇒(iii). We derive from (ii) that for any clause C at p the set $C \setminus \{p\}$ is a state. Using the implication (iii) ⇒ (i) in Theorem 5.4.1, we derive that the space $(Q, \mathcal{K})$ is a learning space.

(iii)⇒(i). By the implication (i) ⇒ (iii) in Theorem 5.4.1, we have $p \in K^{\mathcal{I}}$ for any clause K for p. Condition (i) now follows from Equation (16.3). □

The next result is a straightforward consequence of Theorems 16.1.6 and 16.2.1. It shows how we can verify on the surmise function of a learning space whether the implementation of a positive response to a query yields a learning space, in other words whether the query is hanging-safe.

16.2.2 Theorem. *Let $(Q, \mathcal{L})$ be any learning space, and let (A, q) be any query. The query (A, q) is hanging-safe for $\mathcal{L}$ if and only if there is no clause C for some item r such that $A \cap C = \{r\}$ and $q \in C$.*

PROOF. The statement is a reformulation of the second sentence in Theorem 16.1.6, using Condition (ii) in Theorem 16.2.1. □

16.2.3 Notation. When $(Q, \mathcal{K})$ is a knowledge space with surmise function σ and (A, q) is a query, we denote by $\sigma_{A,q}$ the surmise function of the knowledge space $\mathcal{K} \setminus \mathcal{D}_{\mathcal{K}}(A, q)$ (remember from Theorem 16.1.6 that $\mathcal{K} \setminus \mathcal{D}_{\mathcal{K}}(A, q)$ is a knowledge space).

16.2.4 Example. We again use the learning space from Example 16.1.2:

$$\mathcal{L} = \{\varnothing, \{a\}, \{b\}, \{a, b\}, \{a, c\}, \{a, d\}, \{a, b, c\}, \{a, b, d\}, \{a, c, d\}, Q\}.$$

Its surmise function is:

$$\begin{aligned} \sigma(a) &= \{\{a\}\}, & \sigma(b) &= \{\{b\}\}, \\ \sigma(c) &= \{\{a, c\}\}, & \sigma(d) &= \{\{a, d\}\}. \end{aligned} \tag{16.4}$$

We examine the same three queries $(\{a\}, b)$, $(\{c\}, a)$, $(\{b\}, a)$ as in Example 16.1.7. If we observe a positive response to the query $(\{a\}, b)$, a clause C satisfying the condition of Theorem 16.2.2 would be a clause for a with $b \in C$. As there is no such clause, $(\{a\}, b)$ is hanging-safe and $\mathcal{L} \setminus \mathcal{D}_{\mathcal{L}}(\{a\}, b)$ a learning space. The response $(\{a\}, b)$ is operative for $\mathcal{L}$ and its implementation removes the single state $\{b\}$. Note that the query $(\{a\}, b)$ is still hanging-safe for $\mathcal{L} \setminus \{b\}$, but no longer operative (for it). The surmise function of $\mathcal{L} \setminus \mathcal{D}_{\mathcal{L}}(\{a\}, b)$ is

$$\begin{aligned} \sigma_{\{a\},b}(a) &= \{\{a\}\}, & \sigma_{\{a\},b}(b) &= \{\{a,b\}\}, \\ \sigma_{\{a\},b}(c) &= \{\{a,c\}\}, & \sigma_{\{a\},b}(d) &= \{\{a,d\}\}. \end{aligned} \tag{16.5}$$

We see that the implementation of the query $(\{a\}, b)$ resulted in the removal of the unique clause $\{b\}$ for b, which was replaced by the new clause $\{a, b\} \supset \{b\}$. The state $\{a, b\}$ was not removed, but had to be identified as the new clause for b. We will comment on this replacement in Remarks 16.2.5.

If we observe a positive response to the query $(\{c\}, a)$, we inspect the clauses for c that contain a. There is only one such clause, namely $\{a, c\}$. By Theorem 16.2.2, $\mathcal{L} \setminus \mathcal{D}_{\mathcal{L}}(\{c\}, a)$ is not a learning space. The response $(\{c\}, a)$ is operative but not hanging-safe. In fact, we get

$$\mathcal{L} \setminus \mathcal{D}_{\mathcal{L}}(\{c\}, a) = \{\varnothing, \{b\}, \{a, c\}, \{a, b, c\}, \{a, c, d\}, Q\}.$$

Finally, if we observe a positive response to the query $(\{b\}, a)$, only the clause $\{b\}$ is to be inspected. The query $(\{b\}, a)$ is hanging-safe. Its implementation leads to the removal of the four states $\{a\}$, $\{a, c\}$, $\{a, d\}$, and $\{a, c, d\}$, yielding the learning space

$$\mathcal{L} \setminus \mathcal{D}_{\mathcal{L}}(\{b\}, a) = \{\varnothing, \{b\}, \{a, b\}, \{a, b, c\}, \{a, b, d\}, Q\},$$

with surmise function

$$\begin{aligned} \sigma_{\{b\},a}(a) &= \{\{a,b\}\}, & \sigma_{\{b\},a}(b) &= \{\{b\}\}, \\ \sigma_{\{b\},a}(c) &= \{\{a,b,c\}\}, & \sigma_{\{b\},a}(d) &= \{\{a,b,d\}\}. \end{aligned} \tag{16.6}$$

16.2.5 Remarks. a) As this example shows, the implementation of a hanging-safe query may result in the removal of clauses, which in some cases have to be replaced by new clauses. We encountered two cases of such a replacement in the example.

Implementing the query $(\{a\}, b)$ on the learning space $\mathcal{L}$ of Example 16.2.4 removed the unique clause $\{b\}$ of b. The replacement clause was $\{a, b\}$, which can be seen as the union of the removed clause $\{b\}$ with another, remaining clause $\{a\}$.

The effect of the hanging-safe query $(\{b\}, a)$ was similar, but more complex in that its implementation resulted in the removal of the four states $\{a\}$, $\{a, c\}$, $\{a, d\}$, and $\{a, c, d\}$, three of which are clauses, namely $\{a, c\}$, $\{a, d\}$ and $\{a, d\}$. The replacement clauses are given in the last column of Table 16.2.

Table 16.2. Clauses added or deleted when implementing the query $(\{b\}, a)$ in Remark 16.2.5 a).

Item	clauses deleted	clauses added
a	$\{a\}$	$\{a, b\} = \{a\} \cup \{b\}$
c	$\{a, c\}$	$\{a, b, c\} = \{a, c\} \cup \{b\}$
d	$\{a, d\}$	$\{a, b, d\} = \{a, d\} \cup \{b\}$

We see that in all three cases, the replacement clause could be seen as the union of the removed clause with some remaining clause. In this simple example, finding the new clauses was easy and could be done by a cursory inspection of the new learning space. Needless to say, a more systematic approach, a general algorithm, is required in realistic cases involving very large structures with several hundred items. Actually, we plan a reverse course of action, in that the new surmise function is computed first. It is repeatedly updated by successive implementations of hanging-safe queries. At the end, the new learning space produced by the observed queries is computed as the span of all the final clauses. Nevertheless, this little example was useful in suggesting how the new surmise function might be computed from a given one modified by the implementation of a hanging-safe query[5]. The next example is more complex and more revealing (Theorem 16.2.10 will state the corresponding result).

b) Note that the clauses that were not removed by either $(\{a\}, b)$ or $(\{b\}, a)$ remained clauses in $\mathcal{L} \setminus \mathcal{D}_{\mathcal{L}}(\{a\}, b)$ and $\mathcal{L} \setminus \mathcal{D}_{\mathcal{L}}(\{b\}, a)$ (respectively). This observation can be generalized. For any learning space $\mathcal{G}$ and hanging-safe query (A, q), any clause for some item r is a minimal state of $\mathcal{G}$ containing r. If the clause C still belongs to $\mathcal{L} \setminus \mathcal{D}_{\mathcal{L}}(A, q)$, it will clearly be a minimal state of $\mathcal{L} \setminus \mathcal{D}_{\mathcal{L}}(A, q)$ containing r. So, it will be a clause for r in $\mathcal{L} \setminus \mathcal{D}_{\mathcal{L}}(A, q)$. The difficulty is the replacement of removed clauses, in particular because not all such removed clauses have to be replaced (see Problem 3).

16.2.6 Example. Consider the learning space with surmise function

$$\sigma(a) = \{\{a, b, c, e\}, \{a, d, f\}, \{a, c, d\}\}, \qquad \sigma(d) = \{\{d\}\},$$
$$\sigma(b) = \{\{a, b, d, f\}, \{b, c, e\}\}, \qquad \sigma(e) = \{\{e\}\},$$
$$\sigma(c) = \{\{c, e\}, \{c, d\}\}, \qquad \sigma(f) = \{\{f\}\}.$$
$$\sigma(g) = \{\{c, e, g\}, \{d, e, g\}, \{a, d, f, g\}\}.$$

Thus, the domain of $\mathcal{L}$ is $Q = \{a, b, c, d, e, f, g\}$ and its base is $\mathcal{B} = \cup_{r \in Q} \sigma(r)$. We can verify that none of the clauses is a clause for more than one item. By Condition (ii) in Theorem 5.4.1, the above surmise function σ indeed defines a learning space, which we denote by $\mathcal{L}$.

[5] Such a modification is not necessarily a reduction in size (cf. Example 16.1.7).

Suppose that we observe the positive response $\{c, e\}\mathcal{P}f$. It is easily checked that the query $(\{c, e\}, f)$ is hanging-safe, and so $\mathcal{L} \setminus \mathcal{D}_{\mathcal{L}}(\{c, e\}, f)$ is a learning space. To build its surmise function, we first search for the clauses to be removed from $\mathcal{L}$. There are four such clauses, which form the family

$$\mathcal{D}_{\mathcal{B}}(\{c, e\}, f) = \{\{f\}, \{a, d, f\}, \{a, b, d, f\}, \{a, d, f, g\}\}$$

(in the notation of Definition 16.1.5). It turns out that each of these clauses is the only one removed for a particular item. We have

$$\mathcal{D}_{\sigma(a)}(\{c, e\}, f) = \{\{a, d, f\}\}, \qquad \mathcal{D}_{\sigma(b)}(\{c, e\}, f) = \{\{a, b, d, f\}\},$$
$$\mathcal{D}_{\sigma(g)}(\{c, e\}, f) = \{\{a, d, f, g\}\}, \qquad \mathcal{D}_{\sigma(f)}(\{c, e\}, f) = \{\{f\}\}.$$

To find the replacement clauses and the surmise function of $\mathcal{L} \setminus \mathcal{D}_{\mathcal{L}}(\{c, e\}, f)$, it is impractical, as we argued earlier, to build the new learning space and then to compute its surmise function. In the next subsection, we lay out general principles for building the surmise function of the learning space $\mathcal{L} \setminus \mathcal{D}_{\mathcal{L}}(\{c, e\}, f)$ directly from σ and the hanging-safe query $(\{c, e\}, f)$. We then go back to this example in 16.2.9 and apply these principles to the building of the new clauses.

16.2.7 Building the new clauses: general principles. We consider a learning space $(Q, \mathcal{L})$ and a hanging-safe query (A, q). Suppose that E is a clause for an item r in $\mathcal{L} \setminus \mathcal{D}_{\mathcal{L}}(A, q)$ which was not a clause for r in $\mathcal{L}$. We want to characterize such clauses. As E is a state of $\mathcal{L}$ containing r, there is a clause C for r in $\mathcal{L}$ with $C \subset E$. For the set E to be a clause in $\mathcal{L} \setminus \mathcal{D}_{\mathcal{L}}(A, q)$, the clause C for r in $\mathcal{L}$ must have been removed by the query (A, q). Thus we have $A \cap C = \varnothing$ and $q \in C$ (with possibly $q = r$). Accordingly, we have $q \in E$ and moreover, as E is not removed, there is some item p in $A \cap E$. There is some clause D for p in $\mathcal{L}$ with $D \subseteq E$. In fact, we must have $A \cap D = \{p\}$. Indeed, assume there were s in $A \cap D \setminus \{p\}$. As $D \setminus \{p\}$ is state in $\mathcal{L}$, so is $C \cup (D \setminus \{p\})$. Then $C \cup (D \setminus \{p\})$ is a state in $\mathcal{L} \setminus \mathcal{D}_{\mathcal{L}}(A, q)$ which contains r, contradicting the minimality of E as a clause for r in $\mathcal{L} \setminus \mathcal{D}_{\mathcal{L}}(A, q)$. So, we have $A \cap D = \{p\}$. Again by the minimality of E, we conclude $E = C \cup D$ with C and D clauses in $\mathcal{L}$ belonging to two specific families defined below.

16.2.8 Definition. Suppose that $(\mathcal{L}, Q)$ is a learning space with surmise function σ, and let (A, q) be a query which is hanging-safe for $\mathcal{L}$. Two families of clauses in $\mathcal{L}$ were met in the previous paragraph during the analysis of the new clauses appearing in $\mathcal{L} \setminus \mathcal{D}_{\mathcal{L}}(A, q)$. The first one is the collection of clauses for r removed by (A, q). We recall that it is defined by

$$\mathcal{D}_{\sigma(r)}(A, q) = \{C \in \sigma(r) \,|\, A \cap C = \varnothing, q \in C\}. \tag{16.7}$$

The second family of clauses will be denoted by

$$\mathcal{H}_A = \bigcup_{p \in A} \{D \in \sigma(p) \,|\, A \cap D = \{p\}\}. \tag{16.8}$$

Our discussion in 16.2.7 indicates that any new clause E for r must be equal to some union $C \cup D$, with $C \in \mathcal{D}_{\sigma(r)}(A, q)$ and $D \in \mathcal{H}_A$. Note, however, that not all such unions are necessarily clauses in $\mathcal{L} \setminus \mathcal{D}_{\mathcal{L}}(A, q)$. Example 16.2.9, a continuation of Example 16.2.6, examines several cases.

16.2.9 Example. We are dealing with the learning space $(Q, \mathcal{L})$ whose surmise function σ is defined by

$$\begin{aligned} \sigma(a) &= \{\{a,b,c,e\}, \{a,d,f\}, \{a,c,d\}\}, & \sigma(d) &= \{\{d\}\}, \\ \sigma(b) &= \{\{a,b,d,f\}, \{b,c,e\}\}, & \sigma(e) &= \{\{e\}\}, \\ \sigma(c) &= \{\{c,e\}, \{c,d\}\}, & \sigma(f) &= \{\{f\}\}. \end{aligned}$$
$$\sigma(g) = \{\{c,e,g\}, \{d,e,g\}, \{a,d,f,g\}\}.$$

Considering the (hanging-safe) query $(\{c,e\}, f)$, we want to build the surmise function of the new (learning) space $\mathcal{L} \setminus \mathcal{D}_{\mathcal{L}}(\{c,e\}, f)$. From Example 16.2.6 we know that four clauses are to be removed, which are contained in the four families

$$\begin{aligned} &\mathcal{D}_{\sigma(a)}(\{c,e\}, f) = \{\{a,d,f\}\}, \quad \mathcal{D}_{\sigma(b)}(\{c,e\}, f) = \{\{a,b,d,f\}\}, \\ &\mathcal{D}_{\sigma(g)}(\{c,e\}, f) = \{\{a,d,f,g\}\}, \quad \mathcal{D}_{\sigma(f)}(\{c,e\}, f) = \{\{f\}\}. \end{aligned} \tag{16.9}$$

According to our discussion in 16.2.7, the new clauses to be added for the item r are among the unions of a state in $\mathcal{D}_{\sigma(r)}(\{c,e\}, f)$ with a state in $\mathcal{H}_{\{c,e\}}$. Here we have

$$\begin{aligned} \mathcal{H}_{\{c,e\}} &= \cup_{p \in \{c,e\}} \{D \in \sigma(p) \mid \{c,e\} \cap D = \{p\}\} \\ &= \{D \in \sigma(c) \mid \{c,e\} \cap D = \{c\}\} \cup \{D \in \sigma(e) \mid \{c,e\} \cap D = \{e\}\} \\ &= \{\{c,d\}\} \cup \{\{e\}\} \\ &= \{\{c,d\}, \{e\}\}. \end{aligned} \tag{16.10}$$

To get the new clauses for item a, say, we first combine the states from the family in (16.9) with the states in $\mathcal{H}_{\{c,e\}}$ from (16.10). The resulting unions are potential clauses for a. However, we must reject any such union that either contains a clause for a in $\mathcal{L}$ that is maintained in $\mathcal{L} \setminus \mathcal{D}_{\sigma(r)}(\{c,e\}, f)$, or contains another union of the same type.

Table 16.3 gathers the relevant information for four items.

The following result generalizes this example. We recall that the collection $\mathcal{L} \setminus \mathcal{D}_{\mathcal{L}}(A, q)$ resulting from the implementation of a positive response $A\mathcal{P}q$ to the learning space $\mathcal{L}$ is always a knowledge space (Theorem 16.1.6).

16.2.10 Theorem. *Let $(Q, \mathcal{L})$ be a learning space with surmise function σ. Suppose that $A\mathcal{P}q$ is a positive response to the query (A, q) (that is, $A \subset Q$ and $q \in Q \setminus A$). For any item r in Q, the clauses for r in the space $\mathcal{L} \setminus \mathcal{D}_{\mathcal{L}}(A, q)$ are the states which are minimal for the property of containing r in the collection*

$$(\sigma(r) \setminus \mathcal{D}_{\sigma(r)}(A, q)) \cup \{C \cup D \mid C \in \mathcal{D}_{\sigma(r)}(A, q) \text{ and } D \in \mathcal{H}_A\}. \tag{16.11}$$

Table 16.3. Potential new clauses resulting from the implementation of the hanging-safe query $(\{c, e\}, f)$, reasons for rejecting some of them, and the new sets of clauses for the items (with the new clauses in red). The state given in the third column is included in the potential new clause on the left, and is responsible for the rejection.

Item	Potential New Clauses	Included Clause	Clauses in $\mathcal{L} \setminus \mathcal{D}_{\mathcal{L}}(\{c, e\}, f)$
a	$\{a, d, f\} \cup \{c, d\}$ $\{a, d, f\} \cup \{e\}$	$\{a, c, d\}$	$\{\{a, c, d\}, \{a, b, c, e\}, \{a, d, e, f\}\}$
b	$\{a, b, d, f\} \cup \{c, d\}$ $\{a, b, d, f\} \cup \{e\}$		$\{\{b, c, e\}, \{a, b, c, d, f\}, \{a, b, d, e, f\}\}$
f	$\{f\} \cup \{c, d\}$ $\{f\} \cup \{e\}$		$\{\{c, d, f\}, \{e, f\}\}$
g	$\{a, d, g, f\} \cup \{c, d\}$ $\{a, d, f, g\} \cup \{e\}\}$	$\{d, e, g\}$	$\{\{c, e, g\}, \{d, e, g\}, \{a, c, d, f, g\}\}$

PROOF. Any clause in $\sigma(r) \setminus \mathcal{D}_{\sigma(r)}(A, q)$ is a state in $\mathcal{L} \setminus \mathcal{D}_{\mathcal{L}}(A, q)$ containing r. The same holds for any $C \cup D$ in the second collection in the union (16.11), for $C \cup D$ belongs to $\mathcal{L}$ and $A \cap (C \cup D) \neq \varnothing$ (because $A \cap D \neq \varnothing$). On the other hand, our discussion in 16.2.7 indicates that all the clauses for r must belong to the collection (16.11). The conclusion follows (cf. Problem 4). □

16.2.11 An algorithm for building a learning space. This algorithm is an adaptation of `QUERY`. It starts with an initial learning space. In practice, this initial learning space may be an ordinal space built by the first block of `QUERY`. (By Theorem 4.1.10, any ordinal space is a learning space.)

The general idea is that whenever a positive response to a query (A, q) is observed, the algorithm prunes the current learning space $\mathcal{L}$ only if the resulting space is a learning space, that is (in view of Theorem 16.1.6), only if (A, q) is hanging-safe in the sense of Definition 16.1.9. By Theorem 16.2.2, this means that the query (A, q) passes the

> `HS-test`: in the current learning space $\mathcal{L}$, there is no clause C for any item r in A such that $A \cap C = \{r\}$ and $q \in C$.

If the query (A, q) passes this test, the learning space $\mathcal{L}$ is replaced with the learning space $\mathcal{L} \setminus \mathcal{D}_{\mathcal{L}}(A, q)$. However, a failure of the `HS-test` does not lead to a final rejection of the query. Instead, (A, q) is temporarily put in a buffer, for reconsideration at a later stage. Note that the `HS-test` only requires the verification of a property of the surmise function, so that there is no need to fully store any of the successive learning spaces evolving through the procedure. Rather, the surmise function of $\mathcal{L} \setminus \mathcal{D}_{\mathcal{L}}(A, q)$ is built from the surmise function of the current learning space $\mathcal{L}$ and the response $A\mathcal{P}q$; Theorem 16.2.10 indicates how this can be done. As in the regular `QUERY` procedure, we want to avoid asking queries whose responses could be derived from those of other queries. So, we also store the entailment derived from all accepted positive responses by making inferences such as those listed in Subsection 15.2.1 and Table 15.2. Moreover, as in Koppen's Algorithm (cf. Section 15.2), we also compute negative responses and inferences. This is consistent with the hypothesis that the responses to the queries are dictated by a latent learning space, thus a knowledge space. In this conception, the positive responses to queries that ultimately fail the `HS-test` are regarded as human or statistical errors. By 'ultimately', we mean that the relevant queries have repeatedly failed the test, and are discarded at the end of the procedure. (Clearly, this conception makes sense only in cases in which the number of such 'errors' is small relative to the total number positive responses to queries.) We now go into more details.

The algorithm starts with an initialization step and then proceeds in two main stages. As mentioned earlier, we take the ordinal space obtained from the first block of the standard `QUERY` procedure as the initial learning space, which is represented in the algorithm by its surmise function and its entailment.

The first main stage begins, during which the queries are collected or inferred (see below), and then subjected to the `HS-test`. The algorithm enters the second main stage when all the positive queries have been observed or inferred. In the course of this stage, the algorithm also draws inferences from the responses to the queries, that is, it relies on the relations $\mathcal{P}^{\text{yes}}$ and $\mathcal{P}^{\text{no}}$ of Section 15.2. The management of the inferences is modified, however. Applying the rules from Section 15.2 produces both positive and negative inferences, that is, pairs which could be added to $\mathcal{P}^{\text{yes}}$ and $\mathcal{P}^{\text{no}}$, respectively. In our case, all the negative inferences are accepted, but a positive inference is accepted only if it passes the `HS-test`. When a new positive response $A\mathcal{P}q$ is collected or inferred, the algorithm checks whether (A, q) passes the `HS-test` on the current learning space $\mathcal{L}$. If it does, the algorithm implements (A, q) on $\mathcal{L}$, which produces the next learning space $\mathcal{L} \setminus \mathcal{D}_{\mathcal{L}}(A, q)$. If (A, q) fails the `HS-test`, then (A, q) is added to the pending table. The first stage terminates when all the queries with positive responses have been either implemented or stored in the pending table. Note that, at that time, as illustrated by our Example 16.1.13, some of the queries in the pending table may have become hanging-safe. Indeed, the almost hanging states for which they were critical may have been removed by later queries. Taking care of those queries is the function of the second stage.

In the second stage, the queries in the pending table are successively tested until none of them passes the `HS-test` or the table is empty.

To achieve the operations of the first and second stages, the algorithm relies on two buffers:

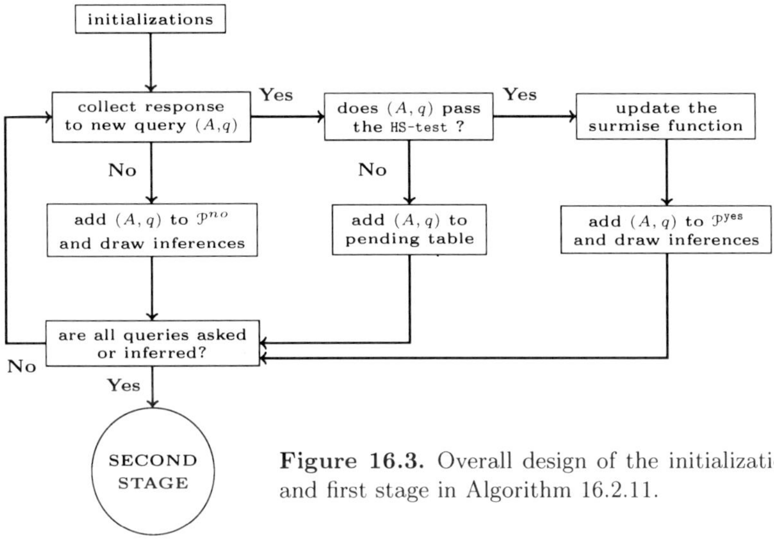

Figure 16.3. Overall design of the initializations and first stage in Algorithm 16.2.11.

- the `Pending-Table` where the queries having failed the `HS-test` are collected awaiting for further use;
- the `R-Store` in which the queries stored in the `Pending-Table` are copied at the beginning, and possibly during other passes of the second stage.

The algorithm is sketched in Figures 16.3 and 16.4. Its basic step are as follows.

1. Initializations: the algorithm computes the surmise function and the entailment of a starting learning space (typically, an ordinal space generated by the first block of `QUERY`).
2. In the first stage (Figure 16.3), the algorithm collects responses to queries and computes the related inferences. The principle is that the negative responses and inferences are always accepted, while the positive responses and inferences must past the `HS-test` before being accepted. A rejected positive response or inference is added to the `Pending-Table`. Notice that the description of the algorithm does not spell out all details covered by the instruction "draw inferences." These are covered in Section 15.2.
3. The second stage begins by checking whether the `Pending-Table` is empty. If so, the algorithm outputs the current surmise function which is the final one. The corresponding base spans the final learning space. Otherwise, all the pairs from the `Pending-Table` are moved to the `R-store`. Then, those pairs are examined one by one and subjected to the `HS-test` to check whether they are implementable. If yes, a new learning space is computed and the pair deleted; if not, the pair is moved into the `Pending-Table` (thus it becomes available for the next pass). Of course, inferences can also be computed along the second stage; we do not mention this in Figure 16.4.

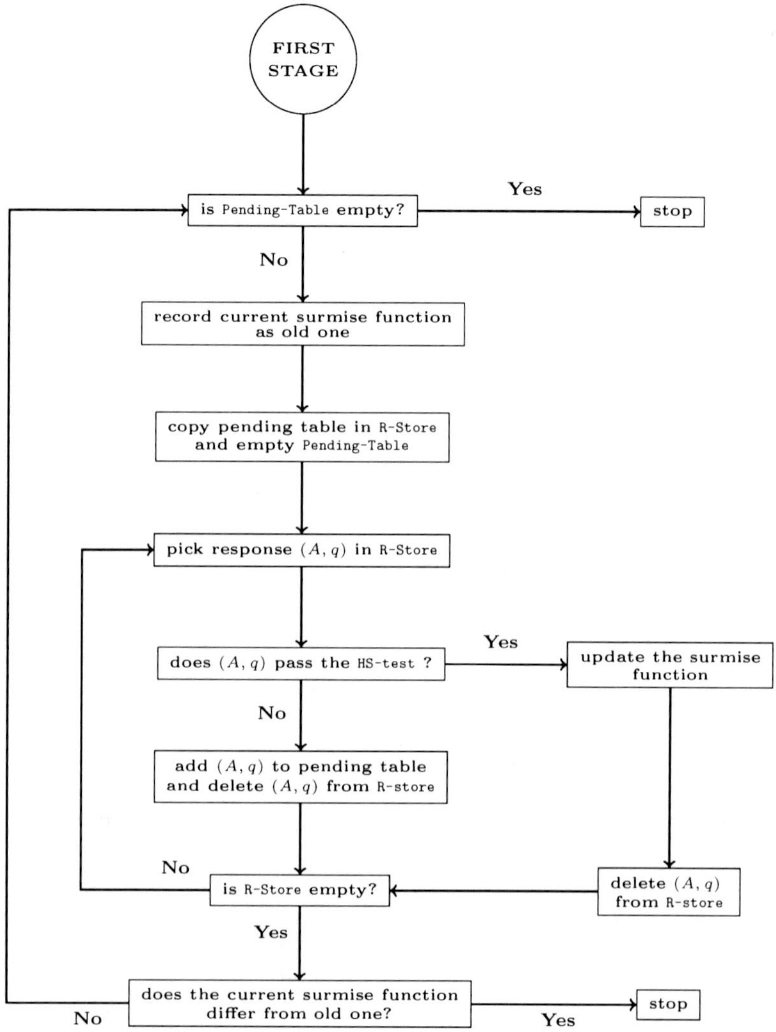

Figure 16.4. Overall design of the second stage of Algorithm 16.2.11.

16.2.12 Comments on Algorithm. a) Notice that in the first stage, we augment the pending table with positive responses and inferences that cannot be implemented; however, we make no use of the information in the table. Another version of the algorithm could specify that whenever the surmise function is modified in Stage 1, the `Pending-Table` is searched for queries whose status are now changed to implementable. (As we saw in Example 16.1.13, a positive response which failed the `HS-test` at some point may later pass that test because some critical state has been removed.)

b) Other features of `QUERY` could be taken into account here, as for instance the block structure or the subtable (cf. Section 15.2.8).

c) It is also clear that `PS-QUERY` rather than the straightforward `QUERY` could be taken as the basic routine. An actual implementation could integrate such features in the routine.

16.3 Engineering a Learning Space

David Eppstein proposes a very different solution to the problem of constructing a learning space through an elaboration of `QUERY` (Eppstein et al., 2009; Eppstein, 2010). He starts with a knowledge space $\mathcal{K}$ constructed by `QUERY` in the usual way, and then asks two questions:

1. How can we test whether the knowledge space $\mathcal{K}$ is well-graded?
2. If it is not, how can we in some optimal sense add states to $\mathcal{K}$ in order to achieve wellgradedness? (This is what is meant by 'engineering' in the title of this section.)

The first question has been considered earlier in this book (see Section 4.5), but Eppstein has obtained new results concerning in particular the complexity of the algorithm. The second question is new and originated with the two papers cited above. We only give a summary of this work here, without proofs.

Eppstein assumes a standard random-access-machine[6] model of computation performing the simple steps in constant time. The input to the algorithms is a family of sets $\mathcal{B}$, with each element of the sets in $\mathcal{B}$ taking a constant amount of computer storage. Typically, the family $\mathcal{B}$ is the purported base of a knowledge space or a learning space. As usual, the O-notation is used to represent the time bound for the algorithms. The results are expressed in terms of the following parameters.

16.3.1 Definition. We denote by

- b the number of sets in $\mathcal{B}$,
- a the size of the largest set in $\mathcal{B}$,
- c the sum of cardinalities of the sets in $\mathcal{B}$.

Clearly, we have $a \leq c \leq ba$.

If $\mathcal{B}$ is the base of a knowledge space $(Q, \mathcal{K})$ with surmise function σ, we have $b = |\cup_{q \in Q} \sigma(q)|$. Moreover, $\mathcal{K}$ is a learning space if and only if $b = \sum_{q \in Q} |\sigma(q)|$ (cf. Theorem 5.4.1).

[6] Random access enables the retrieval of stored data items directly, rather that sequentially, as in sequential access machines (cf. for example Shmoys and Tardos, 1995; Aho, Hoperoft, and Ullman, 1974).

16.3.2 Theorem. *The following two facts regarding a family $\mathcal{B}$ can be determined in time $O(bc)$.*

(i) *$\mathcal{B}$ is the base of a given $\cup$-closed family $\mathcal{F}$.*
(ii) *$\mathcal{B}$ is the base of a learning space.*

Next, Eppstein considers the situation in which a family $\mathcal{B}$ is not the base of a well-graded family. He asks: how can we, by the addition of some suitable states, extend $\mathcal{B}$ so as to produce a well-graded family that is as close as possible, in some sense, to the span of $\mathcal{B}$? The next definition introduces the relevant concept.

16.3.3 Definition. Suppose that $\mathcal{F}$ is a family of sets that is not well-graded. A *minimal well-graded extension* of $\mathcal{F}$ is a well-graded $\cup$-closed set family $\mathcal{H}$ such that:

(i) $\mathcal{F} \subset \mathcal{H}$;
(ii) there is no $\cup$-closed, well-graded family $\mathcal{H}'$ satisfying $\mathcal{F} \subset \mathcal{H}' \subset \mathcal{H}$.

16.3.4 Theorem. *Any family of sets $\mathcal{F}$ has a minimal well-graded extension which can be found in time $O(bca + b^3c)$.*

16.3.5 Remark. This result can of course be applied in a situation where $\mathcal{F} = \mathcal{B}$ is the base of a $\cup$-closed family. Eppstein points out that, in such a case, the completion algorithm used in the proof does not ensure that every set in the original base $\mathcal{B}$ is also a set in the new base, which may be regarded as a flaw. From our standpoint, however, this is not a defect. We would use the completion algorithm in situations in which a knowledge space $\mathcal{K}$ with base $\mathcal{B}$ has been constructed by the `QUERY` routine. What is important is that all the states of $\mathcal{K}$ are also states of the well-graded space constructed by the algorithm, which they are by definition of a well-graded extension.

In any event, the question leads Eppstein to ask: for a base $\mathcal{B}$ of a $\cup$-closed family, can we find a minimal well-graded extension $\mathcal{F}$ of $\mathcal{B}$ such that any set in $\mathcal{B}$ belongs to the base of $\mathcal{F}$? This problem is intractable.

16.3.6 Theorem. *Given a base $\mathcal{B}$ of a $\cup$-closed family, it is NP-complete to determine whether there exists a minimal well-graded extension $\mathcal{F}$ of $\mathcal{B}$ such that $\mathcal{B}$ is a subset of the base of $\mathcal{F}$.*

16.4 Original Sources and Related Works

The results of Section 16.3 are due to David Eppstein and taken from his recently published joint article with Jean-Claude Falmagne and Hasan Uzun (Eppstein et al., 2009).

The results and algorithms described in the rest of this chapter are new. We thank Jeff Matayoshi, Fangyun Yang, and especially Eric Cosyn for some useful discussions on these matters.

Problems

1. In an ordinal space on a set of n items, what is the number of hanging states? of almost hanging states?
2. Prove Lemma 16.1.3. Does the statement of the theorem still holds for knowledge spaces? For discriminative knowledge spaces? If it does not, provide a counterexample.
3. Construct an example of a learning space $\mathcal{L}$ with surmise function σ, and a query (A, q) which is hanging-safe for $\mathcal{L}$ and also satisfying the following condition: implementing (A, q) results in the removal of a clause C for item p, and $\sigma(A, q)(p) = \sigma(p) \setminus \{C\}$ (thus no 'new' clause is added).
4. Complete our proof of Theorem 16.2.10, making the arguments more explicit.
5. For the learning space of Example 16.2.6 (continued in 16.2.9), with domain $Q = \{a, b, c, d, e, f, g\}$ and surmise function

$$\begin{aligned}
\sigma(a) &= \{\{a, b, c, e\}, \{a, d, f\}, \{a, c, d\}\}, & \sigma(d) &= \{\{d\}\}, \\
\sigma(b) &= \{\{a, b, d, f\}, \{b, c, e\}\}, & \sigma(e) &= \{\{e\}\}, \\
\sigma(c) &= \{\{c, e\}, \{c, d\}\}, & \sigma(f) &= \{\{f\}\}.
\end{aligned}$$

$$\sigma(g) = \{\{c, e, g\}, \{d, e, g\}, \{a, d, f, g\}\}.$$

verify that the query $(\{a, b\}, g)$ is hanging-safe and, if so, build the surmise function of the learning space $\mathcal{L} \setminus \mathcal{D}_{\mathcal{L}}(\{a, b\}, g)$.
6. Write $(\mathcal{L}, Q)$ and $\mathcal{B}$ for a learning space and its base. If $\mathcal{L} = 2^Q$, we have $|\mathcal{B}| = |Q|$. Is there an example in which $|\mathcal{B}| = |Q| = |\mathcal{L}|$?
7. For a latent learning space which is a chain, describe the queries that will generate a positive responses.
8. Some infinite $\cup$-closed families have a base, and also satisfy Axioms [MA] and [L3]. Is Theorem 16.2.10 still true for such families?
9. Modify Example 16.2.9 so that the query $(\{c, e\}, f)$ is still hanging-safe but produces a case in which all the potential new clauses are rejected because they fail the minimality condition.

17

Analyzing the Validity of an Assessment

The theory described in this monograph has led to a number of applications, the most prominent ones being the `ALEKS` and the `RATH` educational softwares[1,2]. The focus of this chapter is on the `ALEKS` system. A large scale statistical analysis of the validity of its assessments was recently reported by Cosyn et al. (2010). We summarize these results which we regard as exemplary by the depth and details of the analysis. Using the term 'validity' in this context deserves a discussion.

17.1 The Concept of Validity for an Assessment

In formal logic, 'validity' refers, approximately, to the correctness of formulas or derivations (see, for example, Suppes, 1957). Its meaning in psychometrics, the theoretical basis of standardized testing, is quite different. In fact, 'validity' has several related meanings in this field. We review them briefly in our next subsection in order to clarify, from that standpoint, the similarities and differences between the psychometric and learning spaces methodologies.

17.1.1 On the validity and reliability of psychometric tests. The aim of a psychometric test is to provide a numerical score[3] indicative of the competence of a student in a scholarly subject. Generally speaking, a psychometric test is regarded as valid if its result correlates well with a relevant criterion. For example, a standardized test of quantitative abilities, taken before the beginning of a college education, would be regarded as valid if the correlation between the results of the test and the grades obtained in a mathematics course taken during the first semester is sufficiently high. This particular concept of validity is paramount for psychometricians in view of the methods used for the construction of such tests, which are based on a criterion of homogeneity of the items: an item whose response is poorly correlated with the overall result of the test may be rejected. The rationale behind such a procedure is that the test is regarded as a measurement instrument. The items are assumed to vary along a continuum of competence. Items which cannot be

[1] For the `RATH` system, see Hockemeyer (1997). Other relevant references and systems were mentioned on page 11.

[2] Some other applications are in chemical education, for example (see Arasasingham et al., 2004, 2005; Taagepera et al., 1997; Taagepera and Noori, 2000; Taagepera et al., 2002, 2008).

[3] Or, in some cases, a numerical vector with a small number of dimension.

J.-C. Falmagne, J.-P. Doignon, *Learning Spaces*,
DOI 10.1007/978-3-642-01039-2_17, © Springer-Verlag Berlin Heidelberg 2011

placed somewhere in that continuum are eliminated, even though they may be an integral part of the relevant scholarly curriculum[4].

As a consequence, a psychometric test is not automatically endowed with either 'face validity' or 'content validity.' These two related, but somewhat different concepts attempt to capture the connection between the test score and what the test is supposed to measure. For an introduction to these psychometric concepts, see for example Anastasi and Urbina (1997). In this monograph, which is a classic in this field, the authors explain 'content validity' as a concept involving

> "the systematic examination of the test content to determine whether it covers a representative sample of the behaviour domain to be measured" (Anastasi and Urbina, 1997, p. 114).

'Face validity' has a similar meaning, but is less methodical and relies essentially on the intuition of the experts concerning the relationship between the items of the tests and the variable intended to be measured by it.

The concept of 'reliability' is distinct from that of validity and applies to the replicability of the test results. The Oxford English Dictionary defines reliability as "*The extent to which a measurement made repeatedly in the same circumstances will yield concordant results.*"[5] In psychometrics, a test is regarded as reliable if the correlation between two different but similar versions of the same test is sufficiently high. In our terminology, 'similar versions' would mean that the two versions of the test contain different instances of the same items. It is clear that, while a psychometric test can be reliable without being valid, the reverse implication does not hold. (An extended technical discussion of the concept of 'reliability' can be found, for example, in Crocker and Algina, 1986, Chapters 6-9.)

17.1.2 The validity/reliabity of an assessment in a learning space. In principle, the situation is quite different in the case of the learning space because the collection of all the items potentially used in an assessment is, by design, a fully comprehensive coverage of a particular curriculum. To wit, the items in the learning space are stock features of standard textbooks on the subject, and no important concepts are missing. Asserting then that such an assessment, if it is reliable, is also automatically endowed with a corresponding amount of validity is plausible. In other words, assuming that the database of problem types is a faithful representation of the curriculum, the measurement of reliability is confounded with that of validity. This remains arguably true even in the case of a placement test manufactured by selecting a subset of the full set of items[6], at least if the selected items are chosen to be a representative sample of the curriculum.

---

[4] In other words, the test data must fit a particular unidimensional statistical model. Items contributing to a poor fit of the model are dropped from the test.

[5] O.E.D. 2000 Edition.

[6] This subset of items defines thus a projection in the sense of Definition 2.4.2.

Cosyn et al. (2010) use the following method to evaluate the reliability/validity of the results of an assessment performed in the framework of a learning space. At some point in each assessment, an *extra problem*[7] $\boldsymbol{p}$ is randomly selected in a uniform distribution on the set of all problems. Then, an instance of $\boldsymbol{p}$, also randomly selected, is given to the student, whose response is not taken into account in assessing the student's state. At the end of the assessment, the algorithm chooses one knowledge state, among all those in the learning space, representing the student's competence in the scholarly field. A prediction can thus be made regarding the student's response to the extra problem $\boldsymbol{p}$: if the selected state contains $\boldsymbol{p}$, then the the student response should be correct[8]; otherwise, the student's response should be incorrect. Cosyn et al. (2010) investigate the accuracy of such a prediction on a very large set of assessment data. The authors also examine the evolution of the accuracy of this prediction in the course of the assessment. Note that the authors assume that the probability of a correct response to $\boldsymbol{p}$ does not vary in the course of the assessment. This assumption seems reasonable since no learning is taking place at that time.

In the rest of this chapter, we summarize the results of Cosyn and his collaborators, which are based on the specific learning space built for elementary algebra[9] and on the assessment algorithm used at that time by the ALEKS system.

17.2 The ALEKS Assessment Algorithm

The assessment algorithm in the ALEKS system is a stochastic assessment procedure in the sense of Definition 13.3.4, which uses the parametrized multiplicative updating rule and the half-split questioning rule (cf. Definitions 13.4.4 and 13.4.7, respectively). We recall its main features, referring the reader to Chapter 13 for a detailed exposition. Each trial of the assessment consists in a triple (r_n, q_n, L_n) where n is the trial number, r_n is the response given (coded as 0 or 1 for incorrect or correct), q_n stands for the item asked, and L_n is the likelihood function —or probability distribution—on the set $\mathcal{K}$ of all the knowledge states on trial n.

Writing as usual $\mathcal{K}_q$ for the subcollection of $\mathcal{K}$ containing all the states containing item q,

$$L_n(\mathcal{K}_q) = \sum_{K \in \mathcal{K}_q} L_n(K)$$

[7] This is the terminology used by Cosyn et al. (2010) and we use it in the sequel. Thus, problem and item are synonyms in this chapter.

[8] Or should be corrected with probability $(1 - \beta_{\boldsymbol{p}})$, where $\beta_{\boldsymbol{p}}$ denotes the careless error probability to problem $\boldsymbol{p}$ (cf. Subsection 17.4.4).

[9] In the U.S., this particular mathematics curriculum is often called 'Beginning Algebra' or 'Algebra 1.'

is the likelihood, when sampling a state from $\mathcal{K}$ on trial n, that such a state contains item q. If there are no careless errors[10], $L_n(\mathcal{K}_q)$ can be regarded as the probability of a correct response to item q on trial n from the standpoint of the assessment algorithm.

The likelihood of a state increases or decreases on every trial depending on the event on that trial. If the response of the student to item q is correct, the probability of all the states containing q is increased and the probability of all the states not containing q is decreased. In the case of the multiplicative updating rule, the operator modifying the distribution L_n is commutative, which implies that the order of the student's responses to the problems does not matter: the distribution L_n is the same regardless of the order of the item-response pairs (r_1, q_1), ..., (r_{n-1}, q_{n-1}). The multiplicative updating rule operator is defined by Equations (13.9) and (13.10). On each trial, the item presented to the student is selected according to the half-split rule (cf. Definition 13.4.7). This means that item q may be presented to the student on trial n if the probability that the student has mastered item q is as close to .5 as possible, according to the current probability distribution L_n; that is, $|L_n(\mathcal{K}_q) - .5|$ is minimal. If two or more items gives the same minimal value, the algorithm chooses randomly (from a uniform distribution) between them.

In the `ALEKS` system, for the data relevant to the analysis reported below, the stopping rule was to end the assessment as soon as $L_n(\mathcal{K}_q)$ lies outside the interval $[.2, .8]$ for all items q.

17.3 The Methods

17.3.1 Outline. Cosyn and his colleagues performed three different types of statistical analysis on their data, which are based on more than one hundred thousand assessments taken from January 1, 2004 to July 1, 2007.

1) The first method quantifies the evolution of the information gathered by the algorithm in the course of the assessment. Suppose that the sequence of likelihood functions L_1, ..., L_n, ... has been kept in memory for each assessment. Each of the likelihood values $L_n(\mathcal{K}_{\boldsymbol{p}})$ subsumes the information concerning $\boldsymbol{p}$ accumulated by the algorithm up to trial n. Suppose temporarily that there are no careless errors and that all the assessments have the same length. (These assumptions are unrealistic and will be amended later on.) From the standpoint of the assessment algorithm, $L_n(\mathcal{K}_{\boldsymbol{p}})$ is thus the probability of a correct response to the extra problem $\boldsymbol{p}$ computed on trial n, regardless of when $\boldsymbol{p}$ has actually been presented during that assessment. For any trial n of every assessment, we thus have a pair $(L_n(\mathcal{K}_{\boldsymbol{p}}), r_{\boldsymbol{p}})$, where $r_{\boldsymbol{p}}$ is coded as 0 or 1 depending whether the response to the extra problem is incorrect or is correct. Fixing the trial number, and varying the assessment across the sample of students, the correlation between the $L_n(\mathcal{K}_{\boldsymbol{p}})$ and $r_{\boldsymbol{p}}$

[10] The data analyses reviewed in 17.3.4 and 17.4.4 do not rely on that assumption.

values can be computed. The authors of the study use the point biserial coefficient for this purpose, a choice dictated by the fact that $L_n(\mathcal{K}_{\boldsymbol{p}})$ can be regarded as a continuous variable, while $r_{\boldsymbol{p}}$ is a discrete one (for details about this coefficient, see 17.8).

In reality, the length of the assessment may vary considerably across the students tested. In the actual analysis, the correlation method sketched above is adapted by aligning all the assessments appropriately, using the 'Vincent curves' method (cf. Vincent, 1912, see Subsection 17.4.2). The results are the first ones reported in the next section.

2) The second method is the obvious one. At the end of the assessment, we can predict the student's actual response (correct or incorrect) to the extra problem $\boldsymbol{p}$ by checking whether or not $\boldsymbol{p}$ appears in the student's knowledge state selected by the assessment algorithm at the end of the test. We thus have two dichotomic variables: (1) $\boldsymbol{p}$ is or is not in the student assessed state; (2) the student's reponse is or is not correct. The effectiveness of such predictions can be evaluated using common measures of correlation between two dichotomous variables, such as the tetrachoric coefficient or the phi-correlation coefficient, two standard correlation indices. This analysis does not take possible careless errors into account.

A variant of the above method, described in Subsection 17.4.4, uses the same type of data, but corrects the predictions by a factor depending of the probability that the student commits a careless error in responding to a particular problem. The correlation coefficient used is the point biserial. We shall see that this results in a slight improvement of the correlations.

3) The third method is based on a different idea. At the end of most assessments, the student may initiate learning by choosing an item in the outer fringe of the state assigned by the assessment engine. In practice the student is presented with a display in the form of a pie chart, the slices of which correspond to the different parts of the scholarly material. Moving the computer mouse on one of the slices prompts the opening of a window listing those items of the outer fringe which concern this part of the material. The student chooses an item by clicking on the appropriate location of the window. The student may also select an item to learn by this method in a different situation, that is, not just after an assessment, but in the course of learning the subject matter. Suppose, for example, that K is the knowledge state of the student, as determined by the assessment algorithm. The student chooses some item $\boldsymbol{q}$ by the above method, and masters it. The new knowledge state is then $K \cup \{\boldsymbol{q}\}$. If this new state is not the domain, it has an outer fringe, and the student can choose a new item to learn, again by the same method.

If the knowledge state assigned to the student is the true one or at least strongly resembles the true one, then the prediction of what the student is capable of learning at that time should be sound. Accordingly, we can gauge the validity of the assessments by the probability that the student successfully masters an item chosen in the outer fringe of the assessed state. Cosyn and his collaborators have estimated such probabilities on the basis of a considerable number of assessment/learning trials in elementary algebra. Subsection 17.4.5 contains the results.

17.3.2 Notation. We review and complete our notation, which slightly differs from, but is consistent with, those used by Cosyn et al. (2010). We recall that, to each assessment corresponds one particular extra problem type.

We write:

Q	for the set of items, or domain;
$\mathcal{K}$	for the collection of states of the learning space; thus $\mathcal{K} \subseteq 2^Q$;
$\mathcal{A}$	for the set of all the assessments in the sample; $\mathcal{A} = \{\boldsymbol{a}, \boldsymbol{b}, \ldots, \boldsymbol{x}, \ldots\}$;
$\boldsymbol{x}$	for a variable denoting an assessment in the set $\mathcal{A}$;
$L_{\boldsymbol{x},n}$	for the probability distribution on $\mathcal{K}$ on trial n of assessment $\boldsymbol{x}$;
$\boldsymbol{p_x}$	for the extra problem asked in assessment $\boldsymbol{x}$;
$N_{\boldsymbol{x}}$	for the last trial number in assessment $\boldsymbol{x}$.

Note that the last trial number and the extra problem depend upon the assessment rather than on the student because some students in the sample may have taken several assessments[11]. They also define the collection of random variables

$$\mathbf{R}_{\boldsymbol{x}} = \begin{cases} 0 & \text{if the student's response to problem } \boldsymbol{p_x} \text{ is incorrect} \\ 1 & \text{otherwise} \end{cases} \tag{17.1}$$

with $\boldsymbol{x}$ varying in $\mathcal{A}$. Writing $\mathbf{R}_{\boldsymbol{x}} = \mathbf{R}_{\boldsymbol{p_x}}$ does not involve any ambiguity since any assessment $\boldsymbol{x}$ defines a unique extra problem $\boldsymbol{p_x}$. A careless error probability is attached to each problem. We denoted by

$\beta_{\boldsymbol{q}}$ the probability of committing a careless error to problem $\boldsymbol{q}$.

Thus, $\beta_{\boldsymbol{q}}$ is the conditional probability that a student whose knowledge state contains $\boldsymbol{q}$ commits a careless error in attempting to solve that problem. It is assumed that the parameter $\beta_{\boldsymbol{q}}$ only depends on the problem $\boldsymbol{q}$ and does not vary in the course of an assessment. In accordance with the convention used above for $\mathbf{R}_{\boldsymbol{x}}$, we use from now on the abbreviations

$$\beta_{\boldsymbol{x}} = \beta_{\boldsymbol{p_x}}, \qquad \mathcal{K}_{\boldsymbol{x}} = \mathcal{K}_{\boldsymbol{p_x}}.$$

We denote by $\mathbf{P}_{\boldsymbol{x},n}$ the probability that a student correctly solves the extra problem $\boldsymbol{p_x}$, based on the information accumulated by the assessment algorithm up to and including trial n of assessment $\boldsymbol{x}$. Assuming that there are no lucky guesses, this probability satisfies thus, for the nth trial of assessment $\boldsymbol{x}$, the equation:

$$\mathbf{P}_{\boldsymbol{x},n} = (1 - \beta_{\boldsymbol{x}}) L_{\boldsymbol{x},n}(\mathcal{K}_{\boldsymbol{x}}). \tag{17.2}$$

[11] The authors do not take into account this aspect of the data. This is reasonable considering the large size of their sample of assessments.

17.3.3 Estimating the careless error parameters. The same item is occasionally presented more than once in the course of an assessment: once as the extra problem, and also as one or more regular items of the assessment[12]. This makes it possible to estimate, for each item $\boldsymbol{q}$, the careless error parameter $\beta_{\boldsymbol{q}}$ without relying on the hypotheses of the learning space model. When an item is presented more than once as a regular item of the assessment, only the first presentation is retained for the analysis. For each item $\boldsymbol{q}$, the data takes the form of a 2×2 matrix

		Regular item 0	Regular item 1
Extra problem	0	x	y
	1	z	w

in which 0 and 1 stand for 'incorrect' and 'correct' respectively. (Thus, z is the number of cases in which the response to the extra problem was correct, and that to the regular item was incorrect.) The estimation model used by Cosyn et al. (2010) has two parameters, the probability $\beta_{\boldsymbol{q}}$ of a careless error to item $\boldsymbol{q}$, and the probability $\kappa_{\boldsymbol{q}}$ that the knowledge state of the student belongs to $\mathcal{K}_{\boldsymbol{q}}$. Denoting by $p_{\boldsymbol{q}}(i,j)$ the probability of the case (i,j) in the above table, with $i,j \in \{0,1\}$, the model is defined by the four equations[13]:

$$p_{\boldsymbol{q}}(0,0) = \beta_{\boldsymbol{q}}^2\kappa_{\boldsymbol{q}} + (1-\kappa_{\boldsymbol{q}}) \tag{17.3}$$
$$p_{\boldsymbol{q}}(0,1) = \beta_{\boldsymbol{q}}(1-\beta_{\boldsymbol{q}})\kappa_{\boldsymbol{q}} \tag{17.4}$$
$$p_{\boldsymbol{q}}(1,0) = (1-\beta_{\boldsymbol{q}})\beta_{\boldsymbol{q}}\kappa_{\boldsymbol{q}} \tag{17.5}$$
$$p_{\boldsymbol{q}}(1,1) = (1-\beta_{\boldsymbol{q}})^2\kappa_{\boldsymbol{q}}. \tag{17.6}$$

We thus have

$$\sum_{i,j} p_{\boldsymbol{q}}(i,j) = 1.$$

We now write $N_{\boldsymbol{q}}$ for the number of assessments having at least two presentations of item $\boldsymbol{q}$, with one of them as the extra problem, and $N_{\boldsymbol{q}}(i,j)$ for the number of times (i,j) is realized among the $N_{\boldsymbol{q}}$ assessments. Cosyn et al. (2010) obtain the estimated values of $\beta_{\boldsymbol{q}}$ (and also of $\kappa_{\boldsymbol{q}}$, but this is of lesser interest) by minimizing the Chi-square statistic

$$\mathtt{Chi}_{\boldsymbol{q}}(\beta_{\boldsymbol{q}},\kappa_{\boldsymbol{q}}) = \sum_{i,j} \frac{\left(N_{\boldsymbol{q}}(i,j) - N_{\boldsymbol{q}}p_{\boldsymbol{q}}(i,j)\right)^2}{N_{\boldsymbol{q}}p_{\boldsymbol{q}}(i,j)}. \tag{17.7}$$

We leave to the reader to work out the details (see Problems 1 and 3).

[12] In such cases, different instances of the item are almost always presented.

[13] We recall the the probability of a lucky guess is assumed to be zero.

17.3.4 Aligning the assessments: The Vincent curves. Cosyn and his collaborators have analyzed the temporal course of the assessments by tracking down the correlation between the probability $\mathbf{P}_{\boldsymbol{x},n} = (1 - \beta_{\boldsymbol{x}})L_{\boldsymbol{x},n}(\mathcal{K}_{\boldsymbol{x}})$ that the response to the extra problem $\boldsymbol{p}_{\boldsymbol{x}}$ is correct and the 0-1 variable $\mathbf{R}_{\boldsymbol{x}}$ coding the actual response of the student to that problem (cf. Subsection 17.3.2). Each correlation is thus computed by keeping the trial number n constant and varying $\boldsymbol{x}$, which denotes the assessment, in the pairs $(\mathbf{P}_{\boldsymbol{x},n}, \mathbf{R}_{\boldsymbol{x}})$. If all the assessments had the same length N, these calculations would be straighforward. However, the length of the assessments vary considerably. The solution adopted by Cosyn et al. (2010) to deal with this difficulty is a classical one: they have split each assessment into 10 parts, or 'deciles', of (approximately) equal length, and aligned the assessments on the last trial of each part. To be precise: the trial number retained for computing the correlation in decile i $(1 \leq i \leq 10)$ is the smallest integer not smaller than $i \times N_{\boldsymbol{x}}/10$. They also included the initial trial in each assessment. As an illustration, the trial numbers retained for the computation of the correlations are given in the table below for three assessments $\boldsymbol{a}$, $\boldsymbol{b}$ and $\boldsymbol{c}$ of respective lengths 17, 25 and 30. The 9th correlation (the correlation for the 8th decile) for some item $\boldsymbol{p}$ would thus be based on trials 14, 20 and 24 of assessments $\boldsymbol{a}$, $\boldsymbol{b}$ and $\boldsymbol{c}$ respectively (the red column of the table) in which $\boldsymbol{p}$ has been presented as the extra problem. For each extra problem, the evolution of the correlation is thus traced throughout 11 trials, which are numbered 0, 1, ..., 10 in the second row of the table. The corresponding graph is usually referred as a *Vincent curve* (from the original article implementing this method, which is due to Vincent, 1912). Note that only .01% of the assessment were shorter than 11 trials. There were dealt with by a special rule. We omit the description.

Table 17.1. Trial numbers retained for the Vincent curve analysis for three assessments $\boldsymbol{a}$, $\boldsymbol{b}$, and $\boldsymbol{c}$ of respective lengths 17, 25 and 30. The first column, headed by the letter 'A', lists the assessments. Number 1 in rows 3, 4 and 5 denotes the initial trials in each assessment; the other numbers are those of the last trials of each decile.

	Trial numbers retained for the correlation analysis											
A	0	1	2	3	4	5	6	7	8	9	10	$N_{\boldsymbol{x}}$
$\boldsymbol{a}$	1	2	4	6	7	9	11	12	14	16	17	17
$\boldsymbol{b}$	1	3	5	8	10	13	15	18	20	23	25	25
$\boldsymbol{c}$	1	3	6	9	12	15	18	21	24	27	30	30

17.3.5 Correlation coefficients used. The index chosen to compute the correlation between the variables $\mathbf{P}$ and $\mathbf{R}$ is the *point biserial coefficient*

$$r_{pbis} = \frac{M_1 - M_0}{s_n}\sqrt{\frac{n_1 n_0}{n^2}} \tag{17.8}$$

where

n	is the number of pairs $(\mathbf{P}, \mathbf{R})$,
s_n	is the standard deviation of the continuous variable $\mathbf{P}$,
n_1, n_0	are the numbers of cases $\mathbf{R} = 1$, $\mathbf{R} = 0$, respectively,
M_1, M_0	are the conditional means of $\mathbf{P}$ given $\mathbf{R} = 1$ and $\mathbf{R} = 0$.

This coefficient is frequently used when one of the variables is continuous and the other one discrete, as is the case here. It gives an estimate of a Pearson correlation coefficient under some hypotheses regarding the joint distribution of the two random variables involved (for details, see Tate, 1954, or any other of the standard psychometric texts[14]).

Two different correlation coefficients have also been used for other parts of the study, namely, the 'tetrachoric' and the 'phi' coefficients. We postpone their introduction for the moment.

17.4 Data Analysis

17.4.1 The participants. The assessments were taken via the internet by college or high school students (85% and 15%, respectively) typically in the framework of a course, with some of the student taking more that one assessment during the course. The numbers of students and of assessments differs depending on the part of the study, as will be indicated.

17.4.2 Temporal evolution of the prediction. The Vincent curve analysis of the correlation coefficient r_{pbis} is limited to the initial part of the assessment[15], which uses 82 items selected from the 262 items forming the beginning algebra course in the `ALEKS` system[16]. Of these 82 items, 12 items were discarded because the relevant data was too meager for a reliable estimate of the correlation coefficient. The data considered here concern the remaining 70 items and are based on 78,815 assessments from 42,857 students.

About 17 to 18 questions are asked during the first part, which may be regarded as a placement test (after which the assessment is pursued without the student being aware of any hiatus). This small number of questions must be kept in mind when pondering the values reported for the correlations. As mentioned in Subsection 17.3.4, the point biserial coefficient was used to compute the correlation between the probability of a correct response to the extra problem on trial n, which is specified by Equation (17.2), and the actual response to that problem coded as 0 or 1. Two Vincent curves are displayed in Figure 17.1. The blue curve traces the evolution of the medians

[14] The Wikipedia entry for the concept 'point-biserial coefficient' is a good start.

[15] The split of the assessment into two part is due to storage limitation in the PC's. For technical reasons, a substantial part of an assessment was taking place in the client's computer. (This is no longer true.) The extensive data required for the Vincent curve analysis was only available for the initial part of the assessment.

[16] At that time. This number is now somewhat larger.

of the distributions of the point biserial coefficient, during the first part of the assessment, for items having a careless error probability smaller than .25. The red curve is similar, and concerns items with a careless error probability exceeding .25.

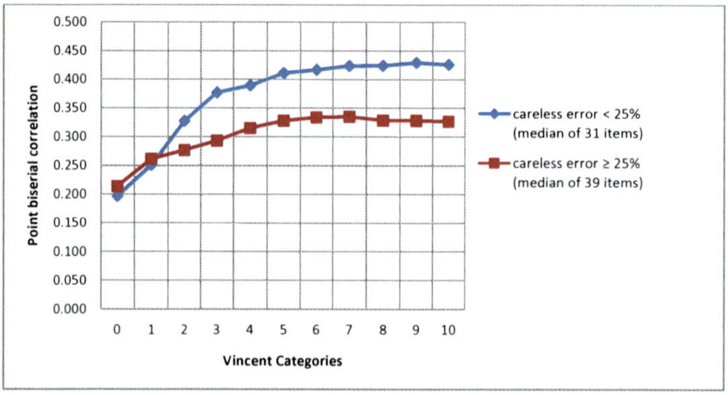

Figure 17.1. Two Vincent curves of the median r_{pbis} values correlating the probability of a correct response to the extra problem predicted by Equation (17.2) and the variable $\mathbf{R}_x$ coding the response to that problem as 0 or 1 for 'incorrect' or 'correct'. The number 0 on the horizontal axis marks the first trial of the assessment. The number 1,..., 10 indicate the ten Vincent categories. The blue Vincent curve describe the evolution of the median correlation for the 31 items having a careless error probability smaller that .25. The red curve is similar and concerns the remaining 39 items. (Reproduced with permission.)

The difference between the two curves stresses the importance of the careless errors and is remarkable. Cosyn et al. (2010) also compute the Vincent curves of a few exemplary items with a careless error probability varying even more widely. The two extreme cases were .13 and .49. While the first item reaches a correlation value around .65 at the end of the assessment, the correlation values for the other never exceed .2 and do not increase through the assessment. Needless to say, as the authors recognize, such an item deserves improvement. (They insistently make the point that items, being an integral part of a curriculum, are rarely rejected, if at all.)

17.4.3 Prediction based on the final state. The Vincent curve analysis from Cosyn et al. (2010) that we just summarized only concerns the initial part of the assessment. The authors consider then the ultimate result of the assessments, and investigate how well the final knowledge state selected by the assessment engine predicts the response to the extra problem. For each item p, the relevant data take the form of a 2×2 matrix, with the two variables:

1. the extra problem p is or is not in the final state selected;
2. the response to the extra problem is correct or incorrect.

Note that this does not take the careless errors into account (see the next paragraph in this regard). These data are substantial, and based on 240,003 assessments performed between January 1, 2004 and July 1, 2007. They are analyzed in terms of two correlation coefficients, the tetrachoric and the phi. Both of them are adequate in that they were designed to deal with such double dichotomies. However, neither of them is ideal because the hypotheses underlying their use do not fit the situation well (as the authors aknowledge[17]).

We reproduce below the covariation graph of the values of the tetrachoric and the phi coefficient for each of the 250 items. Each point of the graph represents an item, the coordinates of which are the values of the two correlation coefficients for that item. The median correlations for both indices are indicated on the graph. They are: .67 for the tetrachoric, and .35 for the phi coefficient. In both cases, the correlations for the grouped data, obtained from the 2×2 matrix resulting from adding the corresponding numbers in each of the four cells in the 250 individual item matrices, are much higher, namely .81 and .57. Note that these results were obtained without consideration of the careless errors, which we already know to be substantial for some items.

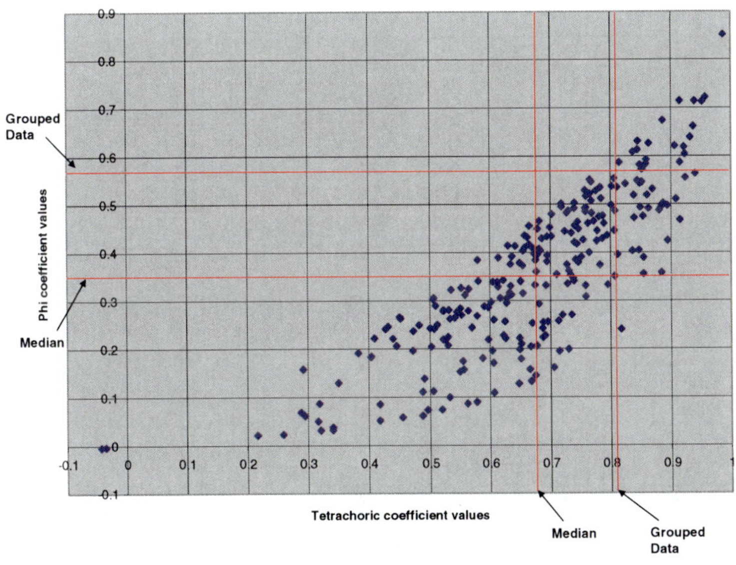

Figure 17.2. Covariation graph of the values of the tetrachoric and the phi coefficients for the 250 items in beginning algebra. (Reproduced with permission.)

[17] The tetrachoric index is an approximation of the Pearson correlation coefficient. It is based on the hypothesis of an underlying pair of random variables, with a joint Gaussian distribution. This is hardly satisfied here. The use of the phi coefficient is predicated on similar assumptions. (Chedzoy, 1983; Harris, 1983, in Volumes 6 and 9, respectively, of the Encyclopedia of Statistical Sciences).

All the representing points of items are below the diagonal, showing that the values of the tetrachoric are inflated in comparison with those the phi coefficient, which is a standard finding. The substantially higher tetrachoric and phi correlation values obtained for the grouped data is not an artifact. An illustration of this phenomenon is given in Table 17.2 in the case of the grouping of the data matrices of two items, one being easy and the other one difficult. The three matrices are displayed in the table. We can see that the grouping results in putting relatively high numbers in the two $(0,0)$–$(1,1)$ cells and thus boosting the correlation.

Table 17.2. The correlation matrices, one pertaining to an easy item on the left, and the other to difficult item on the right. The 'In' and 'Out' labeling the columns refer to the 'In the state' and 'Not in the state' cases. The third matrix below combines the data of the two top matrices.

DIFFICULT

	In	Out	
Correct	7	18	Tetra: .751
Incorrect	4	280	Phi: .359

EASY

	In	Out	
Correct	267	4	Tetra: .742
Incorrect	22	8	Phi: .385

GROUPED

	In	Out	
Correct	274	22	Tetra: .969
Incorrect	26	288	Phi: .842

17.4.4 Adjusting for careless errors. The authors further refine their analysis by taking the careless errors into account as a weighting factor. They introduce, for each item $\boldsymbol{x}$, a variable

$$\mathbf{S}_{\boldsymbol{x}} = \begin{cases} 1 - \beta_{\boldsymbol{x}} & \text{if the final state contains the extra question } \boldsymbol{x} \\ 0 & \text{otherwise.} \end{cases}$$

Using the point biserial coefficient, they compute for the grouped data the correlation between the variables $\mathbf{S}_{\boldsymbol{x}}$ and $\mathbf{R}_{\boldsymbol{x}}$. The value reported is .61, thus slightly higher than the .57 obtained for the phi coefficient for the same grouped data.

The .61 value obtained for the grouped data of the point biserial coefficient is noteworthy in comparison with the point biserial values reported by the Educational Testing Services (ETS) (2008) report[18] for the Algebra I

[18] Produced for the California Department of Education (Test and Assessment Division). See `http://www.cde.ca.gov/ta/tg/sr/documents/csttechrpt07.pdf`.

California Standards Test (CST), which covers roughly the same curriculum as the **ALEKS** assessment for elementary algebra and is given to more than 100 000 students each year. The Algebra I CST is comprised of 65 multiple choice questions (items) and is constructed and scored using Item Response Theory (IRT), for which the point biserial coefficient is a standard measure. In particular, for each of the 65 items, a point biserial item-test correlation was computed, which measured the relationship between the dichotomous variable giving the 1/0 item score (correct/incorrect) and the continuous variable giving the total test score (see p. 397 of the ETS report referenced above). For the 2007 administration of the Algebra I CST, the mean point biserial coefficient for the 65 items was .36, and the median was .38 (see Table 7.2, p. 397 of the ETS report). The minimum coefficient obtained for an item was .10 and the maximum was .53 (Table 7.A.4, pp. 408–9, of the ETS report). The averages for preceding years on the test were similar, namely, the mean point biserial coefficients were .38 in 2005 and .36 in 2006 (see Table 10.B.3, pp. 553, of the same report).

The average correlation obtained by Cosyn et al. (2010) for the ALEKS assessments is only slightly above that reported in the ETS report. However, Cosyn et al. (2010) argue that no selection of items took place in the ALEKS case. By contrast, the items having a point-biserial correlation below .19 in a trial run were removed from the test in the ETS study. Moreover, only around 25–35 questions were asked in the ALEKS assessments, which is roughly half of the number of questions asked in the ETS test.

Finally, and most importantly, it must be recalled that in the case of the ALEKS system, the choice of the extra problem is actually the choice of an instance for that problem. In other words, this involves the random choice of a specific question to ask in a very large set, the size of which is the sum of all the instances in each of the possible extra problems. According to Cosyn et al. (2010), the size of this set is of the order of 100,000 different instances. In the case of the Educational Testing Services (ETS) (2008) study, the comparable random choice is made in a set of 65 items (that is, instances in the ALEKS terminology[19]).

17.4.5 Learning success. In the ALEKS system, at the end of an assessment, the student is offered to start learning by choosing an item in the outer fringe of his state. In the framework of learning space theory, the student should be ready to learn such an item at that time. This choice initiates a process during which the student works on various instances of the item and studies the explanations given. This process is actually a random walk with two absorbing barriers. The walk moves right or left depending on whether the student solves or fails to solve an instance. Hitting the left barrier signifies a failure to solve the item, while hitting the right one means success. In either case, the student's knowledge state is readjusted, and an item in the new outer fringe is then

[19] Remember that, in psychometrics, the term 'item' corresponds to what is called an instance in Knowledge Space Theory.

chosen by the student to continue learning (for details, see Cosyn et al., 2010, Chapter 2).

The probability that a student masters an item—that is, hits the right barrier of the corresponding random walk—provides an indirect way of gauging the validity of an assessment. Presumably, if the assessment is on target, then such probabilities should be high.

Figure 17.3[20] displays the distribution of the estimated probabilities of successfully master 256 items of beginning algebra (when chosen in the outer fringe of a student's state). It is clear that most items are satisfactorily handled. The graph shows that 80% of the items have a probability of success of at least .8, the median of the distribution being around .92. However, the left tail of the distribution indicates that some problems are not learned easily and that adjustments deserve to be made. The data analyzed are based on 1,564,296 such random walks.

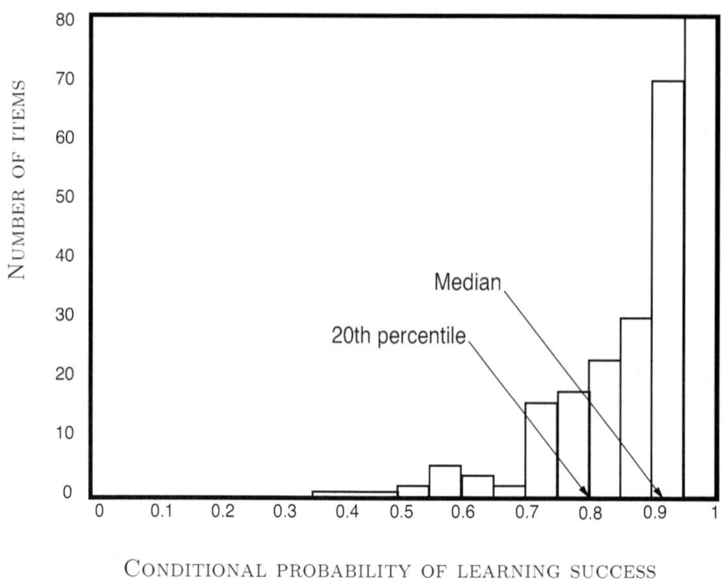

Figure 17.3. For the 256 items in elementary algebra (out of 262), the distribution of the estimated values of the conditional probabilities that a student, having chosen an item in the outer fringe of his state, hits the right barrier of the random walk. The problem is then regarded as having been mastered.

[20] This figure is reproduced from Chapter 2 in Cosyn et al. (2010), with permission.

17.5 Summary

This chapter reviews the results of an application of learning space theory reported by Cosyn et al. (2010). The focus of the study is the validity (or predictive power) of the assessment. The scholarly subject is beginning algebra. The data concern a large number of assessments taken by college and high school students from 2004 to 2007. Three methods are used in the analysis. The first two rely on correlation indices and are based on the extra problem procedure: in each assessment, and extra problem is proposed to the student, the response to which is not used for assessment purposes but can be predicted on the basis of the assessment results.

1) Using Vincentized data, the first method analyzes the evolution of the correlation between: (i) the probability of a correct response to the extra problem based on the information accumulated by the assessment up to a point varying in 10 Vincent deciles categories plus the initial trial; and (ii) the actual response to that extra problem. Two different Vincent curves are computed, each involving the same 11 categories. The first curve concerns the 31 items having a careless error probability smaller .25. Figure 17.1 shows that the median correlation for these items evolve from around .2 on trial 1, to about .425 in the last Vincent category. The second curve of Figure 17.1 traces the evolution of these correlations for the remaining 39 items. The median correlation, which is initially roughly the same as that of the previous curve, increases then smoothly up to about .325. These statistics are based on the first phase of the assessment, which only uses 82 out of the 262 items of the elementary algebra domain of the `ALEKS` system at that time[21]. It is noteworthy that, in any actual assessment, only about 17-18 items, on the average, are proposed to the students during this initial phase of the assessment.

2) The second method analyzes the correlation between the response to the extra problem and the prediction made on the basis of the final knowledge state obtained at the end of the assessment. Two variants are considered. The first one does not take the possibility of careless errors in computing the prediction. For each item, the data take to form of a 2×2 matrix with the two variables: 1. the extra problem $\boldsymbol{p}$ is or is not in the final state selected; 2. the response to the extra problem is correct or incorrect. Two correlation indices, the tetrachoric and the phi, are used for 250 items out of 262. (According to the authors, the data for the remaining 12 items were too scarce to provide reliable estimates of the coefficients.) The covariation graph of the values of the two coefficient reveals the following most salient facts.

(i) The median correlation values are .67 for the tetrachoric, and .35 for the phi coefficient. The much higher values obtained for the tetrachoric coefficient is not surprising in view the literature on the two indices.

[21] Today, the number of items for the elementary algebra domain used in the `ALEKS` system is around 350.

(ii) In both cases, the correlations for the grouped data, obtained from the 2×2 matrix resulting from adding the corresponding numbers in each of the four cells in the 250 individual item matrices, are much higher, namely .81 and .57. An example shows that such an increase should not be surprising.

(iii) For both indices, the values of the coefficients vary considerably across items; for example from 0 to about .86 for the the phi coefficient. A much needed improvement is thus required for some items. Overall, it appears that the correlation values compare quite favorably with a similar analysis of a psychometric test, for the same elementary algebra curriculum, reported in Educational Testing Services (ETS) (2008).

These results were obtained without consideration of the careless errors, which we already know to be substantial for some items. When careless error probabilities estimated from the data are brought into play, the correlation values for both indices are slightly larger.

3) The third method relies on a different type of data, namely, the probability for a student to master an item chosen in his outer fringe. These probabilities have been computed by Cosyn and his colleagues on a very large sample of cases, involving $1,564,296$ learning occasions. The data shows that 80% of the items have a probability of learning success of at least .8, the median of the distribution being around .92. However, they also reveal that a few problems are not learned easily and that adjustments deserve to be made.

Problems

1. The Chi-square statistic defined by Equation (17.7) can be minimized for the values of the parameters β_q and κ_q by a standard optimization algorithm, but Cosyn et al. (2010) proceed differently, and directly. Find the estimation equation for β_q and κ_q analytically. (Hint: you may want to use the Lagrange multipliers method.)

2. What are the weaknesses of the model defined by Equations (17.3)-(17.6), if any, with respect to the problem considered, that is, the estimation of the careless errors? Another possibility for such estimations is to use the final state, and estimate the conditional probability of an error to the extra problem $\boldsymbol{q}$ given that $\boldsymbol{q}$ belongs to the final state. Work out the details of this method. What are its drawbacks, if any?

3. Assume that the probability of lucky guess to problem $\boldsymbol{q}$ is equal to some number $\gamma_q > 0$, a parameter. Thus, γ_q is the probability of a correct response to item $\boldsymbol{q}$, even though this item does not belong to the student state. Can the model defined by Equations (17.3)–(17.6) be adapted to estimate γ_q? What are the drawbacks, if any?

18

Open Problems

We have gathered a number of problems in the course of our investigations, which we have left unsolved. We list them below for the interested reader, with relevant references when available.

18.1 Knowledge Spaces and $\cup$-Closed Families

18.1.1 On large bases. Let Q be a domain with a finite number m of items. Consider all the bases of knowledge spaces on Q. What is the largest cardinality of a base as a function of m ? What are all the knowledge bases on Q having a base of that cardinality? A manuscript of R.T. Johnson and T.P. Vaughan contains results for small values of m. (It is the extended preprint version of Johnson and Vaughan, 1998).

18.1.2 Defining a knowledge spaces by a language. Let $\mathcal{K}$ be a discriminating knowledge space. Does there always exist some assessment language (in the sense of 9.2.3) describing $\mathcal{K}$ but no other knowledge space? Or better, describing no other knowledge structure?

18.1.3 Projections and bases. Let $\mathcal{K}$ be a knowledge space on the domain Q and for any nonempty subset Q' of Q denote by $\mathcal{K}'$ the projection of $\mathcal{K}$ on Q' (cf. Theorem 13.7.4). Find necessary and sufficient (interesting) conditions on the knowledge space $\mathcal{K}$ implying that the space $\mathcal{K}'$ always has a base. Conditions that are only sufficient would also be of interest if they cover a wide variety of examples (among which the finitary spaces).

18.1.4 Uniqueness of a Hasse system. Characterize efficiently the granular knowledge spaces that admit a unique Hasse system (see end of Section 5.5).

J.-C. Falmagne, J.-P. Doignon, *Learning Spaces*,
DOI 10.1007/978-3-642-01039-2_18,

18.1.5 Frankl's Conjecture. A famously difficult problem asks whether for any finite union-closed family $\mathcal{K}$ with $\cup\mathcal{K}$ finite and $\mathcal{K} \neq \{\varnothing\}$, there always exists some element in $\cup\mathcal{K}$ that belongs to at least half of the subsets in $\mathcal{K}$. This is often referred to as Frankl's conjecture. A good access to the literature is the Wikipedia entry for "Union-closed sets conjecture." It should be complemented with Johnson and Vaughan (1998). (Warning: the risk of wasting precious research time is high.)

18.2 Wellgradedness and the Fringes

18.2.1 Strengthening [L1] and [L2]. Lemma 2.2.7 asserts that any well-graded partially union-closed family is a partial learning space, and that the converse implication does not hold. Find axioms that strengthen (or at least are in the spirit of) Axioms [L1] and [L2] that characterize well-graded partially union-closed families.

18.2.2 About the fringes economy. The fringes of a state in a learning space were defined in 4.1.6. The concept was introduced informally earlier, in Section 1.1.5, and presented there as a device permitting an economical representation of the states. Consider the following parameter for measuring the overall economy realized by such a representation: the sum of the sizes of all the states minus the sum of the sizes of all the fringes. (Another parameter is obtained when we divide that difference by the number of states.) It is not difficult to find examples in which the fringe representation of states is not economical at all (that is, in which the parameter takes a negative value). On the other hand, for what kind of learning space is the economy:

(i) maximal in the sense that for a fixed number of items, the parameter takes the largest possible value)?
(ii) minimal (for a fixed number of items, the parameter takes the smallest possible value)?

18.2.3 Characterize fringes mappings. Characterize those mappings $K \mapsto (K^{\mathcal{I}}, K^{\mathcal{O}})$, $K \mapsto K^{\mathcal{I}} \cup K^{\mathcal{O}}$, $K \mapsto K^{\mathcal{I}}$, etc. arising from learning spaces. Which learning spaces are completely specified by any such mappings (one or more of them)? The same problem can be raised for well-graded families.

18.2.4 Characterize well-graded spans. Theorem 4.5.8 characterizes those families whose span† is well-graded. However the characterization refers explicitly to the span†. Find a characterization solely in terms of the spanning family.

18.3 About Granularity

18.3.1 Dropping the granularity assumption. Does the conclusion of Theorem 5.5.6 still hold when granularity is not assumed? (cf. Remark 5.5.7(a)).

18.3.2 Characterize granular attributions. In 8.5.2, we defined the concept of a granular attribution by the property that such an attribution produces a granular knowledge space. So far, we do not have a direct characterization of this concept.

18.4 Miscellaneous

18.4.1 The width and the dimension of a surmise system. Surmise systems and AND/OR graphs (cf. Definitions 5.1.2 and 5.3.1) are two aspects of a same generalization of partially ordered sets. Concepts which are classical for partially ordered sets can, in principle, be extended to surmise systems. This generates a large collection of problems. For instance, what would be appropriate extensions of classical concepts such as the 'width', the 'dimension', etc. of a partial order? Do central theorems about these concepts remain true for the extended situation? (A first pass at some of these problems was made in Doignon and Falmagne, 1988.)

18.4.2 About the set differences for projections. Under wich conditions on a knowledge structure $(\mathring{Q}, \mathring{\mathcal{K}})$ are all the differences $S(a, K) \setminus S(a, K \cup \{b\})$ empty, for all the projections $(Q, \mathcal{K})$ of $(\mathring{Q}, \mathring{\mathcal{K}})$? (Cf. Example 12.7.2).

Glossary

STANDARD SYMBOLS

$\overset{d}{\approx}$	approximately distributed as
$\Leftrightarrow$, iff	the logical equivalence standing for 'if and only if'
$\wedge$, $\neg$	the logical conjunction and the logical negation 'not'
$\wedge_{i=1}^{n}$	the logical conjunction of n statements indexed $1, 2, \ldots, n$
$\exists, \forall$	existential and universal quantifiers
$\varnothing$	empty set
$\in, \subseteq, \subset, \supset, \supseteq$	set membership, set inclusion, proper (or strict) set inclusion, and the reverse inclusions
$\cup$, $\cap$, $\setminus$, $\triangle$	union, intersection, difference, and symmetric difference of sets
$+$, $\sum$	may stand for the ordinary addition or for the union of disjoint sets
$f(B)$	if B is a set and f a function, the image of B by f
$\lvert X \rvert$	number of elements (or cardinal number) of a set X
2^X	power set of the set X (i.e., the set of all subsets of X)
$X_1 \times X_2 \times \cdots \times X_n$	Cartesian product of the sets X_1,X_2, $\ldots$, X_n
$\mathbb{N}$	the set of all natural numbers (excluding 0)
$\mathbb{N}_0$	the set of nonnegative integers
$\mathbb{Q}$	the set of all rational numbers
$\mathbb{R}$	the set of all real numbers
$\mathbb{R}_+$	the set $[0, \infty[$ of all nonnegative real numbers
$\mathbb{Z}$	the set of all integers
$\mathbb{P}$	a probability measure
$]x, y[$	open interval of real numbers $\{z \in \mathbb{R} \mid x < z < y\}$
$[x, y]$	closed interval of real numbers $\{z \in \mathbb{R} \mid x \leq z \leq y\}$
$]x, y]$, $[x, y[$	real, half open intervals
$\breve{R}$	Hasse diagram of a partial order R
$t(R)$	transitive closure of a relation R
$\square$	marks the end of a proof
$\Diamond$	marks the end the proof of a lemma inserted in the proof of a theorem
l.h.s., r.h.s.	abbreviations for 'left-hand side' and 'right-hand side' (of a formula)
r.v.	abbreviation for 'random variable'
w.r.t.	abbreviation for 'with respect to'

J.-C. Falmagne, J.-P. Doignon, *Learning Spaces*,
DOI 10.1007/978-3-642-01039-2, © Springer-Verlag Berlin Heidelberg 2011

Non Standard Technical Terms[1]

In the notation '⟨k → j⟩' appearing in an entry, 'k' refers to the numerical marker of the relevant chapter, section or subsection and 'j' to the corresponding page number. Words or phrases set in blue appear elsewhere as entries in this glossary. The ◁ superscript signifies that the entry refers to media theory.

accessible. A family of sets $\mathcal{K}$ satisfying Axiom [MA] of an antimatroid. Such a family is also said to be downgradable. ⟨2.2.2 → 27⟩

acyclic attribution. An attribution σ is acyclic when the relation $\mathcal{R}_\sigma$ defined by the equivalence $q\mathcal{R}_\sigma q' \iff \exists C \in \sigma(q') : q \in C$ (for $q, q' \in Q$) is acyclic. ⟨5.6.10 → 98⟩

adjacent$^{\triangleleft}$**.** Two distinct states S and V in a medium are adjacent if there exists a token τ such that $S\tau = V$. ⟨10.1.2 → 166⟩

alphabet. In Chapter 9, a set Q of symbols pertaining to a language. The elements of Q are called positive literals. The negative literals are the elements of Q marked with an overbar. Both positive and negative literals are used to write the words of the language. ⟨9.2.1 → 155⟩

antecedent set. The set A appearing as the first component of a query (A, p). ⟨Section 15.1 → 298⟩

antimatroid. Another name for a learning space used in the combinatoric literature and defined there by different axioms. The dual structures are also called 'antimatroid.' Thus the union-closed antimatroids are the well-graded knowledge spaces ⟨2.2.2 → 27⟩ and the intersection-closed antimatroids are the duals. The latter are also called 'convex geometries.'

apex$^{\triangleleft}$**.** A state S is the apex of an oriented medium $(\mathcal{S}, \mathcal{T})$ if $\widehat{S} = \widehat{\mathcal{T}^+}$, that is, if the content of S is made of all the positive tokens in $\mathcal{T}$. ⟨10.5.11 → 182⟩

ascendent family. Consider a knowledge structure $(Q, \mathcal{K})$, a subset Q' of Q and a state W in the projection $\mathcal{K}_{|Q'}$ of $\mathcal{K}$ on Q'. The ascendent family $\mathcal{K}(Q', W)$ of W is the subfamily of all the states of $\mathcal{K}$ whose trace on Q' is W. Formally, we have $\mathcal{K}(Q', W) = \{K \in \mathcal{K} \,|\, K \cap Q' = W\}$. ⟨13.7.1 → 259⟩

assessment. The process of uncovering the competence of an individual in a domain of information ⟨Section 1.3 → 10⟩.

assessment language for a collection $\mathcal{K} \subseteq 2^Q$**.** ⟨9.2.3 → 155⟩ A language L over the alphabet Q that is empty if $|\mathcal{K}| = 0$, has only the word 1 if $|\mathcal{K}| = 1$, and otherwise satisfies $L = qL_1 \cup \bar{q}L_2$, for some q in Q, where

[A1] L_1 is an assessment language for the trace $(\mathcal{K}_q)_{|Q\setminus\{q\}}$;
[A2] L_2 is an assessment language for the collection $\mathcal{K}_{\bar{q}}$ with domain $Q\setminus\{q\}$.

[1] Defined terms in the theory expounded in this book which do not belong to the standard mathematical lingo.

assessment module. The part of a computer educational software, such as the ALEKS system, that is devoted to the assessment of students' competence in a subject. ⟨Section 1.3 → 10⟩

assigned. For any item q in Q, the subset $\tau(q)$ of S is referred to as the set of skills assigned to q by the skill map. ⟨6.2.1 → 106⟩

association (relation). A mapping from the collection $2^Q \setminus \{\varnothing\}$ of all nonempty subsets of a domain Q to Q ⟨8.6.1 → 147⟩. If $\mathcal{K}$ is a knowledge space, then its derived entailment is an example of association ⟨7.1.6 → 123⟩.

atom. A set B in a family of sets $\mathcal{F}$ is an atom at some q in $\cup\mathcal{F}$ if B is a minimal set in $\mathcal{F}$ for the property of containing q ⟨3.4.5 → 48⟩. ('Minimal' is to be understood with respect to set inclusion.) In a knowledge space equipped with a base, the atoms are exactly the elements of the base ⟨3.4.8 → 48⟩.

attribution. A function σ mapping a domain Q into 2^{2^Q} (thus linking any element of Q to some family of subsets of Q) satisfying the condition that $\sigma(q) \neq \varnothing$ for any $q \in Q$. ⟨5.1.2 → 83⟩

attribution order. ⟨5.5.1 → 92⟩ A relation $\precsim$ on the collection $\mathfrak{F}$ of all attributions on a nonempty set Q defined by the equivalence

$$\sigma' \precsim \sigma \iff \forall q \in Q, \forall C \in \sigma(q), \exists C' \in \sigma'(q) : C' \subseteq C \qquad (\sigma, \sigma' \in \mathfrak{F}).$$

ball. For a state K in a knowledge structure $(\mathcal{K}, Q)$, the set of all states whose distance from K is at most h is called the ball of radius h centered at K. It is denoted by $\mathcal{N}(K,h)$. We have thus $\mathcal{N}(K,h) = \{L \in \mathcal{K} \,|\, d(K,L) \leq h\}$. This set is sometimes referred to as the h-neighborhood of K. ⟨4.1.6 → 63⟩

base of a $\cup$-closed family $\mathcal{F}$. A minimal subfamily $\mathcal{B}$ of $\mathcal{F}$ spanning $\mathcal{F}$, where 'minimal' is meant with respect to set inclusion: if $\mathcal{H}$ also spans $\mathcal{F}$ for some $\mathcal{H} \subseteq \mathcal{B}$, then $\mathcal{H} = \mathcal{B}$. ⟨3.4.1 → 47⟩

base†. A concept similar to the base that is instrumental for partial knowledge spaces ⟨Page 262⟩.

basic local independence model. A basic probabilistic model satisfying local independence ⟨11.1.2 → 189⟩. The added qualifier "with no guessing" means that, in such a model, all the lucky guess probabilities are assumed to be equal to zero ⟨11.3.6 → 198⟩.

basic probabilistic model. A quadruple $(Q, \mathcal{K}, p, r)$, in which $(Q, \mathcal{K}, p)$ is a probabilistic knowledge structure and r its response function. ⟨11.1.2 → 189⟩

binary classification language (over a finite alphabet Q). A language L which either consists of the empty word alone or satisfies the two following conditions ⟨9.2.4 → 156⟩:

[B1] a letter may not appear more than once in a word;
[B2] if π is a proper prefix of L, then there exist exactly two prefixes of the form $\pi\alpha$ and $\pi\beta$, where α and β are literals; moreover $\bar{\alpha} = \beta$.

block. The `QUERY` routine proceeds by 'blocks': first Block 1, then Block 2, etc. The responses $A\mathcal{P}q$ with $|A| = k$ appear in Block k. $\langle 15.1.2 \to 299\rangle$

bounded path. In a knowledge structure, a family of states connecting two states and satisfying certain conditions stated in $\langle 4.3.3 \to 70\rangle$.

canonical$^{\triangleleft}$**.** A message $\boldsymbol{m}$ in an oriented medium is called canonical if it is concise and satisfies one of the following three conditions $\langle 10.5.5 \to 180\rangle$:

(i) it is positive, that is, contains only positive tokens;
(ii) it is negative, that is, contains only negative tokens;
(iii) it is mixed, that is, of the form $\boldsymbol{m} = \boldsymbol{n}\boldsymbol{n}'$ where $\boldsymbol{n}$ is a positive message and $\boldsymbol{n}'$ a negative one.

careless error probability. The probability that a student in state K makes an error in responding to an instance of an item in K $\langle$pages 188, 362-364$\rangle$.

cast as. Used in the sense of "assigned the role of", as in: "The relation $\mathcal{R}$ is cast as the attribution σ." $\langle 5.1.4 \to 84\rangle$

child, Q'-child, plus child. The children of the partial knowledge structure $(Q, \mathcal{K})$ are specified by a proper subset Q' of Q. Define the equivalence relation $\sim_{Q'}$ on $\mathcal{K}$ by the formula $L \sim_{Q'} K \iff L \cap Q' = K \cap Q'$. Denote by $[K]$ the equivalence class containing the state K. With $K \in \mathcal{K}$, the set

$$\mathcal{K}_{[K]} = \{M \subseteq Q \,|\, \exists L \in [K], M = L \setminus (\cap[K])\}$$

is a child, or a Q'-child, of $\mathcal{K}$ $\langle 2.4.2 \to 32\rangle$. For any non trivial child $\mathcal{K}_{[K]}$ of $\mathcal{K}$, we call $\mathcal{K}^+_{[K]} = \mathcal{K}_{[K]} \cup \{\varnothing\}$ a plus child of $\mathcal{K}$ $\langle 2.4.11 \to 37\rangle$.

classification. A nomenclature $\{\mathcal{K}_{|Q_i} \,|\, 1 \le i \le k\}$ of a knowledge structure $(Q, \mathcal{K})$ is a classification if $\{Q_1, \ldots, Q_k\}$ is a partition of the domain Q. $\langle 11.8.1 \to 209\rangle$

clause for an item q. Any $C \in \sigma(q)$ where σ is an attribution. A clause for q is also called a foundation of q. $\langle 5.1.2 \to 83\rangle$

closed$^{\triangleleft}$**.** An oriented medium is closed if for any state S and any two distinct tokens τ and μ, both effective for S, we have $S\tau\mu = S\mu\tau$. $\langle 10.5.1 \to 179\rangle$

closed under intersection. See intersection-closed family.

closed under union, closed under finite union. See union-closed family.

$\cup$–closure. Property of being union-closed $\langle 2.2.2 \to 27\rangle$. (See also partial $\cup$–closure.)

closure space. A family of subsets of a set which is closed under intersection. $\langle 3.3.1 \to 46\rangle$

closure of a set. In the context of a closure space $(Q, \mathcal{L})$, the closure of a set $A \subseteq Q$ is the unique set A' in $\mathcal{L}$ including A that is minimal for inclusion in $\mathcal{L}$. For any $A, B \subseteq Q$, we have: (i) $A \subseteq A'$; (ii) $A' \subseteq B'$ when $A \subseteq B$; (iii) $A'' = A'$. $\langle 3.3.4 \to 46\rangle$

collection. A family of sets (or of other specified objects). $\langle 3.3.1 \to 46\rangle$

compatible knowledge structures. A knowledge structure $(Y, \mathcal{F})$ is compatible with a knowledge structure $(Z, \mathcal{G})$ if, for any $F \in \mathcal{F}$, the intersection $F \cap Z$ is the trace on Y of some state of $\mathcal{G}$. $\langle 7.3.5 \to 126\rangle$

competency for an item q in a skill multimap $(Q, S; \mu)$. Any set belonging to $\mu(q)$. $\langle 6.5.1 \to 112\rangle$

concise$^{\triangleleft}$ message. A message in a medium is concise if it is stepwise effective, consistent, and has no token occuring more than once. $\langle 10.1.3 \to 167\rangle$

conjunctive model. See delineated.

1-connected. A finite knowledge structure $(Q, \mathcal{K})$ is 1-connected if there is a stepwise path between any two of its (distinct) states. $\langle 4.1.3 \to 62\rangle$

consistent$^{\triangleleft}$ message. In a medium, a message is consistent if it does not contain both a token and its reverse. $\langle 10.1.3 \to 167\rangle$

content$^{\triangleleft}$ of a message. The set containing all the distinct tokens in that message. $\langle 10.3.1 \to 169\rangle$

convex updating rule. A special kind of non permutable updating rule. $\langle 13.4.2 \to 250 \text{ to } 250\rangle$

critical state. In a knowledge structure, a state K is critical for a state L if the inner fringe of L is some singleton $\{q\}$ and $K = L \setminus \{q\}$. $\langle 16.1.1 \to 336\rangle$

delineated knowledge state. Let (Q, S, τ) be a skill map with T a subset of S. A knowledge state K is delineated by T (via the disjunctive model) if $K = \{q \in Q \,|\, \tau(q) \cap T \neq \varnothing\}$ $\langle 6.2.1 \to 106\rangle$. Such a state K is delineated via the conjunctive model if $K = \{q \in Q \,|\, \tau(q) \subseteq T\}$ $\langle 6.4.1 \to 110\rangle$.

derived quasi ordinal space. $\langle 3.8.6 \to 58\rangle$ In the context of Theorem 3.8.5, the quasi ordinal space $\mathcal{K}$ defined from a relation $\mathcal{Q}$ on a domain Q by the equivalence $K \in \mathcal{K} \iff (\forall (p,q) \in \mathcal{Q} : \; q \in K \Rightarrow p \in K)$. The term "derived" is also used in a similar sense in the context of surmise systems and surmise functions $\langle 5.2.1 \to 85\rangle$, and in Chapter 8 $\langle 8.4.5 \to 144$ and $8.6.1 \to 147\rangle$.

describe (a word describes a state). In Chapter 9, a word is a string belonging to a language. Such a word describes a knowledge state in a knowledge structure if it specifies the state exactly $\langle 9.2.7 \to 157\rangle$. For instance, the word $\bar{a}ed$ specifies the state $\{b, c, d, e\}$ in the knowledge structure

$$\begin{aligned}\mathcal{G} = \big\{&\varnothing, \{a\}, \{b,d\}, \{a,b,c\}, \{b,c,e\}, \{a,b,d\},\\ &\{a,b,c,d\}, \{a,b,c,e\}, \{b,c,d,e\}, \{a,b,c,d,e\}\big\}.\end{aligned}$$

descriptive language. $\langle 9.2.7 \to 157\rangle$ A language L is a descriptive language for a partial knowledge structure $\mathcal{K}$ when

[D1] any word of L describes a unique state in K;

[D2] any state in $\mathcal{K}$ is described by at least one word of L.

discrepancy distribution. For two spaces $\mathcal{K}$ and $\mathcal{K}'$, the distribution $f_{\mathcal{K},\mathcal{K}'}$ of the (minimum) distances from the states in $\mathcal{K}$ to the states in $\mathcal{K}'$ is called the discrepancy distribution from $\mathcal{K}$ to $\mathcal{K}'$. ⟨15.4.5 → 318 to 320⟩

discrepancy index. The discrepancy index from a knowledge structure $\mathcal{K}$ to a knowledge structure $\mathcal{K}'$ is defined by the mean

$$di(\mathcal{K},\mathcal{K}') = \frac{1}{|\mathcal{K}|}\sum_{k=0}^{h(Q)} k f_{\mathcal{K},\mathcal{K}'}(k), \qquad (h(Q) = \lfloor \tfrac{1}{2}|Q| \rfloor),$$

of the discrepancy distribution $f_{\mathcal{K},\mathcal{K}'}$ from $\mathcal{K}$ to $\mathcal{K}'$, where Q is the common domain of $\mathcal{K}$ and $\mathcal{K}'$. ⟨15.4.5 → 318 to 320⟩

discriminative. A knowledge structure is discriminative if every notion is a singleton ⟨2.1.5 → 24⟩. A surmise system (Q,σ) is discriminative if whenever $\sigma(q) = \sigma(q')$ for some $q, q' \in Q$, then $q = q'$. In such a case, the surmise function σ is also called discriminative ⟨5.1.2 → 83⟩.

discriminative reduction. See reduction.

disjunctive model. See delineated.

domain of a (partial) knowledge structure. The set of all its items. If $\mathcal{K}$ is a (partial) knowledge structure, the domain of $\mathcal{K}$ is $\cup\mathcal{K}$. ⟨2.1.2 → 23⟩

downgradable. A synonym of accessible. A nonempty family of sets $\mathcal{F}$ is downgradable if for every nonempty $S \in \mathcal{F}$, there exists $T \in \mathcal{F}$ such that $S \setminus \{q\} = T$ for some $q \in S$. ⟨2.2.2 → 27⟩

dual. The dual of a knowledge structure $\mathcal{K}$ is the knowledge structure $\bar{\mathcal{K}}$ containing all the complements of the states of $\mathcal{K}$. Here, 'complement' is to be understood with respect to the domain $\cup\mathcal{K}$ (in the set-theoretical meaning). ⟨2.2.2 → 27⟩

effective$^{\triangleleft}$**.** (See also 'stepwise effective.') In a medium, a message $\boldsymbol{m}$ is effective for a state S if $S\boldsymbol{m} \neq S$. ⟨10.1.3 → 167⟩

entail relation for a nonempty set Q. A relation $\mathcal{Q}$ on $2^Q \setminus \{\varnothing\}$ satisfying Conditions (i), (ii) and (iii) in Theorem 7.1.5. ⟨7.2.2 → 124⟩

entailment for a nonempty set Q. Any relation $\mathcal{P}$ from $2^Q \setminus \{\varnothing\}$ to Q that satisfies Conditions (i) and (ii) in Theorem 7.1.3. ⟨7.1.4 → 122⟩

essential distance between two states. This concept applies to non necessarily discriminative knowledge structures. The essential distance between two states K and L is defined by $e(K,L) = |K^* \triangle L^*|$, where K^* (resp. L^*) is the set of all notions q^* with q in K (resp. q in L). ⟨2.3.1 → 30⟩

fair stochastic assessment process. In the Markov chain procedure of Chapter 14, the case in which the probabilities of lucky guesses are all equal to zero; thus $\eta_q = 0$ for any item q. ⟨14.2.2 → 279⟩

finitary. A knowledge structure is finitary if the intersection of any chain of states is a state ⟨3.6.1 → 52⟩. A closure space is $\cap$-finitary if the union of any chain of states is a state ⟨Problem 13 on page 60⟩.

finitely learnable $\langle 4.4.2 \to 72\rangle$. A discriminative structure is finitely learnable if there is a positive integer l such that, for any state K and any item $q \notin K$, there exists a positive integer h and a chain of states $K = K_0 \subset K_1 \subset \cdots \subset K_h$ satisfying the two conditions:

(i) $q \in K_h$;
(ii) $d(K_i, K_{i+1}) \leq l, \quad$ for $0 \leq i \leq h-1$.

foundation of an item q**.** See clause.

fringe. The fringe of a state K in a knowledge structure is the union of the inner and outer fringes of K, that is, the set $K^{\mathcal{F}} = K^{\mathcal{I}} \cup K^{\mathcal{O}}$. $\langle 4.1.6 \to 63\rangle$

gradation, ∞**-gradation.** In a finite knowledge structure, a gradation is a tight path from the empty set to the domain $\langle 4.1.3 \to 62\rangle$. A similar concept applies in the infinite case where it is called an ∞-gradation $\langle 4.3.1 \to 69\rangle$.

granular. A knowledge structure is granular if for every state K, any item q in K has an atom at q included in K. $\langle 3.6.1 \to 52\rangle$

guessing probability. See lucky guess probability.

half-split questioning rule. An item q selected by this rule on trial n of an assessment minimizes the quantity $|2L_n(\mathcal{K}_q) - 1|$, where L_n is the current probability distribution on the set $\mathcal{K}$ of states. If two or more items satisfy this condition, the algorithm chooses randomly between them. $\langle 13.4.7 \to 252\rangle$

ε**-half-split.** A special case of the questioning function in the Questioning Rule Axiom [QM]. $\langle 14.3.2 \to 280\rangle$

hanging, almost hanging. $\langle 16.1.1 \to 336\rangle$ In a knowledge structure, a nonempty state K having an empty inner fringe is called a hanging state. A state K is almost hanging if it has at least two items and its inner fringe is a singleton.

hanging-safe. A query (A, q) is hanging-safe (for a learning space) if there is no clause C for some item r with $A \cap C = \{r\}$ and $q \in C$. $\langle 16.1.9 \to 338\rangle$

Hasse system. $\langle 5.5.8 \to 94\rangle$ Generalization of a Hasse diagram $\langle 1.6.8 \to 15\rangle$. This concept applies to granular knowledge spaces.

height of an item. The height of an item q is the number $h(q) = k - 1$, where k is the size of a minimal state containing the item q. A height of zero for an item q means that $\{q\}$ is a state. $\langle 15.4.4 \to 317\rangle$

incidence matrix. $\langle 14.7.1 \to 293\rangle$ For a collection $\mathcal{K}$ of subsets of the finite domain Q of items, the matrix $\mathbf{M} = (M_{q,K})$ whose rows are indexed by items q in Q, columns are indexed by states K in $\mathcal{K}$, and

$$M_{q,K} = \begin{cases} 1 & \text{if } q \in K, \\ 0 & \text{otherwise.} \end{cases}$$

inclusive mesh. A mesh $\mathcal{K}$ of two knowledge structures $\mathcal{F}$ and $\mathcal{G}$ is called (union) inclusive if $F \cup G \in \mathcal{K}$ for any $F \in \mathcal{F}$ and $G \in \mathcal{G}$. ⟨7.4.5 → 128⟩

inconsistent$^{\triangleleft}$**.** A message in a medium is inconsistent if it contains both some token and its reverse. ⟨10.1.3 → 167⟩

ineffective$^{\triangleleft}$**.** A message is ineffective for a state if it is not effective for that state. ⟨10.1.3 → 167⟩

independent states, projections. Let $(Q, \mathcal{K}, p)$ be a probabilistic knowledge structure, and let $\mathbb{P}$ be the induced probability measure on the power set of $\mathcal{K}$. With $Q', Q'' \subset Q$, consider the probabilistic projections $(Q', \mathcal{K}', p')$ and $(Q'', \mathcal{K}'', p'')$. Two states $J \in \mathcal{K}'$, $L \in \mathcal{K}''$ are independent if the events $J^{\diamond} = \{K \in \mathcal{K} \,|\, K \cap Q' = J\}$ and $L^{\diamond} = \{K \in \mathcal{K} \,|\, K \cap Q'' = L\}$ are independent in the probability space $(\mathcal{K}, 2^{\mathcal{K}}, \mathbb{P})$. The projections $(Q', \mathcal{K}', p')$ and $(Q'', \mathcal{K}'', p'')$ are independent if any state $K \in \mathcal{K}$ has independent traces on Q' and Q''. ⟨11.9.1 → 210⟩

informative (equally). Two items belonging to the same notion are called equally informative. ⟨2.1.5 → 24⟩

informative questioning rule. An item selected on trial n according to this rule and presented to the subject minimizes the expected entropy of the likelihood distribution on the set of states on trial $n + 1$. ⟨13.4.8 → 253⟩

inner fringe. In a knowledge structure $(Q, \mathcal{K})$, the inner fringe of a state K is the set $K^{\mathcal{I}} = \{q \in K \,|\, K \setminus \{q\} \in \mathcal{K}\}$. ⟨4.1.6 → 63⟩

inner questioning rule. An abstract constraint on the questioning rule requiring that an item q be chosen so that the likelihood of a correct response to q is as far as possible from 0 or 1. ⟨13.6.4 → 257⟩

instance. ⟨1.1.1 → 2⟩ A particular case of an item that can be used in an assessment, for example by replacing the abstract parameters of an equation by numerical values. This is a pedagogical concept, with no formal definition.

intersection-closed family, $\cap$-closed family. A family $\mathcal{F}$ of subsets of a set X which is closed under intersection: for any subfamily $\mathcal{G}$ of $\mathcal{F}$, we have $\cap\mathcal{G} \in \mathcal{F}$. As we can have $\mathcal{G} = \varnothing \subseteq \mathcal{F}$, we get $\cap\varnothing = X \in \mathcal{F}$. ⟨2.2.2 → 27⟩

item, item indicator. An item is an element in the domain of a knowledge structure or partial knowledge structure ⟨2.1.2 → 23⟩. The term item is also used in the context of skill maps ⟨6.2.1 → 106⟩. An item indicator random variable for an item q is a 0-1 random variable taking value 1 if the response to item q is correct, and 0 otherwise ⟨11.9.4 → 212⟩.

jointly consistent$^{\triangleleft}$**.** Two messages $\boldsymbol{n}$ and $\boldsymbol{m}$ are jointly consistent if $\boldsymbol{nm}$ is consistent. ⟨10.1.3 → 167⟩

knowledge space. A pair $(Q, \mathcal{K})$ where Q is a nonempty set and $\mathcal{K}$ is a family of subsets of Q closed under union and containing the empty set and the set $Q = \cup\mathcal{K}$, which is the domain of the knowledge space. Thus, a knowledge space is a $\cup$-closed knowledge structure. The pair $(Q, \mathcal{K})$ can also be referred to as a space. The family $\mathcal{K}$ itself is often called a space. $\langle 2.2.2 \to 27\rangle$

knowledge state. Any set in a knowledge structure or partial knowledge structure $\langle 2.1.2 \to 23\rangle$. Can be abbreviated as state.

knowledge structure. A pair $(Q, \mathcal{K})$ where Q is a nonempty set and $\mathcal{K}$ is a family of subsets of Q containing both the empty set $\varnothing$ and the set Q, which is called the domain of the knowledge structure. The sets in $\mathcal{K}$ are referred to as the states of the knowledge structure. The family $\mathcal{K}$ itself is also called a knowledge structure. $\langle 2.1.2 \to 23\rangle$

L1-chain in a learning space. Let $K \subset L$ be two knowledge states with $|K \setminus L| = p$ and $L_0 = L \subset L_1 \subset \ldots \subset L_p = K$ a chain of states; so, $|L_i \setminus L_{i-1}| = 1$ for $1 \leq i \leq p$. Such a chain is called a L1-chain. $\langle 2.2.1 \to 26\rangle$

language. A distinguished set of words comprising the positive and negative literals. $\langle 9.2.1 \to 155\rangle$

latent. The term 'latent state' is non-technical and has several meanings. It may refer informally to the hypothetical knowledge state determining the subject's responses to the questions in an assessment $\langle 13.3.1 \to 246\rangle$. The term 'latent structure' is used in the psychometric literature with a germane meaning. Finally, it may also qualify the hypothetical knowledge space or learning space governing the responses in the application of the `QUERY` procedure. $\langle$Section 15.1 $\to 298\rangle$

learning function, learning rate. The function ℓe of Axiom [L] in a system of stochastic learning paths, formalizing the idea that the probability of a state at a given time only depends upon the last state recorded, the time elapsed, the learning rate and the gradation $\langle 12.2.3 \to 218$–$12.2.4 \to 220\rangle$. In a special case, the learning rate is a gamma distributed r.v. $\langle 12.4.1 \to 224\rangle$

learning path. A maximal chain in a knowledge structure. $\langle 4.1.1 \to 61\rangle$

learning space. A knowledge structure satisfying Axioms [L1] and [L2] $\langle 2.2.1 \to 26\rangle$; equivalently, a $\cup$-closed knowledge structure which is either well-graded, or finite and downgradable. $\langle 2.2.4 \to 28\rangle$

learnstep number (of a finitely learnable knowledge structure). The smallest number l in the definition of finite learnability. $\langle 4.4.2 \to 72\rangle$

length$^{\triangleleft}$. The length of a message $\boldsymbol{m} = \tau_1 \ldots \tau_n$ is the number of its (non necessarily distinct) tokens. We write then $\ell(\boldsymbol{m}) = n$. $\langle 10.1.3 \to 167\rangle$.

literals, positive, negative. The symbols used to write the words of a language. $\langle 9.2.1 \to 155\rangle$

local independence. ⟨11.1.2 → 189⟩ The response function r of a basic probabilistic model $(Q, \mathcal{K}, p, r)$ satisfies local independence if for all $K \in \mathcal{K}$ and $R \subseteq Q$, we have

$$r(R, K) = \left(\prod_{q \in K \setminus R} \beta_q\right) \left(\prod_{q \in K \cap R}(1 - \beta_q)\right) \left(\prod_{q \in R \setminus K} \eta_q\right) \left(\prod_{q \in \overline{R \cup K}}(1 - \eta_q)\right)$$

in which, for each item $q \in Q$, the symbols β_q and η_q denote two parameters measuring a careless error probability and a lucky guess probability, respectively, for that item.

lucky guess probability. Probability of a correct response to an instance of an item which does not belong to the student's knowledge state (cf. for example the Local Independence Axiom [N]). ⟨12.4.1 → 224⟩

m-states. ⟨14.4.1 → 283⟩ The states of a Markov chain. The term is used in Chapter 14 to avoid a possible confusion with the states of a knowledge structure.

marking function. In the Markov chain procedure of Chapter 14, the function μ of the Marking Rule Axiom [M]. ⟨14.2.1 → 278⟩

marked states. In the Markov chain procedure of Chapter 14 those states retained as feasible on a given trial. ⟨14.1 → 273 to 278⟩

medium$^{\triangleleft}$**.** A pair $(\mathcal{S}, \mathcal{T})$ in which $\mathcal{S}$ is a set of states, $\mathcal{T}$ is a set of transformations on $\mathcal{S}$ and the two axioms [Ma] and [Mb] are satisfied ⟨10.1.4 → 167⟩. Any discriminative, well-graded family of sets gives rise in a natural manner to a medium; in turn, any medium is obtainable in this way in the sense that the sets of the family are in a one-to-one correspondence with the states of the medium. ⟨10.4.11 → 178–10.5.12 → 182⟩

mesh, meshable, maximal mesh. A knowledge structure $(X, \mathcal{K})$ is a mesh of two knowledge structures $(Y, \mathcal{F})$ and $(Z, \mathcal{G})$ if: (i) $X = Y \cup Z$; and (ii) $\mathcal{F}$ and $\mathcal{G}$ are the projections of $\mathcal{K}$ on Y and Z, respectively ⟨7.3.1 → 126⟩. Two knowledge structures having a mesh are said to be meshable ⟨7.3.1 → 126⟩. For two compatible knowledge structures $(Y, \mathcal{F})$ and $(Z, \mathcal{G})$, the knowledge structure $(Y \cup Z, \mathcal{F} \star \mathcal{G})$ defined by the equation

$$\mathcal{F} \star \mathcal{G} = \{K \in 2^{Y \cup Z} \mid K \cap Y \in \mathcal{F},\ K \cap Z \in \mathcal{G}\}$$

is the maximal mesh of $\mathcal{F}$ and $\mathcal{G}$ ⟨7.4.1 → 127⟩.

message$^{\triangleleft}$**.** A string $\boldsymbol{m} = \tau_1 \dots \tau_n$ of tokens in a medium. ⟨10.1.3 → 167⟩

minimal well-graded extension. ⟨16.3.3 → 356⟩ A minimal well-graded extension of a non well-graded family $\mathcal{F}$ is a well-graded $\cup$-closed family $\mathcal{H}$ such that:

(i) $\mathcal{F} \subset \mathcal{H}$;
(ii) there is no $\cup$-closed, well-graded family $\mathcal{H}'$ satisfying $\mathcal{F} \subset \mathcal{H}' \subset \mathcal{H}$.

mixed$^{\triangleleft}$**.** A message $\boldsymbol{m}$ in an oriented medium is called mixed if it is concise and of the form $\boldsymbol{m} = \boldsymbol{n}\boldsymbol{n}'$, where $\boldsymbol{n}$ is a positive message and $\boldsymbol{n}'$ a negative one (see 'canonical message'). $\langle 10.5.5 \rightarrow 180\rangle$

multiplicative updating rule. A special case of the permutable updating rule. $\langle 13.4.4 \rightarrow 251\rangle$

negative token$^{\triangleleft}$**.** In an orientation $\{\mathcal{T}^+, \mathcal{T}^-\}$ of a medium, any token in $\mathcal{T}^-$. $\langle 10.4.2 \rightarrow 174$ to $175\rangle$

neighborhood, neighbor. The subfamily $\mathcal{N}(K, h)$ of all the states at distance at most h from a state K in a knowledge structure $\mathcal{K}$ is referred to as the h-neighborhood of K, or sometimes as the ball of radius h centered at the state K; thus, $\mathcal{N}(K, h) = \{K' \in \mathcal{K} \,|\, d(K, K') \leq h\}$ $\langle 4.1.6 \rightarrow 63\rangle$.

The term 'neighborhood' is also used with a different, but related meaning pertaining to a family rather than a state. The ε-neighborhood of a subfamily $\mathcal{F}$ of a knowledge structure $\mathcal{K}$ is the subfamily of $\mathcal{F}$ defined by

$$\mathcal{N}(\mathcal{F}, \varepsilon) = \{K' \in \mathcal{K} \,|\, d(K, K') \leq \varepsilon, \text{ for some } K \text{ in } \mathcal{F}\}.$$

The (q, ε)-neighborhood and the $(\bar{q}, \varepsilon)$-neighborhood of $\mathcal{F}$ are respectively $\mathcal{N}_q(\mathcal{F}, \varepsilon) = \mathcal{N}(\mathcal{F}, \varepsilon) \cap \mathcal{K}_q$ and $\mathcal{N}_{\bar{q}}(\mathcal{F}, \varepsilon) = \mathcal{N}(\mathcal{F}, \varepsilon) \cap \mathcal{K}_{\bar{q}}$. The ε-neighbors are the states in a ε-neighborhood; the terms '(q, ε)-neighbors' and '$(\bar{q}, \varepsilon)$-neighbors' have similar meanings. $\langle 14.3.1 \rightarrow 279\rangle$.

nomenclature. Let $Q_1, \ldots, Q_k$ be a collection of nonempty subsets of the domain Q of a knowledge structure $\mathcal{K}$. The collection $\mathcal{K}_{|Q_1}, \ldots, \mathcal{K}_{|Q_k}$ of projections is a nomenclature if the collection $(Q_i)_{1 \leq i \leq n}$ covers Q, that is, if $\cup_{i=1}^n Q_i = Q$. $\langle 11.8.1 \rightarrow 209\rangle$

notion. If q is an item in a knowledge structure $(Q, \mathcal{K})$, then the set q^* of all the items contained in exactly the same states as q is a notion. We have thus $q^* = \{s \in Q \,|\, \forall K \in \mathcal{K}, \, s \in K \Leftrightarrow q \in K\}$. $\langle 2.1.5 \rightarrow 24\rangle$

operative. A query (A, q) is said to be operative for a learning space $\mathcal{L}$ if $\mathcal{L} \setminus \mathcal{D}_{\mathcal{L}}(A, q) \subset \mathcal{L}$. $\langle 16.1.9 \rightarrow 338\rangle$

ordinal space. A discriminative knowledge space closed under intersection is a (partially) ordinal space (with 'partially' referring to the corresponding partial order). $\langle 3.8.1 \rightarrow 56$–$3.8.3 \rightarrow 57\rangle$

orientation$^{\triangleleft}$**.** An orientation of a medium $(\mathcal{S}, \mathcal{T})$ is a partition $\{\mathcal{T}^+, \mathcal{T}^-\}$ of its set of tokens $\mathcal{T}$ such that, for any token τ, we have $\tau \in \mathcal{T}^+ \Leftrightarrow \tilde{\tau} \in \mathcal{T}^-$, with $\tilde{\tau}$ the reverse of τ. By convention, the tokens in $\mathcal{T}^+$ (resp. $\mathcal{T}^-$) are called 'positive' (resp. negative). $\langle 10.4.2 \rightarrow 174\rangle$

oriented medium$^{\triangleleft}$**.** A medium equipped with an orientation is called an oriented medium. $\langle 10.4.2 \rightarrow 174\rangle$

outer fringe. In a knowledge structure $(Q, \mathcal{K})$, the outer fringe of a state K is the set $K^{\mathcal{O}} = \{q \in Q \setminus K \,|\, K + \{q\} \in \mathcal{K}\}$. $\langle 4.1.6 \rightarrow 63\rangle$

parent family. Let $\mathcal{K}_{|Q'}$ be projection of a knowledge structure $(Q, \mathcal{K})$ on a proper subset Q' of Q. For any $J \in \mathcal{K}_{|Q'}$, the parent family $J^\diamond$ of J is defined by $J^\diamond = \{K \in \mathcal{K} \,|\, K \cap Q' = J\}$; so, $\cup_{J \in \mathcal{K}_{|Q'}} J^\diamond = \mathcal{K}$. ⟨11.7.2 → 207⟩

partial knowledge structure. ⟨2.2.6 → 29⟩ A family of sets $\mathcal{K}$ containing the set $\cup\mathcal{K}$. Partial knowledge spaces and partial learning spaces are defined similarly.

partially ordinal space. A discriminative quasi ordinal space ⟨3.8.1 → 56⟩. In other words, a space which is derived from a partial order ⟨3.8.3 → 57⟩.

partially union-closed. A family $\mathcal{F}$ is partially union-closed (or partially $\cup$-closed) if for any nonempty subfamily $\mathcal{G}$ of $\mathcal{F}$, we have $\cup\mathcal{G} \in \mathcal{F}$. (Contrary to the $\cup$-closure condition, partial $\cup$-closure does not imply that the empty set belongs to the family.) ⟨2.2.6 → 29⟩

path in a knowledge structure. See stepwise path or tight path.

Pending-Table. Buffer collecting all the responses $A\mathcal{P}q$ having failed the `HS-test` in the algorithm for building a learning space. ⟨16.2.11 → 351⟩

permutable updating rule. An updating rule satisfying the permutability condition $F\big(F(l, \xi), \xi'\big) = F\big(F(l, \xi'), \xi\big)$. This condition implies that the order of the questions asked during the assessment is irrelevant. ⟨13.4.3 → 251⟩

positive token$^{\triangleleft}$**.** In a medium equipped with an orientation $\{\mathcal{T}^+, \mathcal{T}^-\}$ of a medium, any token in the set $\mathcal{T}^+$. ⟨10.4.2 → 174 to 175⟩

precede. Let $\precsim$ be the surmise relation of a knowledge structure. When $r \precsim q$, we say that r precedes q, or equivalently that r is surmisable from q. If, moreover, $q \precsim r$ does not hold, we say that r strictly precedes q. ⟨3.7.1 → 54⟩

prefix, prefix$^{\triangleleft}$**.** The initial segment of a word ⟨9.2.1 → 155⟩, or of a message. ⟨10.5.4 → 180⟩

probabilistic knowledge structure. A finite partial knowledge structure equipped with a probability distribution on its set of states. ⟨11.1.2 → 189⟩

probabilistic projection. ⟨11.7.3 → 208⟩ Suppose that $(Q, \mathcal{K}, p)$ is a probabilistic knowledge structure, and let $\mathcal{K}'$ be a projection of $(\mathcal{K}, Q)$ on a proper subset Q' of Q. For any $J \in \mathcal{K}_{|Q'}$, write $J^\diamond = \{K \in \mathcal{K} \,|\, K \cap Q' = J\}$. The triple $(Q', \mathcal{K}', p')$ is the probabilistic projection induced by Q', if for all $J \in \mathcal{K}'$, we have $p'(J) = \sum_{K \in J^\diamond} p(K)$.

produce, produce$^{\triangleleft}$**.** Any attribution σ on a set Q produces a knowledge space $(Q, \mathcal{K})$ via the equivalence $K \in \mathcal{K} \iff \forall q \in K, \exists C \in \sigma(q) : C \subseteq K$ ⟨5.2.3 → 86⟩. In media theory, we say that a message $\boldsymbol{m}$ from a state S produces some state $V \neq S$ if $S\boldsymbol{m} = V$ ⟨10.1.3 → 167⟩.

progressive. A system of stochastic learning paths is progressive if it does not allow for any forgetting. ⟨12.2.4 → 220⟩

projection. Let $(Q, \mathcal{K})$ be a partial knowledge structure. For any nonempty proper subset Q' of Q, the family $\mathcal{K}_{|Q'} = \{W \subset Q' \mid \exists K \in \mathcal{K}, W = K \cap Q'\}$ is the projection of $\mathcal{K}$ on Q'. We thus have $\mathcal{K}_{|Q'} \subseteq 2^{Q'}$. If $W = K \cap Q'$ for some $K \in \mathcal{K}$ and so $W \in \mathcal{K}_{|Q'}$, then W is called the trace of K on Q'. ⟨2.4.2 → 32⟩

prolong. The skill map (Q', S', τ') prolongs the skill map (Q, S, τ) if $Q = Q'$, $S \subseteq S'$, and $\tau(q) = \tau'(q) \cap S$ for all $q \in Q$. ⟨6.3.6 → 109⟩.

PS-QUERY. ⟨15.5.1 → 324 to 331⟩ An extension of the `QUERY` routine in which a response to a query (A, p) is not implemented immediately. Instead, it is put in a buffer, awaiting for a confirmation or a contradiction from a later response.

quadratic discrepancy index. For two knowledge structures $\mathcal{K}$ and $\mathcal{K}'$, this index is the quadratic mean $\overline{di}(\mathcal{K}, \mathcal{K}') = \sqrt{di^2(\mathcal{K}, \mathcal{K}') + di^2(\mathcal{K}', \mathcal{K}')}$ between the two discrepancy indices for $\mathcal{K}$ and $\mathcal{K}'$. ⟨15.6.2 → 330 to 330⟩

quasi learning space. A knowledge structure satisfying quasi learning smoothness and quasi learning consistency. This is a variant of the concept of learning spaces for nondiscriminative structure. ⟨2.3.2 → 31⟩.

quasi ordinal space. A knowledge space which is closed under intersection ⟨3.8.1 → 56–3.8.3 → 57⟩. Equivalently, a space derived from a quasi order ⟨3.8.3 → 57⟩.

quasi well-graded family, or qwg-family. A family $\mathcal{F}$ such that, for any two distinct states $K, L \in \mathcal{F}$, there exists a finite sequence of states $K = K_0$, $K_1, \ldots, K_p = L$ satisfying $e(K_{i-1}, K_i) = 1$ for $1 \leq i \leq p$ and moreover $p = e(K, L)$ (where e denotes the essential distance). ⟨2.3.3 → 31⟩

query. ⟨Section 3.2 → 44⟩ ⟨Section 7.1 → 120–Section 7.6 → 123⟩ A question symbolized as (A, p), posed to experts in scholalrly topic, with the following interpretation: "*Will any student failing all the items in the set A also fail item q?*" If Q is the domain, we thus have $A \subseteq Q$ and $p \in Q$. The response data may also be gathered from assessment statistics. ⟨3.2.3 → 45⟩

QUERY (typeset as `QUERY`). A routine for constructing a knowledge space, based on responses to queries of the type (A, q) "*Does failing all the items in the set A entails failing also item q?*" ⟨Chapter 15 → 297 to 323⟩. These responses can either be obtained from questioning an expert, or can be derived from assessment statistics ⟨15.4.7 → 323⟩.

questioning function. The function τ of the Questioning Rule Axiom [QM] of the Markov chain procedure. ⟨14.2.1 → 278–14.2.2 → 279⟩

questioning rule (general). A function $\Psi : (q, L_n) \mapsto \Phi(q, L_n)$ providing a framework for rules governing the choice of the item to be asked on trial n of an assessment, based on the probability distribution L_n on the set of states on that trial ⟨13.3.3 → 248⟩. Special cases of the function Ψ provide actual questioning rules ⟨Section 13.4 → 249⟩.

qwg-family. Abbreviation for quasi well-graded family.

reduction (discriminative). The discriminative reduction of a knowledge structure $(Q, \mathcal{K})$ is the knowledge structure $(Q^*, \mathcal{K}^*)$ constructed by replacing all the items q by the corresponding notions q^*. We have $\mathcal{K}^* = \{K^* \mid K \in \mathcal{K}\}$ where $K^* = \{q^* \mid q \in K\}$. ⟨2.1.5 → 24 to 25⟩

regular updating rule. An abstract constraint on the updating rule generalizing the convex and the multiplicative updating rules. ⟨13.6.2 → 256⟩

resoluble attribution, resolution order. An attribution σ on a domain Q is called resoluble if there exists a linear order $\mathcal{T}$ on Q such that for any item q in Q: (a) there is some $C \in \sigma(q)$ satisfying $C \subseteq \mathcal{T}^{-1}(q)$; (b) $\mathcal{T}^{-1}(q)$ is finite. The order $\mathcal{T}$ is then a resolution order. ⟨5.6.2 → 97⟩

resoluble knowledge space. A knowledge space is resoluble when it is produced by at least one resoluble attribution. ⟨5.6.5 → 97⟩

response function. For a probabilistic knowledge structure $(Q, \mathcal{K}, p)$, a function $r : (R, K) \mapsto r(R, K)$, with $R \subseteq Q$ and $K \in \mathcal{K}$, specifying the probability of the response pattern R for a subject in state K ⟨11.1.2 → 189⟩. This term, denoted by the same symbol r, is also used with a germane meaning in a system of stochastic learning paths ⟨12.2.4 → 220⟩.

response rule. The name of Axiom [R] of a stochastic assessment procedure, which states formally that the probability of a correct response to the question asked on trial n is equal to 1 if the question belong to the latent state of the subject, and to 0 otherwise. (In that context, the careless errors and lucky guesses probabilities are a assumed to be 0.) ⟨13.3.3 → 248⟩

response pattern. The subset R of the domain Q containing all the questions correctly solved by the subject in the course of the assessment. There are thus $2^{|Q|}$ possible response patterns. ⟨11.1.2 → 189⟩

return message$^{\triangleleft}$**.** A message which is both stepwise effective and ineffective for some state is called a return message or, more briefly, a return (for that state). ⟨10.1.3 → 167⟩

reverse$^{\triangleleft}$**.** ⟨10.1.2 → 166–10.1.3 → 167⟩ The reverse of a token τ is a token $\tilde{\tau}$ that annuls the effect of τ. More precisely, for any two adjacent states S and V we have $S\tau = V \Leftrightarrow V\tilde{\tau} = S$. The reverse of a message is defined similarly.

root$^{\triangleleft}$**, rooted**$^{\triangleleft}$ **medium.** The root of an oriented medium is a state R such that any concise message from R producing any other state is positive. An oriented medium having a root is said to be rooted. ⟨10.4.6 → 176⟩

R-store. One of the two buffers of Algorithm 16.2.11. In the second stage of the algorithm, the responses are copied from the `Pending-Table` into the `R-Store` prior to evaluation by the `HS-test`. ⟨16.2.11 → 351 to 16.3 → 353⟩

segment$^{\lhd}$. Suppose that $\boldsymbol{n} = \boldsymbol{mpm'}$ is a message in a medium, with $\boldsymbol{m}$ and $\boldsymbol{m'}$ two possibly (but not necessarily) ineffective messages, and $\boldsymbol{p}$ an effective one. Then $\boldsymbol{p}$ is called a segment of $\boldsymbol{n}$. $\langle 10.5.4 \to 180\rangle$

simple (when applied to a closure space). A closure space $(Q, \mathcal{L})$ is simple when $\varnothing$ is in $\mathcal{L}$. $\langle 3.3.1 \to 46\rangle$

simple learning model. A quadruple $(Q, \mathcal{K}, p, r)$ in which $(Q, \mathcal{K})$ a discriminative knowledge structure, r is a response function and $p : \mathcal{K} \to [0, 1]$ is defined by by the equation

$$p(K) = \prod_{q\in K} g_q \prod_{q'\in K^{\complement}} (1 - g_{q'})$$

in which the g_q's and $g_{q'}$'s are parameters. Moreover, the function p is a probability distribution on $\mathcal{K}$. $\langle 11.4.1 \to 199\rangle$

skill, skill map. $\langle 6.2.1 \to 106\rangle$ A triple (Q, S, τ), where Q is a nonempty set of items, S is a nonempty set of skills, and τ is a mapping from Q to $2^S \setminus \{\varnothing\}$. When the sets Q and S are specified by the context, the function τ itself is called the skill map. Any element of the set S is a skill.

skill multimap. A triple $(Q, S; \mu)$, where Q is a nonempty set of items, S is a nonempty set of skills, and μ is a mapping that associates to any item q a nonempty family $\mu(q)$ of nonempty subsets of S. $\langle 6.5.1 \to 112\rangle$

selective with parameter δ (marking function). In the Markov chain procedure of Chapter 14, a special case of the marking function $\mathfrak{m}$ of the Marking Rule Axiom [M]. $\langle 14.3.4 \to 282\rangle$

space. Abbreviation for knowledge space. $\langle 2.2.2 \to 27\rangle$

span of a family of sets $\mathcal{G}$. The family $\mathbb{S}(\mathcal{G})$ containing any set that is the union of any subfamily of sets in $\mathcal{G}$. We say then that $\mathcal{G}$ spans $\mathbb{S}(\mathcal{G})$. We always have $\varnothing \in \mathbb{S}(\mathcal{G})$ because, by convention the union of the empty family equals the empty set. $\langle 3.4.1 \to 47\rangle$

span† of a family of sets $\mathcal{G}$. The family $\mathbb{S}^{\dagger}(\mathcal{G})$ containing any set that is the union of a nonempty subfamily of sets in $\mathcal{G}$. We thus have $\varnothing \in \mathbb{S}^{\dagger}(\mathcal{G})$ if and only if $\varnothing \in \mathcal{G}$. $\langle 4.5.1 \to 74\rangle$

state, state$^{\lhd}$. Shorthand for knowledge state, that is, a set in a knowledge structure or partial knowledge structure $\langle 2.1.2 \to 23$–$2.2.6 \to 29\rangle$. In media theory, an element in the set $\mathcal{S}$ of a token system $(\mathcal{S}, \mathcal{T})$ $\langle 10.1.2 \to 166\rangle$.

stepwise effective$^{\lhd}$. (See also 'effective'.) A message $\boldsymbol{m} = \tau_1 \ldots \tau_n$ is stepwise effective for a state S if, for $0 \le i \le n-1$ and with $S_i = S\tau_0 \ldots \tau_i$, we have $S_i \ne S_{i+1}$. (We recall that τ_0 is the identity function, which is not a token.) A message $\boldsymbol{m}$ can be stepwise effective for a state without being effective for that state: we may have $S\boldsymbol{m} = S$. $\langle 10.1.3 \to 167\rangle$

stepwise path. A stepwise path between two sets F and G in a family of sets $\mathcal{F}$ is a sequence $F = F_0, F_1, \ldots, F_p = G$ of sets in $\mathcal{F}$ such that $d(F_{i-1}, F_i) = 1$ for $1 \leq i \leq p$, with d denoting the symmetric difference distance between sets. $\langle 4.1.3 \to 62 \rangle$

stochastic assessment process. This phrase is used in Chapters 13 and 14, with different meanings distinguished by the qualifiers "continuous" and "discrete", respectively. In Chapter 13 a stochastic process satisfying Axioms [U], [Q] and [R] $\langle 13.3.3 \to 248$–$13.3.4 \to 249 \rangle$. In the Markov chain procedure of Chapter 14, a stochastic process $(\mathbf{R}_n, \mathbf{Q}_n, \mathbf{K}_n, \mathbf{M}_n)$ satisfying Axioms [K], [QM], [RM], and [M] $\langle 14.2.1 \to 278$–$14.2.2 \to 279 \rangle$.

stochastic learning paths (system of). A stochastic process satisfying the Axioms [B], [R], [I] and [L], modeling the successive mastery of items over time. $\langle 12.2.3 \to 218$–$12.2.4 \to 220 \rangle$

straight process. The case of a discrete stochastic assessment process in which the probabilities of careless errors and lucky guesses are both equal to zero; thus $\beta_q = \eta_q = 0$ for any item q. $\langle 14.2.2 \to 279 \rangle$

substructure. See projection.

suffix, suffix$^{\triangleleft}$. The terminal segment of a word $\langle 9.2.1 \to 155 \rangle$, or of a message $\langle 10.5.4 \to 180 \rangle$.

superfluous. A query is superfluous if its response can be inferred from the responses to other preceding queries. $\langle 15.1.3 \to 302 \rangle$

support. In the Markov chain procedure of Chapter 14, the set $supp(\pi)$ of all the knowledge states K having a positive probability, that is, $\pi(K) > 0$. This subset of states is called the support of π. $\langle 14.2.2 \to 279 \rangle$

surmise function, system. Let σ be an attribution on a nonempty set Q of items. The attribution σ is a surmise function if the following three conditions are satisfied for all $q, q' \in Q$, and $C, C' \subseteq Q$:

(i) if $C \in \sigma(q)$, then $q \in C$;
(ii) if $q' \in C \in \sigma(q)$, then $C' \subseteq C$ for some $C' \in \sigma(q')$;
(iii) if $C, C' \in \sigma(q)$ and $C' \subseteq C$, then $C = C'$.

The pair (Q, σ) is then called a surmise system. $\langle 5.1.2 \to 83 \rangle$

surmise relation. The surmise relation of a knowledge structure $(Q, \mathcal{K})$ is the relation $\precsim$ on Q defined by the equivalence $r \precsim q \Leftrightarrow r \in \cap\mathcal{K}_q$. In such a case, we sometimes say that r is surmisable from q or that r precedes q. $\langle 3.7.1 \to 54 \rangle$

system of stochastic learning paths See stochastic learning paths.

tense. $\langle 5.5.4 \to 93\rangle$ An attribution σ on a nonempty set is tense when for any item q and any clause C for q, there is a state K (in the knowledge space derived from σ) which contains q and includes C but no other clause for q.

tight path. A tight path between two sets F and G in a family of sets $\mathcal{F}$ is a sequence $F = F_0, F_1, \ldots, F_p = G$ of sets in $\mathcal{F}$ such that $d(F_{i-1}, F_i) = 1$ for $1 \leq i \leq p$ with moreover $p = d(F, G)$. (Here, d denotes the symmetric difference distance between sets). $\langle 2.2.2 \to 27$ and $4.1.3 \to 62\rangle$

token$^{\triangleleft}$**.** In a medium $(\mathcal{S}, \mathcal{T})$, an element of $\mathcal{T}$, thus a transformation on the set of states of the medium. $\langle 10.1.2 \to 166\rangle$

token system$^{\triangleleft}$**.** A pair $(\mathcal{S}, \mathcal{T})$, with $\mathcal{S}$ a set of states and $\mathcal{T}$ a collection of functions $\mathcal{T} : \mathcal{S} \to \mathcal{S}$, satisfying the three conditions: (i) $|\mathcal{S}| \geq 2$; (ii) $\mathcal{T} \neq \varnothing$; and (iii) the identity τ_0 is not in $\mathcal{T}$. The elements of $\mathcal{T}$ are called 'tokens.' $\langle 10.1.2 \to 166\rangle$

trace. Let $(Q, \mathcal{K})$ be a partial knowledge structure and Q' a nonempty proper subset Q' of Q. For any state K in $\mathcal{K}$, the intersection $K \cap Q'$ is called the trace of K on Q'. $\langle 2.4.2 \to 32\rangle$.

trivial child. A child of a partial knowledge structure $(Q, \mathcal{K})$ is trivial if it is equal to $\{\varnothing\}$. $\langle 2.4.2 \to 32\rangle$

true state. In the Markov chain procedure of Chapter 14, any state K having a positive probability. The set of those states is the support of the knowledge structure. $\langle 14.2.2 \to 279\rangle$

uniform extension. Let $\mathcal{K}' = \mathcal{K}_{|Q'}$ be the projection induced by a proper subset $Q' \subseteq Q$, and suppose that $(Q', \mathcal{K}', p')$ is a probabilistic knowledge structure. Then $(Q, \mathcal{K}, p)$ is a uniform extension of $(Q', \mathcal{K}', p')$ to $(Q, \mathcal{K})$ if for all $K \in \mathcal{K}$, we have $p(K) = p'(K \cap Q')/|(K \cap Q')^{\circ}|$ $\langle 11.7.5 \to 208\rangle$. We recall that, for any $J \in \mathcal{K}_{|Q'}$, the family $J^{\circ} = \{K \in \mathcal{K} \,|\, K \cap Q' = J\}$ is the parent family of J $\langle 11.7.2 \to 207\rangle$.

union-closed family, closed under finite union. A family $\mathcal{F}$ of sets which is closed under union, that is: for any subfamily $\mathcal{G}$ of $\mathcal{F}$, we have $\cup\mathcal{G} \in \mathcal{F}$. Notice that when $\mathcal{G} = \varnothing$ (the empty subfamily), we get $\cup\mathcal{G} = \varnothing \in \mathcal{F}$. A family $\mathcal{K}$ is closed under finite union when, for any K and L in $\mathcal{K}$, the set $K \cup L$ is also in $\mathcal{K}$. Note that, in such a case, the empty set does not necessarily belong to the family $\mathcal{K}$. $\langle 2.2.2 \to 27\rangle$.

unit support. $\langle 14.2.2 \to 279\rangle$ In the context of a (discrete) stochastic assessment process, the support of a knowledge structure when it contains only one set.

unitary. A special case of a stochastic assessment process with a ε-half-split questioning function and a marking function which is selective with parameter δ. The knowledge structure is assumed to be well-graded and the parameters ε and δ satisfy particular conditions. $\langle 14.5.3 \to 285\rangle$

updating rule. A function $u : (K, r_n, q_n, L_n) \mapsto u_K(r_n, q_n, L_n)$ modifying the probability of state K on trial n on the basis of the subject's response r_n to the question q_n on that trial and on the current distribution L_n on the set of states on that trial ⟨13.3.3 → 248⟩. The function u computes the probability L_{n+1} of the states on trial $n + 1$. Several special cases of the function u are considered ⟨13.4.2 → 250–13.4.4 → 251⟩.

vacuous$^{\triangleleft}$**.** A message $\boldsymbol{m} = \tau_1 \cdots \tau_n$ in a medium is vacuous if its set of indices $\{1, \ldots, n\}$ can be partitioned into unordered pairs $\{i, j\}$, such τ_i and τ_j are mutual reverses. ⟨10.1.3 → 167⟩

well-gradedness, ∞-wellgradedness. ⟨2.2.2 → 27 and 4.3.3 → 70⟩ A family of sets $\mathcal{F}$ is well-graded if, for any two sets K and L in $\mathcal{F}$, there is a tight path in $\mathcal{F}$ from K to L. For knowledge structures, the wellgradedness condition implies the finiteness of the family. A generalized version of this concept, called ∞-wellgradedness, applies to the infinite case ⟨4.3.3 → 70⟩.

word. A string belonging to a language ⟨9.2.1 → 155⟩. See also describes.

yielding. A subset $Q' \subset Q$ of a partial knowledge space $(Q, \mathcal{K})$ is yielding if for any state L of $\mathcal{K}$ that is minimal for inclusion in some equivalence class $[K]$ of the partition induced by Q', we have $|L \setminus \cap[K]| \leq 1$. ⟨2.4.11 → 37⟩

Bibliography

J. Aczél. *Lectures on Functional Equations and their Applications*. Academic Press, New York and San Diego, 1966.

S.R. Adke and S.M. Manshunath, editors. *Introduction to Finite Markov Processes*. Wiley, New York, 1984.

A.V. Aho, J.E. Hoperoft, and J.D. Ullman, editors. *The Design and Analysis of Computer Algorithms*. Addison-Wesley, Reading, MA, 1974.

D. Albert, editor. *Knowledge Structures*. Springer-Verlag, Berlin–Heidelberg, 1994.

D. Albert and Th. Held. Establishing knowledge spaces by systematical problem construction. In D. Albert, editor, *Knowledge Structures*, volume 1, pages 78–112. Springer–Verlag, Berlin–Heidelberg, 1994.

D. Albert and C. Hockemeyer. Dynamic and adaptive hypertext tutoring systems based on knowledge space theory. In B. du Boulay and R. Mizoguchi, editors, *Artificial Intelligence in Education: Knowledge and Media in Learning Systems*, page Amsterdam. IOS Press, Berlin–Heidelberg, 1997. Frontiers in Artificial Intelligence and Applications, Vol. 39.

D. Albert and J. Lukas, editors. *Knowledge Spaces: Theories, Empirical Research, Applications*. Lawrence Erlbaum Associates, Mahwah, NJ, 1999.

D. Albert, M. Schrepp, and Th. Held. Construction of knowledge spaces for problem solving in chess. In G.H. Fischer and D. Laming, editors, *Contributions to Mathematical Psychology, Psychometrics, and Methodology*, pages 123–135. Springer–Verlag, New York, 1992.

D. Albert, A. Nussbaumer, and C. Steiner. Using visual guidance and feedback based on competence structures for personalising e-learning experience. In *ICCE 2008 — The 16th International Conference on Computers in Education*, pages 3–10. Asia–Pacific Society for Computers in Education, 2008.

A. Anastasi and S. Urbina. *Psychological Testing (seventh edition)*. Prentice Hall, Upper Saddle River, NJ, 1997.

R. D. Arasasingham, M. Taagepera, F. Potter, and S. Lonjers. Using Knowledge Space Theory To Assess Student Understanding of Stoichiometry. *Journal of Chemical Education*, 81:1517, 2004.

R.D. Arasasingham, M. Taagepera, F. Potter, I. Martorell, and S. Lonjers. Assessing the Effect of Web-Based Learning Tools on Student Understanding of Stoichiometry Using Knowledge Space Theory. *Journal of Chemical Education*, 82:1251, 2005.

W.W. Armstrong. Dependency structures of database relationships. *Information Processing*, 74:580–583, 1974.

K.C. Arnold. Screen updating and and cursor movement optimization: a library package. In *UNIX Programmer's Supplementary Documents*, volume 1. University of California, Computer Science Division, Berkeley, 1986.

M. Barbut and B. Monjardet. *Ordre et Classification*. Collection Hachette Université, Paris, 1970.

A. Barr and E.A. Feigenbaum. *The Handbook of Artificial Intelligence*. Pittman, London, 1981.

A.T. Barucha-Reid. *Elements of the Theory of Markov Processes and their Applications*. Dover Publications, NYC, New York, 1997.

G. Birkhoff. Rings of sets. *Duke Mathematical Journal*, 3:443–454, 1937.

G. Birkhoff. *Lattice Theory*. American Mathematical Society, Providence, R.I., 1967.

A. Björner, M. Las Vergnas, B. Sturmfels, N. White, and G.M. Ziegler. *Oriented Matroids*. Cambridge University Press, Cambridge, London, and New Haven, second edition, 1999.

T.S. Blyth and M.F. Janowitz. *Residuation Theory*. Pergamon Press, London, 1972.

V.J. Bowman. Permutation polyhedra. *SIAM Journal on Applied Mathematics*, 22:580–589, 1972.

R.P. Brent. *Algorithms for Minimization without Derivatives*. Prentice Hall, Engelwood Cliffs, NJ, 1973.

H.D. Brunk. *An Introduction to Mathematical Statistics*. Blaisdell, Waltham, MA, 1973.

F. Buekenhout. Espaces à fermeture. *Bulletin de la Société Mathématique de Belgique*, 19:147–178, 1967.

R.R. Bush and F. Mosteller. *Stochastic Models for Learning*. John Wiley, New York, 1955.

Nathalie Caspard and Bernard Monjardet. Some lattices of closure systems on a finite set. *Discrete Math. Theor. Comput. Sci.*, 6(2):163–190 (electronic), 2004. fichier pdf et copie papier.

D.R. Cavagnaro. Projection of a medium. To be published in the *Journal of Mathematical Psychology*, 2008.

O.B. Chedzoy. Phi-Coefficient. In S. Kotz and N.L. Johnson, editors, *Encyclopedia of Statistical Sciences*, volume 6. John Wiley & Sons, London and New York, 1983. C.B. Read, associate editor.

K.L. Chung. *Markov Chains with Stationary Transition Probabilities*. Springer-Verlag, Berlin, Heidelberg, and New York, 2nd edition, 1967.

O. Cogis. Ferrers digraphs and threshold graphs. *Discrete Mathematics*, 38: 33–46, 1982.

P.M. Cohn. *Universal Algebra*. Harper and Row, New York, 1965.

O. Conlan, C. Hockemeyer, V. Wade, and D. Albert. Metadata driven approaches to facilitate adaptivity in personalized eLearning systems. *The Journal of Information and Systems in Education*, 1:38–44, 2002.

E. Cosyn. Coarsening a knowledge structure. *Journal of Mathematical Psychology*, 46:123–139, 2002.

E. Cosyn and N. Thiéry. A Practical Procedure to Build a Knowledge Structure. *Journal of Mathematical Psychology*, 44:383–407, 2000.

E. Cosyn and H.B. Uzun. Note on two sufficient axioms for a well-graded knowledge space. *Journal of Mathematical Psychology*, 53(1):40–42, 2009.

E. Cosyn, C.W. Doble, J.-Cl. Falmagne, A. Lenoble, N. Thiéry, and H. Uzun. Assessing mathematical knowledge in a learning space. In D. Albert, C.W. Doble, D. Eppstein, J.-Cl. Falmagne, and X. Hu, editors, *Knowledge Spaces: Applications in Education*. 2010. In preparation.

H. Cramér. *Mathematical Methods of Statistics*. Princeton University Press, Princeton, NJ, 1963.

L. Crocker and J. Algina. *Introduction to Classical & Modern Test Theory*. Wadsworth—Thomson Learning, 1986.

B.A. Davey and H.A. Priestley. *Introduction to Lattices and Order*. Cambridge University Press, Cambridge, London, and New Haven, 1990.

E. Degreef, J.-P. Doignon, A. Ducamp, and J.-Cl. Falmagne. Languages for the assessment of knowledge. *Journal of Mathematical Psychology*, 30:243–256, 1986.

M.C. Desmarais and X. Pu. A Bayesian student model without hidden nodes and its comparison with item response theory. *International Journal of Artificial Intelligence in Education*, 15:291–323, 2005.

M.C. Desmarais, S. Fu, and X. Pu. Tradeoff analysis between knowledge assessment approaches. volume 125, pages 209–216, 2005.

C.W. Doble, J.-P. Doignon, J.-Cl. Falmagne, and P.C. Fishburn. Almost connected orders. *Order*, 18(4):295–311, 2001.

J.-P. Doignon. Probabilistic assessment of knowledge. In Dietrich Albert, editor, *Knowledge Structures*, pages 1–56. Springer Verlag, New York, 1994a.

J.-P. Doignon. Knowledge spaces and skill assignments. In Gerhard H. Fischer and Donald Laming, editors, *Contributions to Mathematical Psychology, Psychometrics, and Methodology*, pages 111–121. Springer–Verlag, New York, 1994b.

J.-P. Doignon and J.-Cl. Falmagne. Spaces for the Assessment of Knowledge. *International Journal of Man-Machine Studies*, 23:175–196, 1985.

J.-P. Doignon and J.-Cl. Falmagne. Knowledge assessment: A set theoretical framework. In B. Ganter, R. Wille, and K.E. Wolfe, editors, *Beiträge zur Begriffsanalyse: Vorträge der Arbeitstagung Begriffsanalyse, Darmstadt 1986*, pages 129–140, Mannheim, 1987. BI Wissenschaftsverlag.

J.-P. Doignon and J.-Cl. Falmagne. Parametrization of knowledge structures. *Discrete Applied Mathematics*, 21:87–100, 1988.

J.-P. Doignon and J.-Cl. Falmagne. Well-graded families of relations. *Discrete Mathematics*, 173:35–44, 1997.

J.-P. Doignon and J.-Cl. Falmagne. *Knowledge Spaces*. Springer-Verlag, Berlin, Heidelberg, and New York, 1999.

J.-P. Doignon, A. Ducamp, and J.-Cl. Falmagne. On realizable biorders and the biorder dimension of a relation. *Journal of Mathematical Psychology*, 28:73–109, 1984.

J.-P. Doignon, B. Monjardet, M. Roubens, and Ph. Vincke. Biorder families, valued relations and preference modelling. *Journal of Mathematical Psychology*, 30:435–480, 1986.

C.E. Dowling. Constructing knowledge spaces from judgements with differing degrees of certainty. In Jean-Paul Doignon and Jean-Claude Falmagne, editors, *Mathematical Psychology: Current Developments*, pages 221–231. Springer–Verlag, New York, 1991a.

C.E. Dowling. *Constructing Knowledge Structures from the Judgements of Experts*. Habilitationsschrift, Technische Universität Carolo-Wilhelmina, Braunschweig, Germany, 1991b.

C.E. Dowling. Applying the basis of a knowledge space for controlling the questioning of an expert. *Journal of Mathematical Psychology*, 37:21–48, 1993a.

C.E. Dowling. On the irredundant construction of knowledge spaces. *Journal of Mathematical Psychology*, 37:49–62, 1993b.

C.E. Dowling. Integrating different knowledge spaces. In G.H. Fischer and D. Laming, editors, *Contributions to Mathematical Psychology, Psychometrics, and Methodology*, pages 149–158. Springer-Verlag, Berlin, Heidelberg, and New York, 1994.

C.E. Dowling, C. Hockemeyer, and A.H. Ludwig. Adaptive assessment and training using the neighbourhood of knowledge spaces. In C. Frasson, G. Gauthier, and A. Lesgold, editors, *Intelligent Tutoring Systems*, pages 578–586. Springer–Verlag, Berlin, 1996.

F. Drasgow. Polychoric and polyserial correlation. In S. Kotz, N.L. Johnson, and C.B. Read, editors, *Encyclopedia of Statistical Sciences*, volume 7, pages 68–74. Wiley, New York, 1986.

A. Ducamp and J.-Cl. Falmagne. Composite measurement. *Journal of Mathematical Psychology*, 6:359–390, 1969.

R.O. Duda and P.E. Hart. *Pattern Classification and Scene Analysis*. John Wiley, New York, 1973.

J. Dugundji. *Topology*. Allyn and Bacon, Boston, 1966.

I. Düntsch and G. Gediga. Skills and knowledge structures. *British Journal of Mathematical and Statistical Psychology*, 48:9–27, 1995a.

I. Düntsch and G. Gediga. On query procedures to buid knowledge structures. *British Journal of Mathematical and Statistical Psychology*, 48:9–27, 1995b.

J. Durnin and J.M. Scandura. An algorithmic approach to assessing behavioral potential: comparison with item forms and hierarchical technologies. *Journal of Educational*, 65:262–272, 1973.

P.H. Edelman and R. Jamison. The theory of convex geometries. *Geometrica Dedicata*, 19:247–271, 1985.

Educational Testing Services (ETS). California Standard Tests (CSTs), Spring 2007 Administration. Technical report, 2008. Contract 5417, `http://www.cde.ca.gov/ta/tg/sr/documents/csttechrpt07.pdf`.

D. Eppstein. Algorithms for drawing media. In *Graph Drawing: 12th International Symposium, GD 2004, New York, NY, USA, September 29–October 2, 2004*, volume 3383 of *Lecture Notes in Computer Science*, pages 173–183, Berlin, Heidelberg, and New York, 2005. Springer-Verlag.

D. Eppstein. Recognizing partial cubes in quadratic time. Electronic preprint 0705.1025, arXiv.org, 2007.

D. Eppstein. Learning Sequences. In D. Albert, C.W. Doble, D. Eppstein, J.-Cl. Falmagne, and X. Hu, editors, *Knowledge Spaces: Applications in Education*. 2010. In preparation.

D. Eppstein and J.-Cl. Falmagne. Algorithms for media. Electronic preprint cs.DS/0206033, arXiv.org, 2002.

D. Eppstein, J.-Cl. Falmagne, and S. Ovchinnikov. *Media Theory*. Springer-Verlag, Berlin, Heidelberg, and New York, 2008.

D. Eppstein, J.-Cl. Falmagne, and H.B. Uzun. On verifying and engineering the wellgradedness of a union-closed family. *Journal of Mathematical Psychology*, 53(1):34–39, 2009.

J.-Cl. Falmagne. A latent trait theory via stochastic learning theory for a knowledge space. *Psychometrika*, 53:283–303, 1989a.

J.-Cl. Falmagne. Probabilistic knowledge spaces: A review. In Fred Roberts, editor, *Applications of Combinatorics and Graph Theory to the Biological and Social Sciences, IMA Volume 17*. Springer Verlag, New York, 1989b.

J.-Cl. Falmagne. Stochastic learning paths in a knowledge structure. *Journal of Mathematical Psychology*, 37:489–512, 1993.

J.-Cl. Falmagne. Finite markov learning models for knowledge structures. In G.H. Fischer and D. Laming, editors, *Contributions to Mathematical Psychology, Psychometrics, and Methodology*. Springer Verlag, New York, 1994.

J.-Cl. Falmagne. Errata to SLP. *Journal of Mathematical Psychology*, 40: 169–174, 1996.

J.-Cl. Falmagne. Stochastic token theory. *Journal of Mathematical Psychology*, 41(2):129–143, 1997.

J.-Cl. Falmagne. Projections of a learning space. Electronic preprint 0803.0575, arXiv.org, 2008. Submitted.

J.-Cl. Falmagne and J-P. Doignon. A class of stochastic procedures for the assessment of knowledge. *British Journal of Mathematical and Statistical Psychology*, 41:1–23, 1988a.

J.-Cl. Falmagne and J-P. Doignon. A Markovian procedure for assessing the state of a system. *Journal of Mathematical Psychology*, 32:232–258, 1988b.

J.-Cl. Falmagne and J.-P. Doignon. Stochastic evolution of rationality. *Theory and Decision*, 43:107–138, 1997.

J.-Cl. Falmagne and J.-P. Doignon. Meshing knowledge structures. In C.E. Dowling, F.S. Roberts, and P. Peter Theuns, editors, *Recent Progress in Mathematical Psychology, Scientific Psychology Series*. Lawrence Erlbaum Associates Ltd., Mahwah, NJ., 1998.

J.-Cl. Falmagne and K. Lakshminarayan. Stochastic learning paths—estimation and simulation. In G.H. Fischer and D. Laming, editors, *Contributions to Mathematical Psychology, Psychometrics, and Methodology*. Springer Verlag, New York, 1994.

J.-Cl. Falmagne and S. Ovchinnikov. Media theory. *Discrete Applied Mathematics*, 121:83–101, 2002.

J.-Cl. Falmagne and S. Ovchinnikov. Mediatic graphs. In S. Brams, W.V. Gehrlein, and F.S. Roberts, editors, *The Mathematics of Preference, Choice and Order. Essays in Honor of Peter, C. Fishburn*, Studies in Choice and Welfare, pages 325–343. Springer, Berlin–Heidelberg, 2009.

J.-Cl. Falmagne, M. Koppen, M. Villano, J.-P. Doignon, and L. Johannesen. Introduction to knowledge spaces: how to build, test and search them. *Psychological Review*, 97:204–224, 1990.

J.-Cl. Falmagne, E. Cosyn, J.-P. Doignon, and N. Thiéry. The assessment of knowledge, in theory and in practice. In B. Ganter and L. Kwuida, editors, *Formal Concept Analysis, 4th International Conference, ICFCA 2006, Dresden, Germany, February 13–17, 2006*, Lecture Notes in Artificial Intelligence, pages 61–79. Springer-Verlag, Berlin, Heidelberg, and New York, 2006a.

J.-Cl. Falmagne, Y.-F. Hsu, F. Leite, and M. Regenwetter. Stochastic applications of media theory: Random walks on weak orders or partial orders. *Discrete Applied Mathematics*, 2007. doi: 10.1016/j.dam.2007.04.032.

S.E. Feinberg. *The Analysis of Cross-Classified Categorical Data*. MIT Press, Cambridge, MA., 2nd edition, 1981.

W. Feller. *An Introduction to Probability Theory and its Applications*, volume 1. John Wiley & Sons, London and New York, 3rd edition, 1968.

P.C. Fishburn. Intransitive indifference with unequal indifference intervals. *Journal of Mathematical Psychology*, 7:144–149, 1970.

P.C. Fishburn. *Interval orders and interval graphs*. John Wiley & Sons, London and New York, 1985.

Cl. Flament. *L'Analyse Booléenne des Questionnaires*. Mouton, Paris-The Haguen, 1976.

D.A.S. Fraser. *Statistics, an Introduction*. John Wiley & Sons, New York, 1958.

S. Fries. Empirical validation of a markovian learning model for knowledge structures. *Journal of Mathematical Psychology*, 41:65–70, 1997.

K.S. Fu. *Syntactic Methods in Pattern Recognition*. Academic Pres, New York, 1974.

P. Gaiha and S.K Gupta. Adjacent vertices on a permutohedron. *SIAM Journal on Applied Mathematics*, 32(2):323–327, 1977.

B. Ganter. Two basic algorithms in concept analysis. (FB4–Preprint, number 831.) TH Darmstadt, 1984.

B. Ganter. Algorithmen zur Formalen Begriffsanalyse. In B. Ganter, R. Wille, and K.E. Wolfe, editors, *Beiträge zur Begriffsanalyse: Vorträge der Arbeitstagung Begriffsanalyse, Darmstadt 1986*, pages 129–140, Mannheim, 1987. BI Wissenschaftsverlag.

B. Ganter and K. Reuter. Finding all the closed sets: a general approach. *Order*, 8(4):283–290, 1991.

B. Ganter and R. Wille. *Formale Begriffsanalyse: Mathematische Grundlagen*. Springer-Verlag, Berlin-Heidelber, 1996. English translation by C. Franske: *Formal Concept Analysis: Mathematical Foundations*, Springer-Verlag, 1998.

M.R. Garey and D.S. Johnson. *Computers and Intractability: A Guide to the Theory of NP-Completeness*. W. H. Freemann, 1979.

K. Gegenfurtner. PRAXIS: Brent's algorithm for function minimization. *Behavior Research Methods Instruments and Computers*, 24:560–564, 1992.

J.-L. Guigues and V. Duquenne. Familles minimales d'implications informatives résultant d'un tableau de données binaires. *Mathématiques, Informatique et Sciences Humaines*, 97:5–18, 1986.

L. Guttman. A basis for scaling qualitative data. *American Sociological Review*, 9:139–150, 1944.

B. Harris. Tetrachoric Correlation Coefficient. In S. Kotz and N.L. Johnson, editors, *Encyclopedia of Statistical Sciences*, volume 9. John Wiley & Sons, London and New York, 1983. C.B. Read, associate editor.

J. Heller. A formal framework for characterizing querying algorithms. *Journal of Mathematical Psychology*, 48(1):1–8, 2004.

J. Heller and C. Repitsch. Distributed skill functions and the meshing of knowledge structures. *J. Math. Psych.*, 52(3):147–157, 2008. ISSN 0022-2496.

C. Hockemeyer. RATH - a relational adaptive tutoring hypertext environment. Rapport 1997/3, Institut für Psychologie, Karl-Franzens-Universität Graz, Austria, 1997.

C. Hockemeyer. Tools and utilities for knowledge spaces. Unpublished technical report, Institut für Psychologie, Karl–Franzens–Universität Graz, Austria, 2001.

C. Hockemeyer, Th. Held, and D. Albert. RATH — a relational adaptive tutoring hypertext WWW–environment based on knowledge space theory. In Christer Alvegård, editor, *CALISCE'98: Proceedings of the Fourth International Conference on Computer Aided Learning in Science and Engineering*, pages 417–423, Göteborg, Sweden, June 1998. Chalmers University of Technology. ISBN 91-7197-683-3. URL `http://wundt.kfunigraz.ac.at/rath/publications/calisce/`.

C. Hockemeyer, A. Nussbaumer, E. Lövquist, A. Aboulafia, D. Breen, G. Shorten, and D. Albert. Applying a web and simulation–based system for adaptive competence assessment of spinal anaesthesia. In Marc Spaniol,

Qing Li, Ralf Klamma, and Rynson Lau, editors, *Advances in Web–Based learning — ICWL 2009*, pages 182–191, Berlin, 8 2009. Springer Verlag.

L. Hyafill and R.L. Rivest. Constructing optimal decision trees is NP-complete. *Information Processing Letters*, 5:15–17, 1976.

W. Imrich and S. Klavžar. *Product Graphs*. John Wiley & Sons, London and New York, 2000.

K. Jameson. Empirical methods for generative semiotics models: an application to the roman majuscules. In W.C. Watt, editor, *Writing Systems and Cognition*, Neuropsychology and Cognition Series. Kluwer Academic Publishers, Dordrecht, 1992.

R.E. Jamison-Waldner. A perspective on abstract convexity: classifying alignments by varieties (Norman, Oklahoma, 1980. In W.C. Watt, editor, *Lecture Notes in Pure and Applied Mathematics*, volume 76, pages 113–150. Dekker, New York, 1982.

R.T. Johnson and T.P. Vaughan. On union-closed families. I. *Journal of Combinatorial Theory, Series A*, 84:242–249, 1998.

M. Kambouri. *Knowledge assessment: A comparison between human experts and computerized procedure*. PhD thesis, New York University, New York, 1991.

M. Kambouri, M. Koppen, M. Villano, and J.-Cl. Falmagne. Knowledge assessment: Tapping human expertise by the QUERY routine. *International Journal of Human-Computer Studies*, 40:119–151, 1994.

W.T. Kelvin. *Popular Lectures and Addresses. Volume 1-3.* MacMillan, London, 1889. Volume 1: Constitution of Matter (Chapter: Electrical Units of Measurement).

J.G. Kemeny and J.L. Snell. *Finite Markov Chains*. Van Nostrand, Princeton, N.J., 1960.

M.D. Kickmeier-Rust, B. Marte, S.B. Linek, T. Lalonde, and D. Albert. Learning with computer games: Micro level feedback and interventions. In M. E. Auer, editor, *Proceedings of the International Conference on Interactive Computer Aided Learning (ICL)*, Kassel, 2008. Kassel University Press. CD–ROM publication.

R. Kimbal. A self-improving tutor for symbolic integration. In D. Sleeman and J.S. Brown, editors, *Intelligent Tutoring Systems*, volume 1 of *Computer and People Series*. Academic Press, London, 1982.

M. Koppen. *Ordinal Data Analysis: Biorder Representation and Knowledge Spaces*. Doctoral dissertation, Katholieke Universiteit te Nijmegen, Nijmegen, The Netherlands, 1989.

M. Koppen. Extracting human expertise for constructing knowledge spaces: An algorithm. *Journal of Mathematical Psychology*, 37:1–20, 1993.

M. Koppen. On alternative representations for knowledge spaces. *Mathematical Social Sciences*, 36:127–143, 1998.

M. Koppen and J.-P. Doignon. How to build a knowledge space by querying an expert. *Journal of Mathematical Psychology*, 34:311–331, 1990.

B. Korte, L. Lovász, and R. Schrader. *Greedoids*. Number 4 in Algorithms and Combinatorics. Springer-Verlag, 1991.

K. Lakshminarayan. *Theoretical and Empirical Aspects of Some Stochastic Learning Models*. Doctoral dissertation, University of California, Irvine, 1995.

K. Lakshminarayan and F. Gilson. An application of a stochastic knowledge structure model. In C.E. Dowling, F.S. Roberts, and P. Peter Theuns, editors, *Recent Progress in Mathematical Psychology, Scientific Psychology Series*. Lawrence Erlbaum Associates Ltd., Mahwah, NJ., 1998.

M. Landy and R.A. Hummel. A brief survey of knowledge aggregation methods. In *Proceedings, 1986 International Conference on Pattern Recognition*, pages 248–252, 1986.

P. F. Lazarsfeld and N. W. Henry. *Latent Structure Analysis*. Houghton Mifflin, Boston, 1968.

Cl. Le Conte de Poly-Barbut. Le diagramme du treillis permutohèdre est Intersection des diagrammes de deux produits directs d'ordres totaux. *Mathématiques, Informatique et Sciences Humaines*, 112:49–53, 1990.

E.L. Lehman. *Testing Statistical Hypotheses*. John Wiley, New York, 1959.

B.W. Lindgren. *Statistical Theory*. Macmillan, New York, 2nd edition, 1968.

F.M. Lord. Individualized testing and item characteristic curve theory. In D.H. Krantz, R.C. Atkinson, R.D. Luce, and P. Suppes, editors, *Contemporary Developments in Mathematical Psychology, Vol. II. Measurement, Psychophysics and Neural Information Processing*. Freeman, San Francisco, CA, 1974.

F.M. Lord and M.R. Novick. *Statistical Theories of Mental Tests* (2nd. ed.). Addison-Wesley, Reading, MA, 1974.

R.D. Luce. Semiorders and a theory of utility discrimination. *Econometrica*, 24:178–191, 1956.

R.D. Luce. *Individual Choice Behavior*. John Wiley, New York, 1959.

R.D. Luce. Some one-parameter families of learning operators. In R.C. Atkinson, editor, *Studies in Mathematical Psychology*, pages 380–398. Stanford University Press, Stanford, CA, 1964.

J. Lukas and D. Albert. Knowledge assessment based on skill assignment and psychological task analysis. In Gerhard Strube and Karl F. Wender, editors, *The Cognitive Psychology of Knowledge*, volume 101 of *Advances in Psychology*, pages 139–160. North–Holland, Amsterdam, 1993.

A.A.J. Marley. Abstract one-parameter families of learning operators. *Journal of Mathematical Psychology*, 4:414–429, 1967.

S.P. Marshall. Sequential item selection; optimal and heuristic policies. *Journal of Mathematical Psychology*, 23:134–152, 1981.

B. Matalon. *L'Analyse Hiérarchique*. Mouton, Paris, 1965.

B. Monjardet. Tresses, fuseaux, préordres et topologies. *Mathématique et Sciences Humaines*, 30:11–22, 1970.

B. Monjardet. A use for frequently rediscovering a concept. *Order*, 1(4): 415–417, 1985.

C.E. Müller. A procedure for facilitating an expert's judgments on a set of rules. In E.E. Roskam, editor, *Mathematical Psychology in Progress*, Recent Research in Psychology, pages 157–170. Springer-Verlag, Berlin, Heidelberg, and New York, 1989.

M.F. Norman. *Markov Processes and Learning Models.* Academic Press, New York, 1972.

J. Nunnally and I. Bernstein. *Psychometric Theory.* MacGraw-Hill, New York, 1994.

S. Ovchinnikov. Convex geometry and group choice. *Mathematical Social Sciences*, 5:1–16, 1983.

S. Ovchinnikov and A. Dukhovny. Advances in media theory. *International Journal of Uncertainty, Fuzziness and Knowledge-Based Systems*, 8(1):45–71, 2000.

E. Parzen. *Stochastic Processes.* Holden-Day, San Francisco, 1994.

K. Pearson. *The Life, Letters and Labours of Francis Galton.* Cambridge University Press, London, 1924. Volume 2: Researches of Middle Life.

G. Pilato, R. Pirrone, and R. Rizzo. A KST–based system for student tutoring. *Applied Artificial Intelligence*, 22:283–308, 2008.

M. Pirlot and Ph. Vincke. *Semiorders: Properties, Representations, Applications.* Kluwer, Amsterdam, 1997.

M.J.D. Powell. An efficient method for finding the minimum of a function of several variables without calculating derivatives. *Computer Journal*, 7: 155–162, 1964.

M. Regenwetter, J.-Cl. Falmagne, and B. Grofman. A stochastic model of preference change and its application to 1992 presidential election panel data. *Psychological Review*, 106(2):362–384, 1999.

E. Rich. *Artificial Intelligence.* MacGraw-Hill, Singapore, 1983.

J. Riguet. Les relations de Ferrers. *Compte Rendus des Scéances de l'Académie des Sciences (Paris)*, 232:1729–1730, 1951.

F.S. Roberts. *Measurement Theory, with Applications to Decisionmaking, Utility, and the Social Sciences.* Addison-Wesley, Reading, Mass., 1979.

F.S. Roberts. *Applied Combinatorics.* Prentice Hall, Englewood Cliffs, New Jersey, 1984.

M. Roubens and Ph. Vincke. *Preference Modelling.* Lecture Notes in Economics and Mathematical Systems, vol. 250. Springer-Verlag, Berlin-Heidelberg, 1985.

G. Rozenberg and A. Salomaa, editors. *Handbook of Formal Language Theory, Vol. 1-3.* Springer-Verlag, Berlin, Heidelberg, and New York, 1997.

A. Rusch and Rudolf Wille. Knowledge spaces and formal concept analysis. In H.-H. Bock and W. Polasek, editors, *Data Analysis and Information Systems*, pages 427–436, Berlin, Heidelberg, and New York, 1996. Springer-Verlag.

D. Scott and P. Suppes. Foundational aspects of theories of measurement. *Journal of Symbolic Logic*, 23:113–128, 1958.

D.B. Shmoys and É. Tardos. Computational Compexity. In R.L. Graham, M. Grötschel, and L. Lovász, editors, *Handbook of Combinatorics*, volume 2. The M.I.T. Press, Cambridge, MA, 1995.

E.H. Shortliffe. *Computer-Based Medical Consultation: Mycin.* American Elsevier, New York, 1976.

E.H. Shortliffe and D.G. Buchanan. A model of inexact reasoning in medicine. *Mathematical Bioscience*, 23:351–379, 1975.

A.N. Shyryayev. *Probability.* Springer-Verlag, New York, 1960.

G. Sierksma. Convexity on unions of sets. *Compositio Mathematica*, 42: 391–400, 1981.

J. Stern and K. Lakshminarayan. Comments on mathematical and logical aspects of a model of acquisition behavior. *Technical Report 95-05*, 1995.

P. Suppes. *Introduction to Logic.* Van Nostrand, Princeton, N.J., 1957.

P. Suppes. *Axiomatic Set Theory.* Van Nostrand, Princeton, N.J., 1960.

P. Suppes, D.H. Krantz, R.D. Luce, and A. Tversky. *Foundations of Measurement, Volume 2: Geometrical, Threshold, and Probabilistic Representations.* Academic Press, New York and San Diego, 1989.

E. Szpilrajn. Sur l'extension de l'ordre partiel. *Fundamenta Mathematicae*, 16:386–389, 1930.

M. Taagepera and S. Noori. Mapping Students' Thinking Patterns in Learning Organic Chemistry by the Use of the Knowledge Space Theory. *Journal of Chemical Education*, 77:1224, 2000.

M. Taagepera, F. Potter, G.E. Miller, and K. Lakshminarayan. Mapping Students' Thinking Patterns by the Use of Knowledge Space Theory,. *International Journal of Science Education*, 19:283, 1997.

M. Taagepera, R. Arasasingham, F. Potter, A. Soroudi, and G. Lam. Following the Development of the Bonding Concept Using Knowledge Space Theory. *Journal of Chemical Education*, 79:1756, 2002.

M. Taagepera, R.D. Arasasingham, S. King, F. Potter, I. Martorell, D. Ford, J. Wu, and A. M. Kearney. How Students Think About Stereochemistry. Technical report, 2008. Submitted and under revision.

G.M. Tallis. The maximum likelihood estimation of correlation for contingency tables. *Biometrics*, 9:342–353, 1962.

R.F. Tate. Correlation between a discrete and a continuous variable. Point-Biserial Correlation. *The Annals of Mathematical Statistics*, 25(3):603–607, 1954.

W.T. Trotter. *Combinatorics and Partially Ordered Sets: Dimension Theory.* The Johns Hopkins University Press, Baltimore and London, 1992.

M. Van de Vel. *The theory of convex structures.* North-Holland Publishing Co., Amsterdam, 1993.

K. Van Lehn. Student modeling. In M.C. Polson and J.J. Richardson, editors, *Foundations of Intelligent Tutoring Systems.* Erlbaum, Hillsdale, N.J., 1988.

M. Villano. *Computerized knowledge assessment: Building the knowledge structure and calibrating the assessment routine.* PhD thesis, New York

University, New York, 1991. In *Dissertation Abstracts International*, vol. 552, p. 12B.

M. Villano, J.-Cl. Falmagne, L. Johannesen, and J.-P. Doignon. Stochastic procedures for assessing an individual's state of knowledge. In *Proceedings of the International Conference on Computer-assisted Learning in Post-Secondary Education, Calgary 1987*, pages 369–371, Calgary, 1987. University of Calgary Press.

S.B. Vincent. The function of the vibrissae in the behavior of the white rat. *Behavioral Monographs*, 1(5), 1912.

H. Wainer and S. Messick. *Principles of Modern Psychological Development: A Festschrift for Frederic Lord.* Lawrence Erlbaum Associates, Hillsdale, N.J., 1983.

H. Wainer, N.J. Dorans, D. Eignor, R. Flaugher, B.F. Green, R.J. Mislevy, L. Steinberg, and D. Thissen. *Computerized Adaptive Testing: A Primer.* Lawrence Erlbaum Associates, New Jersey and London, 2000.

D.J. Weiss. *New Horizons in Testing: Latent Trait Theory and Computerized Testing Theory.* Academic Press, New York, 1983.

D.J.A. Welsh. Matroids: Fundamental concepts. In R.L. Graham, M. Grötschel, and L. Lovász, editors, *Handbook of Combinatorics*, volume 1. The M.I.T. Press, Cambridge, MA, 1995.

M. Wild. A theory of finite closure spaces based on implications. *Advances in Mathematics*, 108:118–139, 1994.

G. M. Ziegler. *Lectures on polytopes*, volume 152 of *Graduate Texts in Mathematics.* Springer-Verlag, New York, 1995. ISBN 0-387-94365-X.

Index

Printed by Publishers' Graphics LLC
SO20130116.19.19.40